Information Systems and Ser

This book is due for return on or b
period of lo

COMPREHENSIVE ANALYTICAL CHEMISTRY

ELSEVIER SCIENCE B.V.
Sara Burgerhartstraat 25
P.O. Box 211, 1000 AE Amsterdam, The Netherlands

ISBN 0-444-89033-5

This book is printed on acid-free paper.

Printed in The Netherlands

Wilson & Wilson's

COMPREHENSIVE ANALYTICAL CHEMISTRY

Edited by

S.G. WEBER

Department of Chemistry, University of Pittsburgh, Pittsburgh, PA, U.S.A.

COMPREHENSIVE ANALYTICAL CHEMISTRY

Wilson & Wilson's

COMPREHENSIVE ANALYTICAL CHEMISTRY

VOLUME XXXI

SURFACTANTS IN ANALYTICAL CHEMISTRY
APPLICATIONS OF ORGANIZED AMPHIPHILIC MEDIA

by

E. PRAMAURO and E. PELEZETTI

1996
ELSEVIER
AMSTERDAM – LAUSANNE – OXFORD – NEW YORK – SHANNON – TOKYO

CONTRIBUTORS TO VOLUME XXXI

Dr. E. Pramauro and Dr. E. Pelezetti
University of Turin, Department of Analytical Chemistry, Via Petro Giuria 5, 10125 Turin, Italy

WILSON AND WILSON'S

COMPREHENSIVE ANALYTICAL CHEMISTRY

VOLUMES IN THE SERIES

Vol. IA	Analytical Processes
	Gas Analysis
	Inorganic Qualitative Analysis
	Organic Qualitative Analysis
	Inorganic Gravimetric Analysis
Vol. IB	Inorganic Titrimetric Analysis
	Organic Quantitative Analysis
Vol. IC	Analytical Chemistry of the Elements
Vol. IIA	Electrochemical Analysis
	Electrodeposition
	Potentiometric Titrations
	Conductometric Titrations
	High-Frequency Titrations
Vol. IIB	Liquid Chromatography in Columns
	Gas Chromatography
	Ion Exchangers
	Distillation
Vol. IIC	Paper and Thin Layer Chromatography
	Radiochemical Methods
	Nuclear Magnetic Resonance and Electron Spin Resonance Methods
	X-Ray Spectrometry
Vol. IID	Coulometric Analysis
Vol. III	Elemental Analysis with Minute Sample
	Standards and Standardization
	Separation by Liquid Amalgams
	Vacuum Fusion Analysis of Gases in Metals
	Electroanalysis in Molten Salts
Vol. IV	Instrumentation for Spectroscopy
	Atomic Absorption and Fluorescence Spectroscopy
	Diffuse Reflectance Spectroscopy
Vol. V	Emission Spectroscopy
	Analytical Microwave Spectroscopy
	Analytical Applications of Electron Microscopy
Vol. VI	Analytical Infrared Spectroscopy
Vol. VII	Thermal Methods in Analytical Chemistry
	Substoichiometric Analytical Methods
Vol. VIII	Enzyme Electrodes in Analytical Chemistry
	Molecular Fluorescence Spectroscopy
	Photometric Titrations
	Analytical Applications of Interferometry

Vol. IX	Ultraviolet Photoelectron and Photoion Spectroscopy
	Auger Electron Spectroscopy
	Plasma Excitation in Spectrochemical Analysis
Vol. X	Organic Spot Tests Analysis
	The History of Analytical Chemistry
Vol. XI	The Application of Mathematical Statistics in Analytical Chemistry
	Mass Spectrometry
	Ion Selective Electrodes
Vol. XII	Thermal Analysis
	Part A. Simultaneous Thermoanalytical Examination by Means of the Derivatograph
	Part B. Biochemical and Clinical Applications of Thermometric and Thermal Analysis
	Part C. Emanation Thermal Analysis and other Radiometric Emanation Methods
	Part D. Thermophysical Properties of Solids
	Part E. Pulse Method of Measuring Thermophysical Parameters
Vol. XIII	Analysis of Complex Hydrocarbons
	Part A. Separation Methods
	Part B. Group Analysis and Detailed Analysis
Vol. XIV	Ion-Exchangers in Analytical Chemistry
Vol. XV	Methods of Organic Analysis
Vol. XVI	Chemical Microscopy
	Thermomicroscopy of Organic Compounds
Vol. XVII	Gas and Liquid Analysers
Vol. XVIII	Kinetic Methods in Chemical Analysis
	Application of Computers in Analytical Chemistry
Vol. XIX	Analytical Visible and Ultraviolet Spectrometry
Vol. XX	Photometric Methods in Inorganic Trace Analysis
Vol. XXI	New Developments in Conductometric and Oscillometric Analysis
Vol. XXII	Titrimetric Analysis in Organic Solvents
Vol. XXIII	Analytical and Biomedical Applications of Ion-Selective Field-Effect Transistors
Vol. XXIV	Energy Dispersive A-Ray Fluorescence Analysis
Vol. XXV	Preconcentration of Trace Elements
Vol. XXVI	Radionucleide X-Ray Fluorescence Analysis
Vol. XXVII	Voltammetry
Vol. XXVIII	Analysis of Substances in the Gaseous Phase
Vol. XXIX	Chemiluminescence Immunoassay
Vol. XXX	Spectrochemical Trace Analysis for Metals and Metalloids
Vol. XXXI	Surfactants in Analytical Chemistry

Contents

Editor's Preface

Analytical and physical chemists have tamed the marvellous, rich, and surprising chemistry of surfactants. By augmenting solvent power, creating new phases, by providing a measure of size and shape selectivity in binding reactions, surfactants have allowed chemists to be very creative in solving analytical problems.

Here, in this volume, Dr. Pramauro and Dr. Pelizetti have gathered and critically discussed an impressively broad subject matter. Because of the breadth of the material, and by the use of a large number of enlightening examples and tables of data, chemists of many disciplines will enjoy and learn from this valuable contribution.

S.G. Weber

Authors' Preface

For several years now surfactant molecules and their aggregates have been increasingly used in analytical techniques, in part as new solvent systems and in part to provide a very particular environment able to alter the analytes' properties and their reactivity. As a consequence of such surfactant–analyte interactions, new determination methods have been developed and a new generation of separation processes has been established which also finds interesting applications in other important areas such as biotechnology and pollution control.

In order to obtain the maximum advantage from the coupling of surfactants and chemical analysis, the knowledge of amphiphilic structures and their physico-chemical properties, the quantitative description of solute–surfactant interactions and the effects of such interactions on chemical equilibria and reactivity must be carefully considered.

The main objective of this book is to provide a comprehensive assessment of the state of surfactants and surfactant aggregates in analytical chemistry following the general approach indicated above, examining the application of surfactant structures in all the analytical areas where they can offer significant advantages in respect to previous methods and describing the features of recently developed surfactant-based systems designed to solve specific analytical problems.

The book is aimed at academic and industrial researchers working with surfactants in various fields of chemistry and separation science where the interactions between the organized media and substrates is investigated and developed using analytical methods, as well as at people studying chemical, biochemical and biological phenomena occurring in surfactant biomimetic systems who are interested in understanding the general aspects of reaction equilibria and kinetics in these microheterogeneous media.

The preparation of this book has been stimulated by many useful discussions with a group of colleagues whose well known scientific contributions to surfactant chemistry have greatly expanded our knowledge and interest in this field. We are, in particular, grateful to Prof. Michael Gratzel, Dr. Dan Meisel, Prof. Miguel de la Guardia and Prof. Willie L. Hinze who, in recent years, were our partners in several international research projects.

We are also grateful to the authors and publishers of books and journals for the permission to reproduce original illustrations and, as far as the financial support of our own work is concerned, we are grateful to CNR and MURST (Rome) and to EniRicerche (Milan) for their long-term contribution to our investigations.

We also thank the Editor of this series for his comments and suggestions and Elsevier staff for their help with this undertaking and for their patience.

E. Pramauro
E. Pelizetti

Chapter 1

Behaviour and properties of amphiphilic molecules and their assemblies

Dilute solutions, electrolyte solutions and regular solutions, although differing from each other according to types of solvents and the nature of interactional forces, are all adequately described on the basis of random mixing by thermal motion of their solute and solvent molecules. They also constitute the basis of classical analytical chemistry. Such ideal solutions represent exceptions amongst the cases which are encountered in the real world.

In many biological and environmental solutions, amphiphilic molecules and irregular polymers give rise to formation of structures, designated as "organized assemblies", exhibiting several characteristic properties.

The analytical chemist has to be aware of the influence of these organized assemblies on the chemical equilibrium, and can exploit their unique properties for improving analytical methodologies and for the development of entirely new concepts in analytical chemistry.

Organized assemblies may exhibit several peculiar properties, such as:

(a) solubilize, concentrate and compartmentalize chemical species;
(b) alter dissociation constants, oxidation and reduction properties;
(c) provide a unique reaction media, affecting chemical pathways and rates;
(d) influence transport properties.

Obviously, a proper selection of a given system for a given application is necessary, and this requires a sufficient knowledge of the properties of the organized systems and those of the substrate interactions.

In this chapter we will briefly describe the characteristics of aggregate forming agents in aqueous and non-aqueous media, which render them useful in many analytical applications. Our attention will be focused on aggregate formation and on the effect of physical and chemical

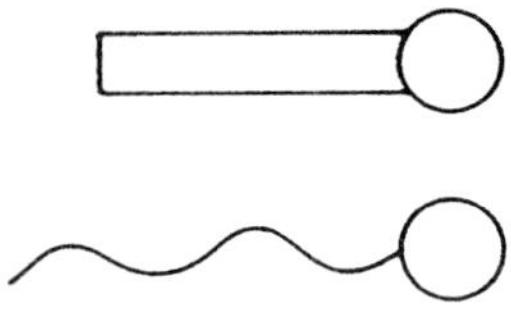

Fig. 1.1. Conventional representation of a surfactant molecule as a flexible tail or rod (the hydrocarbon chain) and a head group.

variables, which can drastically alter the properties of amphiphilic assemblies and their interaction with the analytes.

1.1. AMPHIPHILIC MOLECULES

Amphiphile surfactants or detergents are molecules with distinct hydrophobic and hydrophilic regions (schematically represented in Fig. 1.1). Depending on their chemical structures they can be neutral, cationic, anionic, and zwitterionic. The hydrophobic part of the surfactant can be of various lengths (typically 8–20 carbon atoms), contain multiple bonds, or consist of two or more hydrocarbon chains. Surfactants have also been functionalized to contain desired reactive and/or reporter groups. Aggregation behaviour depends on the nature and the concentration of the surfactants, the nature of the solvent, and the method of preparation.

The name and structure of some common amphiphiles (synthetic and naturally occurring) are summarized in Table 1.1.

TABLE 1.1

Summary of name(s) (trivial names), structures and abbreviations of some amphiphilic compounds

Surfactant	Molecular structure	Abbreviation
Anionic		
Sodium dodecylsulfate (sodium laurylsulfate)	$CH_3(CH_2)_{11}$–O–$SO_3^-Na^+$	SDS
Sodium decylsulfate	$CH_3(CH_2)_9$–$OSO_3^-Na^+$	SDeS
Sodium hexadecylsulfate (sodium cetylsulfate)	$CH_3(CH_2)_{15}$-O-$SO_3^-Na^+$	SHS
Sodium dodecanoate (sodium laurate)	$CH_3(CH_2)_{10}$–COO^-Na^+	NaL

Surfactant	Molecular structure	Abbreviation
Potassium dodecanoate (potassium laurate)	$CH_3(CH_2)_{10}{-}COO^-K^+$	KL
Sodium dodecylsulfonate	$CH_3(CH_2)_{11}{-}SO_3^-Na^+$	SDDS
Sodium dodecylbenzene-sulfonate	$CH_3(CH_2)_{11}{-}C_6H_4{-}SO_3^-Na^+$	SDBS
Sodium 3,7,12-trihydroxy-5-cholanate (sodium cholate)	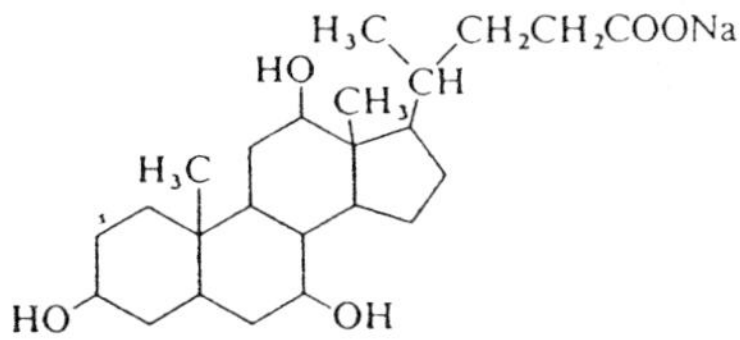	NaC
Sodium bis(2-ethylhexyl)sulfosuccinate (Aerosol OT)[a]	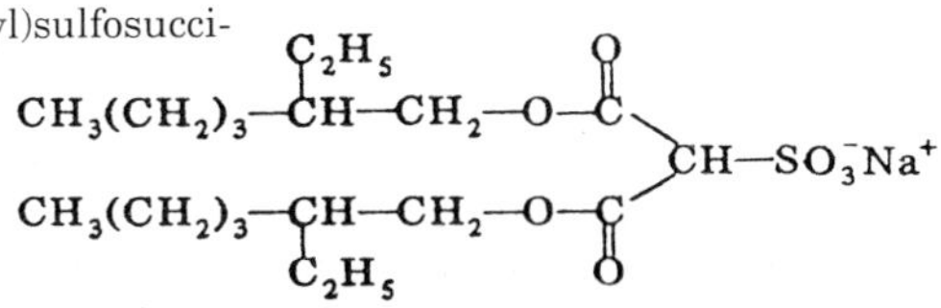	AOT
Sodium dihexadecylphosphate[b]	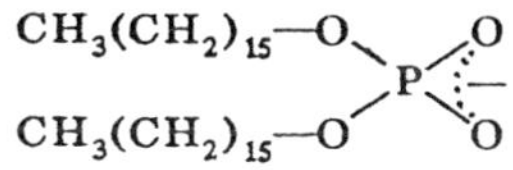	DHP
Cationic		
Hexadecyltrimethylammonium bromide (cetyltrimethyl-ammonium bromide)	$CH_3(CH_2)_{15}N^+(CH_3)_3Br^-$	HTAB
Hexadecyltrimethylammonium chloride (cetyltrimethylammonium-chloride)	$CH_3(CH_2)_{15}N^+(CH_3)_3\ Cl^-$	HTAC
Dodecyltrimethylammonium bromide (lauryltrimethyl-ammonium bromide)	$CH_3(CH_2)_{11}N^+(CH_3)_3\ Br^-$	DTAB
Decyltrimethylammonium bromide	$CH_3(CH_2)_9N^+(CH_3)_3\ Br^-$	DeTAB
Octyltrimethylammonium bromide	$CH_3(CH_2)_7N^+(CH_3)_3\ Br^-$	OTAB
Hexadecylpiridinium chloride (cetylpiridinium chloride)	$CH_3(CH_2)_{15}{-}\ C_6H_4N^+\ Cl^-$	HPC
Dodecylpyridinium chloride	$CH_3(CH_2)_{11}{-}C_6H_4N^+Cl^-$	DPC
Hexadecyldimethylbenzyl-ammonium chloride	$CH_3(CH_2)_{15}N^+(CH_3)_2CH_2C_6H_5\ Cl^-$	HDBAC
Dioctadecyldimethylammonium chloride[b]	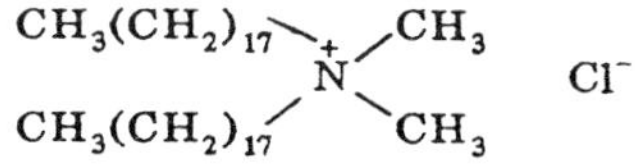	DODAC

continuation

TABLE 1.1 *continued*

Surfactant	Molecular structure	Abbreviation
Non-ionic		
Polyoxyethylene(4)dodecanol	$CH_3(CH_2)_{11}$–O–$(CH_2$–$CH_2O)_4H$	$C_{12}E_4$
Polyoxyethylene(6)dodecanol	$CH_3(CH_2)_{11}$–O–$(CH_2$–$CH_2O)_6H$	$C_{12}E_6$
Polyoxyethylene(8)dodecanol	$CH_3(CH_2)_{11}$–O–$(CH_2$–$CH_2O)_8H$	$C_{12}E_8$
Polyoxyethylene(6)hexadecanol	$CH_3(CH_2)_{15}$–O–$(CH_2$–$CH_2O)_6H$	$C_{16}E_6$
Polyoxyethylene(9)hexadecanol	$CH_3(CH_2)_{15}$–O–$(CH_2$–$CH_2O)_9H$	$C_{16}E_9$
Polyoxyethylene(23)dodecanol	$CH_3(CH_2)_{11}$–O–$(CH_2$–$CH_2O)_{23}H$	Brij 35
Polyoxyethylene(9.5)-t-octyl-phenol	$(CH_3)_3C$–CH_2–$C(CH_3)_2$–C_6H_4––$(OCH_2CH_2)_{9.5}H$	Triton X100
Polyoxyethylene(7.5)-t-octyl-phenol	$(CH_3)_3C$–CH_2–$C(CH_3)_2$–C_6H_4––$(OCH_2CH_2)_{7.5}H$	Triton X114
Polyoxyethylene(7.5) nonyl-phenyl ether	$CH_3(CH_2)_8$–C_6H_4–$(OCH_2CH_2)_{7.5}H$	PONPE 7.5
Polyoxyethylene sorbitan esters of fatty acids[c]	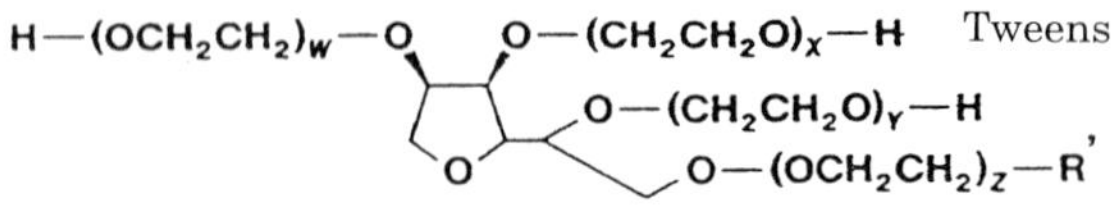	Tweens
Polyoxyethylene(20)sorbitol monolaurate		Tween 20
Polyoxyethylene(20)sorbitol mono-palmitate		Tween 40
Polyoxyethylene(20)sorbitol mono-stearate		Tween 60
Polyoxyethylene(20)sorbitol mono-oleate		Tween 80
Octylglucoside	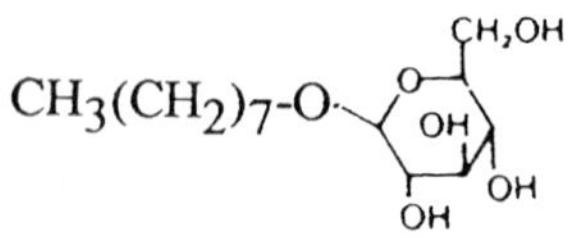	OG
Dodecylglucoside	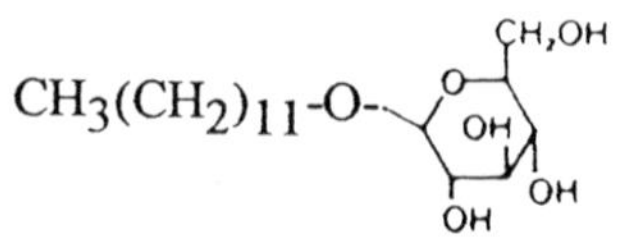	DG
Dodecylmaltoside (laurylmaltoside)	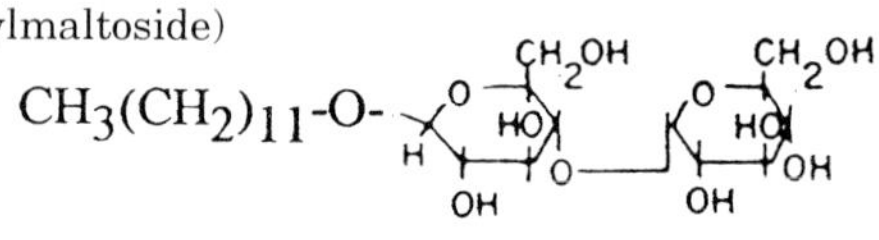	LM
Zwitterionic		
Dodecyldimethylammonium acetate (n-dodecylbetaine)	$CH_3(CH_2)_{11}(CH_3)_2N^+CH_2COO^-$	DDAA

Surfactant	Molecular structure	Abbreviation
Decyldimethylammonium acetate (n-decylbetaine)	$CH_3(CH_2)_9(CH_3)_2N^+CH_2COO^-$	DeDAA
Hexadecyldimethylammonium acetate (n-hexadecylbetaine)	$CH_3(CH_2)_{15}(CH_3)_2N^+CH_2COO^-$	HDAA
N-Decyl-N,N-dimethylammonium-3-propane-1-sulfonate (n-decylsultaine)	$CH_3(CH_2)_9(CH_3)_2N^+(CH_2)_3SO_3^-$	SB-10
N-Dodecyl-N,N-dimethylammonium-3-propane-1-sulfonate (n-dodecylsultaine)	$CH_3(CH_2)_{11}(CH_3)_2N^+(CH_2)_3SO_3^-$	SB-12
N-Hexadecyl-N,N-dimethylammonium-3-propane-1-sulfonate (n-hexadecylsultaine)	$CH_3(CH_2)_{15}(CH_3)_2N^+(CH_2)_3SO_3^-$	SB-16

Lecithins[d]

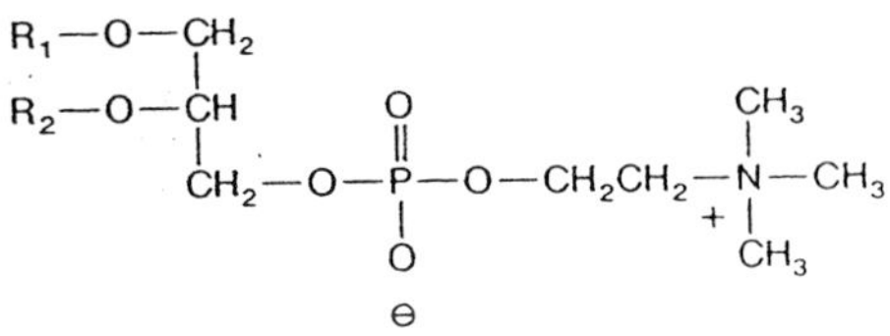

a: Reverse micelle forming; *b*: vesicle forming; *c*: the total number of polyoxyethylene units (indicated between parenthesis) is given by the sum: x + y + w + z; *d*: R_1 and R_2 represent alkanoyl groups.

1.1.1. Functionalized amphiphilic molecules

In respect to the conventional structure of surfactants shown in Fig. 1.1, a series of modifications can change the properties of such molecules and/or introduce a specific chemical function. These modifications may involve either the hydrophilic head-groups or the hydrocarbon chains.

Modification of the polar head: This is a very important method of surfactant functionalization. In particular it is possible to introduce suitable polar groups possessing acid-base, redox or chelating capabilities, which can form aggregates that are reactive towards target analytes present in the system.

The introduction of proper reactive counterions in solutions of ionic amphiphiles can also be adequately exploited to perform chemical reactions with selected substrate components.

Modification of the hydrophobic tail: This is another typical functionalization which includes: (i) the partial or total substitution of the hydrogen atoms of the hydrocarbon chains by fluorine; (ii) the introduction of a double hydrocarbon chain attached to the same hydrophilic moiety; (iii) the presence of a mixed double chain composed of a hydrocarbon part and a fluorinated one; (iv) the formation of "gemini" structures in which two amphiphilic units are connected through their hydrocarbon tails; (v) the formation of "bolaform" structures containing two terminal hydrophilic groups bounded to the same long hydrocarbon chain.

The modification of hydrophobic tails dramatically changes the properties of the amphiphile. For example, fluorinated surfactants are not attracted to the oil phase in an oil/water system and, thus, they are poor emulsifiers for those systems. Other important properties of these compounds are the large reduction in surface tension and their chemical and thermal stability. In fact, fluorinated surfactants can operate at higher temperatures and are able to reduce the surface tension of chromic acid solutions, where conventional hydrocarbon amphiphiles are usually destroyed.

Selected examples of functionalized reactive surfactants possessing different chemical groups, together with some schematic examples of amphiphile structures obtained through modification of the hydrophobic part of the molecule, are shown in Table 1.2.

TABLE 1.2

Typical structures of some functionalized amphiphiles

Functionalized polar head		
Redox function		
1-Dodecyl-1′-methyl-4,4′-bipyridine dichloride (dodecylmethylviologen dichloride)	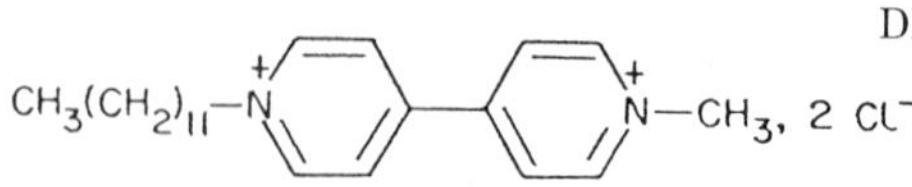	DMV
Ligand function		
5-n-Tetradecyl-1,10-diaza-4,7,13,16-tetraoxacyclooctadecane	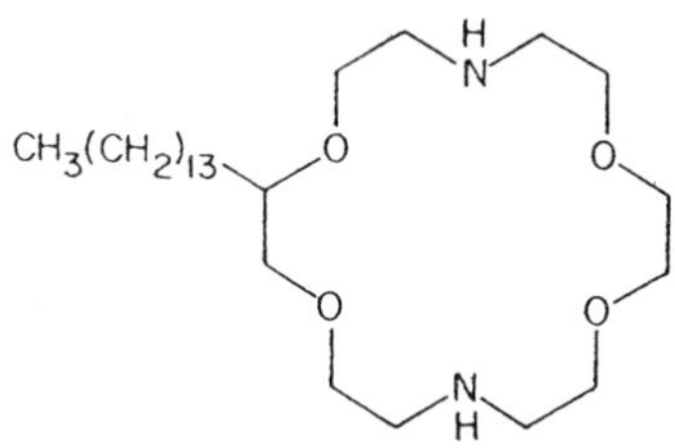	

4-Alkylamidosalicylic acid derivatives:

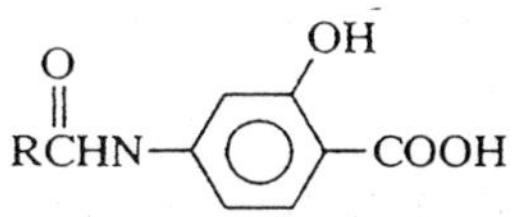

PAS-C_n [a]

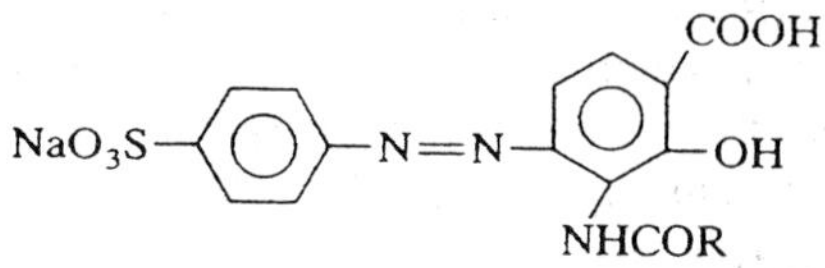

Y-PAS-C_n [a]

Amphiphiles forming ionic or non-ionic micelles as a function of pH

$R\text{-}(CH_2\text{-}CH_2\text{-}O)_n\text{-}CH_2COOH$

Reactive counterions of anionic and cationic surfactants

$-N^+R_3$, Cu^{2+}, Ag^+, Tl^+

$-SO_3^-$, $S_2O_8^=$, $SO_3^=$

Functionalized hydrophobic tail(s)[b]

Fluorinated: 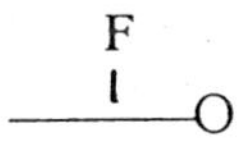

Mixed fluoro- and hydrocarbon chain: 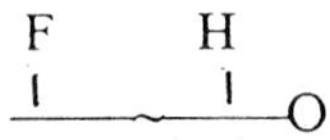

Double hydrocarbon chain: 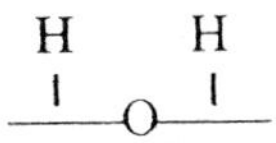

Hybrid double chain: 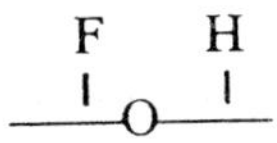

Gemini surfactants: 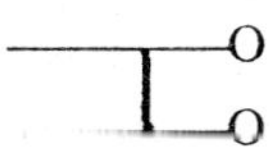

Bolaform surfactants: $O\text{–}C_{22}H_{44}\text{–}O$

a: $C_n = C_{n-1}H_{2n-1}CO$; *b*: the symbol represents a polar group.

1.2. PREPARATION AND PURIFICATION OF SYNTHETIC AND NATURALLY OCCURRING SURFACTANTS

Commercially available synthetic surfactants are often impure, containing either starting materials and/or mixtures of homologs often in unspecified proportions. Depending on the analytical applications, different degrees of purity may be requested. Isomerically pure surfactants are usually expensive to produce and generally have only a limited potential advantage in performance over the less expensive commercially available surfactants. However, accurate physical data are most reliable and comparable for highly purified surfactants.

Preparation and purification of the starting materials or of the surfactants (from unreacted initial compounds) include several techniques: recrystallization and washing with organic solvents (methanol, ethanol, acetone, diethyl ether), distillation, column chromatography, etc. Removal of water may be important and is accomplished by pulverization and drying in vacuo over phosphorus pentoxide.

The industrial preparation of non-ionic amphiphiles containing polyoxyethylene chains leads to polydisperse products, generally containing a Poisson distribution of ethylene oxide units around a mean which depends on the condensation conditions. They exhibit properties very similar to those of the pure surfactant corresponding to the mean and thus can be used without detriment in physicochemical investigations.

Pure homogeneous surfactants are preferable in some particular cases, but their synthesis and purification are tedious and arduous. They are commercially available, although costly.

The most important naturally occurring amphiphiles include simple and complex lipids, fatty acid esters (containing phosphorus, nitrogen bases or sugars) and bile acids. These surfactants may be of interest in some peculiar bioanalytical applications.

The isolation of pure homogeneous compounds from the starting complex mixtures is generally extremely difficult. Column, paper and thin-layer chromatography have been widely used to purify some of these materials, such as lipids. Details on preparation and purification procedures are reported in reference books [1,2].

1.3. SURFACE ACTIVITY IN AQUEOUS SOLUTION

The most significant characteristic of the amphiphilic molecules is the tendency to adsorb very strongly at the interface between air and

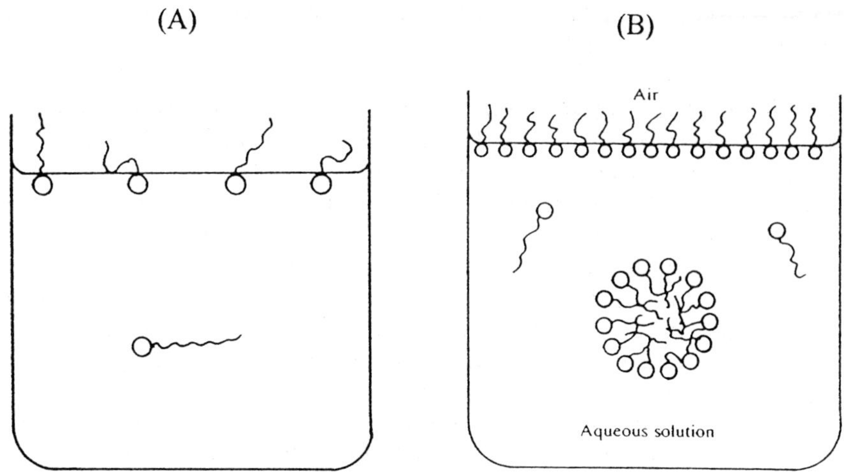

Fig. 1.2. Schematic arrangement of amphiphile (soap or detergent) molecules at low concentration in water (A), and above the CMC (B).

water (Fig. 1.2A). Such substances are said to be strongly surface-active because they lower the surface (or interfacial) tension, γ. They therefore favour the formation of a new surface and are widely used as dispersing agents.

The ability of a given surfactant to reduce the surface tension of the solvent is usually measured by its C_{20} value, where C_{20} represents the amphiphile concentration (in the solution phase) which lowers the surface tension of the pure solvent of 20 dynes cm^{-1}. "Gemini" surfactants [3,4] are among the most active, having a great tendency to adsorb at the liquid–air interface. For example, the amphiphilic molecule shown below exhibits a very low C_{20} value (ca. 1×10^{-6} M), whereas much higher concentrations of usual anionic surfactants are necessary ($C_{20} \approx 3 \times 10^{-3}$ M for SDS).

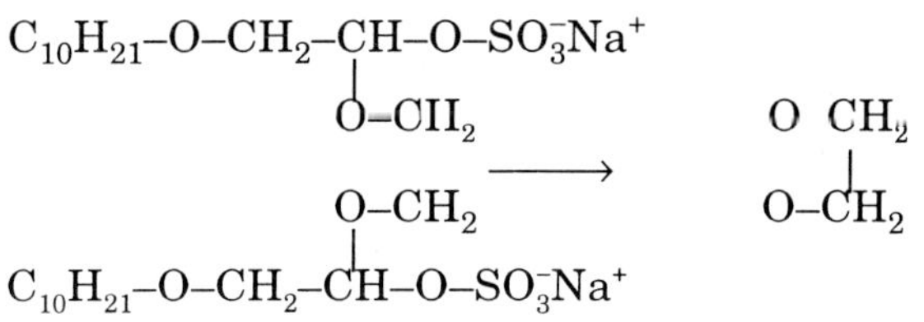

At very low concentrations (e.g. $^{-4}$ M) many surface active agents are soluble in water, forming simple solutions; if they are ionic, like the fatty acid soaps or the alkyl sulphate detergent, they will be dissociated as weak or strong electrolytes. As the concentration increases, the adsorption at the air–solution interface becomes stronger. Saturation is reached when the molecules are packed close together, with strong lateral interactions occurring between the hydrophobic chains, which tend to stick up out of the water (see Fig. 1.2B).

Another aggregation process, which often occurs at about the same surfactant concentration, is the formation of micelles. The concentration at which micelles first form in the solution is called the critical micellization concentration (CMC). The normal micelle aggregate consists of from 40 to 200 monomers (aggregation number, N_{AG}). In these structures the hydrophobic portions of the surfactant molecule associate together to form regions from which the solvent (water) is practically excluded.

The hydrophilic head groups remain on the outer surface to maximize their interaction with water and, for ionic amphiphiles, with the oppositely charged ions (counterions). A significant fraction of the

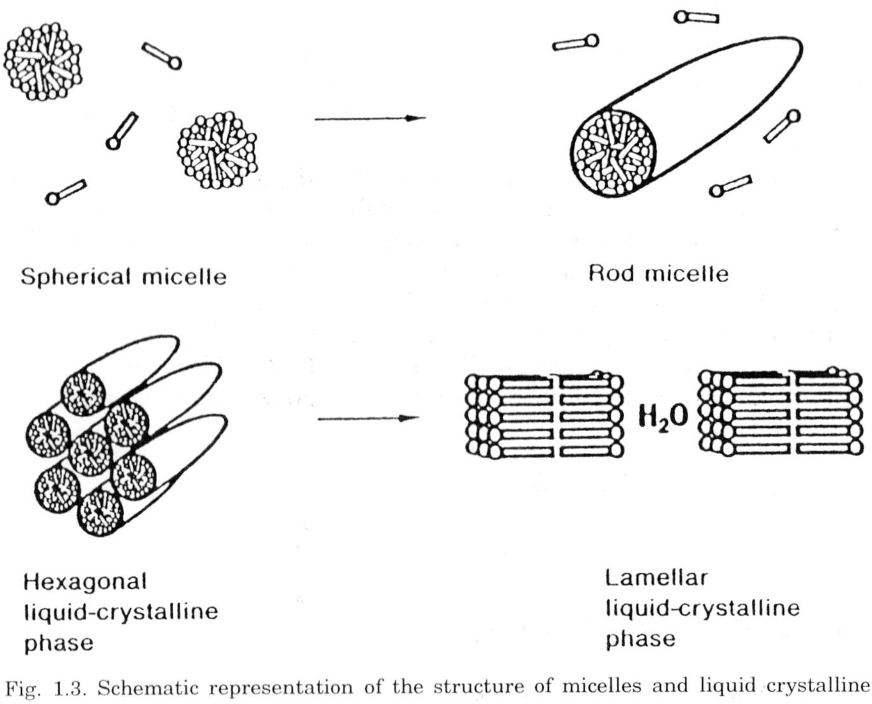

Fig. 1.3. Schematic representation of the structure of micelles and liquid crystalline phases.

counterions remains strongly bound to the head groups so that the lateral repulsive force between those groups is greatly reduced.

The precise structure of the micelle depends upon the temperature and concentration but also on the molecular structure: size of head group, length and number of hydrocarbon chains, presence of branches, double bonds or aromatic rings, etc. Increasing the concentration of the surfactant leads to the formation of rodlike micelles and, subsequently, to liquid crystals (see Fig. 1.3).

1.4. THE CRITICAL MICELLE CONCENTRATION (CMC)

As noted in the previous section, the amphiphiles are able to form aggregates called micelles in aqueous solution, provided their concentration is sufficiently high. The concentration at which this micelle formation occurs is usually fairly sharply defined and it can be identified by observing the behaviour of any one of a number of equilibrium or transport properties of the solution (see Fig. 1.4), each of which undergoes a rather abrupt change in concentration dependence at much the same point.

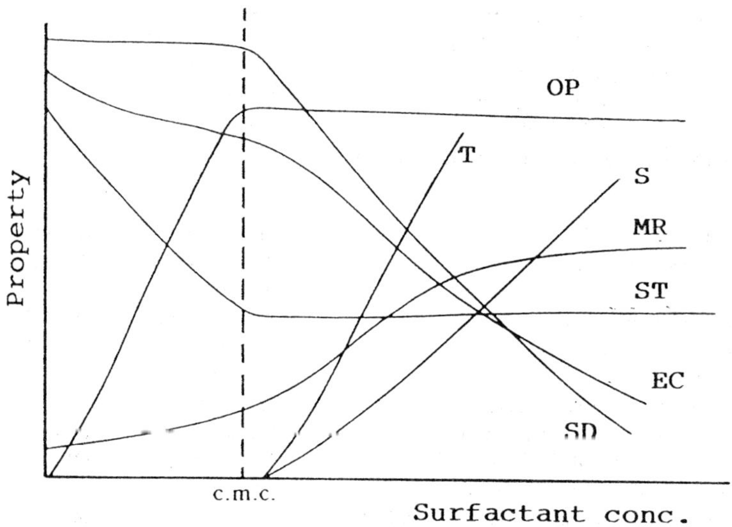

Fig. 1.4. Concentration dependence of some physical properties for solutions of micelle-forming amphiphiles. OP: osmotic pressure; T: turbidity; S: solubilization; MR: magnetic resonance; ST: surface tension; EC: equivalent conductivity; SD: self diffusion.

A very large and valuable collection of data on the CMC of various surfactants has been compiled by Mukerjee and Mysels [7]. Table 1.3 lists representative values of the CMC and aggregation numbers of some amphiphiles in aqueous solution.

TABLE 1.3

Critical micelle concentrations and aggregation numbers of selected amphiphiles[a]

Amphiphile	cmc (M)	N_{AG}
Anionic		
Sodium decylsulfate (SDeS)	3.3×10^{-2}	50
Sodium dodecylsulfate (SDS)	8.1×10^{-3}	62
Sodium hexadecylsulfate (SHS)	5.2×10^{-4}	100
Sodium dodecanoate (NaL)	2.4×10^{-2}	56
Potassium dodecanoate (KL)	1.2×10^{-2}	48
Sodium dodecylsulfonate (SDDS)	9.8×10^{-3}	54
Sodium cholate (NaC)	$(3.0–6.4)\times10^{-3}$	4–5
Sodium deoxycholate (NaDOC)	2.8×10^{-3}	
Cationic		
Octyltrimethylammonium bromide	1.8×10^{-1}	
Decyltrimethylammonium bromide (DeTAB)	6.5×10^{-2}	47
Dodecyltrimethylammonium bromide (DTAB)	1.5×10^{-2}	50
Hexadecyltrimethylammonium bromide (HTAB)	9.0×10^{-4}	61
Hexadecyltrimethylammonium chloride (HTAC)	1.3×10^{-3}	78
Hexadecylpyridinium chloride (HPC)	9.0×10^{-4}	95
Hexadecyldimethylbenzylammonium chloride (HDBAC)	2.7×10^{-4}	
Non-ionic		
Polyoxyethylene(4)octanol (C_8E_4)	$(6.5–8.5)\times10^{-3}$	82–147
Polyoxyethylene(8)octanol (C_8E_8)	$(9.3–10.4)\times10^{-3}$	41–72
Polyoxyethylene(4)decanol ($C_{10}E_4$)	$(6.4–8.1)\times10^{-4}$	30
Polyoxyethylene(4)dodecanol ($C_{12}E_4$)	$(2.3–6.4)\times10^{-5}$	40
Polyoxyethylene(6)dodecanol ($C_{12}E_6$)	6.0×10^{-5}	40–110
Polyoxyethylene(8)dodecanol ($C_{12}E_8$)	7.0×10^{-5}	120
Polyoxyethylene(8)hexadecanol ($C_{16}E_8$)	$(5–6)\times10^{-7}$	
Polyoxyethylene(9.5)-t-octylphenol (Triton X100)	2.3×10^{-4}	140
Polyoxyethylene(23)dodecanol (Brij 35)	1.0×10^{-4}	40
Polyoxyethylene(10)hexadecanol (Brij 56)	6.0×10^{-7}	620
Polyoxyethylene(20)sorbitol monooleate (Tween 80)	1.2×10^{-5}	60
Polyoxyethylene(20)sorbitol monolaurate (Tween 20)	$(4.0–5.9)\times10^{-5}$	
Polyoxyethylene(15)nonylphenol (Igepal CO-730)	2.8×10^{-4}	52–80
Polyoxyethylene(20)nonylphenol (Igepal CO-850)	$(1.3–1.8)\times10^{-4}$	62

Octylglucoside (OG)	2.5×10^{-2}	27
Dodecylglucoside (DG)	1.9×10^{-4}	
Dodecylmaltoside (LM)	2.0×10^{-4}	98
Zwitterionic		
n-Decylsultaine (SB-10)	3.9×10^{-3}	
n-Dodecylsultaine (SB-12)	1.2×10^{-3}	55
n-Hexadecylsultaine (SB-16)	1.0×10^{-4}	
n-Decylbetaine (DeDAA)	$(1.0–2.1)\times10^{-2}$	34
n-Dodecylbetaine (DDAA)	1.5×10^{-3}	73
n-Hexadecylbetaine (HDAA)	2.0×10^{-5}	
Dioctanoylphosphatidylcholine (C_8-lecithin)		500
Functionalized		
Gemini		
$C_{10}H_{21}–O–CH_2–CH–OSO_3^-Na^+$ $O–CH_2$ $O–CH_2$ $C_{10}H_{21}–O–CH_2–CH–OSO_3^-Na^+$		
Fluorinated		
Sodium perfluorohexanoate[b]	1.1×10^{-1}	
Sodium perfluorooctanoate	9.3×10^{-3}	
Sodium perfluorooctylsulfonate[c]	8.5×10^{-3}	
C_7H_{15} $\rangle CH–O–SO_3^-Na^+$ C_7F_{15}	4.8×10^{-4}	
Complex forming		
5-n-Tetradecyl-1,10-diaza-4,7,13,16-tetraoxacyclo-octadecane	1.3×10^{-4}	
1-Methyl-10-n-dodecyl-1,10-diaza-4,7,13,16-tetraoxa-cyclooctadecane	3.7×10^{-5}	110
	9.2×10^{-5d}	60^d
Sodium 4-octylamidosalicylate (PAS-C_8)[e]	8.5×10^{-3}	25
Sodium 4-dodecylamidosalicylate (PAS-C_{12})[e]	4.2×10^{-4}	164
Surfactants bearing reactive counterions		
Silver dodecylsulfate[f]	7.3×10^{-3}	
Copper(II) dodecylsulfate[g]	1.2×10^{-3}	85

Cmc values taken from Refs. [2,7–21]. *a*: Aqueous solutions at 25°C; *b*: 18°C; *c*: 75°C; *d*: data of the 1:1 complex with Ba^{2+}; *e*: in NaOH 0.1 M, at 45°C; *f*: 35°C; *g*: 30°C.

1.4.1. The measurement of cmc

The critical micelle concentration is a crucial parameter in the characterization of the amphiphiles. Surface tension, optical turbidity, electric conductivity, osmotic coefficient, density, sound velocity, diffusion, viscosity, solubilization, ultracentrifugal sedimentation, NMR chemical shifts, are among the properties that have been used in determining the cmc.

However, it is not to be expected that all will be equally convenient or even give the same results. Mukerjee and Mysels give a detailed discussion of most of the applied methods, together with tabulated values for more than seven hundred compounds under various conditions [7].

Usually, the selected property (or some function of it) is plotted against amphiphile concentration; the functions are selected so that the graph consists as nearly as possible of a pair of straight lines whose intersection can be taken as the CMC. Extreme care and caution are needed for a correct CMC evaluation.

A common method, principally for convenience, is surface tension measurement. The preferred plot is of γ versus the logarithm of the concentration, since the relation is usually virtually linear before the cmc is reached, with a very sharp change to near constancy at the cmc. Lowering is of the order of tens of dynes per centimeter below the cmc and dynes per centimetre or less above the cmc per tenfold concentration change.

The determination of a cmc value from changes in colour, absorption spectrum, or the amount of solubilizate, is popular in routine commercial work and was once widely used also in fundamental studies. This method, although convenient if discreetly used, may suffer pitfalls stemming from the effect of the indicating additives on the micellization process and from chemical reactions between indicator and amphiphile. In many cases an insoluble salt may form between dye and amphiphile ions, which is insoluble below cmc but is solubilized in the micelles above it. Solubilized materials generally lower cmc values significantly (see below).

Conductance, suitable for strong electrolyte amphiphiles, has the advantage that very high precision is not too difficult to achieve (from 10 to 100 ppm in the individual conductance measurement). However, obtaining and maintaining sufficient purity of materials to warrant such precision is often difficult. For cmc determination the preferred plot is of equivalent conductance against square root of molality, because this has a strictly linear form (Onsager slope) at high dilution, with a large change as micelles form.

Of the colligative properties, vapour pressure lowering is the most common measure of particle solution concentration in micellar systems. It is a sensitive but not very accurate measurement, since it is difficult to maintain saturation pressure without change of concentration in the minute samples employed.

1.4.2. Factors affecting the cmc

(i) *Nature of the hydrophobic group:* The large majority of amphiphiles, either ionics or non-ionics, have hydrophobic regions composed of hydrocarbon chains. For ionic amphiphiles, increase in the number of carbon atoms in unbranched hydrocarbon chains leads to a decrease in the cmc. Figure 1.5 shows a linear relationship between log cmc and the number of carbon atoms in the chain, m, for compounds with the same head group.

The dependence of cmc on m may be expressed by the empirical equation:

$$\log \text{cmc} = A - B\,m \qquad (1)$$

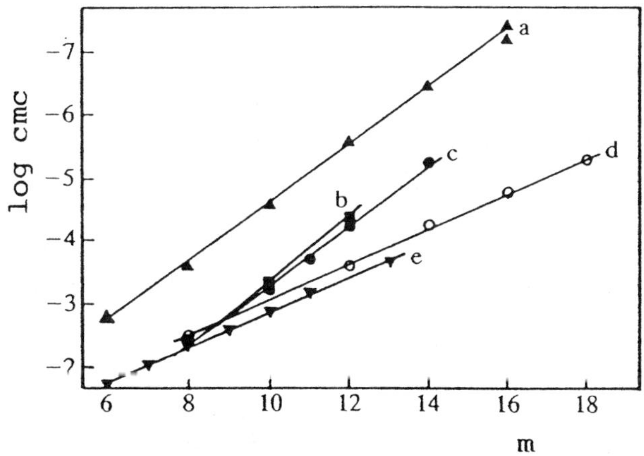

Fig. 1.5. Plots of log cmc (in mole fraction units) versus m. Temperature is generally 25°C. (a) Alkyl hexaoxyethylene glycol monoethers. (b) Alkyl trimethylammonium bromides in 0.5 M NaBr. (c) N-alkyl betaines. (d) Sodium alkyl sulfates (40°C). (e) Sodium alkylcarboxylates. Taken from Ref. [6], reproduced with permission.

where A and B are constants for a homologous series. As a general rule for ionic surfactants, the cmc decreases by roughly a factor of 2 when the length of the straight hydrocarbon chain is increased by one methylene group. For chains of greater length than 16 carbon atoms this relationship no longer holds and further increase in chain length often has no appreciable effect on the cmc, possibly due to the coiling of the long chains in solution.

Modifications to the hydrocarbon chain (such as introducing branching, or double bonds, or polar functional groups along the chain) usually lead to increases in the cmc. The introduction of a benzene ring is equivalent to adding about 3.5 methylene groups to the chain length.

Introducing fluorine in place of hydrogen has a quite marked effect, at first increasing and then decreasing the cmc (ultimately to less than 10% of the hydrocarbon value), as the proportion of fluorine atoms is increased towards saturation [22].

(ii) *Nature of the hydrophilic group:* There is a pronounced difference between the cmc's of ionic and non-ionic surfactants with identical hydrophobic moieties. The lower cmc's of the non-ionic surfactants are a consequence of the lack of electrical work necessary in forming the micelles.

Most ionics of given chain length show very similar values for cmc. For example for the straight chain C_{12} (dodecyl) surfactants the following values (with the counterion in brackets) are reported in the literature:

Anionics: 8×10^{-3} M for $-O-SO_3^-(Na^+)$; 10×10^{-3} M for $-SO_3^-(Na^+)$; 12×10^{-3} M for $-COO^-(K^+)$.

Cationics: $20-10^{-3}$ M for $-N^+(CH_3)_3(Cl^-)$; 15×10^{-3} M for $-NH_3^+(Cl^-)$.

As might be expected, the more ionized groups there are present in the surfactants, the higher the cmc due to the increase in electrical work required to form the micelle. The position of the ionic group also affects the micellar properties.

For the polyoxyethylated ether type of non-ionic surfactant, increase of the length of the polyoxyethylene chain makes the monomer more hydrophilic and the cmc increases.

(iii) *Nature of the counterion:* In conventional ionic surfactants a change in counterion to one of greater polarizability or valence leads to a decrease in cmc. The size of the counterion is also a determining factor, an increase of cmc is noted with increase in hydrated radius. Counterion effects on the cmc of two typical surfactants are shown in Table 1.4.

The analytical applications based on the effect of transition metal ions, either aquoions and complexes (e.g. with phenanthroline-like

TABLE 1.4

Variation of cmc with different counterions

Dodecyl sulfates	cmc (M×10^3)^a	Dodecyltrimethyl-ammonium salts	cmc (M×10^3)^b
Li^+	8.9	F^-	8.4
Na^+	8.3	Cl^-	3.8
K^+	7.2^c	Br^-	1.9
Cs^+	6.1	NO_3^-	0.8
$(CH_3)_4N^+$	5.5	BrO_3^-	3.3
$(C_2H_5)_4N^+$	3.8$_5$	IO_3^-	5.1
$(n\text{-}C_3H_7)_4N^+$	2.2	CH_3COO^-	6.0
$(n\text{-}C_4H_9)(CH_3)_3N^+$	2.4		

a: In water at 25°C; *b*: in 0.5 M sodium salt, at 31–32°C; *c*: at 32°C. Data taken from Ref. [23], reproduced with permission.

ligands or crown ethers) are of particular interest. These systems will be described later.

(iv) *Effect of organic molecules:* Quite small amounts or organic material can have a significant influence on the cmc, and the properties of micellar solutions. The classical example of this behaviour occurs in aqueous solutions of sodium dodecyl sulfate (SDS), where the presence of dodecanol (an impurity or a hydrolysis product) causes a minimum in the surface tension due to the competing effects of adsorption of dodecanol at the aqueous solution–air interface and solubilization in the SDS micelles. Such anomalous behaviour is to be expected between supposedly similar industrial surfactants, as a consequence of the presence of impurities or manufacturing byproducts.

It is an important aspect of the behaviour of micelles that they are able to act as sites for the dissolution of lipophilic (i.e. fat-soluble) molecules. The use of surfactants as detergents, stabilizers and dispersing agents depends on this property, known as solubilization, as we will see in Chapter 2. It is characterized by a dramatic increase in the solubility of the lipophilic material at concentrations of the surfactant above the cmc.

It is common practice to divide organic materials into two main groups, depending on their mode of action in influencing the cmc. The first group is composed of molecules (like alcohols with moderate to long hydrocarbon chains) that appear to be adsorbed in the outer regions of the micelle, forming a palisade structure with the surfactant molecules. This lowers the free energy of micellization to more negative values and so reduces the cmc. Moreover, such molecules can also influence the micelle shape.

Straight chain molecules have the most marked effect, the latter reaching a maximum when the length of the hydrophobic chain of the additive is about the same as that of the surfactant. A decreased electrostatic repulsion between ionized head groups, and reduction in steric hindrance for non-ionic surfactants have been proposed as likely explanations for these observed effects. The compounds of the first group are generally effective at quite low bulk concentrations.

The second group of compounds alter the cmc at substantially higher bulk concentrations and probably exert their influence through modification of the bulk water structure. The effect is usually discussed in terms of whether the additive is a (water) structure-maker or a structure-breaker.

Typical structure-makers are xylose and fructose [24], whereas urea and formamide are examples of structure-breakers [25]. The short-chain alcohols, like ethanol, can act in both groups.

Structure-breakers increase the cmc of surfactants in aqueous solution, exerting their strongest influence on non-ionic surfactants of the polyoxyethylene type. The variations of the thermodynamic parameters with temperature and with composition are often so complex and difficult to explain that the dichotomy into makers and breakers of structure may appear too simplistic.

The above classification remains, however, a useful notion in the extreme case to distinguish the increased interactions that occur between water molecules because a hydrocarbon moiety is present, and those that occur between solute and water (usually hydrogen bonding) that disturb the normal (very strong) hydrogen bonding of water itself.

An interesting example is represented by the effect of n-alcohols on the cmc of ionic surfactants. A linear decrease of cmc with increasing alcohol concentration has been observed (see Fig. 1.6).

In contrast to the extensive study of the effect of added alcohols to ionic micelles, only a few studies have been reported for non-ionic systems. Whereas methanol and ethanol cause a cmc increase, the higher alcohols butanol and pentanol cause a decrease in this property. Propanol exhibits an intermediate effect, low concentrations causing a decrease in cmc, higher concentrations (e.g. >1 mol l^{-1}) causing an increase.

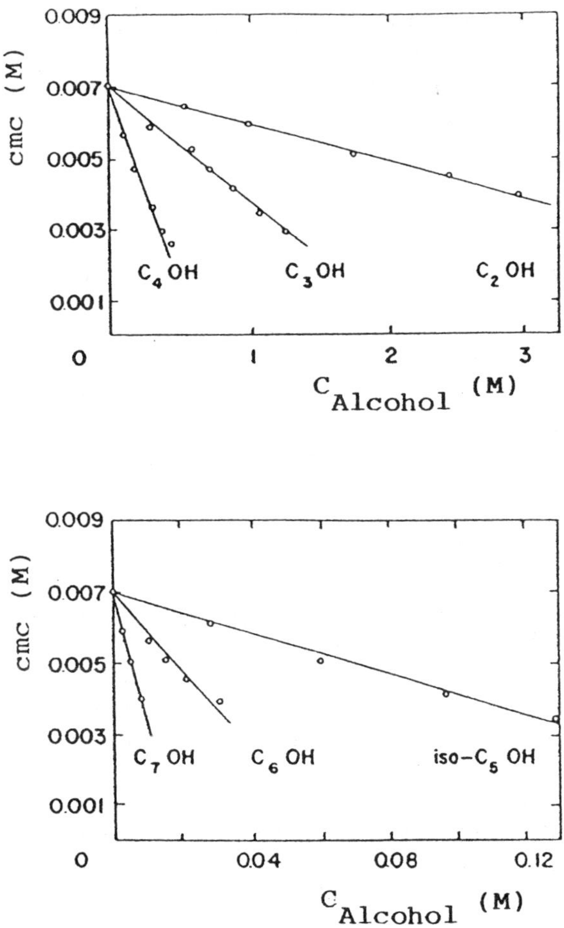

Fig. 1.6. Effect of different alcohols on the cmc of potassium tetradecanoate at 18°C. Reproduced from Ref. [26] with permission.

(v) *Effect of temperature:* The most drastic effect of temperature on surfactants is the effect on solubility. Figure 1.7 shows the variation of the cmc with temperature for an ionized surfactant (sodium dodecyl sulphate) and a non-ionic surfactant ($C_{10}E_5$). The minimum in the curve for ionic surfactants occurs typically between 20 and 30°C. Isolated examples exist of minima in cmc-temperature curves for non-ionic surfactants, e.g. minima were noted at approximately 50°C in a series of octylphenoxyethanols with oxyethylene chain lengths of between 6 and 10 [27].

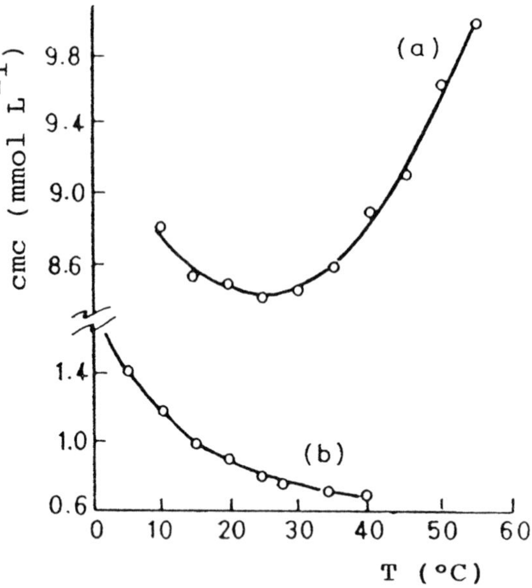

Fig. 1.7. Variation of cmc with temperature for: (a) sodium dodecyl sulfate; (b) $CH_3(CH_2)_9$–$(OCH_2CH_2)_5OH$. Reproduced from Refs. [13,28] with permission.

The general failure to detect minima in curves for non-ionics could conceivably be a consequence of a lack of data at sufficiently high temperatures. In many cases such measurements would not be feasible due to phase separation at elevated temperatures. The decrease in the cmc of ionic surfactants with temperature increase at lower temperatures is possibly due to dehydration of the monomers, whilst further temperature increase causes disruption of the structured water around the hydrophobic groups which opposes micellization.

(vi) *Effect of pressure:* The effect of pressure on the cmc of a series of alkyltrimethylammonium bromides and on sodium dodecyl sulfate has been studied by different techniques. An increase in cmc with pressure increase was reported up to pressure of about 150 MPa, followed by a cmc decrease at higher pressures. Evidence for a maximum in the cmc-pressure plot around 150 MPa seems to emerge. More recent studies have, however, shown a monotonic increase in the cmc of SDS with pressure up to 550 MPa and it has been suggested that the reversal of behaviour reported by earlier workers may be an artefact arising from the conductivity technique.

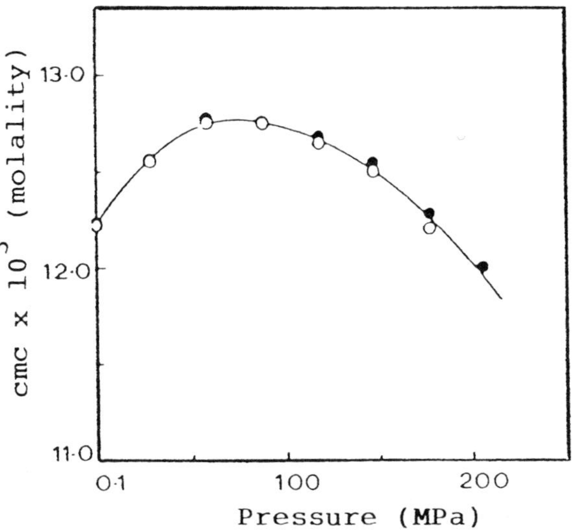

Fig. 1.8. Variation of cmc of dodecylpyridinium bromide as a function of pressure at 30°C: (O) optical method; (●) electroconductivity method. Reproduced from Ref. [29] with permission.

The differential absorbance measurements involved the use of an optical probe (naphthalene) which was solubilized within the SDS micelles. This method has been criticized on the grounds that the cmc's determined by these workers were those for the formation of mixed SDS–naphthalene micelles, and thus related to the effect of pressure on the solubilization process. Further evidence for a maximum in the cmc–pressure plot was presented [29], obtaining contrasting conclusions. Figure 1.8 compares cmc values from this method and those previously determined using the electroconductivity approach.

1.5. EQUILIBRIUM TREATMENT OF MICELLE FORMATION

The formation of micelles can be treated in a formally rigorous way in terms of all of the possible equilibria:

$$D + D \overset{K_2}{\rightleftarrows} D_2 + D \overset{K_3}{\rightleftarrows} D_3 \; \ldots\ldots \; D_{n-1} + D \overset{K_n}{\rightleftarrows} D_n \qquad (2)$$

with equilibrium constants K_n for $n = 2 - \infty$. Unfortunately, it is not possible to measure the individual equilibrium constants, and recourse must be had to one of a number of models to simplify the situation.

Observations of the size of more or less spherical micelles near the cmc (such as those of SDS) suggest that the size range is very limited. The simplest assumption to make in treating Eqn. (2) is, therefore, that only one of the K_n values is important. For SDS it would be about K_{60} at 25°C and, in that case, the micelle formation is represented as:

$$nD \rightleftarrows M \tag{3}$$

for which the equilibrium constant, K, is:

$$K + \frac{[\text{micelles}]}{[\text{monomers}]^n} = \frac{C_M}{C_D^n} \tag{4}$$

The inherent assumption here is that the activities may be replaced by concentrations. For the monomer this implies the assumption that the only departure from ideal behaviour is the aggregation process. It could, in principle, be removed by estimating other activity corrections from solution theory. Assuming ideal behaviour for the micelles is more problematical because of the large size difference between monomers and micelles.

The micelles will also interact strongly. For ionics the interaction will become very significant as soon as the mean separation is less than about (8–10)/***K***, where ***K*** is the Debye–Hückel parameter, and that occurs at surfactant concentrations not far above the cmc. It must also be noted that when ionic micelles are formed there is a strong tendency for the counterions to be associated closely with the head groups, because of the high electrostatic potential in that region. This is a further source of non-ideality.

From Eqn. (4), the following expression can be derived:

$$\Delta G^o = -RT \ln K = -RT \ln C_M + n\,RT \ln C_D \tag{5}$$

and thus:

$$\frac{-\Delta G^o}{n} = -\,\Delta G^o = \frac{RT}{n} \ln C_M - RT \ln C_D \tag{6}$$

At the cmc:

$$C_D = C_0 \tag{7}$$

and the total surfactant concentration, C_T, above this point is given by:

$$C_T = C_0 + n\, C_M \tag{8}$$

From Eqns. (4) and (8):

$$K = \frac{C_M}{(C_T - nC_M)^n} \tag{9}$$

It has been shown that Eqn. (9), with a K value of unity and n = 100 gives rise to a sharp transition from a system in which all of the surfactant is present as monomer to one in which the monomer concentration remains essentially constant above the cmc and all additional surfactant goes into micelle formation. As an alternative approach, Eqn. (9) can be differentiated:

$$\frac{dC_m}{dC_T} = \frac{K^{1/n}}{[nK^{1/n} + 1/n\, C_m^{((1-n)/n)}]} \tag{10}$$

and the plot of dC_M/dC_T against concentration for K = 1 and various values of n is shown in Fig. 1.9.

When it is realized that for most of the commonly used surfactants the value of n is at least 50, it becomes clear that the concept of a critical micellization concentration is a reasonable one.

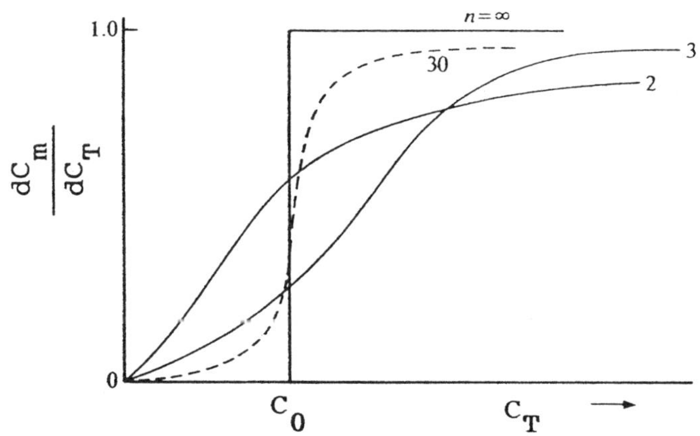

Fig. 1.9. Variation of dC_M/dC_T with total surfactant concentration for different values of the aggregation number, n. C_0 is the critical micellization concentration and C_M the concentration of micelles. Taken from Ref. [5], reproduced with permission.

As $n \rightarrow \infty$ the transition becomes sharper and ultimately approaches the behaviour expected of a first-order phase transition. In this limit case, it is possible to treat the micelle formation as a phase separation.

The association model just described is not, however, physically appealing. If, for example, n = 60, it is difficult to see why the addition of one extra monomer to the aggregate of 59 should drastically reduce the free energy of the aggregate. Moreover, why should it be difficult, or impossible, to add an additional monomer? In order to obtain a physically reasonable model is necessary to write out the full equilibrium between monomer and micelles of all sizes and then to seek a physical basis for defining a relationship between the equilibrium constants. Such a relationship should predict the observed facts of micelle size (or size distribution) and the thermodynamic parameters, while still having only a small number of adjustable parameters. The treatment here follows that of Ref. [30], with some modifications.

Consider again the equilibria in Eqn. (2). For any n-mer, the stepwise association constant is:

$$X_{n-1} + X_1 \overset{K_n}{\rightleftarrows} X_n \tag{11}$$

The overall association constant for the formation of X_n from the monomer X_1 (i.e. n X_1 X_n) is:

$$\beta_n = \frac{[X_n]}{[X_1]^n} \tag{12}$$

where:

$$\beta_n = \prod_2^n K_n \tag{13}$$

(i.e. β_n is the product of all stepwise association constants K_2, K_3 ... up to K_n). The constant β_n is defined in terms of concentrations (instead of activities), with all of the uncertainties related to this simplification.

The concentration of all solute species, S, (in mol l^{-1}) is given by:

$$S = \Sigma[X_n] \tag{14}$$

and, expressed in terms of the total monomer concentration, M_1, Eqn. (14) becomes:

$$M_1 = \Sigma n[X_n] \tag{15}$$

The number average degree of association, N_n, of all species (including the monomer) is then given by:

$$N_n = \frac{\Sigma n[X_n]}{\Sigma[X_n]} = \frac{M_1}{S} \tag{16}$$

The corresponding values of number average for the associated species are:

$$N_n = \frac{\sum_2 n[X_n]}{\sum_2 [X_n]} = \frac{M_1 - [X_1]}{S - [X_1]} \tag{17}$$

It is evident from Eqn. (12) that, as $[X_1]$ increases, the concentration of each associated species also increases. The value of n influences each associated species and the percentage increase of each increases with n; moreover, the number average (N_n) must increase as the total concentration, M_1, increases. On the other hand, with increasing dilution: $M_1 \to 0$, $N_n \to 1$, and $M_1 \approx [X_1]$.

The individual K_n values define the concentrations at which particular n-mers become most important and, hence, they control the value of N_n at any concentration. The products of association and their size distribution depend on concentration and on $K_n(n)$, i.e. how K_n depends upon n, which reflects the molecular architecture of the associated species.

Amongst the different types of association behaviour which may be identified (dimerization, aggregates with a wide range of aggregation numbers) the case of formation of micelles with a narrow range is the most relevant to the present purposes.

This is the situation that was examined in the closed association model. The delicate force balance involved in micelle formation leads not to an increase or decrease in K_n with n, but rather to a value of n for which K_n is a maximum. The values of K_{n-1} and K_{n+1} will be of comparable magnitude but one will expect to find the resulting size distribution to be fairly narrow.

As mentioned earlier, the existence of a preferred micelle size is easily explained on the basis of the competing requirements for bringing the hydrocarbon chains into intimate contact, away from the aqueous environment, while maintaining the head groups as far apart as possible. There is very little energy change in separating the hydrocarbon chains from the water and the negative free energy associated with that process stems largely from the concomitant entropy increase in the water. In the case of an ionic surfactant, as each additional monomer is added to the micelle, the contribution to the free energy change becomes less negative because the developing micellar charge causes an increasing (positive) free energy change, reflecting the repulsion between the head groups.

Even when n is large, a relatively broad maximum in the values of K_n may produce a narrow size distribution. The free energy change involved in the formation of the n-mer from the monomer is, from Eqns. (5) and (12):

$$\Delta G_n^o = -RT \ln \beta_n \tag{18}$$

and only a slight minimum in the plot of $\Delta G_n^o/n$ against n is required to produce narrow size distributions of micelles [31]. Mukerjee [30] gives a striking illustration of this point.

Using the following empirical expression for $\ln \beta_n$:

$$\ln \beta_n = 2(n-1)\ln(n-1) - 0.02\,(n-1)^2 + 2.7896(n-1) \tag{19}$$

which exhibits a very broad maximum (see Fig. 1.10), the values of $n[X_n]$ were calculated from Eqn. (12) (i.e. the concentration of n-mers expressed in terms of monomer concentration) for a monomer concentration of 4.11×10^{-5} mol l^{-1} (i.e. $\ln [X_1] = -10$). The result is a narrow size distribution peaking at n = 97 and with a half-width of less than 10, even though $\Delta G_n^o/n$ changes by less than 2% over the whole range from 70–120 for n.

Figure 1.10 clearly shows why the calculation of the anticipated size distribution in a particular case requires accurately measured β_n values near the maximum or good estimates of ΔG_n^o.

When the monomer chains are very long, very large polydisperse aggregates apparently form, even in dilute solution [32]. These micelles are thought to be flexible cylinders and may be described by a self-association model similar to that discussed above [33]. The molecular requirements for micelles of various geometries will be discussed later.

The estimate of the value of ΔG_n^o as accurately as possible should allow the determination of the micelle size distribution and of the other properties.

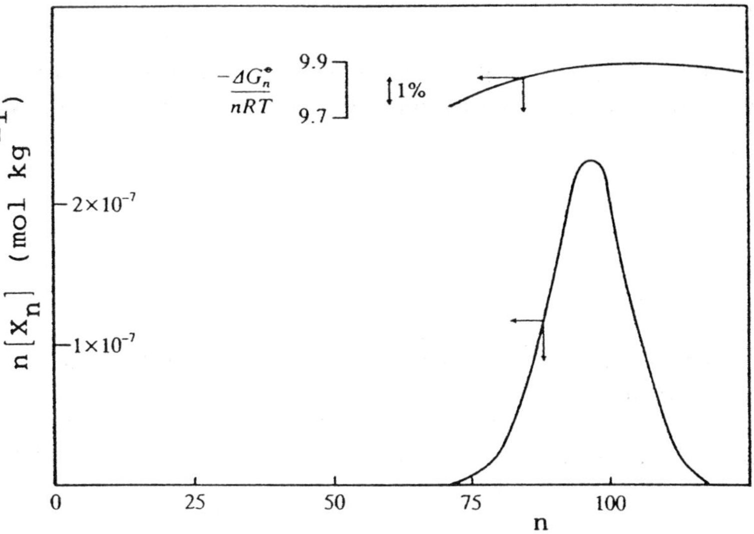

Fig. 1.10. Variation of $n[X_n]$ as a function of the number of monomers per micelle, n, for an assumed free energy profile, $\Delta G_n^o/n$. Figure taken from Ref. [30], reproduced with permission.

1.6. THERMODYNAMICS OF MICELLE FORMATION

Another approach to micellar equilibria is to try to deduce the free energy of micellization from considerations of molecular geometry and interactions. The use of the relation $\Delta G^o = -RT \ln K$, requires a definition of the standard states. For monomers the infinite dilution in water with concentration expressed in mole fraction X_1 is usually adopted, writing for its chemical potential:

$$\mu_1 = \mu_1^o + RT \ln X_1 \tag{20}$$

For a micelle that contains n monomers, the corresponding expression is:

$$\mu_{mic} = \mu_{mic}^o + \frac{RT}{n} \ln \frac{X_{mic}}{n} \tag{21}$$

where the standard state at infinite dilution and the free energy of incorporating one mole of amphiphile into micelles of size n is considered. The relation of the chemical potentials (again neglecting activity coefficients) is:

$$n(\mu^{o}_{mic} - \mu^{o}_{1}) = -RT \ln \beta_n \tag{22}$$

where β_n is defined by Eqn. (12). If this equation is written in logarithmic form (remembering that concentrations are expressed as mole fractions):

$$\Delta G^{o} = -RT \ln X_n + nRT \ln X_1 \tag{23}$$

and:

$$\Delta G^{o} = -RT \ln (X_{q+1}/X_q) + RT \ln X_1 \tag{24}$$

It can be expected for either of the above relations that the first term will be substantially smaller than the second. Furthermore, from surface tension results, the monomer concentration can be assumed as substantially equal to the cmc at higher concentrations. Accordingly, the following expression holds:

$$\Delta G^{o} = RT \ln \text{cmc} \tag{25}$$

interpreting ΔG^{o} as either the free energy change per monomer of forming a micelle of size n or the free energy of adding one monomer to such a micelle.

1.6.1. Estimation of $\Delta\overline{G}^{o}$

It seems reasonable to assume that the free energy change associated with the transfer of one mole of monomer from free solution into a micelle could be broken into various contributions:

$$\Delta\overline{G}^{o} = \Delta\overline{G}^{o}_{hc} + \Delta\overline{G}^{o}_{H} \tag{26}$$

where the subscripts hc and *H* represent the contribution of the hydrophobic moiety and that of the hydrophilic head group, respectively.

This kind of breakdown is suggested by the solubility and vapour pressure behaviour of homologous series of organic molecules, in which a constant increment or decrement is noted for each additional $-CH_2$ group (see Eqn.(1)).

On the other hand the contribution of the hydrophilic head group ΔG^{o}_{H} to the free energy of micellization is invariably positive and so opposes the process. Very little progress has been made on the calculation of $\Delta\overline{G}^o_H$ for non-ionic molecules. It is assumed to arise from steric interactions as the large head groups crowd the surface but beyond that there is little that can be said from a theoretical point of view; the contributions of different non-ionic head groups can, however, be evaluated from experimental data.

By contrast, more progress has been made on the calculation of $\Delta\overline{G}^o_H$ for ionic surfactants, assuming that it is dominated by the electrostatic effects predictable from the Gouy–Chapman theory of the double layer. The procedure is analogous to that used in the estimation of the electrical part of the free energy of other colloidal particles, except that instead of working in terms of unit area we consider the energy change involved in establishing an n-mer (of charge $q = ne$). It has been shown [34], however, that a more exact calculation, using the computer-calculated relation between the potential and q for spherical particles, can satisfactorily account for the variation of cmc with salt concentration in SDS solutions at 25°C over the range from 0 to 0.2 M NaCl.

At somewhat higher salt concentrations (above about 0.45 M) the aggregation number of SDS increases dramatically to over 1000 and a more elaborate model is required.

1.6.2. Enthalphy and entropy of micelle formation

The variation of entropy, $\Delta\overline{S}^o$ can be obtained from: $\Delta\overline{S}^o = (\Delta\overline{H}^o - \Delta\overline{G}^{o)}/T$. The resulting ΔH^o and $\Delta\overline{S}^o$ data have been used to develop a deeper insight into the micelle formation process.

The most striking feature is the relation between $\Delta\overline{H}^o$ and $T\Delta\overline{S}^o$, which is shown in Fig. 1.11. Taking T = 27°C a slope of unity is obtained, so that this is called the compensation temperature. It is significant that this compensation (which makes $\Delta\overline{G}^o = 0$) should occur so close to room temperature, emphasizing how delicate the balance between energy and entropy must be in the micellization process.

As noted earlier it has, until very recently, been assumed that the major factor driving the surfactant molecules into aggregation in water is a positive entropy change, presumably associated with breakdown of the structured water which surrounds the hydrocarbon chain in the unassociated species. Such an interpretation carries over readily to the formamide system in which some structuring by dissolved hydrocarbon also occurs.

Extensive studies [35] on the alkyltrimethylammonium bromides in water (from 25°C to ca. 160°C) (see Fig. 1.12) and in hydrazine [36] suggest that this interpretation is misleading. Despite large changes in

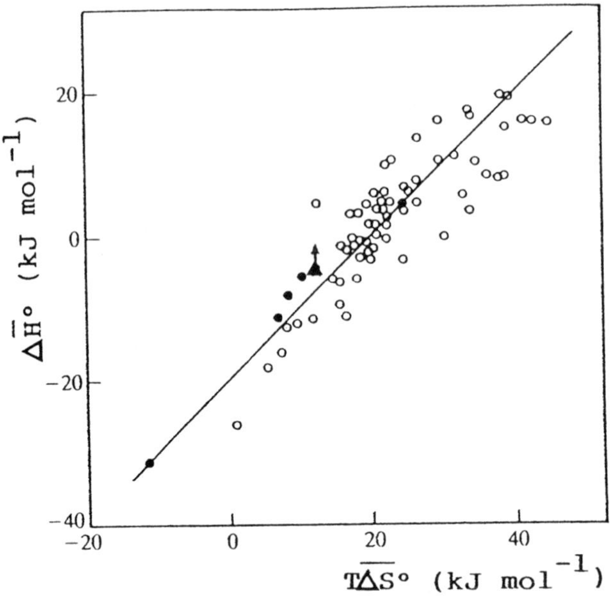

Fig. 1.11. Compensation plot of data for a variety of ionic and non-ionic surfactants in various liquids. (○): water, with or without additives; (●): benzene; (▲) formamide. Taken from Ref. [5], reproduced with permission.

ΔH and TΔS, the free energy of micellization (ΔG) remains almost constant across the entire temperature range from 25 to 160°C.

At high temperatures (>90°C) water loses most of its peculiar structural properties and the formation of structured water in the walls of the hydrocarbon cavities is no longer possible. Neither is it expected to occur in hydrazine and yet both of these systems exhibit micellization phenomena.

The difference is that ΔS° for the process is now negative, as is also ΔH°. It was argued that, in such circumstances, it is not sensible to attribute the micellization at room temperature to the positive entropy change. That change is made up of two parts: a large positive part due to removal of water from around the hydrocarbon and a (smaller) negative part due to transfer of the hydrocarbon (and counterions) into the micelle. Since the first water structure part is not present in hydrazine or hot water it cannot be the general driving force for micellization. Rather it is the second part (for which ΔH° is also negative, due to some extent to the re-establishment of the hydrogen bonds in the solvent) that must be the usual driving force [38].

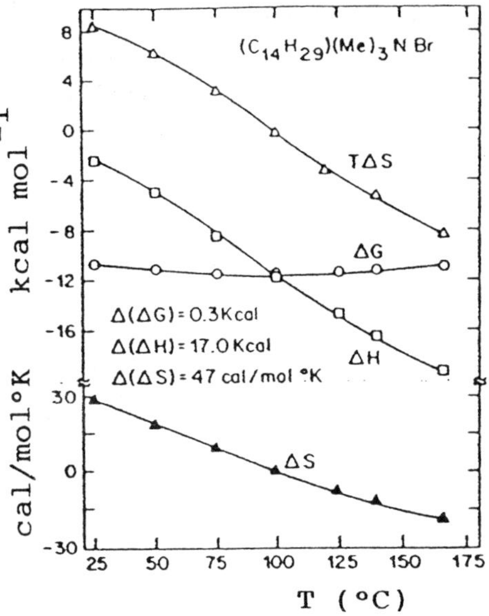

Fig. 1.12. Thermodynamics of micellization of tetradecyltrimethylammonium bromide obtained by using the dressed micelle model. Taken from Ref. [37], reproduced with permission.

The limited amount of data on other solvents also presents a complex picture. Reported data (given below) allow the comparison of the behaviour of dodecylammonium octanoate in benzene and cyclohexane [39]

$\Delta\bar{H}^\circ$ (kJ mol^{-1}): –5.4 (benzene); –32 (cyclohexane)

$T\Delta\bar{S}^\circ$ (kJ mol^{-1}): + 10 (benzene); –12.1 (cyclohexane)

In benzene, $T\Delta S^\circ$ is in fact positive and smaller than the corresponding values in water. Although the above data were derived at different temperatures (26°C and 40°C, respectively), no simple interpretation can explain such a difference. It is important to recognize that the surfactant molecules, when dissolved in benzene at low concentrations, are present as ion pairs, so the starting point for the aggregation processes is very different from that in aqueous solution. Indeed, the question concerning the whole concept of micelle formation as applied to solutions in aprotic non-aqueous media (like benzene) has to be carefully considered. It should also be noted that the presence of trace amounts of water has a relevant effect on the micellization process.

From these considerations it is clear that the very simple behaviour of the free energy function often masks a very complex shift in ΔS and ΔH values as temperature changes. Sometimes the study of these detailed shifts provides deeper insights into the process. In the present case it may well have led to an overemphasis of the role of structured water in the phenomenon referred to as "the hydrophobic bond".

1.7. THE DEGREE OF DISSOCIATION OF MICELLES

The delicate balance between the various forces which leads to the formation of the micellar aggregate includes partial neutralization of the charged micellar surface. Neutralization of the charged head groups, by binding counterions, considerably reduces the electrostatic repulsion between the head groups which counteracts the attractive hydrophobic forces. The binding of counterions may be represented for an anionic surfactant by:

$$m\,D^- + p\,X^+ \rightleftarrows D_mX_p^{(m-p)-} \tag{27}$$

where D^-, X^+ and D_mX_p correspond to the monomer surfactant, counterion and micelle respectively; m is the aggregation number and (m–p) is the charge of the micelle. The degree of dissociation, α, is given by the ratio (m–p)/m. This binding of counterions does not indicate any site specific binding. Rather, it can be envisaged as a continuous radial distribution of counterions corresponding to high concentration near the micellar surface and continuously decreasing as the distance from the micelle increases.

There is therefore no clear and sharp distinction between bound and free counterions. Consequently, no unambiguous method for measuring the degree of freedom (or binding) can be identified. Experimental results have thus been interpreted in a variety of degrees of sophistication and complexities. The resulting α is therefore not a very well-defined experimental quantity and may be expected to vary with the experimental method of determination.

The degree of dissociation is an important parameter to be considered when competing counterions are present in the same aggregate solution and this aspect will be examined in detail.

Several methods and approaches have been utilized for α determination. These include light scattering (static or dynamic), small-angle neutron scattering (SANS), ultracentrifugation, colligative properties measurements, ion activity measurements, electrophoretic mobility, conductivity, variation of cmc with added salts, and tracer diffusion techniques [40].

Spectroscopic methods which monitor the amount of ions whose spectroscopic properties are affected by the immediate microenvironment have also been utilized. These include, in addition to optical spectroscopy, NMR chemical shifts or relaxation studies [41] and epr techniques [42,43].

Table 1.5 provides an extensive list of measured dissociation degrees of several surfactant aggregates in different experimental conditions and using different methods of determination. A lot of data concerning further examples can be also found in Ref. [44].

In the following, attention will be focused on different approaches to the same technique in determining the degree of ionization. Moreover, the commonly utilized and easily accessible conductivity technique has been chosen to illustrate the ambiguity involved in this concept.

TABLE 1.5

Summary of values of α for some selected surfactants

Surfactant	Method	T (°C)	Additive	α	Ref.
Anionic					
$C_{12}H_{25}SO_4Na$	EMF (Na^+)	25	—	0.14	45
	SANS	30	—	0.14	46
	LS[a]	25	—	0.15	47
	EMF (D^-)[b]	25	—	0.15	48
	EMF (D^-)	25	—	0.15	49
	EMF (Na^+)	25	—	0.16	50
	Mass action law	25	—	0.18	51
	Dialysis	25	—	0.18	52
	Osmometry	25	—	0.20	53
	EMF (Na^+)	25	—	0.22	54
	EMF (Na^+)	25	—	0.23	55
	Mass action law	25	—	0.25	43
	Conductivity	25	—	0.25	56
	EMF (Na^+)	40	—	0.256	57
	Osmometry	25	—	0.26	53
	EMF (Na^+,D^-)	25	—	0.27	58

continued

TABLE 1.5 *continuation*

Surfactant	Method	T (°C)	Additive	α	Ref.
	Conductivity	25	—	0.275	59
	Electrophoresis	25	—	0.28	60
	EMF (Na^+,D^-)	25	—	0.37	61
	Conductivity	25	—	0.38	62
	ζ potential	25	—	0.50	63
	Mass action law	25	—	0.54	64
$C_{12}H_{25}SO_4Na$	LS	25	NaCl 0.1 M	0.39	65
			NaCl 0.2 M	0.38	
			NaCl 0.5 M	0.33	
	SANS	30	NaCl 0.2 M (SDS 0.04 M)	0.20	46
		30	NaCl 0.2 M (SDS 0.8 M)	0.03	
	Conductivity	25	NaCl 0.03 M	0.267	59
			NaCl 0.1 M	0.338	
$C_{12}H_{25}SO_4Li$[c]	SANS	37	—	0.39	66
			LiCl 0.2 M	0.30	
		50	LiCl 0.2 M	0.28	
		37	LiCl 0.4 M	0.29	
			LiCl 1.0 M	0.26	
$C_{12}H_{25}SO_4Li$			LiDS 0.074 M	0.40	
			LiDS 0.147 M	0.40	
			LiDS 0.294 M	0.41	
			LiDS 0.325 M	0.35	
			LiDS 0.845 M	0.24	
			LiDS 1.107 M	0.12	
$C_{12}H_{25}SO_4Na$	LS	25	NaCl 0.1 M	0.39	65
	LS	40	NaCl 0.1 M	0.43	
$C_{12}H_{25}SO_4Li$	Conductivity	25	—	0.30	59
$C_{12}H_{25}SO_4Cs$			—	0.24	
$C_{12}H_{25}SO_4TM$[d]			—	0.26	
$C_{12}H_{25}SO_4TE$[e]			—	0.24	
$C_{12}H_{25}SO_4TP$[f]			—	0.205	
$C_{12}H_{25}SO_4Co/2$	Conductivity	25	—	0.14	62
$C_{12}H_{25}SO_4Ni/2$			—	0.12	
$C_8H_{17}SO_4Na$	Conductivity	40	—	0.315	57
$C_{14}H_{29}SO_4Na$			—	0.234	
$C_{16}H_{33}SO_4Na$			—	0.209	

Surfactant	Method	T (°C)	Additive	α	Ref.
$C_8H_{17}SO_4Na$	Conductivity	25	—	0.45	56
$C_{10}H_{21}SO_4Na$				0.28	
$C_{14}H_{29}SO_4Na$		25		0.22	
$C_{12}H_{25}$–CH–CH_3 SO_4Na	Conductivity	40	—	0.250	57
$C_{11}H_{23}$–CH–C_2H_5 SO_4Na	Conductivity	40	—	0.333	
$C_{10}H_{21}$–CH–C_3H_7 SO_4Na			—	0.412	
C_9H_{19}–CH–C_4H_9 SO_4Na			—	0.468	
C_5H_{11})CH–Ø–SO_3Na C_5H_{11}	Conductivity	25	—	0.69	67
			Alcohol	0.69–0.95	
Cationic					
$C_{16}H_{33}N^+(CH_3)_3Br^-$	EMF	25	—	0.23	68
	Conductivity	25	—	0.17	69
	Conductivity	25	—	0.27	
	EMF	25	—	0.18–0.21	70
	Potentiometry	25	—	0.22	71
	Conductivity	25	—	0.16	72
	EMF	25	NaBr 0.1 M	0.33	68
	LS	40	NaBr 0.02–1 M	0.22	73
	EMF	25	CCl_4 0.1 M	0.23	68
	EMF	25	Alcohols (C_1–C_4)	0.18–0.94	70
$C_{12}H_{25}N^+(CH_3)_3Br^-$	LS	25	NaBr 0–0.2 M	0.04–0.14	74
$C_{10}H_{21}N^+H_3Br^-$		25	NaBr 0.5 M	0.07	75
$C_{10}H_{21}N^+H_2(CH_3)Br^-$				0.07	
$C_{10}H_{21}N^+H(CH_3)_2Br^-$				0.08	
$C_{10}H_{21}N^+(CH_3)_3Br^-$				0.13	
$C_{10}H_{21}N^+H(C_2H_5)_2Br^-$				0.13	
$C_{10}H_{21}N^+(C_2H_5)_3Br^-$				0.23	
$C_{12}H_{25}N^+(CH_3)_2$–R					
R = –$C_2H_5Br^-$	Conductivity	25	—	0.28	72

continued

TABLE 1.5 *continuation*

Surfactant	Method	T (°C)	Additive	α	Ref.
$-C_3H_7Br^-$			—	0.32	
$-C_4H_9Br^-$			—	0.44	
R = $-C_8H_{17}Br^-$	Conductivity	25		0.62	
$-C_{10}H_{21}Br^-$				0.74	
$C_{14}H_{29}N^+(C_2H_5)_3Br^-$				0.35	
$(C_3H_7)_3Br^-$				0.42	
$(C_4H_9)_3Br^-$				0.48	
$C_{16}H_{33}N^+(CH_3)_3Cl^-$	EMF	25	—	0.24	68
	EMF	25	NaCl 0.1 M	0.18	
	LS	40	NaCl 0.02–1 M	0.27	73
$C_{10}H_{21}N^+(CH_3)_3$					
$HCOO^-$	LS	25	HCOONa 0.5 M	0.01	76
CH_3COO^-			CH_3COONa 0.5 M	0.00	
C_2H_5COO–			C_2H_5COONa 0.5 M	0.09	
n-C_3H_7COO–			C_3H_7COONa 0.5 M	0.17	
n-C_4H_9COO–			C_4H_9COONa 0.5 M	0.24	
n-$C_5H_{11}COO$–			$C_5H_{11}COONa$ 0.5 M	0.29	
n-$C_6H_{13}COO$–			$C_6H_{13}COONa$ 0.5 M	0.36	
$C_8H_{17}N^+(CH_3)_3\ Br^-$	Conductivity	25	—	0.36	72
$C_{10}H_{21}N^+(CH_3)_3\ Br^-$			—	0.27	
$C_{12}H_{25}N^+(CH_3)_3\ Br^-$				0.23	
$C_{14}H_{29}N^+(CH_3)_3Br^-$				0.20	

a: Light scattering; *b*: EMF measurement of D^- activity through surfactant ion electrodes; *c*: LiDS concentration = 0.037 M; *d*: TM = tetramethylammonium; *e*: TE = tetraethylammonium; *f*: TP = tetrapropylammonium.

When the conductance of a surfactant solution is measured as a function of the surfactant concentration, the usual break in conductance, marking the cmc, is observed. The ratio of the slopes above and below the cmc in the curves of conductivity vs. concentration provides an estimate for α if it is assumed that the conductivity of the micelle is negligibly small compared to that of the counterions and the monomeric surfactant [77]. The mobility of the micelle using Stoke's law was calculated [78] as:

$$1000\ S_2 = \frac{(m-p)^2}{m^{4/3}}(1000\ S_1 - \Lambda_{Na}) + \frac{m-p}{m}\Lambda_{Na} \tag{28}$$

where Λ_{Na} is the equivalent conductance of Na^+ and S_1 and S_2 are the slopes below and above the cmc. As an example of the effect of the modification outlined by Zana [42], α for HTAB is calculated to be 0.27 by the simple approach while it reduces to 0.17 with the latter one [79]. If the mobility of the micelles is measured directly, or calculated from diffuse double-layer theory, its value is smaller than that obtained by the previous model. Consequently, the value of α will be higher [80]. For SDS, α= 0.256 is reported [78] whereas a value of 0.275 is reported in Ref. [80].

As can be seen in Table 1.5, quantitative differences exist between values of α given for the same compound when studied by different methods. Even when studied by the same method, differences due to different interpretations are evident. In the following comparative discussion, we try to adhere to a single experimental method and, when possible, to the same assumptions used in the analysis. With this cautionary comment, some general conclusions may be drawn from the examination of Table 1.5:

(a) Changes in temperature are not likely to produce large changes in α, as can be seen for SDS at 25 and 40°C (NaCl 0.1 M) and for LiDS at 37 and 50°C (LiCl 0.2 M) [65,66]. However, as is shown in Fig. 1.13, detailed investigations of the effect of temperature on the cmc and α of $C_{14}H_{29}N^+(CH_3)_3$ Br^- reveal only small changes in cmc, whereas α changes from 0.2 to 0.3 [72].

For the same cationic surfactant, a similar increase of α has also been reported [81]. The increase indicates that the dissociation of Br^- from the micellar surface is an endothermic reaction.

(b) α is only slightly dependent on surfactant concentration as long as the micellar shape remains unchanged (see the data for LiDS in the surfactant concentration range 0.037–0.294 M)[66]. Increasing surfactant concentration may cause a sphere-to-rod transition, as in the case of SDS passing from 0.04 M to 0.8 M (NaCl 0.2 M) [46]. Such transitions are accompanied by increased binding of counterions (decreased α). In the case of lithium dodecylsulfate, the decrease in α observed by increasing the surfactant concentration from 0.325 M to 1.107 M is not a result of a phase transition. Rather, the intermicellar distances become comparable to the micellar size. Then the counterion has no other alternative but to remain in the vicinity of the micelles.

(c) Increasing the alkyl chain leads to a decrease in α (see for example the case of sodium octyl-, tetradecyl-, and hexadecylsulfate at 40°C) [57]. Another example in the table concerns different alkyltrimethylam-

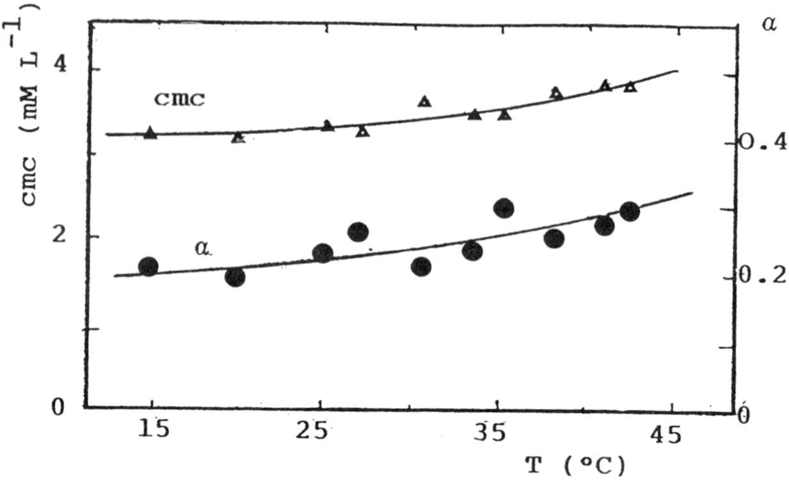

Fig. 1.13. Changes of α and cmc of tetradecyltrimethylammonium bromide with temperature. Adapted from Ref. [72], with permission.

monium bromides at 25°C [72], also reported in the table. A similar trend has also been reported for sodium alkanoates [82].

This effect of increased counterion binding can be rationalized in terms of an increase in the micellar size for the longer chain surfactants. The larger size (larger aggregation number) results in higher charge density at the micelle surface.

A similar argument may be invoked to explain the effect of the position of the charged head group along the alkyl chain on α (e.g., see the linear tetradecyl sulfates having the sulfate group attached to different carbon atoms) [57]. The closer the head group is to the chain end, the larger the micelle and the charge density is higher, resulting in smaller α. An extreme case is the micelle of $(C_5H_{11})_2CH–C_6H_4–SO_3Na$ [67]. The small aggregation number of this micelle (N_{AG} = 32) leads to a large surface area for each surfactant head group which in turn results in low charge density.

Basically the same effect may be observed on increasing the size of the head group, as for cationic surfactants obtained from decyl-, dodecyl- and tetradecylammonium bromide by introduction of different alkyl substituents on the N atom [72,75]. A larger degree of dissociation is observed for the larger head groups.

(d) The identity of the counterion is of utmost importance to the degree of dissociation, as can be seen in Table 1.5 for dodecylsulfates of different cations at 25°C, investigated using the same technique (conductivity) [59]. The smaller the hydrated counterion the stronger is the attractive force (Cs^+ is smaller than Li^+ and thus α is smaller for the former). For alkali ion dissociation the sequence Li > Na K Rb Cs has been observed [83]. However, when the counterions contain hydrophobic moieties, hydrophobic interactions may contribute to the binding (note the decrease in α on changing TM to the larger TP) [59].

Of course, divalent ions are stronger bound (per unit charge) than monovalent ions. NMR has proven to be a powerful tool in studies of binding of alkali ions. In particular, this technique has revealed head-group specificity towards $^{23}Na^+$ for $-OSO_3^-$ and $-SO_3^-$ as compared to $-COO^-$ [41]. Counterion specificity is even more pronounced for cationic surfactants. The degree of dissociation of halides from dodecylamino-ethanol salts obtained by the conductivity technique [84] follows the sequence:

Cl^- ($\alpha = 0.27$) > Br^- ($\alpha = 0.20$) > I^- ($\alpha = 0.16$) > NO_3^- ($\alpha = 0.11$)

From the variation of cmc with counterion and from ion activity measurements it was inferred that the degree of dissociation from $-N^+(CH_3)_3$ and $-N^+H_3$ head groups follows the sequence Cl^- > Br^- > NO_3^- > I^-. A similar sequence has been found for 1-methyl-4-dodecyl-pyridinium micelles [85]. While increasing the size of hydrophobic counterions increases its binding to cationic micelles [86], lengthening the chain of n-alkyl carboxylates seems to increase the degree of dissociation [76].

(e) The effect of added salts has been a subject of controversial assumptions and interpretations. Mass-action considerations predict a decrease in α on increasing ionic strength. Such an effect may be seen in the table for SDS at 25°C [65] and for LiDS at 37°C [66]. Opposite trends, however, are reported for SDS in the presence of 0.03–0.1 M NaCl [59].

In other cases, α seems to remain constant (see for example HTAB at 40°C, in the presence of NaBr 0.02–1 M) [73]. Such a behaviour corresponds to ion condensation, often assumed in polyelectrolyte chemistry [87].

(f) The solubilization of organic solutes (e.g. CCl_4) has only a limited effect on α (see the effect on HTAB) [68]. Non-polar additives, such as benzene, cyclohexane and n-hexane do not alter the degree of dissociation of SDS. Conductometric studies show that, while the cmc is reduced, the slopes in the conductivity vs. concentration curves are not

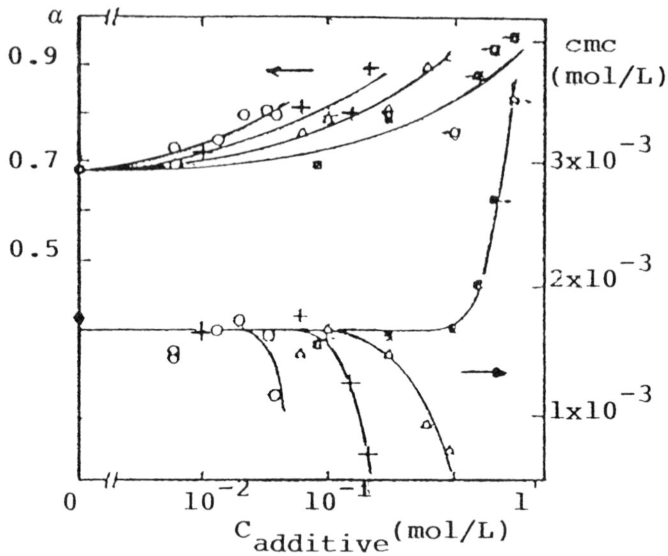

Fig. 1.14. Effect of alcohol concentration on cmc and α of: $(C_5H_{11})_2CH–C_6H_4–SO_3Na$. (□) 1-propanol; (Δ) 1-butanol; (+) 1-pentanol; (O) 1-hexanol. Taken from Ref. [67], reproduced with permission.

affected by such additives. This means that either the substrates are solubilized predominantly in the interior of the micelles or, if a significant fraction is adsorbed near the micellar surface, α nevertheless remains unaffected [76]. On the other hand, amphiphilic additives (e.g., diols, phenols) decrease the cmc and concomitantly increase α. Such an effect of alcohols is shown in Fig. 1.14 and in Table 1.5 for the anionic surfactant $(C_5H_{11})_2CH–C_6H_4–SO_3Na$ [67] and for HTAB [70]. In these cases, the alcohol molecules intercalate between the surfactant molecules with their hydroxyl groups close to the head groups. The surface area of the micellar surface thus increases and the charge density decreases.

1.8. MICELLAR DYNAMICS

As reported earlier in this chapter micellar aggregates are dynamic units, constantly forming and dissociating on a time-scale in the microsecond to millisecond range. The kinetics of micelle formation and breakdown have mainly been studied by fast relaxation methods (temperature-jump, pressure-jump, ultrasonic absorption, shock-tube

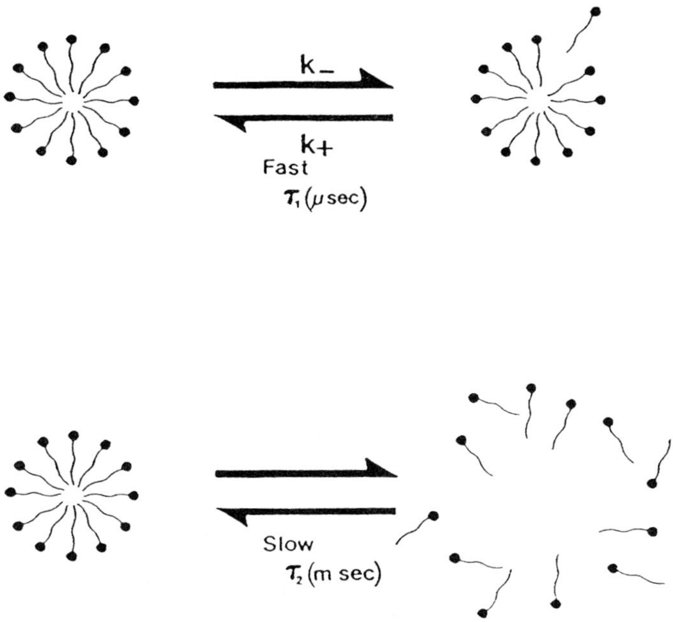

Fig. 1.15. Two relaxation mechanisms are proposed for dilute micellar solutions of pure surfactants. The slow process is thought to take place through a sequence of increasingly smaller aggregates and not suddenly as shown.

methods, etc. [88,89]). In these methods a system is subjected to a sudden perturbation and it is then monitored as it returns to equilibrium or departs from it. Such measurements reveal the presence of two relaxation processes in the perturbed system: a fast process (nanosecond to microsecond time-scale) related to the exchange of monomers between the bulk aqueous phase and micelles of different sizes, and a slower relaxation (microsecond to millisecond) related to the formation or break-up of micelles (see Fig. 1.15).

The relaxation times are, of course, related to the rate constants for association (k_+) and dissociation (k_-) as represented by reaction:

$$M_{n-1} + S \overset{K_+}{\rightleftarrows} M_n \tag{29}$$

where S represents a surfactant monomer.

The rate constants for a series of sodium alkyl sulfates are given in

TABLE 1.6

Kinetic parameters of association and dissociation of some micelle-forming alkyl sulfates[a]

Compound	cmc (M)	N_{AG}	k_- (s^{-1})	k_+ ($M^{-1}\ s^{-1}$)	K_{eq} [b]
C_6SO_4Na	4.2×10^{-1}	17	1.3×10^{9}	3.2×10^{9}	40.5
C_7SO_4Na	2.2×10^{-1}	22	7.3×10^{8}	3.3×10^{9}	100
C_8SO_4Na	1.3×10^{-1}	27	1.0×10^{8}	7.7×10^{8}	207
C_9SO_4Na	6.0×10^{-2}	33	1.4×10^{8}	2.3×10^{9}	550
$C_{11}SO_4Na$	1.6×10^{-2}	52	4.0×10^{7}	2.6×10^{9}	3250
$C_{12}SO_4Na$	8.2×10^{-3}	64	1.0×10^{7}	1.2×10^{9}	7800
$C_{14}SO_4Na$	2.0×10^{-3}	80	9.6×10^{5}	4.7×10^{8}	39000

a: Data taken from Ref. [88]; *b*: $K_{eq} = k_+ N_{AG}/k_-$.

Table 1.6. As the alkyl chain becomes larger, the exit rate of monomer becomes appreciably slower, as is expected on the basis of its increased hydrophobic character. The association rate constant also decreases with increasing chain length (by about a factor of six from C_6 to C_{14}), but only slightly.

The above observations can be explained assuming the existence of a diffusion-controlled rate step with an electrostatic repulsion term due to the interaction between the charged micelle and the anionic monomer [88].

From Table 1.6 it appears that, whereas the entry rate constant is not dependent on surfactant structure, the exit rate is proportional to 1/cmc, thus the exit time of the monomer is very high when the cmc is low.

1.9. MICELLAR STRUCTURE

1.9.1. General structure of ionic and non-ionic micelles

The cross-section of a typical ionic micelle containing n molecules may be thought of as having a liquid core formed by the n associated hydrocarbon chains with the fully ionized head groups projecting out into the water. Immediately surrounding the core is the Stern layer which contains not only the ionic head groups but also $n(1-\alpha)$ counterions.

The Stern layer constitutes the inner part of the electrical double layer surrounding the micelle. The outermost boundary of the Stern layer corresponds with the hydrodynamic shear surface of the micelle. The core and the Stern layer form the kinetic micelle with a total charge αne where e is the elementary charge. The surface potential of the kinetic micelle is the electrophoretic zeta potential (ζ). The outer, more diffuse layer, which contains the remaining $n\alpha$ counterions, is termed the Gouy–Chapman layer.

In polyoxyethylated non-ionic micelles the core is surrounded by a layer composed of the polyoxyethylene chains to which solvent molecules may be hydrogen bonded. This region of the micelle is often termed the palisade layer.

Recent years have seen a renewed interest in the investigation of the nature of the micellar regions as more sophisticated experimental techniques have become available. An understanding of the detailed nature of the micellar microenvironments is important in understanding the ability of the micelles to act as solubilizing agents and for the prediction of the location of the solubilizate molecules within the micellar entity.

1.9.2. The microenvironment in, on, and near the micellar aggregate

1.9.2.1. The core

Starting from the solubilization in micellar solutions of water-insoluble hydrocarbon-soluble materials it seems evident that the hydrocarbon core of a micelle is very much like that of the corresponding liquid hydrocarbon.

More direct evidence about molecular motion within the micelle core is provided by the ESR spectra of solubilized spin-labelled molecules or by depolarization of the fluorescence of solubilized perylene or similar compounds.

Carbon-13 NMR T_1 measurements are the most conclusive. T_1 measurements on sodium octanoate, in which T_1 is inversely related to the rotation times about the corresponding C–C bond have shown that the rotation is slower near the head of the chain than its $-CH_3$ end, both for monomeric octanoate ion and for the micellized material, by factors of 4 and 10, respectively [90]. It has also been found from the ^{13}C chemical shifts that the proportion of chain methylene pairs in a gauche (cis) as compared with an anti (trans) conformation is decreased by micellization [91]. Studies performed on n-dodecyl octaoxyethyleneglycol micelles in D_2O showed, from T_1 data, that rotation about C–C bonds at the end of the dodecyl chain is nearly the same as in liquid alkanes but decreases toward the head group.

1.9.2.2. The irregular micellar surface

The segmental motions of the micellized amphiphiles are somewhat hindered at the micelle surface, but the rapid equilibrium with monomers suggests that quite turbulent conditions are present in this region. The dissociation of an entire monomer must occur by means of segmental jumps, one CH_2 group at a time, as a kind of diffusion process. For a dodecyl chain the time-scale of protrusion is in the range of picoseconds for a single methylene group, whereas it is several nanoseconds for the whole chain. At any instant perhaps every third monomer would project one unit, every seventh two units, and so exponentially. A calculation performed for monomers containing a charged head group gives similar results, with an average distance of protrusion of about 0.2 nm and similar time scales.

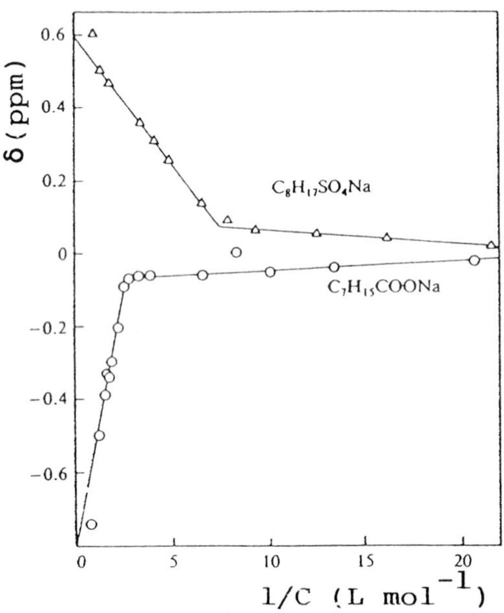

Fig. 1.16. Differences between sodium ion binding to micelles for alkyl sulfate and alkyl carboxylate as shown by ^{23}Na NMR. Taken from Ref. [92], reproduced with permission.

1.9.2.3. The micellar environment

The interfacial region around a micelle is not easy to describe. Often it is represented like a concentrated solution of molecules resembling the head groups, the bound or free counterions, the adsorbed solutes, etc. The counterion binding has been previously discussed (see Section 1.7) from the standpoint of determining its extent and some assumptions appear to be valid. One of them is that bound ions retain their hydration since, if not, a large heat of hydration would be expected when a micellar solution is diluted.

Since anions are usually less hydrated than cations, their binding to cationic micelles must show more specificity than is found for cationic counterions and anionic micelles. Several examples confirm the above expected behaviour. The binding of sodium ion to octanoate and n-octyl sulfate micelles will be examined in some detail.

The ^{23}Na NMR quadrupole relaxation occurs for the alkylsulfate with a chemical shift to higher applied field, whereas it is found for the carboxylate at lower field (see Fig. 1.16). A possible interpretation is that the cation is placed between the three oxygen atoms on the outer face of the sulfate tetrahedron head group, and between two adjacent carboxylate groups in the other case.

1.10. MICELLAR PARAMETERS

In addition to cmc, the crucial parameter for micelle formation, the aggregates are characterized by other properties: micellar size and shape, aggregation number, fraction of charge and interfacial electrostatic potential, competition of ions at the micellar surface, polarity and microviscosity. These different parameters will be separately considered.

1.10.1. Micellar size and shape. Aggregation number

1.10.1.1. Micellar shape

Micelles are considered as dynamic structures with a liquid core and, for this reason, it is difficult to regard them as rigid structures with a precise shape. It is, however, useful to consider an average micellar shape.

The experimental determination of the shape for small micelles, such as those formed by ionic surfactants in the absence of added electrolyte and close to the cmc, has not yet proved feasible. Although for the purposes of interpretation of experimental data it is usual to describe the micelles as spheres, it can be demonstrated from geometrical considerations that most of the common surfactants with a single unbranched hydrocarbon chain cannot form truly spherical micelles.

Because no holes may exist in the centre of the aggregate, the micellar radius is limited by the maximum extension of the hydrocarbon chain and it is possible to establish the maximum aggregation number for a given hydrocarbon chain starting from the experimental density values available for such compounds.

The experimental aggregation numbers of micelles are generally larger than the calculated maximum values estimated for spherical aggregates and, consequently, there is some dispute about the exact shape of these small micelles. Tanford [93] has proposed that the distortion of the micellar shape into an ellipsoid of revolution is the simplest way of incorporating a larger number of molecules into the micelles.

The shape of the larger micelles formed by non-ionic surfactants and ionic surfactants at high concentration or in the presence of added electrolyte is also controversial. From theoretical considerations, except when head group repulsion is very strong, oblate ellipsoids are thermodynamically favoured over prolate ellipsoids for most micelles. Non-ionic surfactants and ionic surfactants in the presence of added electrolyte would be included in this category.

The three different aggregates originated from Triton X100 monomers are shown in Fig. 1.17. In order to obtain a sphere, the intrusion of oxyethylene chains into the micellar core must be assumed.

1.10.1.2. Effects of different parameters on aggregation number

(i) *Nature of the hydrophobic group*: The increased hydrophobicity derived from an increase of chain length also causes an increase in micellar size. In many cases a linear relationship exists between log(micellar weight) and the hydrocarbon chain length. Moreover, a linear relationship between the aggregation numbers of polyoxyethylene non-ionic surfactants and the number of carbon atoms in the hydrocarbon chain has been assessed for surfactants having similar (number of ethylene oxide units)/(number of carbon atoms in hydrocarbon chain) ratios.

(ii) *Nature of the hydrophilic group*: A detailed study of the effect of the nature of the polar group of ionic surfactants on the micellar properties has concluded that an important factor controlling the micellar size was the mean distance of closest approach of a counterion to the

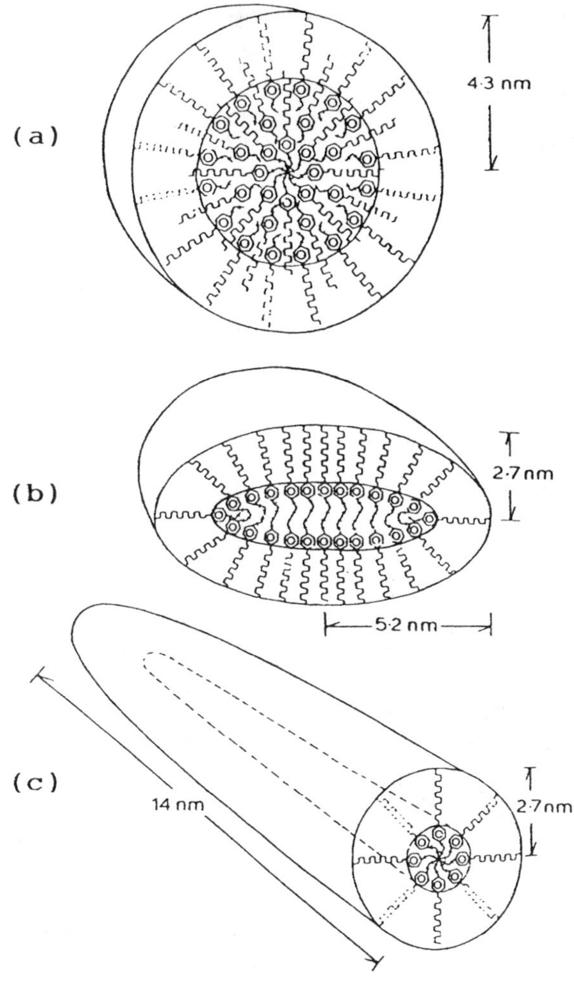

Fig. 1.17. Schematic representation of one-half of the micelle of Triton X-100 obtained from geometrical calculations. (a) spherical, (b) oblate, (c) prolate. Taken from Ref. [162], reproduced with permission.

charge centre of the surfactant [94]. For example, decylammonium bromide forms very much larger micelles than decyltrimethylammonium bromide because the Br^- counterions are able to approach more closely the charged nitrogen atom of decylammonium, shielding the repulsive electrical forces and allowing larger micelles to form.

TABLE 1.7

Variation of cmc, micellar weight and aggregation number of hexadecylpolyoxyethylene ethers as a function of the number of polyoxyethylene groups[a]

Oxyethylene units	cmc ($M \times 10^6$)	Micellar weight (Da)	N_{AG}
6	1.7	1,230,000	2430
7	1.7	327,000	590
9	2.1	140,000	220
12	2.3	117,000	150
15	3.1		
20[b]			82
21	3.9	82,000	70
30[b]			64
40			52

a: Data taken from Ref. [13]; *b*: from Ref. [95].

Solvent interaction may also play a relevant role. The cause of this effect was attributed to changes in the effective dielectric constant produced when the polar head structure is changed, to intermolecular hydrogen bonding and to hydrogen bonding between the polar head group and water.

In most cationic surfactants the charge is localized on a single N atom which is typically located at the end of the hydrocarbon chain. However, when the polar group is a piperazinium or pyridinium ring the mode of delocalization of the charge around the ring may affect the aggregation number. An increase in the positive charge on that particular atom of the ring to which the hydrocarbon chain is attached leads to a decrease in aggregation number.

For the polyoxyethylene-type non-ionic surfactants at constant number of E units, the increase in the alkyl chain decreases the cmc and increases the aggregation number. For a fixed alkyl chain, however, the increase of the length of the polyoxyethylene chain causes an increase in the cmc and a decrease in the micelle size, as can be seen in Table 1.7.

Increasing the polyoxyethylene chain length in fact makes the monomer more hydrophilic and the cmc increases. The same effect may also induce a decrease in micellar size, although other factors (including the type of geometric packing of the monomers within the micelles) can play a role.

(iii) *Nature of the counterion*: In conventional ionic surfactants a change in counterion to one of greater polarizability or valence leads to a decrease in cmc and corresponding increase in aggregation number.

A recent study carried out by small angle neutron scattering has examined in detail the influence of the counterion on the micellar properties of trimethylammonium halide surfactants. It was found that the N_{AG} increases in the order: $NO_3^- > Br^- > CH_3SO_4^- > Cl^- >> OH^-$ and that the aggregate tends to become more prolate elliptical. Fractional charge and size are also influenced. The fractional charge follows the order: $OH^- > Cl^- > CH_3SO_4^- \approx Br^- \approx NO_3^-$ [96].

(iv) *Effect of additives*: Addition of electrolyte causes a reduction in the thickness of the ionic atmosphere surrounding the polar head groups and a consequent decreased repulsion between them. These effects are manifest as a reduction in cmc and an increase in aggregation number.

As an illustration of the effect of electrolyte on micellar size of ionic surfactants, the extensively investigated sodium dodecyl sulfate system will be considered. The literature values for the aggregation number of SDS in the presence of added sodium chloride from a variety of techniques including dynamic and static light scattering, sedimentation equilibrium and luminescence quenching have been critically examined [97].

The shape of SDS micelles exhibits a transition from sphere to rod at a concentration of approximately 0.45 M NaCl at 25°C, and a corresponding great increase of N_{AG} was observed by several workers at a similar salt concentration (see Fig. 1.18).

The studies on the effect of n-alcohols on the aggregation number of ionic surfactants suggest that water-soluble alcohols (methanol to butanol) are predominantly dissolved in the water phase and may increase or decrease the aggregation number depending on the alcohol concentration. Less soluble alcohols, such as pentanol or hexanol, are partitioned between the aqueous and micellar phases and, when present at low concentrations, may increase N_{AG}. Sparingly soluble alcohols, such as heptanol and octanol, are almost entirely solubilized in the micelles and thereby increase N_{AG}.

(v) *Effect of temperature and pressure*: The decrease of the micelle size of ionic surfactants with increase in temperature has been reported, whereas the aggregates of many polyoxyethylene-type non-ionic surfactants increase rapidly in size with rising temperature. In some systems the rapid increase in micelle size is observed only above a characteristic transition temperature and is accompanied by an increase in micelle asymmetry. From viscosity data obtained at higher tempera-

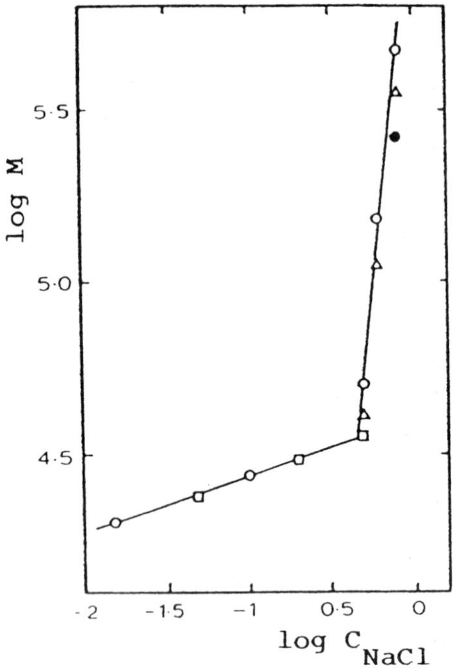

Fig. 1.18. Dependence of log M (M: apparent micellar weight) of SDS on ionic strength: (O) 25°C; (Δ) 30°C; (●) 35°C; (□) data from Ref. [98]. Figure taken from Ref. [163], reproduced with permission.

tures [99], an increase in the extension of the polyoxyethylene chains has been suggested, with a corresponding increase of the amount of water physically trapped by the micelles.

The effect of pressure on the micellar size in aqueous solutions of SDS and $C_{12}E_6$ [100] showed two different effects: a minimum in the aggregation number vs. pressure curve for SDS, which corresponds to the maximum in the cmc–pressure curve (see Fig. 1.19A), and a continuous decrease of the aggregation number of $C_{12}E_6$ with increasing pressure (Fig. 1.19B), the rate of decrease being more significant at lower pressures.

1.10.2. Interfacial potential. Electrostatic interactions

As we have seen, ionic amphiphiles are one of the most important classes of micelle-forming surfactants. Above the cmc their aggregates

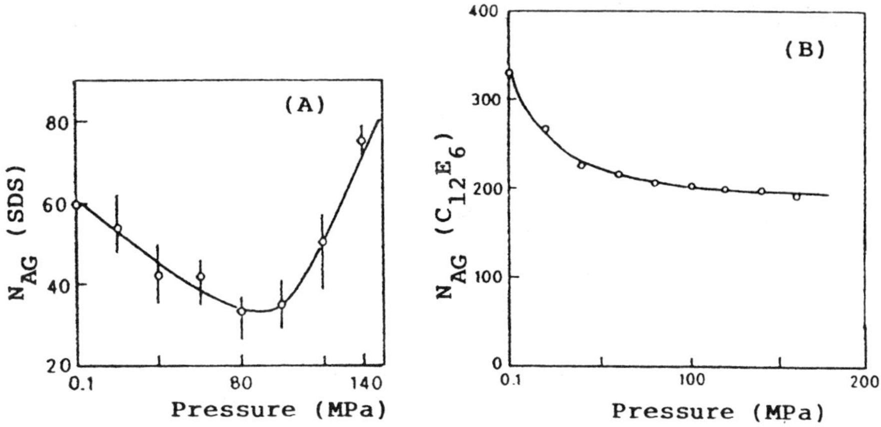

Fig. 1.19. Aggregation number of ionic and non-ionic micelles as a function of the applied pressure. (A) SDS at 30°C; (B) $C_{12}E_6$ at 25°C. Figure taken from Ref. [100], reproduced with permission.

carry high formal charge at the micelle/water interface. The strong repulsion between the head groups, inherent in the aggregation of ions of like charges, is partially compensated by binding of counterions. The charged micelles thus give rise to a high local electrostatic field extending into the solution which influences the distribution and concentration of the other ionic species present in the same solution. A detailed knowledge of the ions' distribution is essential, therefore, in the understanding of micellar effects on reaction equilibria and dynamics, in particular when ionic reactants are involved.

The most straightforward approach to describe binding of ions to the micelle is through Eqn. (28), as an ordinary chemical equilibrium. The fundamental difficulties associated with such a description arise from the fact that the electrostatic interactions are of long-range character. The extension of the electrostatic potential away from the micellar surface into the solution is shown schematically in Fig. 1.20.

It has been shown that when the potential field extends to large distances away from the charge-carrying entity, the use of mass action law, as in Eqn. (28), leads to various difficulties. Due to the long-range nature of the electrostatic potential, the concentration of counterions continuously decreases from its concentration at the Stern layer to the bulk concentrations. As a consequence, different experimental techniques will probe different regions and different degrees of counterion association will be reported. A model which explicitly describes the ions'

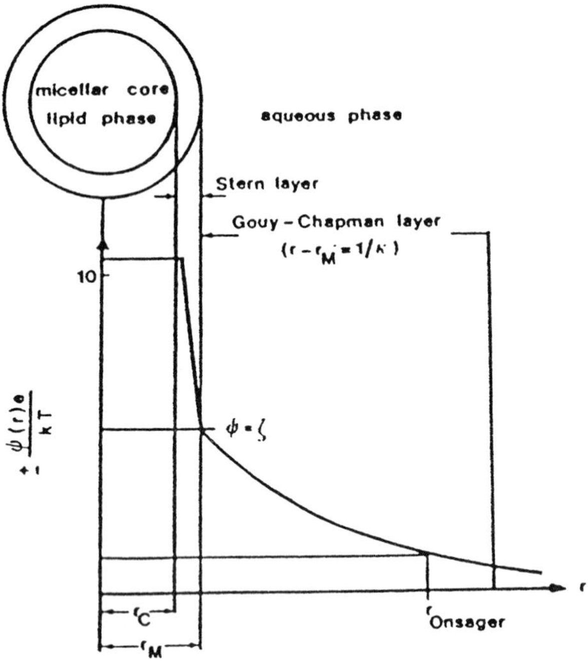

Fig. 1.20. Variation of the electrostatic potential as a function of distance from the micelle centre. Taken from Ref. [101], reproduced with permission.

distribution around the charged micelle seems, therefore, preferable to the simple equilibrium description.

In the following model it is assumed that the micelle is a sphere of radius r_m, and that the counterions are spherically distributed around the micelle. Due to the finite size of the ionic groups (head groups and counterions), the counterions cannot approach the micelle to closer than a sphere of radius r_i (see Fig. 1.20). The ionic groups at the micellar surface will thus give rise to a non-uniform electrostatic potential ψ which will follow Poisson statistics. If the electrostatic effect dominates, the counterions' distribution is determined by ψ and will obey the Poisson–Boltzmann equation:

$$-\varepsilon_r \varepsilon_0 \nabla^2 \psi = \rho = \Sigma_i Z_i C_i e \exp(-Z_i e \psi / kT) \qquad (30)$$

where ε_0 and ε_r are the dielectric constants of free space and of the medium, ρ is the charge distribution, C_i is a constant with the dimensions of concentrations and relates to the average concentration of ion i, Z_i is the charge of ion i and e is the elementary charge. The boundary conditions are obtained from the micellar total charge and from $(d\psi/dr)_{r\to\infty} = 0$.

The Poisson–Boltzmann equation in the region $r > r_i$ can be solved numerically [102,103]. In this calculation, the micelles are assumed to be dissolved in an infinitely dilute electrolyte solution. The micelles are thus assumed to be at a fixed chemical potential of the surrounding water and electrolyte. The electrostatic potential is affected by the micelles concentration. This effect may be viewed as micelle–micelle interaction mediated by the counterions.

A more satisfactory description of the micellar potential is obtained by confining the volume per micelle to an outer radius r_f (see Fig. 1.21). This "cell model", often invoked for polyelectrolyte systems, has been successfully extended to micellar solutions [104–106]. Frahm and Diekmann

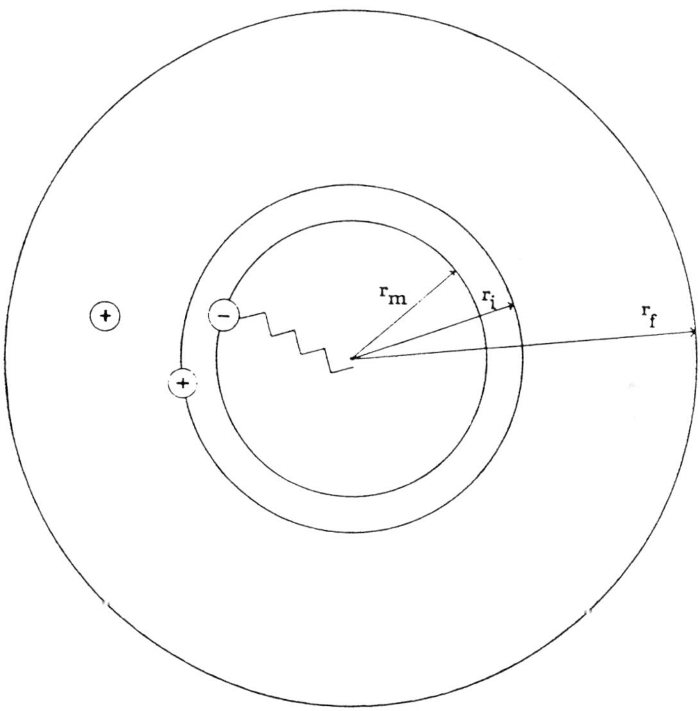

Fig. 1.21. Model for a micellar system used to calculate the electrostatic properties. r_m is the radius of the micelle, r_i the radius of closest approach of the counterions and r_f the radius of a sphere representing the volume per micelle.

[107] applied the cell model to the kinetic micellar effect on ionic reactions.

Monte Carlo computations, using the cell model, were performed to simulate the ionic composition around a spherical micelle and were compared to the results from the Poisson–Boltzmann equation [108]. Up to moderate concentration of added salt there is a good agreement between the two treatments. At high salt concentrations, ion–ion interactions become more important than ion–micelle interactions. Since the Poisson–Boltzman equation neglects the former interactions, the cell model will better describe the system at high ionic strength.

Improvements to the simple Poisson–Boltzmann equation have been extensively discussed [109]. Field dependence of the dielectric constant, finite size of the ions and their polarizability have been taken into account. However, the interaction between ions, hydration forces, compressibility of the solvent and short-range potential fields were all neglected. In spite of these approximations, the solution of Eqn. (30) provides an acceptable qualitative description of the electrostatic effect.

For several examples, the above approach explains the dependence of cmc on salt concentration, alkyl chain length and charge of counterion [106]. In particular, binding of counterions and its relation to thermodynamic, transport and spectroscopic properties. The degree of dissociation calculated is similar to the one measured experimentally.

Ions of higher valency usually bind stronger than the lower valency ions to the micelles. Thus, Gd^{3+} is bound stronger than Mn^{2+} or Cu^{2+} to SDS micellar aggregates [110]. However, effects other than simple electrostatics are sometimes observed to dominate the binding. Such is the case with Tl^{+}, which has a binding constant comparable to that of divalent ions. On the other hand, Cr^{3+} has a rather low binding constant. While for Cr^{3+}, reduced charge due to complexation (e.g., hydroxide formation) may be invoked, the Tl^{+} case still remains unresolved [72].

Another interesting case to examine is the electrostatic behaviour of zwitterionic micelles, which is fundamentally different from that of common ionic aggregates. A specific model was developed to describe the electrostatic properties of these systems, starting from detailed studies performed on 3-(N-hexadecyl-N,N-dimethylammonio)propane sulfonate aggregates using light scattering and conductivity techniques [111]. The model, which is equivalent to a double shell concentric spherical capacitor immersed in an electrolyte solution, is schematically shown in Fig. 1.22(A).

Some peculiar observed effects, such as the sequestering of mobile anions in the dipole layer (positive inner surface, negative outer) and

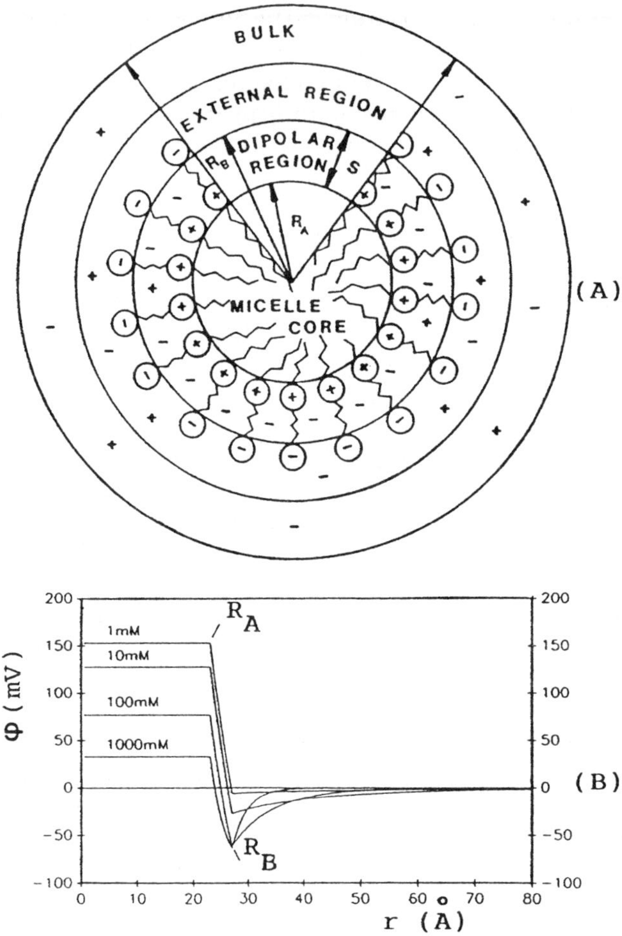

Fig. 1.22. (A) Scheme of a concentric sphere model for zwitterionic micelles. (B) Potential variation for selected salt concentrations. Taken from Ref. [111], reproduced with permission.

the low net mobile ion charge densities outside the aggregate, can be explained using the model. The calculations are based on the Debye–Huckel linearization of the Poisson–Boltzmann equation, assuming that the potentials within and outside the micelle originate from the superposition of solutions for a separate cationic micelle of radius R_A and an anionic micelle of radius R_B. Moreover, each isolated micelle is considered as impermeable to mobile ions.

Using the above approach, the variation of the electrostatic potential as a function of the distance from the zwitterionic micellar core and in the presence of different concentrations of added univalent electrolytes (in mM), has been calculated. The corresponding profile is very different from that obtained for simple ionic micelles, as can be seen in Fig. 1.22(B).

1.10.3. Competition of ions at the micellar surface

The surfactant counterions can be exchanged with other like-charged ions according to the following equilibrium equation (valid for cationic micelles bearing monovalent counterions):

$$D_mX_p^{(m-p)} + q\,Y^- \rightleftarrows D_mX_{(p-q)}Y_q^{(m-p)+} + q\,X^- \qquad (31)$$

The above ion-exchange equilibrium can be described (as for ion-exchange resins) using the pertinent ion-exchange coefficients $K_{x/y}$. The values of $K_{x/y}$ can be estimated by fitting the experimental equilibrium or kinetic results to the equations derived from pseudophase ion-exchange models. Often, however, independent methods are available for these determinations. The applied techniques include absorption spectroscopy [112], fluorescence quenching [113], EMF measurements of activity [108], ultrafiltration [114], conductivity [115], complex formation [114], and chemical trapping [164,165].

In Table 1.8, a series of $K_{Br/X}$ values determined by various techniques are compiled. Similarly to the situation in the determination of α (see Section 1.7), different methods may differently discriminate between "bound" and "free" ions. The extensive set of data, given in Table 1.8, for Br^-/NO_3^- and for Br^-/Cl^- exchange shows an excellent agreement between values determined by the various techniques. Less satisfactory is the agreement for ions which bind inefficiently, such as OH^- or F^-. The discrepancies observed here may be due to errors in mass balance calculations or to the inherent defects of the simple pseudophase model.

An important observation from Table 1.8 is that the complexing ability of the ions and/or their polarizability are more important factors than size in influencing selectivity. Values for $K_{Br/OH}$ may be compared with kinetic estimates which yield values in the range of 10–40. Ion exchange selectivity coefficients for the binding of OH^- to a series of homologous alkyltrimethylammonium bromides and chlorides and to hexadecylpyridinium chloride (HPC) have been determined from kinetic

TABLE 1.8

Selectivity coefficients for bromide/anion exchange ($K_{br/X}$) in HTAB micelles

Anion	Fluorescence		Absorption spectroscopy		UF	cmc variation
	a	*b*	*c*	*d*	*e*	*f*
OH^-	42			21		
I^-		0.08				
Cl^-	4.2	6	5	5.1	4.2	2.25
F^-	42	23		23	19	
NO_3^{-g}	0.8	0.9	0.73	0.9	0.9	0.73
$H_2PO_4^-$		22				
HCO_3^-		22	13			
Ac^{-h}	28	18	13	10	10	
Cl-Ac$^-$		3				
Br-Ac$^-$		2				
I-Ac$^-$		1				
$SO_4^=$				1.6	1.9	
$HPO_4^=$				6.5	7.4	
$B_4O_7^=$				6.4	12	

a: Bromide quenching of naphthalene or pyrene fluorescence [113]; *b*: competitive displacement of the iodoacetate ion (IAC) with pyrene as fluorescent probe ($K_{Br/X} = K_{IAC/X}/K_{IAC/Br}$ is assumed) [113]; *c*: displacement of 2-furoate ion [112] *d*: displacement of 2-benzenesulfonate [117]; *e*: Ref. [114]; *f*: cmc variation in the presence of added salt [118]; *g*: a value of 0.7 is also reported from EMF measurements [108]; *h*: Ac^- = acetate ion.

data on alkaline hydrolysis [116]. The magnitude of the selectivity is determined primarily by the nature of the surfactant counterion (Cl^- or Br^-). Furthermore, the quantitative differences observed in micellar HTAB or HPC could be attributed almost exclusively to somewhat more favourable binding of OH^- to the HPC surface than to HTAB.

Ion-exchange between monovalent and divalent anions in cationic micelles has also been studied through fluorescence quenching and ultrafiltration [119] using the simple "two component" pseudophase formalism. While $S_2O_3^{2-}$ competition with Br^-, Cl^-, NO_3^- is satisfactorily described by this model, SO_4^{2-}/Br^- exchange deviates considerably.

The extent of H^+ binding to alkyl sulfonic acid micelles [120] and to SDS micelles [115] has been studied conductometrically. The effect of Tl^+ ions on fluorescence yields of 9-methylanthracene in SDS provides an estimate for the Tl^+/Na^+ ion-exchange coefficient [121].

Although data for cation affinity on anionic aggregates are less abundant, some studies on dodecyl sulfate and decyl phosphate have been reported. Decyl phosphate has the peculiarity of bearing one or two negative charges (DPH^- and DP^{2-}) as a function of pH, thus allowing us to investigate the effect of double charged head groups.

Specific counterion effects on the apparent pK_a of an indicator bound to micelles of SDS and DPH^- follow the order: $Cs^+ > Rb^+ > K^+ > Na^+$, i.e. for counterions having the same charge, the effect increases with the counterion size (crystal radius) and ease of dehydration of the counterion (Hofmeister series). However, specificity may also depend upon hydrogen-bonding interactions between hydrated counterions and head groups or the partial disruption of the hydration layers of the head groups and counterions and the possibility that a fraction of the counterions are site-bound to a surfactant head group.

Changes in selectivity order probably have not been observed because the counterions that form very strong complexes with surfactants are water-insoluble. For example, divalent metal ions, Li^+ and the two guanidinium ions precipitate decyl phosphate; K^+ and Rb^+ (but not Li^+, Na^+ and Cs^+) form a precipitate with dodecyl sulfate at room temperature; Mg^{2+} and Ca^{2+} form precipitates with long-chain alkylcarboxylates as ClO_4^- precipitates alkyltrimethylammonium surfactants. In acidic solutions ($pH < 1$) the proton precipitates long chain alkyl carboxylates and alkylphosphates because protons bind covalently to their weakly acidic head groups.

Thus, the fact that the affinity order for those alkaline metal ions forming stable solutions with dodecyl sulfate, DPH^- and DP^{2-} follows a Hofmeister series implies that these counterions remain at least partially hydrated, i.e. do not form tight ion-pairs at the micellar surface.

The alkali metal cations show significantly less affinity than tetramethylammonium (TMA^+) and tetraethylammonium(TEA^+) ions for DPH^- micelles and, then, TMA^+ for dodecyl sulfate micelles. The quaternary ammonium ions interact both electrostatically and hydrophobically with the micellar surface.

Surprisingly, TMA^+ and TEA^+ fail to displace Na^+ from the DP^{2-} micellar interface. The alkali metal ion probably interacts much more strongly with divalent than monovalent head groups, perhaps by side-binding to the dianionic phosphate head group through an intervening water molecule [122].

1.10.4. Micropolarity and microviscosity

Traditionally aqueous micelles have been considered to contain a hydrophobic lipid-like core and a polar surface. In the light of more recent evidence and results, this picture should be somewhat modified. The hydrocarbon tails are relatively mobile and polar headgroups are hydrated and in contact with a number of water molecules. The extent of water penetration is still rather controversial.

Several experimental techniques have been used to obtain evidence regarding the properties of aggregates and the availability of water in them. Reactivity, charge transfer and fluorescence spectra, indicator shifts, NMR, and other methods have been proposed. The various methods and probes may measure different properties of a solvent aggregate, especially in micelles or other aggregates where solutes have considerable mobility and migrate rapidly between the various regions of the aggregate and between it and the bulk solvent.

For example, a substrate may reside mostly in one region of a droplet but react in another, so that the estimated polarity depends on the rate of measurement and on the sensing process, which may be rapid (as in fluorescence) or relatively slow (as in kinetics of hydrolysis or decarboxylation). In addition rates of decarboxylation (see later) are not related in any simple way to water activity or solvent parameters (such as E_T [123] or Z [124]). They are dependent on the hydrogen-bonding ability of the medium.

Among the kinetic methods and probes, a few unimolecular reactions have been selected, since they avoid complications due to the incorporation of more than one reactant into the aggregate. Decarboxylation of 6-nitrobenzisoxazole-3-carboxylate (NBIOC, see the structure below) is a useful indicator of reaction medium polarity since the rate increases as water is replaced by less polar solvents and it is almost free from acid and base catalysis.

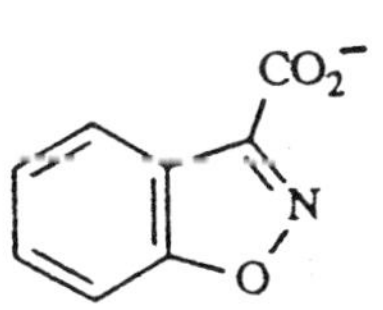

NBIOC

An apolar, non-hydrogen bonding environment evidently stabilises the charge delocalised transition state of the benzisoxazole carboxylate anion thereby accelerating the rate of decarboxylation [125].

Solvent and aggregate effects on the decarboxylation of NBIOC will be examined in detail. It has been suggested that the factors which influence the rate of decarboxylation are the carboxylate ion hydrogen bond formation in protic solvents (which inhibits the reaction by selectively stabilising the starting material) and the transition-state stabilisation in dipolar aprotic solvents (which accelerates the reaction) [126].

Table 1.9 reports the kinetic data in homogeneous solvents. It is interesting to note that poor correlation can be observed between the kinetic data and solvent effects on electronic transitions (Fig. 1.23) [127].

TABLE 1.9

Relative rate constants for decarboxylation of 6-nitrobenzisoxazole-3-carboxylate in homogeneous solvents at 30°C

Solvent	k_{rel}^{a}	$\Delta H^{\neq\, b}$	$\Delta S^{\neq\, c}$
Water	1	135	+80
Methanol	34		
Formamide	100		
Chloroform	110		
Ethanol	130		
Benzene	650		
Dimethoxymethane	5.0×10^{3}		
Nitromethane	7.8×10^{4}		
Acetonitrile	4.0×10^{5}	105	+85
Diglyme	6.8×10^{5}		
Dimethylsulphoxide	1.3×10^{6}	100	+75
Acetone	3.2×10^{6}		
Dimethylacetamide	2.1×10^{7}		
Hexamethylphosphoramide	1.0×10^{8}	95	+75

a: Rate constant = 3×10^{-6} s^{-1} at 25°C (the rate increases to 7.4×10^{-6} s^{-1} at 30°C); *b*: kJ mol^{-1}; *c*: JK^{-1} mol^{-1}. Data taken from Ref. [128], with permission.

Micellar effects are listed in Table 1.10 for cationic, zwitterionic and non-ionic aggregates (NBIOC does not interact with anionic micelles). When the substrate is incorporated into the micelle, the change in concentration of ions or other solutes in the Stern layer can influence the reaction rate only by changing the structure of the micelle and its water interface and therefore changing the interaction of the initial and transition states with the medium.

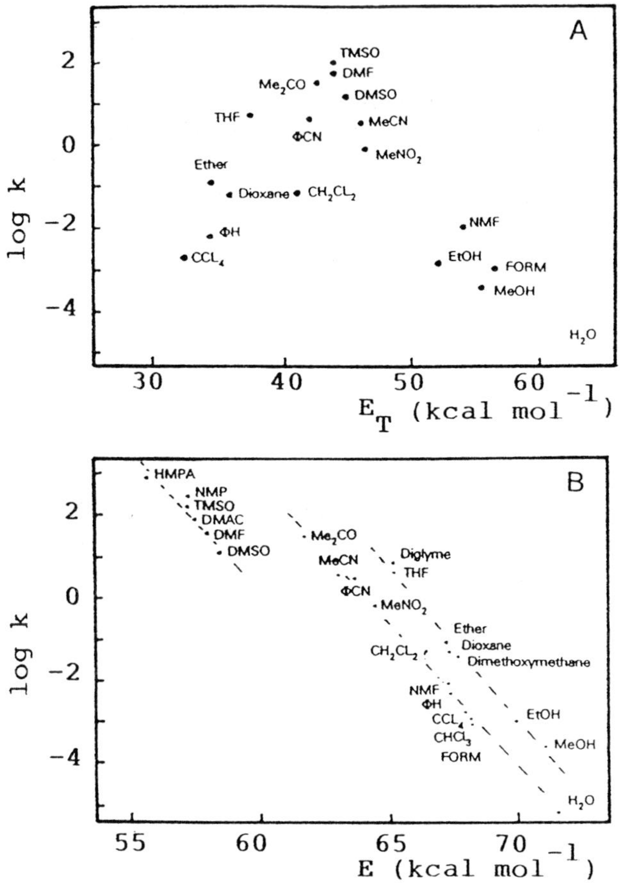

Fig. 1.23. Variation of log k (for decarboxylation of NBIOC) with the Reichardt E_T parameter (A), and with the transition energy of the long wavelength absorption maximum of 2-cyano-5-nitrophenolate anion (B). Taken from Ref. [127], reproduced with permission.

The last entries in Table 1.10 allow the comparison between three distinct types of aqueous aggregate obtained from alkylammonium salts [129]. The decarboxylation reaction is accelerated in the following order; triple chain > double chain > single chain. Although the relative rates cannot be correlated directly with the carbon number or with the aggregate weight, it was observed that the rate enhancement by the dialkylammonium membrane was influenced by its hydrophobicity and

TABLE 1.10

Relative rate constants for decarboxylation of 6-nitrobenzisoxazole-3-carboxylate in surfactant aggregates at 25°C[a]

Amphiphile	k_{rel}	Activation parameters[b]
C_9H_{19}–C_6H_4-p-$(OCH_2CH_2)_{24}OH$[c]	65	
$C_{12}H_{25}N^+(CH_3)_3$ [c]	70	
$C_{16}H_{33}N^+(CH_3)_3$ [c]	95	$\Delta H^{\neq} = 89$; $\Delta S^{\neq} = -11$
$C_{16}H_{33}N^+(CH_3)_2$–$(CH_2)_6$-$N^+(CH_3)_2C_{16}H_{33}$ [c]	400	
$C_{16}H_{33}N^+(CH_3)_3$ [d]	85	
$(C_{16}H_{33})_2N^+(CH_3)_2$ [e]	400	$E_a = 92$ ($<T_c$); 84 ($>T_c$)
$(C_{12}H_{25})_2N^+(CH_3)_2$ [e]	430	$E_a = 117$($<T_c$); 59 ($>T_c$)
$(C_8H_{17})_3N^+CH_3$ [f]	1000	

a: Data taken from Ref. [128] and from references cited therein; *b*: in kJ mol^{-1} or J K^{-1} mol^{-1}; *c*: micelles; *d*: microemulsion (1-butanol:HTAB:n-octane 2:1:1 by weight); *e*: bilayer (T = 30°C); *f*: small tight aggregate (T = 30°C).

fluidity, the fluidity in turn being dependent on the crystal to liquid crystal phase transition.

The activation data suggest that above the phase transition temperature (T_c) the membrane fluidity becomes influential and the most fluid $(C_{12}H_{25})_2N^+(CH_3)_2$ membrane should give the lowest E_a value.

Table 1.10 reports data for this and other aggregates. As can be seen from the kinetic data, despite the difference between the structure of the aggregates, ionic reactions occur in an environment (surface) whose polarity is intermediate between those of water and ethanol, but in which water is available and reactive.

The importance of the hydrophobic environment is also evident in kinetic data obtained in the presence of polyions (see Table 1.11) and, in particular, for polysoaps (compounds 9–17 in the table) [130]. These compounds do not generally have a critical micelle concentration in the same sense as conventional micelles.

A transition from polyelectrolyte to polymer micelle is, however, observed at 10–13 mol% of the lauryl-group content, where a drastic reduction in viscosity and increase in solubilisation ability occurs. Thus polymers with more than 12 mol% of lauryl group would be classified as polymer micelles and below this as typical polyelectrolytes.

The decarboxylation of NBIOC has been assessed only in polymer

TABLE 1.11

Relative rate constants for decarboxylation of NBIOC in the presence of different polyions, at 25°C

Polymer	k_{rel}	Activation parameters[a]
1 PVBC[b]	2300	$\Delta H^{\neq} = 88$; $\Delta S^{\neq} = +10$
2 PVBC/K^+	3700	
3 PVBC/Cs^+	14000	$\Delta H^{\neq} = 67$; $\Delta S^{\neq} = +45$
4 CyHAm[c]	3.5	
Polyethyleneimines[d]:		
5 $R(C_{12}H_{25})_{0.1m}(CH_3)_{1.9m}$	90	
6 $R(C_{12}H_{25})_{0.25m}(CH_3)_{1.75m}$	430	
7 $R(C_{12}H_{25})_{0.25m}(C_2H_5)_{1.75m}$	1300	
8 $R(C_{12}H_{25})_{0.25m}$	220	
Other polyions:		
9 EVI[e,f] -L-3	14	
10 -L-12	20	
11 -L-22	80	
12 -L-29	400	
13 -L-41	270	
14 4-VP[e,g]-Et-85	3.5	
15 -Hx-65	35	
16 -Bz-95	20	
17 DACS[e,h]-LB-22	110	

a: In kJ mol^{-1} or J K^{-1} mol^{-1}; *b*: poly(vinylbenzo-18-crown-6); *c*: cycloheptaamylose, at 60°C; *d*: R = $(C_2H_4N)_m$, with m = 1400; *e*: formulae shown in Fig. 1.24; *f*: L-x means polymer containing x mol% of n-dodecyl groups; *g*: Et = ethyl, Hx = hexyl, Bz = benzyl; *h*: LB = 4-n-dodecylbenzyl. Data from Ref. [128].

micelles. The polyion EVI-L-29 is a better catalyst than EVI-L-41 due to the fact that an excess of lauryl groups stiffens the polymer chain. The hydrophobicity of the polymer micelle is, in fact, closely related to the conformational compactness of the polymer chain.

Several investigations on the polarity and microviscosity of surfactant aggregates have appeared in the literature. The published values of the parameters obtained by using different probes and techniques vary within a rather broad range. Such a discrepancy may have its origin in different mechanisms of the response of the various probes,

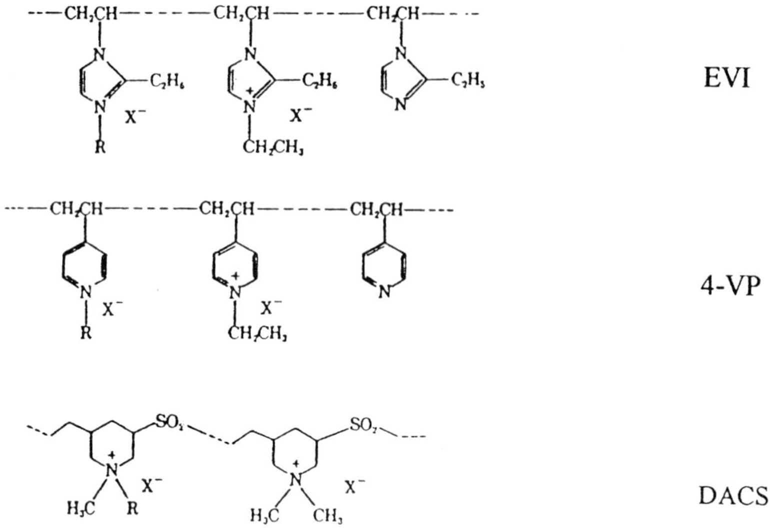

Fig. 1.24. Chemical structures of some polyions reported in Table 1.11.

inappropriate choice of empirical scales for polarity and viscosity, dissimilar location of different probes in the bilayer of vesicles and the different degree of interaction of probes with their environments.

Studies based on the use of $E_{T(30)}$, Z and other related micelle-specific parameters have been reported in several papers. In particular, distinct changes in $E_{T(30)}$ values for aggregates have been observed as a function of the amphiphile nature, chain length, type of counterion, presence of additives, etc. Especially-designed azobenzene derivatives have been proposed in the literature as suitable and sensitive solvatochromic probes for microheterogeneous media [131].

Table 1.12 lists the values of the estimated effective dielectric constants of some amphiphilic aggregates, obtained from different experimental measurements. Dioxane–water mixtures were used as reference media.

As far as microviscosity is concerned, Table 1.13 presents a homogeneous series of data estimated from the monomer–excimer ratio of 1,3-di(α-naphthyl)propane (DNP). It can be shown that the microviscosity of SDS and HTAC micelles in pure water is 14 and 27 Cp, respectively (to be compared with water = 10, n-octanol = 8.9 and glycerol = 875 Cp). The data suggest a model in which the micellar core occupied by the probe contains some water molecules and is comparatively polar.

TABLE 1.12

Effective dielectric constants at the aggregate surface, at 25°C, referred to dioxane–water mixtures

Amphiphilic aggregate	Effective dielectric constant
SDS micelles	32^{a}; 55^{b}
HTAB micelles	32^{a}; 31^{b}
Triton X100 micelles	32^{a}; 30^{b}
Brij 35 micelles	29^{b}
HTAB microemulsion^{c}	27^{b}
SHS microemulsions^{d}	33^{b}
DMP vesicles^{e}	14^{b}

a: From Ref. [132]; *b*: from Ref. [133]; *c*: microemulsion composition 4.6 g HTAB, 10 ml water, 1.1 ml hexadecane and 5.5 ml 1-butanol; *d*: microemulsion composition 1.25 g SHS (sodium hexadecylsulfate), 13.5 ml water, 0.6 ml hexadecane and 3.4 ml 1-pentanol; *e*: DMP = dimyristoylphosphatidylcholine (1.2 mg/ml).

The increase in ionic strength causes a decrease in microviscosity in spite of a considerable increase in the micellar size, which is also consistent with the partial penetration of water into the micellar core. Ethanol addition also decreases the microviscosity, resulting from an increase in the micelle core's polarity and concomitant loosening of the core structure because of the interpenetration of ethanol molecules.

1.11. OCCURRENCE OF ORGANIZED STRUCTURES IN OTHER POLAR SOLVENTS

Suitable models of solutions based on casual mixing of their components can allow the understanding of a variety of chemical systems such as ideal dilute solutions, electrolyte solutions, and regular polymer solutions.

Since self-organizing systems are the exact opposite of randomly mixed solutions, a different approach is necessary. Solute–solvent interactions may thus be better described in terms of lyophilic, lyophobic and lyophilic–phobic interactions.

If interactions between solute and solvents are lyophobic enough, phase separation occurs beyond a well defined solubility limit. As examples, oil–water and water–mercury systems can be mentioned. On

TABLE 1.13

Micellar microviscosities of ionic surfactants sensed by DNP[a] under different experimental conditions

Surfactant	T (°C)	Additives (M)	Microviscosity (cP) at pressure[b]			
			1	450	990	2070
HTAB	25	none	47	66	86	
HTAC	25	none	27	39	53	84
SDS	15	none	14	22	29	50
SDS	25	none	11.5	18	25	39
SDS	40	none	10	14	19	29
SDS	50	none	10	13	17	27
SDS	25	NaCl 0.1	4.0	8.0	12	19
SDS	25	NaCl 0.2	3.1	5.8	10	19
SDS	25	NaCl 0.4	3.5	6.4	11	21
SDS	25	Na_2SO_4 0.1	3.3	6.9	12	24
SDS	25	Na_2SO_4 0.2	3.2	6.3	11	24
SDS	25	Na_2SO_4 0.4	4.7	7.9	13	33
SDS	25	C_2H_5OH 0.1	11	18	23	37
SDS	25	C_2H_5OH 0.3	11	16	23	36
SDS	25	C_2H_5OH 1.0	9.8	14	20	30
SDS	25	C_2H_5OH 2.0	7.6	12	16	27

a: C_{DNP} in cationic micelles = 3×10^{-5} M, C_{DNP} in SDS micelles = 1.2×10^{-5} M; *b*: pressure measured in bars. Surfactant concentration = 0.02 M. Data taken from Ref. [128] (and references cited therein).

the other hand, if solute–solvent interactions are sufficiently hydrophilic, solute and solvent are completely miscible (in all proportions).

A model based on random mixing of components driven by thermal motion is appropriate to describe the above two limit situations, which represent the two limits of the well known Gibbs free-energy expression ($\Delta G = \Delta H - T\Delta S$), where only one of the terms dominates. The enthalpy is predominant in the phase separation model, whereas the entropy is the most important factor in the random mixing model.

The model can be tentatively extended to the more complex amphiphilic solutions in which lyophilic–phobic interactions are simultaneously present, although the formation of a variety of aggregate structures cannot be simply explained by combining the molecular and phase separated systems. As summarized by Shinoda [134], there are four conditions that define an organized solution: (i) low solute solubility,

(ii) swelling of the solute phase by the solvent, (iii) solute present in a liquid or liquid-crystalline state, and (iv) solute aggregation to form a structure having high aggregate weight. In particular, conditions (i) and (ii) require that the solute molecules be amphiphilic, since the lyophobic moiety favours the aggregation and the lyophilic moiety prevents phase separation.

The aggregation in aqueous systems has led to an overemphasis of the role of the unique properties of water in promoting the self-assembly of amphiphiles. In fact, organized amphiphilic assemblies can also be found in other solvents, such as hydrazine [36], glycerol [135], ethyleneglycol [136–138], formamide [139,140], etc. In order to permit the amphiphilic assembly, the solvent must meet some requirements which will be examined.

A high solvent polarity, and hence low solubilization of non-polar solutes, is an essential factor. Two measures of such solvent polarity are the Gordon "cohesiveness" parameter ($\gamma/v^{1/3}$) [141], and the Hildebrand solubility parameter ($\delta = (\Delta H_v/v)^{1/2}$) [142]. The symbols γ, ΔH_v and v represent the surface tension of the solvent, the heat of vaporization, and the molar volume, respectively.

The solvents which allow surfactant aggregation possess Gordon parameters above 12.5 dyn cm^{-2} and Hildebrand parameters greater than 16 (cal $cm^{-3})^{1/2}$, as can be seen in Table 1.14.

Starting from water, the solubility of non-polar molecules increases (and the solvophobic driving force decreases) along the solvent series, with a parallel increase of cmc and decrease of amphiphilic aggregate size. NMR studies performed on SDS/decanol liquid crystals in glycerol showed that the organized structures are much more disordered than those obtained in water and that further decrease in solvent cohesiveness gives rise to the formation of a molecularly dispersed solution.

Hydrogen bonding is at the basis of solvent cohesiveness in several solvents, but other mechanisms may also be invoked. For example, for the highly polar aprotic solvent 3-methylsydnone, having a Gordon parameter of ca. 15 dyn cm^{-2}, the formation of amphiphilic aggregates has been attributed to other types of molecular interactions [144].

Solvents such as water, ethylenglycol and hydrazine, which exhibit strong intermolecular interactions, favour the separation of non-polar molecules into another phase (phase separation model) and the formation of organized amphiphilic structures (self-organizing fluid model). The application of free-energy minimization to surfactant self-assembly allows the prediction of the final size and shape of these aggregates as

TABLE 1.14

Solvent parameters and surfactant aggregation in different media

Solvent	$\gamma/v^{1/3}$ (dyn cm^{-2})	δ (cal cm^{-3})$^{1/2}$	Aggregate formation
Water	27.7	23.4	Yes
Hydrazine	22	19	Yes
Glycerol	15.1	17.7	Yes
Ethylammonium nitrate	13.7		Yes
Ethyleneglycol	12.5	16.1	Yes
Formamide		17.9	Yes
Dimethylsulfoxide	11.2	13	No
Nitromethane	9.8	12.3	No
Methanol	6.5	14.5	No
Ethanol	5.7	13	No
Acetone	5.6	9.8	No
Hexane	4	7.3	No

Data taken from Ref. [143], with permission.

the result of two opposite driving forces: solvophobicity and solvophilicity. The first one tends to drive the amphiphile molecules into aggregates of infinite size, whereas the second tends to molecular dispersity.

The optimal size and shape of a given aggregate is related to a geometric amphiphile parameter dependent on the volume of the solvophobic part of the molecule (v), on the length of the solvophobic chains (l) and on the area of the solvophilic head group (a) [145].

The theory predicts the formation of different types of micelles on the basis of the value of the parameter v/al. Spherical micelles are obtained for values lower than 0.333, while cylindrical micelles and bilayers form when the geometric parameter is in the range 0.333–0.5 and 0.5–1, respectively.

Solvent–solute interactions can be, of course, modified by changing the ionic strength, the counterion (for ionic amphiphiles) and the temperature. Such modifications affect, in particular, the interfacial properties of the aggregate and therefore the area per head group a. For example, the salt addition to ionic micellar solutions reduces the a value by screening the electrostatic repulsion between the charged head groups. As a consequence, the term v/al becomes larger, increasing the micellar size.

1.12. FORMATION OF NON-MICELLAR ORGANIZED ASSEMBLIES IN WATER

1.12.1. Formation of liquid crystalline phases

Different types of organized amphiphilic structures can be obtained in aqueous solutions as a function of surfactant concentration and/or experimental conditions. Spheroidal micelles are the simplest aggregates which form at low surfactant concentrations while at higher concentrations the appearance of cylindrical structures are observed and, above 30–40% of surfactant, liquid crystalline phases can form.

These organized assemblies (see Fig. 1.25) have properties which are intermediate between those of solids and liquids. They result from aggregation of surfactant molecules into large domains, often of hexagonal or lamellar structure, although many other geometric arrangements are possible.

As indicated in the previous paragraphs, not only the increase of the surfactant concentration causes changes in shapes and structures. There are also other methods of reducing interface curvature which originate the aggregate growth:

– the addition of an electrolyte, such as NaCl, reduces the electrostatic repulsion between neighbouring head groups, increasing the radius of curvature of the aggregates;

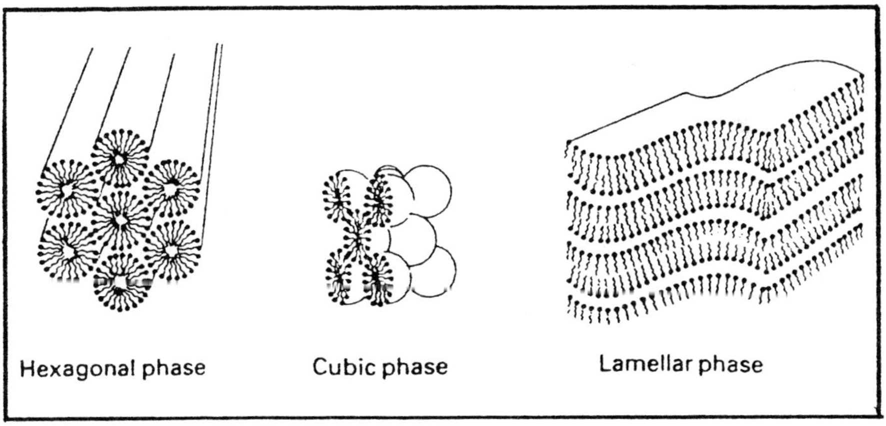

Fig. 1.25. Schematic representation of liquid crystalline structures.

– the addition of a cosurfactant, such as an n-alkanol, induces a similar effect;

– changes in the counterions, e.g. using Mg^{2+} instead of Na^{+} in anionic surfactants favours the aggregate growth because both the hydrated cations are of about the same size, but only half the number of Mg^{2+} are required. In this case, the total volume requirement of each head group is lower;

– the use of surfactant mixtures can also cause micellar growth since the formation of liquid crystalline phases can occur at low surfactant concentration in these systems.

It must be noted that, whereas solutions containing spherical micelles have low viscosity, the liquid crystalline phases (especially the hexagonal phase) exhibit very high viscosity values.

1.12.2. Formation of monolayers, bilayers and vesicles

Spreading an organic solution of an amphiphile on water can lead to monolayer formation at the water–air interface. Spreading is accomplished in a Langmuir film, which controls the surface area and surface pressure. These aggregates represent ideal media for investigating two-dimensional processes.

Bilayer (or black) lipid membranes (BLMs) are formed by brushing an organic solution of surfactants (or lipids) across a pinhole having a diameter of ca. 2–4 mm, separating two aqueous phases. These aggregates have been extensively used in studies concerning the transport mechanisms across biological membranes. In fact, closed bilayer aggregates formed from phospholipids (liposomes) or from surfactants (vesicles) are currently used as models of biomembranes.

Swelling of thin lipid films in water gives rise to the formation of multilamellar vesicles (MLVs) having a large mean diameter (1000–8000 Å). The sonication of these structures at a temperature above the phase-transition limit (at which the MLVs transform from a gel to a liquid phase) leads to the formation of smaller aggregates, termed single bilayer vesicles or small unilamellar vesicles (SUVs). These vesicles have a mean diameter in the range 300–600 Å.

It is important to note that the kinetic stability of the vesicles (typically several weeks) is considerably greater than that of micelles. The vesicles do not exhibit a typical aggregation concentration and their stability can be further enhanced through suitable polymerization reactions.

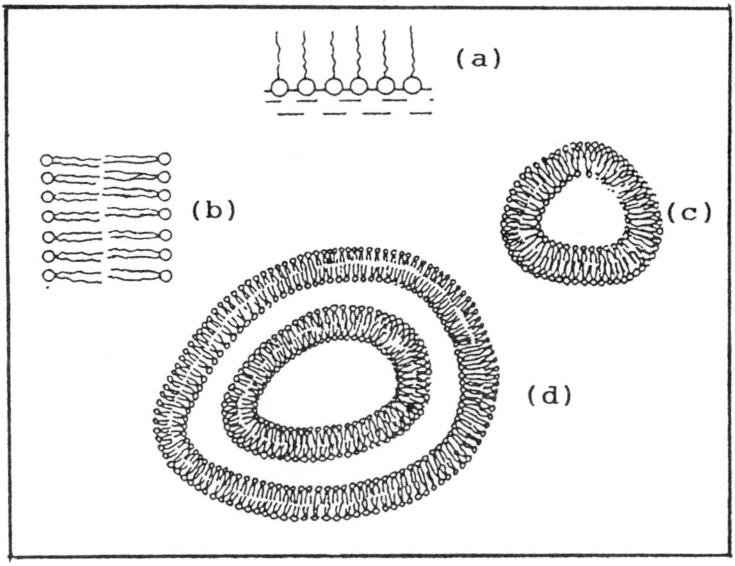

Fig. 1.26. Structures of: (a) monolayers, (b) bilayers, (c) single compartment vesicles, (d) multicompartment vesicles.

These aggregates possess a large number of solubilization sites, with hydrophobic molecules mainly located among the hydrocarbon bilayers and the polar molecules confined in the vesicle-entrapped water pools. Charged polar species, in particular, can be electrostatically repelled from the inner vesicle surface. On the other hand, small cations and anions can be attracted by the oppositely charged outer or inner surfaces, whereas charged species having the same charge of the vesicles can still be attached to the vesicle surface if they possess a suitable moiety (e.g. a long hydrocarbon chain), capable of hydrophobic interactions with the aggregate surface.

Figure 1.26 shows the typical structure of the above mentioned surfactant aggregates, while their main properties are summarized in Table 1.15.

1.12.3. Formation of microemulsions

Mixtures of water, oils and surfactants represent very important systems from fundamental and practical points of view. When a surfactant is added (under particular conditions) to an oil–water mixture, a

TABLE 1.15

Summary of the properties of monolayers, bilayers and vesicles

Property	Monolayers	Bilayers (BLMs)	Vesicles
Mean aggregate weight (Da)	*a*	*a*	10^7
Hydrodynamic diameter (A)	*a*	*a*	300–10000
Stability	days–weeks	hours	weeks–months
Time-scale of aggregate formation and breakdown	min–hours^b	min^c	min–hours^d
Number of solubilized molecules per aggregate	large	large	large
Location of the solubilizates	intercalation and surface	either or both sides of the bilayer, within the bilayer	inner water pool, outer surface, bilayer

a: Dependent on area covered and on the density of coverage; *b*: monomer to subphase; *c*: monomer to plateau border; *d*: monomer to bulk.

single phase system called a microemulsion can be obtained. The presence of polar additives (usually aliphatic alcohols) can help to stabilize the aggregate. These components are termed cosurfactants.

A typical oil-in-water (o/w) microemulsion structure can be represented as an oily drop in which the surfactant (and cosurfactant) molecules are placed with their hydrocarbon tails directed to the drop center and their polar heads oriented to the aqueous bulk phase. Depending on the ratio of the components, the formation of aggregates having an inner hydrophilic (aqueous) core dispersed in a bulk organic phase can be also obtained. These structures, called water-in-oil microemulsions (w/o), are very important from the analytical point of view, either as extraction solvents or as suitable chromatographic mobile phases.

The microemulsion formation and break down are very fast processes (the time-scale is in the range 10^{-4}–10^{-6} s). The obtained aggregates, which are stable for months, can solubilize a quite large number of solute molecules in the inner pool and in the surfactant tail region. The hydrodynamic diameter of microemulsions is usually in the range 50–5000 Å.

1.13.1. Phase diagrams of ionic surfactants close to the Krafft point

The solubility of amphiphilic molecules in water, defined as the concentration of dissolved molecules in equilibrium with a solid (crystalline) surfactant phase, increases significantly with rising temperature. In fact, a sharp bend in the solubility curve is observed, which is due to micelle formation (see Fig. 1.27). The point of intersection of the solubility and C_M curves, plotted as a function of temperature, is indicated as the Krafft point, T_K. At this point, surfactant monomers, micelles and hydrated solid phases coexist.

The micelles exist only at temperatures above the Krafft point, whereas below this temperature the amphiphile is present in molecular form up to the solubility limit. At higher surfactant concentrations, the monomers are in equilibrium with the crystals, and above the Krafft

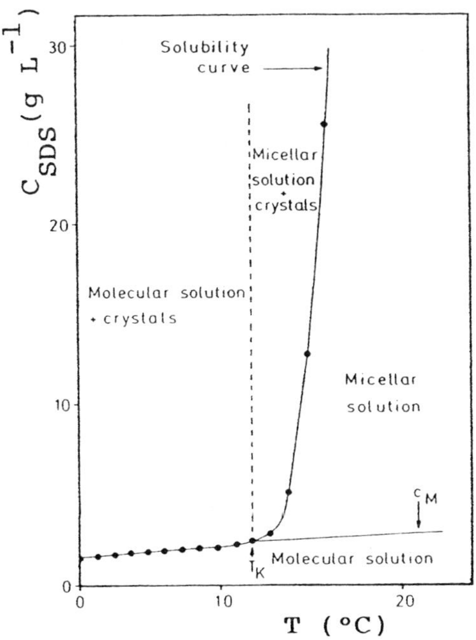

Fig. 1.27. Phase diagram of the SDS–water system close to the Krafft point. Reproduced from Ref. [148], with permission.

TABLE 1.16

Krafft points (°C) of n-dodecyloxyethylene sulfates

Number of oxyethylene groups in the molecule	Counterion			
	Na^+	Ca^{2+}	Sr^{2+}	Ba^{2+}
None	9	50	64	105
1	5	15	32	62
2	–1	<0	—	35
3	<0	<0	<0	12

Data taken from Refs. [148] and [149] with permission.

point the three phases are simultaneously present. The transition from crystals to micellar aggregates is reversible and, upon cooling micellar solutions below T_K, supercooled liquid phases may form.

It is interesting to note that the Krafft point in the homologous surfactant series rises with the increasing alkyl chain length. Moreover, it depends greatly on the counterion. For example, the Krafft points of the calcium salts of anionic surfactants are higher than those of the sodium salts, thus limiting the use of such amphiphiles as detergents in hard water.

Since in most applications surfactant precipitation must be avoided, the reduction of the Krafft point is often required. The introduction of ethoxy groups into the molecules of anionic surfactants is particularly effective. For example, the introduction of one ethoxy group in dodecylsulfate moiety permits the reduction of T_K of the corresponding calcium salts from 50°C to ca. 15°C.

The dependence of Krafft temperatures of different laurylethoxysulfates on the number of ethoxy groups is illustrated in Table 1.16.

The addition of electrolytes to surfactant solutions increases the Krafft point, whereas a lowering is observed upon addition of alcohols or structure-breaking solutes, such as n-methylacetamide.

1.13.2. Phase diagrams of non-ionic amphiphiles

Non-ionic surfactants can also give rise to the formation of different types of mesophases depending on temperature and concentration conditions. Figure 1.28 shows the typical phase diagram of $C_{12}E_6$ in water. The region B represents a single isotropic solution, where the micellar aggregates exhibit gradual changes of shape and size upon changing

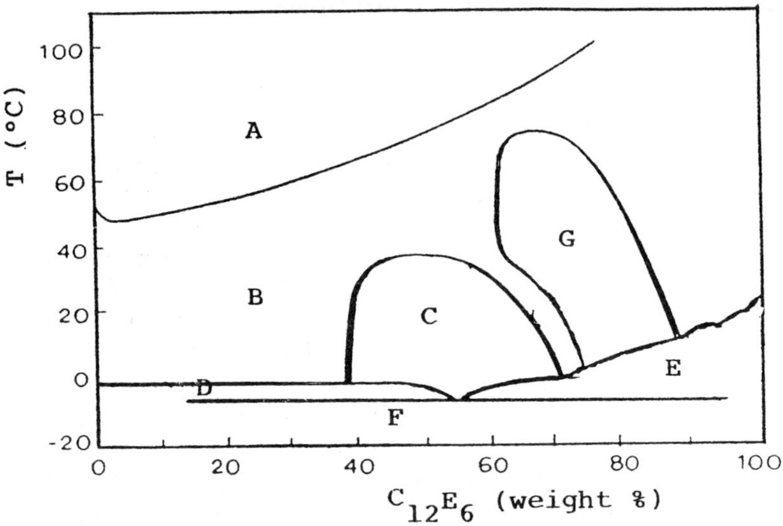

Fig. 1.28. Condensed binary phase diagrams for $C_{12}E_6$-H_2O. (A) two isotropic liquids; (B) one isotropic liquid; (C) middle phase; (D) ice + liquid; (E) crystals + liquid; (F) ice + crystals; (G) neat phase. Reproduced from Ref. [150], with permission.

temperature and/or surfactant concentration.

Above a critical temperature, termed the cloud point (T_C), the solution becomes turbid and two liquid phases separate out (region A), one of which is surfactant-rich and the other contains a monomer solution, with practically no aggregates present. The clouding phenomenon is reversible. The curve which separates regions (A) and (B) is called the consolution curve.

For a homologous series of alkylpolyoxyethylene surfactants, T_C decreases with increasing the hydrocarbon chain length and decreasing the number of oxyethylene units.

The effect of additives on the cloud point may be noticeable. Two categories of additives cause considerable increases in T_C:(*i*) urea and salts containing anions known to break the structure of water (e.g., iodide, thyocyanate, perchlorate); (*ii*) salts of cations capable of forming complexes with ether groups, such as Li^+, Ca^{2+}, Mg^{2+}, Al^{3+}, Ni^{2+}.

Lowering of the cloud point (i.e. salting out) was noted after addition of a few electrolytes containing non-complexing cations (Na^+, K^+, NH_4^+, Cs^+, Rb^+), in association of anions such as nitrate.

A detailed study of the effect of inorganic additives on solutions of non-ionic surfactants can be found in Refs. [151–153].

1.13.3. Phase diagrams of ternary mixtures

When oil, water and surfactant are blended together and allowed to equilibrate, one, two or more phases may appear. In many cases almost the entire inventory of surfactant will reside in one of the phases together with various proportions of oil and water. This phase, which contains the bulk of the surfactant, is often called the micellar phase, although it may also present the structure of a microemulsion.

Phase equilibria within ternary systems are most conveniently represented by triangular phase diagrams which allow the identification of the different phase regions and the combination of the three components in order to obtain a given suitable system. In these diagrams a property of equilateral triangles is exploited, that is: the sum of the lengths of perpendiculars from a given point in the triangle plane to the three sides is equal to the height of the triangle.

Figure 1.29 shows an equilateral triangle in which each side (divided into 100 parts corresponding to percentage composition) is assigned to one of the components. The composition of the ternary system at a given point (e.g. X) is obtained by measuring the perpendicular distances from this point to the three sides (20% A, 30% B and 50% C).

The lines connecting the apex to a point on the opposite side represent all the possible mixtures which have the same relative amounts of the other two components. For example, the line CD represents all possible compositions obtained starting with a mixture containing 70% A and 30% B, further diluted with component C.

Simple phase diagrams are obtained for ternary systems in which two of the three components are completely miscible. For example if components A and B, and A and C are assumed to be completely miscible liquids, whilst B and C are only partially miscible, the region under the curve lmnop (indicated as 2L region) is an area of immiscibility. Thus, any system having a composition within this region will separate into two liquid phases at equilibrium. For example, a ternary system X gives rise to the separation of two liquids having the composition indicated by points o (20% A, 20% B and 60%C) and m (20% A, 60% B and 20% C). Along the tie line mo, all the ternary mixtures form two liquids having the above indicated compositions, but the weight ratio between these two liquids depends on the relative position of the system along the line mo.

In surfactant-containing ternary systems, A is usually the surfactant, B is water and C is an oil. The resulting simplified phase diagram is represented in Fig. 1.30.

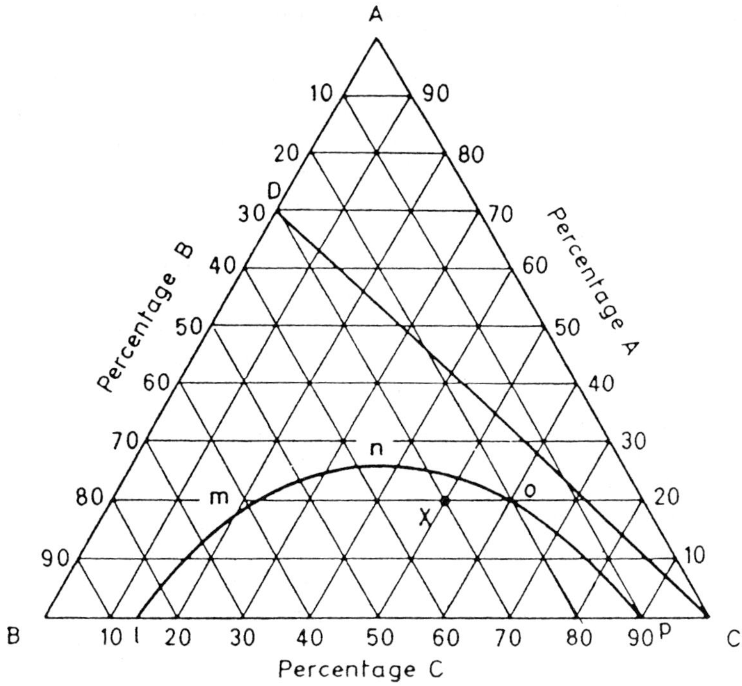

Fig. 1.29. Triangular three-component phase diagram. Reproduced from Ref. [13], with permission.

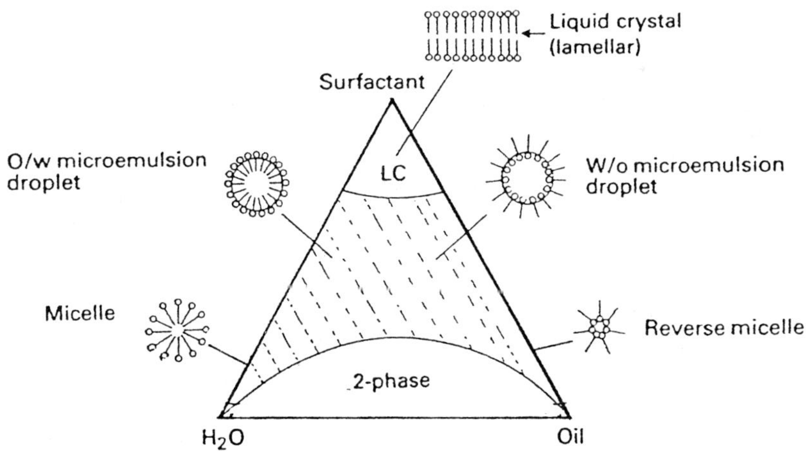

Fig. 1.30. Phase diagram of a surfactant–water–oil system. Reproduced with permission from Ref. [154].

The single-phase region known as the microemulsion shows interesting properties. In the oil-rich domain a w/o microemulsion forms, whereas in the water-rich corner, o/w aggregates are more likely to be formed. When droplets in w/o microemulsions are small, they are sometimes referred to as "reversed micelles". In the intermediate region of the phase diagram, bicontinuous structures are present.

The shaded area of Fig. 1.30 therefore represents a single phase microemulsion domain which is optically transparent and exhibits a low viscosity. This system is thermodynamically stable, unlike emulsion droplets, which are unstable and as such have no place in the phase diagram.

Water may be replaced by other polar solvents, such as glycerol, formamide or short-chain alcohols, forming droplet dispersions in oil.

The molecular structure of the surfactant molecule is a major factor in determining the nature of the dispersion, since droplet shapes can be related to the curvature of the oil–water interface.

For an o/w microemulsion the suitable curvature is positive and the droplets can be regarded as essentially normal micelles in water, swollen with oil in the core. On the contrary, w/o microemulsions exhibit negative curvature and are preferentially formed by wedge-shaped surfactants, such as AerosolOT, didodecyldimethylammonium bromide, $C_{12}E_4$, etc., where the hydrophilic head group represents the thin end of the wedge. These o/w systems are of particular interest for their ability to dissolve water-soluble species present in oily matrices. For example, they have been applied for extracting some biomolecules from complex samples.

Since microemulsions are self-organizing systems, they are easily prepared by gentle shaking of the oil/water/surfactant mixture. A convenient representation of the stability of one-phase microemulsions is given in Fig. 1.31. A suitable stabilizing agent for a microemulsion may also be obtained using a combination of a hydrophobic surfactant (e.g. SDS) and an alcohol of intermediate chain (cosurfactant).

The w_0 term reported in the figure represents the ratio between water and surfactant concentrations. The two full lines CD and AB indicate the phase separation. The size of the droplet core in w/o microemulsions is essentially independent of droplet concentration and temperature and, for AOT-stabilized w/o microemulsions, r_w ¢ $0.18\, w_0$. This implies that droplets of chosen sizes can be prepared by simply controlling the w_0 ratio.

Although the microemulsion system is thermodynamically stable, the droplets are kinetically unstable and, as a consequence, solubilized

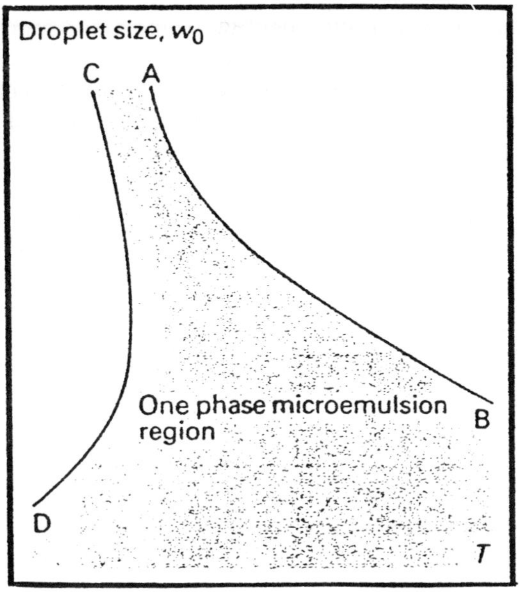

Fig. 1.31. Stability of microemulsions. From Ref. [154], reproduced with permission.

species are not confined within a single droplet but exchange between different droplets occurs on a microsecond to millisecond time-scale.

As indicated before, the molecular design of surfactant components represents the major controlling factor, but other experimental parameters may also play a significant role in the phase behaviour. They include: the nature of the oil matrix, the type of counterions, the ionic strength, temperature and pressure.

By carefully regulating the above parameters, the ternary system can be driven to well defined regions of the phase diagram, particularly useful for different analytical applications. Ternary phase diagrams including supercritical fluids have been also investigated, with interesting applications in analytical techniques. These applications will be examined in detail in other chapters.

1.14. SURFACTANT MIXTURES

Surfactants used in practical applications are often mixtures of surface-active compounds and, as mentioned in Section 1.2, their purification may be necessary in some cases. However, properly designed mixtures of dissimilar surfactants can have peculiar properties, superior to those of the individual components involved.

In solutions containing mixtures of surfactants, the tendency to form aggregates can be different than in solutions containing the pure surfactants. For example, precipitation occurs when some pure surfactants are present in aqueous solutions but it may be inhibited by adding other amphiphiles. It must also be noted that the surfactant composition of the mixed micelles can differ greatly from that of the surfactant monomer and this is important since the applications usually depend only on monomer composition or on aggregate composition. For example, surfactant adsorption on solids depends on monomer composition and concentration, whereas micelles do not play a relevant role. On the contrary, solubilization within the aggregates depends only on micellar composition.

The ability to predict and manipulate the tendency for surfactant mixtures to form aggregates, the aggregate properties and the distribution of the components between monomer solution and aggregates, may offer a wide range of possibilities.

1.14.1. Formation of mixed micelles

Certain surfactants form ideal mixed micelles and there the mixture properties can be predicted on the basis of the component properties. In these cases, the chemical potentials of the surfactants in the micellar pseudophase can be determined as a function of the cmc of the pure components. For a binary mixture of surfactants A and B, the following equation holds:

$$C_M = \{cmc_{(A)}\, cmc_{(B)}\}/\{Y_A cmc_{(B)} + Y_B\, cmc_{(A)}\} \qquad (32)$$

where C_M is the total monomer concentration, $cmc_{(A)}$ and $cmc_{(B)}$ are the cmc values for the individual components, Y_A and Y_B are the monomer mole fractions of A and B, respectively.

The cmc of the mixture SDS/SDeS (given by the C_M value) is satisfactorily described by Eqn. (32), whereas other surfactant mixtures show remarkable deviations [156] (see Fig. 1.32). Regular solution theory can well describe the data of cationic/non-ionic mixtures and, from tabulated values of the interaction parameters, the following conclusions have been reached:

1. The deviation from ideality for binary mixtures of surfactants increases in the order: cationic/non-ionic, anionic/non-ionic, anionic/cationic.

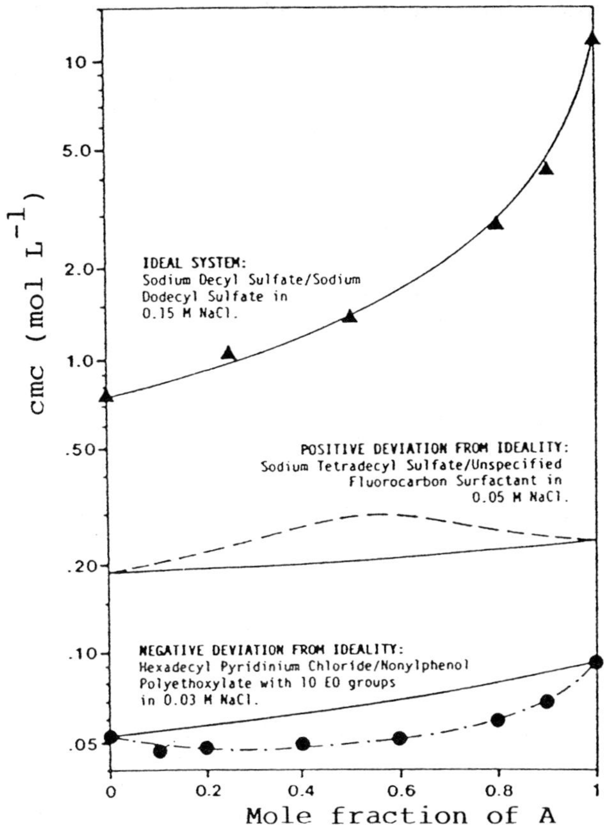

Fig. 1.32. Variation of cmc as a function of composition for different surfactant binary mixtures at 30°C. Surfactant A is the first component listed. Taken from Ref. [156], reproduced with permission.

2. The deviation from ideality decreases as added electrolyte concentration increases.

3. For anionic/non-ionic systems, where the non-ionic is of polyoxyethylene type, the deviation from ideality increases with the number of polyoxyethylene units.

4. The deviation from ideality of a zwitterionic/anionic system can vary widely from a value corresponding to a typical cationic/non-ionic mixture to nearly that of an anionic/cationic system.

5. The deviation from ideality decreases as temperature increases for ionic/non-ionic systems.

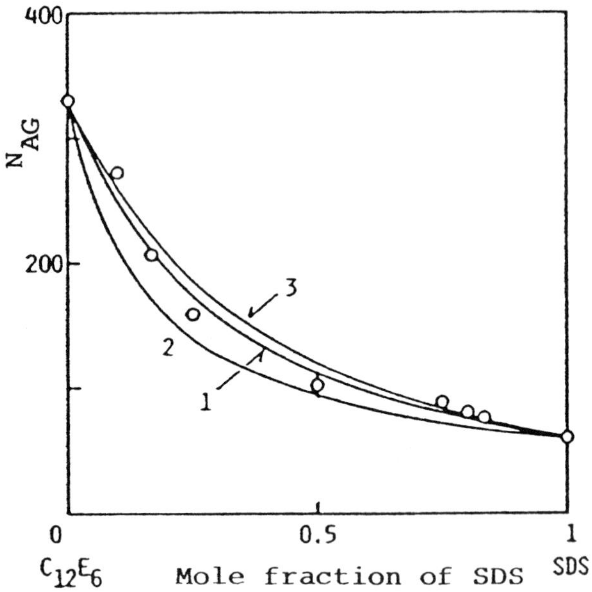

Fig. 1.33. N_{AG} of SDS/$C_{12}E_6$ mixed aggregates as a function of the micelle composition, at 25°C. Calculated curves: (1) oblate ellipsoids; (2) prolate ellipsoids; (3) spherical micelles. (O) experimental points. Taken from Ref. [155], reproduced with permission.

1.14.2. Size and aggregation number of mixed micelles

Several investigations on the size and aggregation number of mixed surfactant systems, performed using different experimental techniques, have been recently reported in the literature. In some particular cases, the properties of the mixed aggregates can be predicted as a function of their composition.

As an example, Fig. 1.33 shows the variation of aggregation numbers for mixed SDS/$C_{12}E_6$ micelles as a function of the SDS mole fraction [155]. The formation of oblate ellipsoids seems to prevail in this case, although the presence of nearly spherical micelles cannot be completely excluded.

For more detailed information concerning this topic see Refs. [156,157].

1.14.3. Counterion binding in mixed aggregates

The fractional counterion binding on charged mixed micelles is of great interest because it gives an indication of surface charge density,

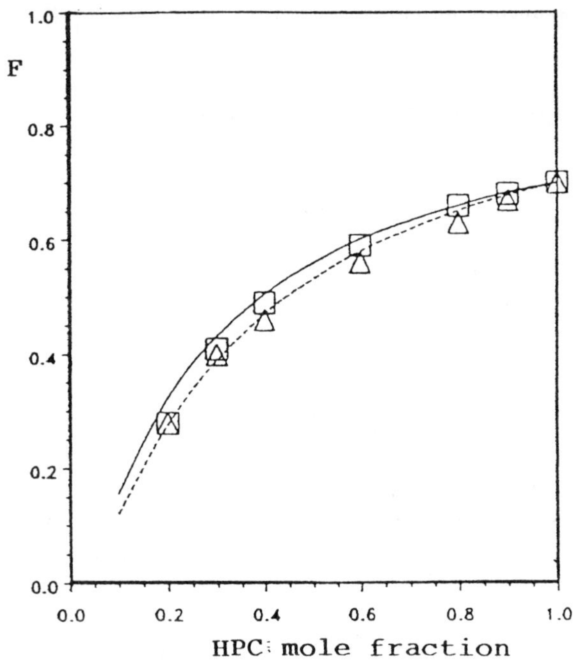

Fig. 1.34. Variation of the fractional counterion binding (F) as a function of the mole fraction of the cationic component in binary non-ionic/cationic surfactant mixtures. (□): HPC/$C_{12}E_6$. (Δ): HPC/nonylphenol-E_{10}. Taken from Ref. [158], reproduced with permission.

which is related to the mechanism of mixing non-ideality in ionic/non-ionic micelles. Moreover, this parameter allows the estimation of the electrical potential at the micelle surface.

The fractional counterion binding on mixed micelles composed of two surfactants having similarly structured aggregates and like charge varies monotonically between the values of the two individual components. For ionic/non-ionic aggregates the aggregation number and counterion binding (or electrical potential) decrease as the non-ionic content of the micelle increases (see Fig. 1.34).

This effect is expected since the intercalation of non-ionic head groups between ionic hydrophilic groups causes a lower density of the charged groups on the micellar surface, thus reducing the micellar electrical potential. Therefore, counterion binding is reduced.

The counterion binding decreases by increasing the number of oxyethylene groups of the non-ionic surfactant because the larger hydrophilic groups occupy a larger fraction of the aggregate surface area, thus reducing the surface charge density even more.

1.14.4. Surfactant precipitation in mixed systems

Precipitation of surfactants from aqueous solutions is generally undesirable in most surfactant-based analytical procedures as well as in other practical applications. Suitable additives are, in fact, often added to pure surfactants or to surfactant mixtures in order to avoid precipitations.

However, in some cases, surfactant precipitation can be useful. For example for surfactant recovery from aqueous solutions after their use in separation and/or enrichment procedures.

1.14.4.1. Precipitation in anionic/anionic and cationic/cationic mixtures

Mixing of like-charged surfactants having similar structures (e.g., close members of an homologous series) and possessing the same counterion, gives rise to a Krafft temperature which is intermediate between the pure components or exhibits only a slight minimum. Mixed crystals containing both surfactants are obtained in this case.

When the surfactants possess dissimilar structures or when the counterion is different, the Krafft temperature of the mixture can be lower than either of the pure components. These systems show the classical eutectic-type behaviour and the crystal phase contains only a type of surfactant (or counterion) in relevant amounts.

1.14.4.2. Precipitation in ionic/non-ionic surfactant mixtures

Since non-ionic surfactants are uncharged, they do not form salt precipitates with counterions. Moreover, a mixture of ionic and non-ionic surfactants gives rise to non-ideal micelles having a reduced ionic surfactant monomer concentration compared to ideal mixed micelles. The above two effects induce either an increase of the salinity/hardness tolerance and a Krafft point decrease, which is proportional to the amount of the non-ionic surfactant added.

1.14.4.3. Precipitation in anionic/cationic surfactant mixtures

Mixed micelles formed between cationic and anionic surfactants exhibit high negative deviation from ideality, as has already been discussed. The electrostatic attractions between the anionic and cationic surfactant can cause ion-pairing formation or coacervate formation in these mixtures. When the precipitation is desirable, these use of these mixtures may be beneficial.

Detailed studies on the phase diagrams of various "catanionic" amphiphiles, including the prediction of the complete precipitation phase boundary using the regular solution theory, have been reported in the literature [159,160].

1.14.5. Cloud point phenomena in mixed surfactant systems

The cloud point of a mixture of non-ionic surfactants is intermediate between those of the pure non-ionic surfactants involved. The cloud point of a dilute non-ionic surfactant solution increases upon addition of ionic surfactants.

Since the coacervate phase formation originates from the attractive forces operating between the aggregates, the addition of an ionic surfactant to non-ionic micelles introduces electrostatic repulsion between the mixed aggregates. This hinders the coacervate formation, thus raising the cloud temperature of the system.

The anionic/non-ionic amphiphile mixing in the coacervate phase has been approximated by equations based on the regular solution theory.

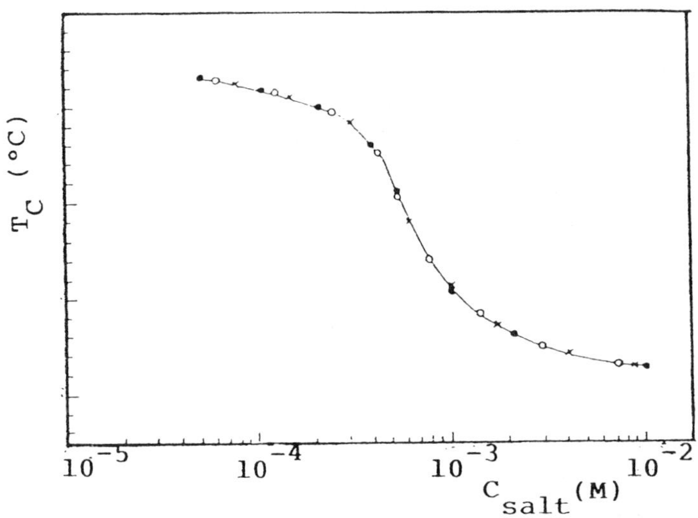

Fig. 1.35. Cloud temperature of a Triton X100 solution (1% w/v) containing one SDS molecule per non-ionic micelle as a function of the concentration of added electrolyte. (O) LiCl; (●) NaCl; (×) KCl. Taken from Ref. [161], reproduced with permission.

The observed non-ideality of mixing is similar to that reported for the corresponding mixed micelles. Studies performed above the cloud point involve the analysis and modelling of monomer–micelle and monomer–coacervate equilibria.

The effect of added electrolytes on the cloud point of ionic/non-ionic mixtures is, as expected, relevant. As a rule, T_C decreases as the ionic strength increases, following a sigmoidal curve. A typical example of such behaviour is shown in Fig. 1.35, where the cloud point of an SDS/Triton X100 mixture is plotted against the concentration of added salt [161].

An increasing number of interesting applications of the cloud-point phenomenon in separation science has been reported in recent years. In particular, the use of functionalized mixed micelles in aqueous media as effective non-traditional extraction systems has been greatly developed. Further details concerning the behaviour of such clouding mixtures will be given in Chapter 7.

REFERENCES

1 J. Falbe (Ed.), *Surfactants in Consumer Products*. Springer Verlag, Berlin, 1987, pp. 547.

2 J.H. Fendler and E.J. Fendler, *Catalysis in Micellar and Macromolecular Systems*. Academic Press, New York, 1975, 545 pp.

3 F.M. Menger and C.A. Littan, *J. Am. Chem. Soc.*, 113 (1991) 1451.

4 M.J. Rosen, Z.H. Zhu and H.Y. Hua, *J. Am. Oil Chem. Soc.*, 69 (1992) 30.

5 R.J. Hunter, *Foundations of Colloid Science. Vol. I.* Oxford University Press, New York, 1986, 673 pp.

6 B. Lindman and H. Wennestrom, *Topics in Current Chemistry, Vol. 87.* Springer, Berlin, 1980, 1–83 pp.

7 P. Mukerjee and K. Mysels, *Critical Micelle Concentrations of Aqueous Surfactant Systems.* National Standards Reference Data Series, Vol. 36, National Bureau of Standards, Washington DC., 1971.

8 C. Tanford, *The Hydrophobic Effect: Formation of Micelles and Biological Membranes.* Wiley, New York, 1973.

9 M.J. Rosen, *Surfactants and Interfacial Phenomena.* Wiley Interscience, New York, 1978.

10 F.M. Menger, *Acc. Chem. Res.,* 12 (1979) 111.

11 J.H. Fendler, *Membrane Mimetic Chemistry.* Wiley Interscience, New York, 1982.

12 B.R. Bluestein and C.L. Hilton, *Amphoteric Surfactants.* Marcel Dekker, New York, 1982.

13 D. Atwood and A. Florence, *Surfactant Systems.* Chapman and Hall, New York, 1983.
14 D.W. Armstrong, *Sep. Purif. Methods*, 14 (1985) 212.
15 E. Pelizzetti and E. Pramauro, *Anal. Chim. Acta,* 169 (1985) 1.
16 E. Pramauro, E. Pelizzetti, C. Minero, E. Barni, P. Savarino and G. Viscardi, *Ann. Chim. (Rome)*, 77 (1987) 209.
17 K. Shinoda, M. Hato and T. Hayashi, *J. Phys. Chem.*, 76 (1972) 909.
18 W. Guo, Z. Li, B.M. Fung, E.A. O'Rear and J.H. Harwell, *J. Phys. Chem.*, 96 (1992) 6738.
19 Y.P. Zhu, A. Masuyama and M. Okahara, *J. Am. Oil Chem. Soc.*, 68 (1991) 268.
20 W.L. Hinze and E. Pramauro, *Crit. Revs. Anal. Chem.*, 24 (1993) 177.
21 M.C. Allison in: D.R. Karson (Ed.), *Industrial Applications of Surfactants.* Royal Chem. Soc., London, 1987, p.333.
22 P. Mukerjee and K.J. Mysels, in: K.L. Mittal (Ed.), *Anomalies of Partially Fluorinated Surfactant Micelles.* ACS Symp. Ser. 9, Am. Chem. Soc., Washington DC, 1975, p. 239.
23 P. Mukerjee, *Adv. Colloid Interface Sci.*,1 (1967) 241.
24 M. Schwuger, *Ber. Bunsenges. Phys. Chem.*, 75 (1971) 167.
25 M.J. Schick (Ed.), *Nonionic Surfactants.* Marcel Dekker, New York, 1967.
26 K. Shinoda, T. Nakagawa, B. Tamamushi and T. Isemura, *Colloidal Surfactants. Some Physico-Chemical Properties.* Academic Press, New York, 1963, pp. 69–70.
27 J. Swarbrick and J. Darawala, *J. Phys. Chem.*, 73 (1969) 2627.
28 E.D. Goddard and G.C. Benson, *Can. J. Chem.*, 35 (1957) 986.
29 N. Nishikido, N. Yoshimura and M. Tanaka, *J. Phys. Chem.*, 84 (1980) 558.
30 P. Mukerjee in: H. van Olphen and K.J. Mysels (Eds.), *Physical Chemistry, Enriching Topics From Colloid and Surface Science.* IUPAC I.6 Commission. Theorex, La Jolla, CA, 1975, Chapter 9.
31 D. Stigter and J.T.G. Overbeek in: *Proceedings of the Second International Congress on Surface Activity. Vol. I*, 1957, p.311.
32 P. Debye and E.W. Anacker, *J. Phys. Colloid Chem.*, 55 (1951) 644.
33 P. Mukerjee, *J. Pharm. Sci.*, 63 (1974) 972.
34 D. Stigter in: H. van Olphen and K.J. Mysels (Eds.), *Physical Chemistry, Enriching Topics From Colloid and Surface Science.* IUPAC I.6 Commission I.6. Theorex, La Jolla, CA, 1975, Chapter 12.
35 D.F. Evans, M. Allen, B.W. Ninham and A. Fouda, *J. Solution Chem.*, 13 (1984) 87.
36 M.S. Ramadan, D.F. Evans and R. Lumry, *J. Phys. Chem.*, 87 (1983) 4538.
37 D.F. Evans and P.J. Wightman, *J. Colloid Interface Sci.*, 86 (1982) 515.
38 D.F. Evans and B.W. Ninham, *J. Phys. Chem.*, 90 (1986) 226.
39 G.C. Kresheck in: F. Franks (Ed.), *Water. A Comprehensive Treatise.* Plenum Press, New York, 1975, Chapter 2.

40 S.V. Lymar, M.I. Khramov, V.N. Parmon and K.I. Zamaraev, *Khim. Fiz.*, 1 (1982) 405.
41 W.J. Gettings, J.E. Rassing and E. Wyn-Jones, in: K.L. Mittal (Ed.), *Micellization, Solubilization and Microemulsions.* Plenum Press, New York, 1977, Vol. I, p. 347.
42 R. Zana, in: K.L. Mittal (Ed.), *Surfactants in Solution.* Plenum Press, New York, 1979, Vol. I, p. 473.
43 I. Satake, T. Tahara and R. Matsuura, *Bull. Chem. Soc. Jpn.*, 42 (1969) 319
44 M. Almgren, *J. Phys. Chem.,* 85 (1981) 3599.
45 M.E. Feinstein and H.L. Rosano, *J. Colloid Interface Sci.*, 24 (1967) 73.
46 J.B. Hayter and J. Penfold, *J. Chem. Soc. Faraday I*, 77 (1981) 1851.
47 J.N. Phillips and K.J. Mysels, *J. Phys. Chem.*, 59 (1955) 325.
48 K.M. Kale, E.L. Cussler and D.F. Evans, *J. Phys. Chem.*, 84 (1980) 593.
49 S.G. Cutler, P. Meares and D.G. Hall, *J. Chem. Soc. Faraday I*, 74 (1978) 1758.
50 C. Botrè, V.L. Crescenzi and A. Mele, *J. Phys. Chem.*, 63 (1959) 650.
51 J.N. Phillips, *Trans. Faraday Soc.*, 54 (1958) 587.
52 M.L. Fishman and E.R. Eirich, *J. Phys. Chem.*, 75 (1971) 3135.
53 H. Huff, J.W. McBain and A.P. Brady, *J. Phys. Colloid Chem.*, 55 (1951) 311.
54 L. Shedlovsky, C.W. Jacob and M.B. Epstein, *J. Phys. Chem.*, 67 (1963) 2075.
55 T. Ingram and M.N. Jones, *Trans. Faraday Soc.*, 65 (1969) 297.
56 N. Nishikido, *J. Colloid Interface Sci.*, 92 (1983) 588.
57 H.C. Evans, *J. Chem. Soc.*, (1956) 579.
58 T. Sasaki, M. Hattori, J. Sasaki and K. Nukina, *Bull. Chem. Soc. Jpn.*, 48 (1975) 1397.
59 P. Mukerjee, K.J. Mysels and P. Kapauan, *J. Phys. Chem.*, 71 (1967) 4166.
60 D. Stigter and K.J. Mysels, *J. Phys. Chem.*, 59 (1955) 45.
61 K. Shirahama, *Kolloid-Z.Z. Polym.*, 250 (1972) 620.
62 W. Baumuller, H. Hoffmann, W. Ulbricht, C. Tondre and R. Zana, *J. Colloid Interface Sci.*, 64 (1978) 418.
63 D. Stigter, *J. Colloid Interface Sci.*, 23 (1967) 379.
64 M.L. Corrin, *J. Colloid Sci.*, 3 (1948) 333.
65 M. Corti and V. Degiorgio, *J. Phys. Chem.*, 85 (1981) 711.
66 D. Bendedouch, S.H. Chen and W.C. Koehler, *J. Phys. Chem.*, 87 (1983) 2621.
67 P. Lianos and J. Lang, *J. Colloid Interface Sci.*, 96 (1983) 222.
68 W.K. Matthews, J.W. Larsen and M.J. Pikal, *Tetrahedron Lett.*, 6 (1972) 513.
69 C.A. Bunton and F. de Buzzaccarini, *J. Phys. Chem.*, 85 (1981) 3142.
70 J.W. Larsen and L.B. Tepley, *J. Colloid Interface Sci.*, 49 (1974) 113.
71 E. Keh, C. Gravach and J. Gustalla, *C. R. Acad. Sci., Ser. C*, 263 (1966) 1488.
72 R. Zana, *J. Coll. Interface Sci.*, 78 (1980) 330.
73 R.B. Dorshow, C.A. Bunton and D.F. Nicoli, *J. Phys. Chem.*, 87 (1983) 1409.

74 A. Rohde and E. Sackmann, *J. Phys. Chem.*, 84 (1980) 1598.
75 E.W. Anacker, R.D. Geer and E.H. Eylar, *J. Phys. Chem.*, 75 (1971) 369.
76 E.W. Anacker and A.L. Underwood, *J. Phys. Chem.*, 85 (1981) 2463.
77 R. Zana and J. Lang, in: K.L. Mittal and E.J. Fendler (Eds.), *Solution Behavior of Surfactants: Theoretical and Applied Aspects.* Plenum Press, New York, 1982, Vol. II, p. 1195.
78 H.C. Evans, *J. Chem. Soc.*, (1956) 579.
79 C.A. Bunton, F.de Buzzaccarini and F.H. Hamed, *J. Org. Chem.*, 48 (1983) 2461.
80 L.L. Melhado and C.D. Gutsche, *J. Am. Chem. Soc.*, 100 (1978) 1850.
81 G.O. Bizzigotti, *J. Org. Chem.*, 48 (1983) 2598.
82 E. Vikingstad and H. Hoiland, *J. Coll. Interface Sci.*, 64 (1978) 510.
83 E. Keh, C. Gavach and J. Gustalla, *C.R. Acad. Sci. (Paris), Ser. C.*, 263 (1966) 1488.
84 C. Ramachandran, R.A. Pyter and P. Mukerjee, *J. Phys. Chem.*, 86 (1982) 3198.
85 E.J.R. Sudholter and J.B.F.N. Engberts, *J. Phys. Chem.*, 83 (1979) 1854.
86 N.R. Hoskin, *Trans. Faraday Soc.*, 49 (1953) 1471.
87 J.W. Larsen and L. Magid, *J. Am. Chem. Soc.*, 96 (1974) 5774.
88 E.G. Aniansson, S.N. Wall, M. Almgren, H. Hoffmann, I. Kielmann, W. Ulbricht, R. Zana, J. Lang and C. Tondre, *J. Phys. Chem.*, 80 (1976) 905.
89 N. Muller, in: K.L. Mittal (Ed.), *Micellization, Solubilization and Microemulsions*. Plenum Press, New York, 1977, Vol. 1, pp. 229–239.
90 H. Wennerstrom and B. Lindman, *Phys. Rep.*, 1 (1979) 1.
91 B. Persson, T. Drakenberg and B. Lindman, *J. Phys. Chem.*, 80 (1976) 2124.
92 H. Gustavsson and B.Lindman, *J. Am. Chem. Soc.*, 97 (1975) 3923.
93 C. Tanford, *J. Phys. Chem.*, 76 (1972) 3020.
94 R.D. Geer, E.H. Eylar and E.W. Anacker, *J. Phys. Chem.*, 75 (1971) 369.
95 M. Abe, H. Uchiyama, T. Yamaguchi, T. Suzuki, K. Ogino, J. Scamehorn and S.D. Christian, *Langmuir*, 8 (1992) 2147.
96 S. Berr, R.R.M. Jones and J.S. Johnson Jr., *J. Phys. Chem.*, 96 (1992) 5611.
97 J.P. Kratochvil, *J. Colloid Interface Sci.*, 75 (1980) 271.
98 M.F. Emerson and A. Holtzer, *J. Phys. Chem.*, 71 (1967) 1898.
99 D. Attwood, *J. Phys. Chem.*, 72 (1968) 339.
100 N. Nishikido, M. Shinozaka, G. Sugihara and M. Tanaka, *J. Colloid Interface Sci.*, 74 (1980) 474.
101 G. Burgess and E. Pelizzetti, *Gazz. Chim. Ital.*, 118 (1988) 803.
102 J. LeMoigne, P. Gramain and J. Simon, *J. Colloid Interface Sci.*, 60 (1977) 565.
103 D. Stigter, *J. Phys. Chem.*, 79 (1975) 1008.
104 G.M. Bell and A.J. Dunning, *Trans. Faraday Soc.*, 66 (1970) 500.
105 M. Mille and G. Vanderkooi, *J. Colloid Interface Sci.*, 59 (1977) 211.

106 G. Gunnarsson, B. Jonsson and H. Wennerstrom, *J. Phys. Chem.*, 84 (1980) 3114.
107 J. Frahm and S. Diekmann, *J. Coll. Surface Sci.*, 70 (1979) 440.
108 P. Linse, G. Gunnarsson and B. Jonsson, *J. Phys. Chem.*, 86 (1982) 413.
109 G.M. Bell and S. Levine, in: B.E. Conway and R.G. Barradas (Eds.), *Chemical Physics of Ionic Solutions*. Wiley, New York, 1966, p. 409.
110 I.D. Robb, *J. Colloid Interface Sci.*, 37 (1971) 251.
111 M. da Silva Baptista, I. Cuccovia, H. Chaimovich, M.J. Politi and W.F. Reed, *J. Phys. Chem.*, 96 (1992) 6442.
112 M. Almgren, F. Grieser and J.K. Thomas, *J. Am. Chem. Soc.*, 101 (1979) 279.
113 M. Almgren and S. Swarup, *J. Phys. Chem.*, 86 (1982) 4212.
114 G.F. Goodeve, *Trans. Faraday Soc.*, 31 (1935) 197.
115 B. Barry and R. Wilson, *Colloid Polym. Sci.*, 256 (1978) 251.
116 C.A. Bunton, J. Frankson and L.S. Romsted, *J. Phys. Chem.*, 84 (1980) 2607.
117 M. Kahlweit and M. Teubner, *Adv. Colloid Interface Sci.*, 13 (1980) 1.
118 R.A. Moss, Y.S. Lee and K.W. Alwis, in: K.L. Mittal and E.J. Fendler (Eds.), *Solution Behavior of Surfactants. Theoretical and Applied Aspects*. Plenum Press, New York, 1982, p. 993.
119 C.M. Link, D.K. Jansen and C.N. Sukenik, *J. Am. Chem. Soc.*, 102 (1980) 7798.
120 C.A. Bunton, L.S. Romsted and G. Savelli, *J. Am. Chem. Soc.*, 101 (1979) 1253.
121 H. Al-Lohedan, C.A. Bunton and M.M. Mhala, *J. Am. Chem. Soc.*, 104 (1982) 6654.
122 L.S. Romsted and C.O. Yoon, *J. Am. Chem. Soc.*, 115 (1993) 989.
123 C. Reichardt, *Mol. Interactions*, 3 (1982) 241.
124 E.M. Kosover, *J. Am. Chem. Soc.*, 80 (1958) 3253.
125 D.S. Kemp and K.G. Paul, *J. Am. Chem. Soc.*, 92 (1970) 2253.
126 D.S. Kemp, D.D. Cox and K.G. Paul, *J. Am. Chem. Soc.*, 97 (1975) 7312.
127 D.S. Kemp and K.G. Paul, *J. Am. Chem. Soc.*, 97 (1975) 7305.
128 J. Burgess and E. Pelizzetti, *Progr. Reaction Kin.*, 17 (1992) 1–70.
129 T. Kunitake, Y. Okahata, R. Ando, S. Shinkai and S. Hizakava, *J. Am. Chem. Soc.*, 94 (1972) 8875.
130 T. Kunitake, S. Shinkai and S. Hirotsu, *J. Org. Chem.*, 42 (1977) 306.
131 Dong-Myung Shin, K.S. Schanze and D. Whitten, *J. Am. Chem. Soc.*, 111 (1989) 8494.
132 M.S. Fernandez and P. Fromherz, *J. Phys. Chem.*, 81 (1975) 1755.
133 K.A. Zachariase, N. Van Phuc and B. Kozankievicz, *J. Phys. Chem.*, 85 (1981) 2676.
134 K. Shinoda, *J. Phys. Chem.*, 89 (1985) 2429.
135 S.E. Friberg, A.J.I. Ward and D.W. Larsen, *Langmuir*, 3 (1987) 735.

136 A. Ray, *Nature*, 231 (1971) 313.
137 N. Moucharafieh and S.E. Friberg, *Mol. Cryst. Liq. Cryst.*, 49 (1979) 231.
138 M.A. El-Nokaly, L.D. Ford, S.E. Friberg and D.W. Larsen, *J. Polym. Sci.*, 19 (1981) 1585.
139 M.A. Thomason, D.M. Bloor and E. Wyn-Jones, *J. Phys. Chem.*, 95 (1991) 6017.
140 S.E. Friberg, G. Rong and A.J.I. Ward, *J. Phys. Chem.*, 92 (1988) 7247.
141 J.E. Gordon, *The Organic Chemistry of Electrolyte Solutions*. Wiley, New York, 1975.
142 A.F.M. Barton, *Chem. Rev.*, 75 (1975) 731.
143 D. Fennell Evans and D. Miller, in: S.E. Friberg and B. Lindman (Eds.), *Organized Solutions*. Marcel Dekker, New York, 1992, pp. 33–45.
144 X. Auvray, C. Petipas, T. Perche, R. Anthore, M.J. Marti, I. Rico and A. Lattes, *J. Phys. Chem.*, 94 (1990) 8604.
145 D.J. Mitchell and B.W. Ninham, *J. Chem. Soc. Faraday Trans. 2*, 77 (1981) 601.
146 J.H. Clint, *Chem. Brit.*, (1990) 333.
147 J.H. Fendler, *Chem. Rev.*, 87 (1987) 877.
148 K. Shinoda, *Pure Appl. Chem.*, 52 (1980) 1195.
149 M.J. Schwuger in: M.J. Rosen (Ed.), *Structure / Performance Relationships in Surfactants.* ACS Symp. Ser. 253, Am. Chem. Soc., Washington DC, 1984, p. 3.
150 J.M. Corkill and F. Goodman, *Adv. Coll. Interface Sci.*, 2 (1969) 297.
151 H. Schott, *J. Coll. Interface Sci.*, 43 (1973) 150.
152 H. Schott and S.K. Han, *J. Pharm. Sci.*, 64 (1975) 658.
153 H. Schott and S.K. Han, *Ibid.* 65 (1976) 975.
154 B.H. Robinson, *Chem. Brit.* (1990) 342.
155 N. Nishikido, *J. Colloid Interface Sci.*, 120 (1987) 495.
156 J.F. Scamehorn (Ed.), *Phenomena in Mixed Surfactant Systems.* ACS Symp. Ser. 311. Am. Chem. Soc., Washington DC, 1986.
157 K. Ogino and M. Abe (Eds.), *Mixed Surfactant Systems.* Marcel Dekker, New York, 1994.
158 J.F. Rathman and J.F. Scamehorn, *Langmuir*, 3 (1987) 372.
159 P. Jokela, B. Jonsson and A. Khan, *J. Phys. Chem.*, 91 (1987) 3291.
160 K.L. Stellner, J.C. Amante, J.F. Scamehorn and J.H. Harwell, *J. Colloid Interface Sci.*, 123 (1988) 186.
161 L.Marszall, *Langmuir*, 4 (1988) 90.
162 R.J. Robson and E.A. Dennis, *J. Phys. Chem.*, 81 (1977) 1075.
163 S. Hayashi and S. Ikeda, *J. Phys. Chem.*, 84 (1980) 744.
164 J.A. Loughlin and L.S. Romsted, *Coll. Surfaces*, 48 (1990) 123.
165 J. Yao and L.S. Romsted, *J. Am. Chem. Soc.*, 116 (1994) 11779.

Chapter 2

Solubilization in micellar systems

An exceptional feature of surfactant solutions is their ability to dissolve (solubilize) substances that are sparingly soluble in water, forming a thermodynamically stable isotropic solution. The solubilization phenomenon is clearly related to the formation of micelles, since a sharp increase in solubility is observed when the cmc is attained. Once micelles are formed, the solubility increases proportionally with increase in surfactant concentration.

In Chapter 1 it has been shown that to the extent that anhydrous amphiphilic molecules dissolve in hydrocarbons, they also aggregate there giving reverse micelles where the polar portions of the amphiphilic molecules are in the interior. Such organized assemblies dissolve water and a variety of water-soluble (hydrocarbon-insoluble) substances become soluble in this four-component system. The term "solubilization" refers to incorporation of water-insoluble materials in systems containing normal micelles and also to incorporation of hydrocarbon-insoluble materials in reverse micelles. Reviews of the early work and much useful data are found in Refs. [1–4].

In the following we will focus attention on several aspects of solubilization including: the evaluation of the amount of material that can be solubilized, the determination of micelle–water distribution equilibria, the thermodynamics and kinetics of solubilization, the location of solubilizates in the micellar regions and the effect of dissolved analytes on the aggregation properties and on the solubilization capacity of the amphiphiles.

2.1. DETERMINATION OF THE MAXIMUM ADDITIVE CONCENTRATION

The study of solubilized systems must obviously start with the determination of the concentration of solubilizate which can be incorporated into a given system with the maintenance of a single isotropic solution. This saturation concentration of solubilizate for a given concentration of surfactant is termed the maximum additive concentration (MAC) and the ratio of moles of solubilizate to mole of micellized amphiphile is often called the "solubilizing power".

Basically, the methods of measuring this concentration are the same as those for the determination of the solubility of any compound in a given solvent but with the additional difficulties due to the presence of a colloidal solution. As these solubilization properties are temperature sensitive, it is important that all these methods should be subject to adequate temperature control; a factor often missing in much experimental work.

A rapid increase of solubilizing power with surfactant chain length is observed for solutes such as ethylbenzene, *n*-heptane and dimethylaminoazobenzene. However, the power is nearly independent of chain length for both 1-octanol and 1-heptanol, except for the case of the very short octanoate aggregates.

For the same amphiphile, the solubilizing power will, of course, to some extent, reflect the solubility of the given solubilizate in the hydrocarbon most closely related to the amphiphile tail. Nevertheless, distinct trends among solubilizates of similar chemical nature are apparent.

Detection of the appearance of a second phase gives a direct measure of the MAC, but methods based on this are not always suitable. Analytical determination of the amount solubilized is then required, spectrophotometry being very useful in certain cases [5,6].

2.2. SOLUBILITY OF GASES IN MICELLAR SYSTEMS

Most common gases are basically hydrophobic in nature in that they are considerably more soluble in oil than in water. Consequently, higher solubility of gases in micellar systems (in respect to water) has been observed, as reported in Table 2.1.

Above the cmc the gas solubility increases linearly with surfactant concentration. The slopes of the plots: moles of gas absorbed (per atmosphere in 1000 g of water) vs. surfactant concentration (mol kg^{-1}) correspond to the molar ratios of solubilized gas molecules to the surfactant in micellar form.

Below the cmc the ionic surfactant decreases the gas solubilities, acting as a salting out agent. This behaviour is particularly important for surfactants with high cmc such as sodium 1-heptanesulfonate (SHS).

TABLE 2.1

Gas solubilities (in mole fraction × 10^4) at 25°C and 1 atm in some surfactant solutions and bulk solvents

Gas	T_C (K)	Investigated solvent system						
		Water	CH_3OH	C_8H_{18}	C_7F_{16}	SHS	SDS	HTAB
Ar	150.8	0.25	4.15	23.36	56	4	11	13
O_2	154.6	0.23	4.11	21.77	60	4	10	10
CO_2	304.1	6.11	56.30	119.80	208.8	42	93	147
CH_4	190.5	0.25	8.67	50.26	73	10	19	29
N_2O	309.6	4.38	65	175	278			
C_2H_6	305.4	0.33	40.5	324	217	45	110	155
C_3H_8	369.8	0.27	130	1200	612	102	330	480

T_C is the critical temperature. Data taken from Ref. [7] and references cited therein.

2.3. RATE OF SOLUBILIZATION

The solubilization rate of hydrocarbons into water by non-ionic surfactants has been found to be directly proportional to the surfactant concentration above the cmc and decreases with the molecular weight of the oil [8]. A correlation between the solubilization rate, the interfacial tension and the size of the surfactant aggregate has been proposed [9].

Investigations on the solubilization of hydrocarbons by mixtures of non-ionics and alkylarene sulfonates showed the dramatic effect of the position of the sulfonate group on the rate of solubilization [2,3].

Amphiphile molecules can also influence the deaggregation and dissolution of sparingly soluble organic compounds in solid form. A theory of solubilization kinetics of a sparingly soluble solid and, based on the analysis of the five steps shown in Fig. 2.1, its relation to the dynamics of the dissolution medium has been proposed [10]. Surfactant molecules diffuse to the surface as micellar species. These molecules are adsorbed on the surface of the solid and form mixed micelles.

The solubilization rate is assumed to be controlled by the two last steps, i.e. detachment and diffusion away into the bulk of the mixed micelle. These predictions were verified for the fatty acid dissolution into sodium dodecyl sulfate solutions. The kinetic analysis of Higuchi [11] predicts that the rate of change of concentration of a solute (dc/dt)

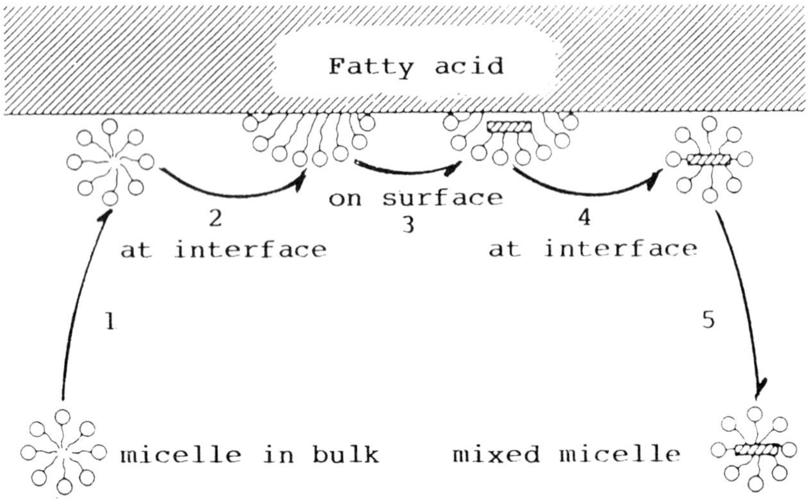

Fig.2.1. Schematic mechanism for initial solubilization. Mixed micelle desorption and diffusion (steps 4 to 5) are assumed to control stearic acid solubilization. Reproduced with permission from Ref. [10].

is related through its surface area, A, to its saturation solubility, c_s, and to the increase in solubility due to the aggregates, c_m, by:

$$dc/dt = A[(D_s c_s/\delta + (D_m c_m/\delta] \tag{1}$$

where D_s is the diffusion coefficient of the solute, D_m is the diffusion coefficient of the solute–micelle adduct and δ the diffusion layer thickness.

2.4. MICELLE–WATER DISTRIBUTION EQUILIBRIA

In solubilized systems in which the solubilizate has a significant water solubility, it is of interest to know not only the distribution ratio of solubilizate between the micelles and water under saturation conditions but also at varying degrees of saturation of the system with solubilizate. Such information cannot, of course, be obtained using the solubility methods.

The association of a solute to a surfactant aggregate can be treated from a thermodynamic or a kinetic point of view. Furthermore, it has also been treated as a statistical problem, by examining the type of distribution of solute molecules among the aggregate [12]. These three approaches are not independent but closely related.

The most realistic model for analyzing solute incorporation is that based on the mass action approach [13] (see Chapter 1). In this respect, the solubilization process is treated in terms of a stepwise addition of solute molecules S to aggregate M, giving MS_i adducts (a micelle containing i solute molecules).

Introducing a set of assumptions (e.g. micellar aggregation number independent of solute; solute entry rate independent of solute molecules in the micelle; solute exit rate directly proportional to the solute molecules in the micelle; "infinite" solubilization capacity of the micelle) and a Poisson distribution for the micelles containing i solutes, solute incorporation can be treated as if it were governed by the pseudo-equilibrium:

$$S_w + M \overset{K_B}{\rightleftarrows} MS \tag{2}$$

with the related expression for K_M:

$$K_M = \frac{\bar{n}}{[S_w]} = \frac{[S_m]}{[S_w][M]} \tag{3}$$

From a practical standpoint, the use of K_M to describe solute incorporation suffers from the inconvenience of requiring a knowledge of the concentration of micelles. Although the micelle concentration [M] can in principle be calculated by dividing the concentration of micellized surfactant (C_D) by the average aggregation number N_{AG}, reliable aggregation numbers (in the absence of solutes) are available for only a few detergents. As a consequence, the more common form of expressing the solute incorporation constant has been as K_s, defined in terms of C_D and trivially related to K_M:

$$K_B = \frac{[S_m]}{[S_w]C_D} = \frac{K_m}{N_{AG}} \tag{4}$$

This definition of K_B essentially reduces the solubilization process to a pseudophase equilibrium of the solute between a micellized surfactant phase and the aqueous phase:

$$S_w + D_m \overset{K_B}{\rightleftarrows} S_m \tag{5}$$

where D_m is the micellized detergent. This approach has played a central role in more recent pseudophase models for analyzing micellar effects on the kinetics of ground state reactions [14].

This same pseudophase equilibrium also serves as the basis for defining two partitioning coefficients, the local concentration scale coefficient K_{MW} and the mole fraction scale coefficient K_x, both of which have been used to obtain thermodynamic parameters associated with the transfer of the solute from the aqueous phase to the micelle. K_{MW} (sometimes referred to as P) is defined as:

$$K_{MW} = \frac{[S_m]}{[S_w]} \tag{6}$$

where $[S_m]$ and $[S_w]$ refer to the local concentration of the solute in moles per litre of the actual volume of either the micellar or aqueous phase, respectively. If $[C_{sm}]$ and $[C_{sw}]$ are the analytical concentrations of micellar and aqueous substrate (based on the total solution volume), the local molar solute concentrations are given by:

$$[S_m] = \frac{[C_{sm}]}{C_D\overline{V}} \tag{7}$$

ar d:

$$[S_w] = \frac{[C_{sw}]}{(1 - C_D\overline{V})} \tag{8}$$

where $\overline{V}$, the appropriate volume of the micellar phase per mole of micellized detergent, is generally taken to be the partial molar volume of the micellized surfactant. Substituting these into the expression for K_{MW} it is trivial to obtain, by assuming that the volume fraction of the solution occupied by the micelles ($C_D\overline{V}$) is usually quite small relative to unity:

$$K_{MW} = K_s/\overline{V} = P \tag{9}$$

Thus, K_{MW} can be calculated from K_s by assuming a value for $\overline{V}$. Alternatively, chromatographic methods for measuring solute partitioning coefficients provide the product $\overline{V}(K_{MW}-1)$ directly, the value of which is approximately equal to $(K_s-\overline{V})$ (see Chapter 6). K_{MW} has been used to describe substrate incorporation in several kinetic models for micellar catalysis (see Chapter 3).

Ben Naim [15] has argued that K_{MW}, which assumes a standard state based on local molar concentrations, is the preferred form of the partitioning coefficient for calculating standard free energies of transfer of solutes from water to the micelle.

The dimensionless mole fraction partitioning constant, K_x, is defined as:

$$K_x = X_m / X_s \tag{10}$$

The expression for the solute mole fraction in the aqueous phase is straightforward:

$$X_w = \frac{[S_w]}{[S_w] + \mathrm{cmc} + 55.5} \tag{11}$$

As for X_m, it has become customary to express the mole fraction of the micelle-incorporated solute in terms of its mole fraction in a "dry" micelle [16], i.e., to ignore water as a possible third micellar component and simply write:

$$X_m = \frac{[S_m]}{[S_m] + C_D} \tag{12}$$

Combining these equations provides an expression for K_x in terms of the analytical concentrations of the detergent and solute:

$$K_x = \frac{55.5[S_m]}{[S_w][S_m] + C_D} = \frac{55.5K_s}{1 + K_s[S_w]} \tag{13}$$

At low degrees of solute incorporation ($K_s[S_w] = [S_m]/C_D << 1$), K_x can be interrelated to K_{MW} and K_s via

$$K_x = 55.5\ K_{MW}\ \overline{V} = 55.5\ K_s \tag{14}$$

The principal application of K_x has been in the calculation of standard free energies of transfer of solutes from water to the micelle, based on the unitary scale (mole fraction standard state). Implicit in this use of K_x is the supposition that the solute is distributed homogeneously within the micelle, forming an ideal mixture with the detergent.

In summary, the following relationships can be established between the different binding constants:

$K_x = 55.5\, K_B$ dimensionless (15)

$K_x = K_{MW}\, 55.5\, \overline{V}_m$ dimensionless (16)

$K_B = K_{MW}\, \overline{V}_m$ M^{-1} (17)

$K_M = K_B\, N_{AG}$ M^{-1} (18)

Thus, if the value of the partition coefficient is known on any given scale the corresponding values on the other scale can be calculated by assigning values to the parameters $\overline{V}_m$ and N_{AG}. The main inconvenience is that these parameters are not always reported in the literature for every type of amphiphilic compound.

However, for the most commonly used detergents the above data is available. For example, for SDS and HTAB, the following values of $\overline{V}_m$ and N_{AG} (in the absence of added electrolytes) are recommended: $\overline{V}_m$ (HTAB) = 0.363 dm^3/mol [17]; $\overline{V}_m$ (SDS) = 0.25 dm^3/mole [18]; N_{AG} (HTAB) = 84 [19]; N_{AG} (SDS) = 58 [20].

2.5. DETERMINATION OF BINDING CONSTANTS

The binding constant of substrates to the micelles repetitively appears in practically all treatments of micellar applications. It is thus clear that information on the value of this parameter is essential in the evaluation of the effect of micelles on reaction equilibria and kinetics. This section is devoted to a short description of the techniques adopted to determine binding constants or partition coefficients.

A commonly utilized technique to determine these coefficients is through the changes in the UV/VIS spectra of the solute upon partitioning into the micellar subphase. When the absorbance of a substrate in water is A_w, and the absorbance of the same concentration of the substrate when completely bound is A_m, then the experimental results when plotted according to Eqn. (19) provide the value of the binding constant, K_B:

$$\frac{1}{A_s - A_w} = \frac{1}{A_m - A_w}\left(1 + \frac{1}{K_B C_D}\right) \quad (19)$$

A_s in Eqn. (19) is the absorbance of the same amount of solute at any surfactant concentration, C_D, and it is assumed that the surfactant does not contribute to the absorption. In fact, Eqn. (19) can be utilized when any physical parameter which is proportional to the concentration of the solute at each of the pseudophases is measured (fluorescence intensity, epr signals, etc.). These techniques, however, can be used only when there are significant spectral changes upon incorporation in the micelles and when the spectra are conveniently accessible.

Another approach to determine binding constants is through solubilization studies, as described in the previous section.

In addition to the above-mentioned techniques, a wide variety of other techniques has been also proposed for measuring K_B. These include dialysis [21], ultrafiltration [22], vapour pressure measurements [23], NMR [24], ESR [25], fluorescence quenching [26], potentiometry [27], polarography [28] and variations of dissociation constants in the micellar systems [29–31]. Generally speaking, these techniques are restricted to selected systems, are often time-consuming and require very accurate measurements.

Kinetic measurements may also provide information on K_B, but they require a detailed knowledge of the mechanism of the reaction studied before information on the binding constant may be extracted.

Recently chromatographic methods have proven useful and of general applicability in determination of binding constants. Gel filtration [39–41] and HPLC [42] provide experimental data from which partition coefficients, P_{MW}, for micelle–water partitioning can be calculated, as discussed in detail in Chapter 6. Inaccuracy in these chromatographic determinations may arise from large partition coefficients between the stationary phase and the aqueous phase (P_{SW}) and this means that the method cannot be applied to very hydrophobic solutes.

Figure 2.2 compares the values of a large number of binding constants of aromatic solutes to SDS and HTAB as determined by a variety of experimental methods. The binding constants reported in this figure are expressed in terms of free energy of transfer. The latter parameter is discussed in detail in the next section.

2.6. FREE ENERGY OF TRANSFER FROM WATER TO MICELLES

The preferential association of non-polar moieties with each other rather than with surrounding polar molecules (water) is generally referred to as hydrophobic interactions. Detailed discussions on the nature and thermodynamics of the hydrophobic interactions that have

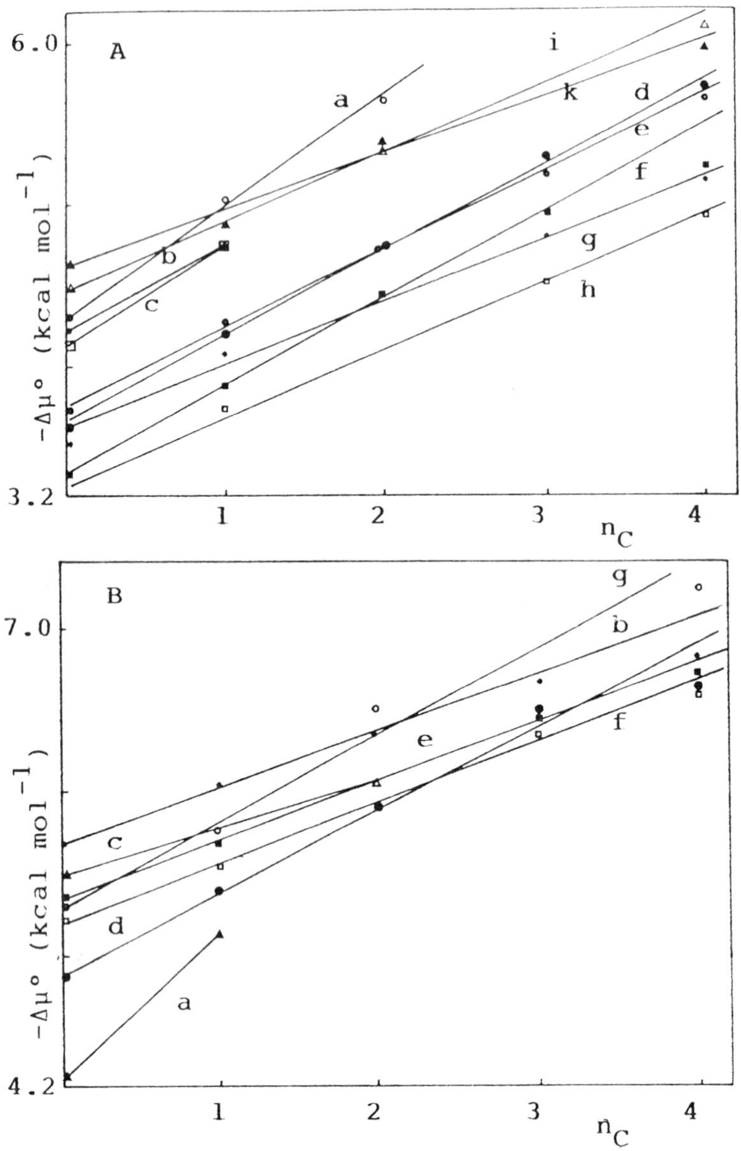

Fig. 2.2. Free energy of transfer of aromatic solutes from water to micelles as a function of the number of carbon atoms of the alkyl chain determined by various techniques. **(A)** SDS micelles: Arenes: (a) solubility [32], (b) isopiestic point [22], (c) chromatography [33]; Phenols: (d) dissociation constants [30,31], (e) chromatography [34,35], (f) ultrafiltration [22]; Catechols: (g) chromatography [34,35], (h) kinetics [34]; Benzoic acids: (i) dissociation constants [36], (k) ultrafiltration [22].

appeared in the literature [15,43,44], have been reported in Chapter 1. The free energy of transfer of one mole of solute from water to the micellar pseudophase is related to the binding constant through

$$\Delta\mu_t^o = \mu_m^o - \mu_w^o = -\,RT \ln K_B \tag{20}$$

When the micellar incorporation is primarily a result of hydrophobic interactions, the free energy of transfer can be correlated with the number of carbon atoms of the solute, n_C, through the linear equation

$$\Delta\mu_t^o = a + b\, n_C \tag{21}$$

For example, measuring the solubilities of aliphatic hydrocarbons from ethane through pentane in SDS micelles, the parameters a = -1934 cal mol^{-1} and b = -771 cal mol^{-1} have been determined [45].

A comparison between these values and the values for the corresponding free energy of transfer to liquid hydrocarbons reveals that the free energy of transfer to the micelles is lower by only ca. 15%. One may, therefore, infer that the interior part of the micelle represents a medium which is quite similar to a liquid hydrocarbon.

These values are also similar to the free energy of transfer of a variety of non-ionic or zwitterionic surfactant monomers into their micelles (b = -700 cal/mole) and for ionic surfactants to their micelles when electrostatic interactions are minimized by high ionic strength [43,46].

When the hydrocarbon molecule contains different moieties, such as aromatic rings, polar substituents (e.g. halides, hydroxy groups, etc.), the free energy of transfer can be expressed as the sum of different contributions of the various groups:

$$\Delta\mu_t^o = \Delta\mu_{Ph}^o + n\Delta\mu_C^o \tag{22}$$

where $\delta\mu_{Ph}^o$ includes the contribution from the aromatic (for arenes) or substituted aromatic (phenols, benzoic acids) groups. From these plots, the contribution of an aliphatic carbon atom in these structures is ~-500 cal mol^{-1} for SDS (see Fig. 2.2A).

For the anionic micelles, this value of free energy of transfer suggests the palisade layer as the locus of solubilization, whereas in cationic

Fig. 2.2 *continued*. **(B)** HTAB micelles: Arenes: (a) isopiestic method [22]; Phenols: (b) absorption spectroscopy [37], (c) fluorescence [38], (d) ultrafiltration [22]; Catechols: (e) absorption spectroscopy [35], (f) kinetics [35]; Benzoic acids: (g) ultrafiltration [22]. Reproduced from Ref. [35] with permission.

micelles (the estimated $\Delta\mu_C^o$ value for HTAB is ~–300 cal mol^{-1}) solubilization close to the micellar surface is more probable.

An interesting effect is also observed for hydrophilic groups, such as –OH or –COOH, when attached to a benzene ring. While these groups decrease the binding constant to SDS when going from benzene to phenol, an increase is observed when bound to HTAB. This observation supports the hypothesis of interaction of the benzene ring with positive head-groups at the micellar interface.

Halides also contribute noticeably to the binding constant, and this contribution increases with increase in atomic number. Plots of free energies of transfer from water to SDS of different halophenols showed a linear dependence on the number of halogen atoms from which the contribution of each halogen substituent has been estimated. The values were: –330 cal mol^{-1} for fluorine, –640 cal mol^{-1} for chlorine, –1030 cal mol^{-1} for bromine, and –1500 cal mol^{-1} for iodine, respectively [47]. Very similar contributions have been estimated for chlorine and bromine atoms when present on naphthols.

While it is possible to attribute a contribution to the binding constant from each group on the molecule (similar to the Rekker hydrophobic fragment constant [48]), the additivity holds only for a given class of solutes with a given surfactant. In fact, the examples mentioned above have demonstrated that the free energy of transfer from water to micelles are highly dependent on both the nature of the solute and the type of surfactant.

For example, studies performed on several polysubstituted phenolic compounds have shown a good agreement between the measured free energies of transfer of such compounds, obtained using different experimental methods, and the values predicted using the additive group contribution method (see Fig. 2.3). These data confirm the utility of this approach when homogeneous classes of hydrophobic solutes are considered.

If a molecule is bound to the micelle mainly through hydrophobic interactions, $\Delta\mu_t^o$ can be also estimated from data on the free energy of transfer from water to a non-aqueous solvent, e.g., n-octanol [50]. Such a correlation between the free energy of transfer to micelles and to n-octanol for a series of halophenols is shown in Fig. 2.4.

While the predictability of this approach has been verified in some selected cases, it should be realized that different solubilization processes will yield different free energies of transfer and thus the correlation may depend on the solubilization system.

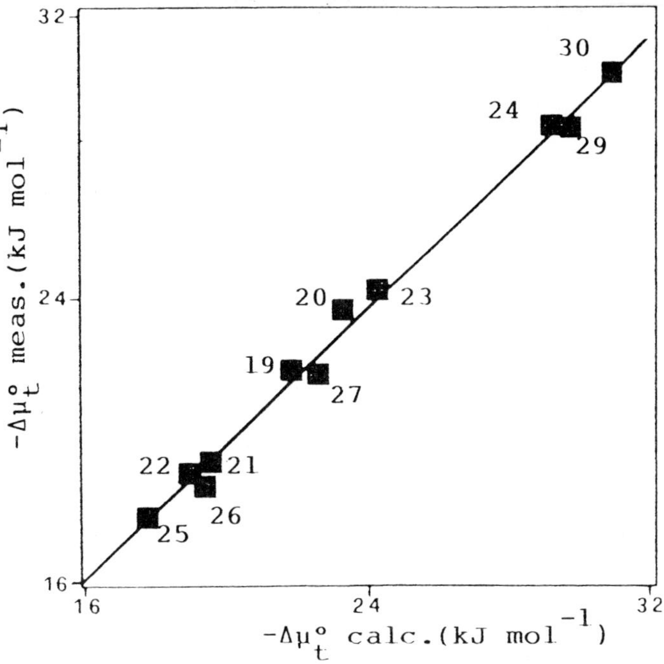

Fig. 2.3. Comparison between calculated and measured free energies of transfer of polysubstituted phenols (correlation coefficient 0.996). (19) 3-trifluoromethylphenol; (20) 2-N-phenylcarboxamidophenol; (21) 2-nitro-4-chlorophenol; (22) 3-nitro-4-methylphenol; (23) 2-chloro-4-bromo-6-methylphenol; (24) 4-nitro-2,6-diiodophenol; (25) 3,5-dimethoxyphenol; (26) 2,6-dimethoxy-4-acetylphenol; (27) 2-acetamido-4,6-dichloro-5-methyl-phenol; (28) 1-naphthol; (29) 2-N-butylcarboxamido-1-naphthol; (30) 2-N-phenylcarboxamido-1-naphthol. Reproduced with permission from Ref. [49].

A group contribution method, based on the structure formula of the solute and on the Henry's law constant in water, has been also proposed to predict the limiting equilibrium constant for transferring the solute from the infinitely diluted solution in water into the micelle [51].

The solubilization equilibrium constant is a type of distribution coefficient defined by

$$K = X_m/C_A \tag{23}$$

where C_A is the concentration of unsolubilized solute in the bulk aqueous solution [52]. Detailed and accurate solubilization isotherms, that is solubilization measurements at varying solute activity and mole

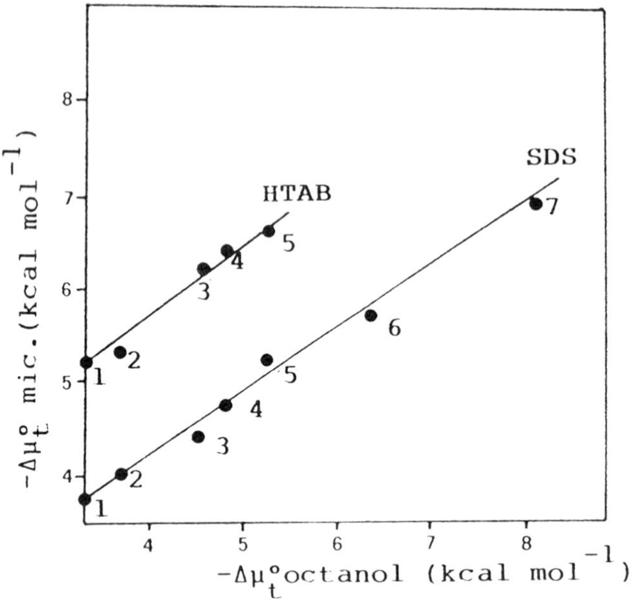

Fig. 2.4. Correlation between free energy of transfer from water to SDS and HTAB micelles and from water to octanol for the following compounds: (1) phenol, (2) 4-fluorophenol, (3) 4-chlorophenol, (4) 4-bromophenol, (5) 4-iodophenol, (6) 2,4,5-trichlorophenol, (7) pentachlorophenol. Taken from Ref. [47]. Reproduced with permission.

fractions of the solute in the micelles, have been obtained by semi-equilibrium dialysis [53] and head-space chromatography [54].

Defining the activity coefficient of an organic solute in the micelle (γ_A) as $f_A/(f_A^o X_m)$, where f_A is the fugacity of the solute and f_A^o the fugacity in the standard state (the pure component), it follows that because f_A is proportional to C_A in the very dilute aqueous solution, K varies inversely with γ_A.

Information about the activity coefficient of solubilizates and the dependence of γ_A on X_m may be useful in determining the environment of these compounds within the micelle. In particular, values of γ_A near unity indicate that the environment of the solute is energetically similar to that of the pure component (liquid or solid). Large values of γ_A are characteristic of compounds that solubilize in environments less favourable than the pure solute state, whereas values of $\gamma_A < 1$ are found when the aggregate strongly attracts the solute.

Aliphatic hydrocarbons are essentially dissolved within the hydrocarbon core of micelles and the corresponding solubilization isotherms show γ_A vs X_m curves which decrease from large values at infinite dilution to lower values as X_m approaches unity. Typical values of γ_A for alkanes in micelles are in the range 5–10 when X_m is very low, whereas the corresponding γ_A values in pure water are much higher (about 10^5).

The location of other important classes of solubilizates, such as the aromatic compounds, in ionic micelles is still controversial. Depending upon the solute property investigated, the location of benzene has been reported at the micelle–water interface [55,56], within the micellar core [57–59] or in intermediate regions [60]. Accurate investigations performed in several micellar systems using the vapour pressure method do not indicate a preferential site of solubilization for aromatic hydrocarbons [52,61]. The activity coefficients of benzene and toluene are greater than unity in several ionic surfactants (at small values of X), but these values are considerably smaller than those measured for the aliphatic hydrocarbons.

Several equations have been introduced to describe the dependence of the activity coefficient (or the solubilization equilibrium constant) on X [62–64]. For micellar solutions of polar organic molecules, at $X <$ 0.2–0.3, K can be expressed as a linear function of X of the form:

$$K = K_0 (1 - bX) \tag{24}$$

However, for higher X values (0.3–0.6), a quadratic or higher order term must be included in the function. The equation

$$K = K_0 (1 - aX + bX^2) \tag{25}$$

better describes the solubilization data of a variety of systems. Moreover, in some particular cases (phenol, chlorophenol, dichlorophenols in hexadecylpyridinium chloride micellar solutions) [65] a simpler quadratic equation holds:

$$K = K_0 (1 - bX)^2 \tag{26}$$

In the case of organic solutes bearing ionic charges, the contribution of the electrostatic interaction is less than that expected from surface potential ($Ze\Psi$). There probably is an opposing effect due to loss of solvation of the ionic moiety upon incorporation into the micelle. Such a situation may be noticed for the association of phenoxide ions with positive micelles as compared to the association of phenols with the same micelles. The difference between the free energy of transfer of phenoxides and phenols in this case is ca. –1150 cal/mol [66], corresponding to a surface potential of 50 mV. The latter is substantially lower than the commonly accepted value for surface potential. A similar trend was observed for the binding of N-methylphenothiazonium radical cations as compared to N-methylphenothiazine bound to anionic micelles. The difference in free energy of transfer is, again, ca. –1000 cal/mol [67].

While binding of substrates to micelles has attracted a great deal of attention, much less effort has been devoted to solutes that do not seem to interact with the micelles. The repulsion of a solute from a micelle is often implied (e.g., micellar inhibition of a reaction) but rarely treated in a rigorous manner. The recent application of chromatographic techniques to micellar systems (see Chapter 6) has proven useful in this respect [68].

That the solutes of charge similar to that of the micelle are in effect excluded from the micellar volume is noticeable in their chromatographic behaviour. Thus, the retention volumes of negatively charged organic substrates (e.g., naphthalene sulfonate) or inorganic complexes (potassium nitroferricyanide, ammonium thiocyanate) are increased by the presence of SDS micelles. This type of interaction is referred to as antibinding and is easily distinguished from nonbinding, in which no change in retention time is observed on changing micellar concentration in the mobile phase.

Using appropriate HPLC equations (described in detail in Chapter 6), it is possible to measure the various micelle–solute interactions and express the antibinding interaction through negative binding constants.

It is interesting to note that the antibinding behaviour may be considered as an excluded volume effect resulting from the repulsion of a charged solute from the similarly charged micelle. This repulsion will exclude the solute not only from the micellar volume, but also from much of the volume of the double layer surrounding the micelle. From strictly electrostatic arguments, one may expect a decrease in antibinding interaction upon increasing the ionic strength. This is, indeed, the observation for most ions studied.

However, when hydrophobic interactions become significant, reduction of the surface potential upon increasing ionic strength leads to binding of the hydrophobic portion of the solute to the micelle. The implications of such reversal of binding behaviour to kinetic effects may be quite important. Changes in micellar properties (caused by such factors as solution pH, buffer concentration, ionic strength, etc.) may be sufficient to convert, for example, micellar catalysis into micellar inhibition.

Ionic amphiphiles constitute one of the most important classes of micelle-forming surfactants. Below the cmc the monomeric amphiphile behaves as a simple electrolyte. Above the cmc, however, the micellar aggregates carry high formal charge at the micelle/water interface. The strong repulsion between the head-groups, inherent in the aggregation of ions of like charge, is partially compensated for by binding of counterions. The charged micelles thus give rise to a high local electrostatic field extending into the solution which influences the distribution and concentration of the other ionic species present in the same solution. A detailed knowledge of the distribution of the ions is essential, therefore, in the understanding of micellar effects on reaction dynamics and equilibria, in particular when ionic reactants are involved.

2.7.1. Interactions with inorganic ions

The partition coefficient, $P_{S,I}$, for charged species can be expressed through the equation:

$$P_{S,I} = P^{o}_{S,I} \exp(-Z\Psi e/kT) \qquad (27)$$

where Ψ is the surface potential of the aggregate, $P^{o}_{S,I}$ is a constant reflecting the contributions of non-electrostatic forces in the ion–micelle interaction, Z is the ionic charge and e is the charge of the electron [69]; a result very close to Eqn. (27), corrected with a factor 1/2 in the exponential term, is obtained from a two-pseudophase model [70].

This approach to inorganic ions is severely limited by the requirement of the knowledge of some empirical values and the partition coefficient shows a marked dependence on ionic strength [70]. In addition, because of the long-range nature of the electrostatic potential, the concentration of ions of charge opposite to that of the surface continuously decreases from the concentration at the aggregate surface (Stern layer) to the bulk concentrations. As a consequence, different experimental techniques will probe different regions and different degrees of counterion association will be reported.

A model which explicitly describes the distribution of the ions around the charged micelles has been developed assuming a spherical distribution of the ions around the micelle, governed by the Poisson–Boltzmann equation and proper boundary conditions [71]. The model assumes that

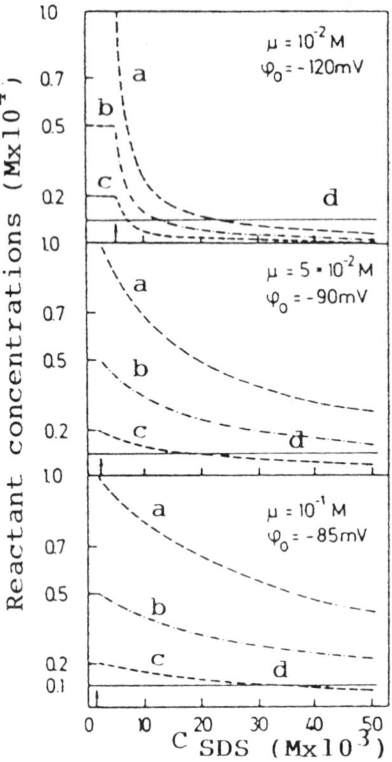

Fig. 2.5. Calculated bulk concentrations of the reactants in the system $IrCl_6^{=}/Fe^{2+}/SDS$ as a function of SDS concentration at different metal concentrations and ionic strengths (μ). Curves (a): $[Fe^{2+}] = 1\times10^{-4}$ M; (b): $[Fe^{2+}] = 5\times10^{-5}$ M; (c): $[Fe^{2+}] = 2\times10^{-5}$ M; (d): $[IrCl_6^{=}] = 1\times10^{-5}$ M. Reproduced with permission from Ref. [73].

the micelle is confined in an outer radius, r_f (see Chapter 1, Fig. 1.21). In spite of some approximations adopted, the model provides an acceptable description of the electrostatic effect. In particular, binding of counterions and its relation to thermodynamic, transport and spectroscopic properties, agrees satisfactorily with experimental observations [72].

The distribution of the concentration of Fe^{2+} in an anionic micellar system is depicted in Fig. 2.5. The data describe very accurately the kinetics of electron transfer reactions involving the $Fe^{3+/2+}$ couple in the presence of anionic aggregates [73,74]. It must be emphasized however that effects other than simple electrostatics may be operating in some cases.

Given the discrete number of micelles present at a given total surfactant concentration, introduction of solutes leads to their solubilization in the available number of micelles. Knowledge of this distribution of solutes among the micelles is essential for the interpretation of various processes [75–77].

When the average number of solutes per micelle n is considered ($\bar{n} = N/[M]$, where N is the total number of solutes introduced and [M] the concentration of micelles) the determination of the probability P_i of finding i solutes in a given micelle can be an interesting statistical exercise. Considering the solutes as simple N entities to be thrown into a given number M of micelles, there are several distribution laws that can be derived under various conditions: random or geometric, binomial, and Poissonian. For indistinguishable entities, the number of ways of arranging them among the micelle is given by (Bose statistics):

$$W = \frac{M!}{\Pi_{i=0} M_i!} \tag{28}$$

where M_i represents micelles with i solutes. Maximizing W under the conditions $\Sigma M_i = M$ and $\Sigma iM_i = N$, we obtain the geometric (or random) distribution:

$$P_i = \frac{\bar{n}^i}{(1+\bar{n})^{i+1}} \tag{29}$$

where P_i is the probability of finding a micelle that contains i solutes. For a Poisson distribution, we have the following equations:

$$W = \frac{N!M!}{\Pi_{i=0}[M_i!(i!)^{Mi}]} \tag{30}$$

$$P_i = (\bar{n}^i e^{-\bar{n}}) / i! \tag{31}$$

On the other hand, binomial distribution is characterized by the following expression:

$$P_i = \begin{pmatrix} i \\ m \end{pmatrix} \begin{pmatrix} n \\ m \end{pmatrix}^i [1 - \bar{n}/m)]^{m-i} \tag{32}$$

where m is the maximum number of solutes that can be solubilized in a given micelle (i = 0,1,2, ..., m). When $m \rightarrow \infty$, the binomial distribution reduces to Poissonian.

Poissonian distribution is the most widely used model for the distribution of solutes and micelles and other aggregate/molecular host systems. The Poisson distribution can be shown to be a normal consequence of the dynamic model of the micellar association and solubilization process. The dynamic equilibrium of the solute S with the micelle can be represented as

$$S_w + M_i \underset{(1+i)k_-}{\overset{k_n}{\rightleftarrows}} M_{i+1} \qquad (33)$$

where S_w represents a solute molecule in water and M_i a micelle containing i solute molecules. The rate k_n is a second order rate constant which describes the rate of entry of solutes in the micelles. It is assumed to be independent of the number of solutes in the micelle. The k_- is the exit rate of one solute and is assumed to be independent of the occupation number i. The rate at which the solutes leave micelles containing several of them is assumed to be linearly dependent on the number of solutes, i.e., rate = $(i + 1)k_-$.

There is no limit to the maximum number of solutes that may occupy a given micelle. Writing equilibrium expressions (Eqn. (33)) for all values of i, it is easy to obtain the fraction of micelles that are occupied by i solute molecules as

$$[M_i]/[M] = (n_i\, e^{-\bar{n}})/i! \qquad (34)$$

where $\bar{n}$ is the average number of solutes per micelle. This equation shows that the distribution of solubilizates in micelles is governed by a Poisson distribution. The same conclusion can also be reached by considering that each micelle is a small fraction v of a total micellar volume V offered to the solubilizates.

Figure 2.6 presents schematically the variation of P_i values for various mean occupancy numbers $\bar{n}$. It is worthwhile examining these P_i values to appreciate the influence of the Poisson distribution. At a low value of $\bar{n}$ (ca. 0.1), most of the micelles are empty and only 10% of the micelles contain one or more solutes ($P_0 = 0.90$, $P_1 = 0.09$, and $P_n = 0$ for $n \geq 2$). When $\bar{n} = 1.0$, i.e., there are equal concentrations of solutes and micelles, 37% of the micelles are still empty but about 26% of them contain two or more solutes ($P_0 = 0.37$, $P_1 = 0.37$ and $P_n = 0.26$ for $n \geq 2$). It has been shown that Poisson distribution adequately explains the results in most cases.

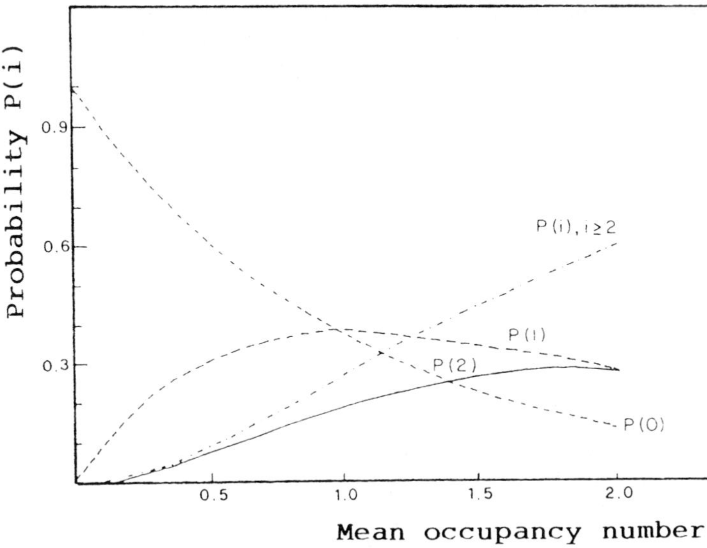

Fig. 2.6. Variation in the probability P_i of finding i solutes in a given micelle with the mean occupancy number N of solutes in a given micellar solution, calculated according to Poisson statistics. Reproduced with permission from Ref. [78].

2.8.1. Exchange of solubilizates

The dynamics of exchange of the solute via entry and exit rates relative to the equilibrium can be expressed with a net solute entry rate of

$$v_+ = k_+ [M] [S_w] \qquad (35)$$

and a net exit rate of

$$v_- = k_- [MS] = k_- \bar{n}[M] \qquad (36)$$

where $\bar{n}$ is the average occupation number of solute in the micelle. Since k_- is a pseudo first order rate constant, $1/k_-$ is in effect the mean lifetime of a given solute molecule inside a micelle, k_+ is a bimolecular rate constant which can in turn be expressed as:

$$k_+ = \beta\, k_{diff} \qquad (37)$$

where k_{diff} is the rate constant for diffusion controlled encounters between the aqueous solute and the micelles and β is the net efficiency of solute incorporation per encounter.

From a simple consideration of relative mobility, the rate constant for bimolecular encounter between a molecule and a micelle of radius R_M in aqueous solution, should be given approximately by [79]:

$$k_{diff} = 4\pi \frac{N_A}{1000} R_M D_S \tag{38}$$

where N_A is Avogadro's number and D_S is the diffusion coefficient of the solute. The values of k_{diff} should then lie in the range 5×10^{10} to 1×10^{11} M^{-1} s^{-1}, depending on the micellar radius and probe diffusion coefficient.

Experimental data range from ca. 1×10^9 to 5×10^{10} M^{-1} s^{-1} with no clear relationship to either micellar or solute properties [4,13]. Although some experimental errors have probably to be taken into account, these data indicate that solute incorporation is slower than solute–micelle encounter. Contrary to the entry rate, micellar exit rates vary in a very large range from 10^8 s^{-1} to less than 10^3 s^{-1} (see Table 2.2) [4].

Furthermore, there is a clear trend; that is k_- decreases as the solute hydrophobicity increases. The dependence of K_M on solute hydrophobicity thus appears to be essentially related to k_- rather than to k_+. The value of K_M is then crucial in controlling the extension of the association of a solute to the colloid and in the rate of release of the solubilizate.

Under normal experimental conditions, this exchange of the probe takes place on a time scale of between 10^{-3} and 10^{-8} s and hence fast detection techniques (stopped flow, relaxation and fast spectroscopic techniques) are required to evaluate the individual rate constants.

Stopped flow techniques, being rather slow, are useful only in selected cases [80]. Although extensively exploited to study the dynamics of monomer–micelle exchange, ultrasonic relaxation has been only sparingly applied to solute exchange. Wyn-Jones et al. [81] did, however, use ultrasonic relaxation to evaluate exchange rates of n-alcohols in micellar HTAB.

Spectroscopic techniques such as electron paramagnetic resonance [82] and transient time-resolved absorption [83,84] and emission spectroscopy [13,85] are particularly suitable for measuring exchange rates on the time scale considered. A time-resolved flash CIDNP (chemically induced dynamic nuclear polarization) method has also been used to evaluate the rate of exit of benzyl radicals from SDS micelles [84]. In at least one system [86], exchange rate constants were extracted from steady state photolysis data.

TABLE 2.2

Entry (k_+) and exit (k_-) rate constants of different solute molecules to and from micelles

Solute	Surfactant	k_+ (10^{-9} M^{-1} s^{-1})	k_- (10^{-6} s^{-1})
Oxygen	SDS	14	350
Oxygen	HTAB	13	—
Acetone	SDS	10	—
Acetone	HTAC	34	—
Propanol	HTAB	3.8	81
Butanol	HTAB	4.4	58
Pentanol	HTAB	6.8	26
Hexanol	HTAB	4.2	9
Naphthalene	SDS	—	>0.05
1-Bromonaphthalene	SDS	5–8	0.025
Anthracene	HTAB	—	<10
Acetophenone	SDS	15.6	7.8
Propiophenone	SDS	14.1	3.0
Isobutyrophenone	SDS	12	1.6
Xanthone	SDS	17	2.7
Methylene iodide	SDS	25	10
di-terbutyl nitroxide	SDS	1.4	—
Cis-1,3-pentadiene	SDS	1.2	8.9
Trans-1,3-pentadiene	SDS	0.95	6.9
1,3-hexadiene	SDS	0.83	2.3
1,3-cyclohexadiene	SDS	1.3	5.6
2,4-dimethylpentadiene	SDS	—	0.37
1,3-cycloheptadiene	SDS	—	1.3
1,3-octadiene	SDS	—	0.13
2,5-dimethyl-2,4-hexadiene	SDS	—	0.19
Cyclohexadienyl radical	SDS	—	4.4×10^6
Benzyl radical	SDS	—	1.4×10^6
Iodine dioxide	SDS	1.5	37×10^7

Data taken with permission from Ref. [4] (and references cited therein).

2.9. LOCATION OF SOLUBILIZATE IN THE MICELLES

The type of interactions which determine the location and orientation of the solute with respect to the micelle varies with the nature of the solubilized species and the surfactant. Information on the site of solu-

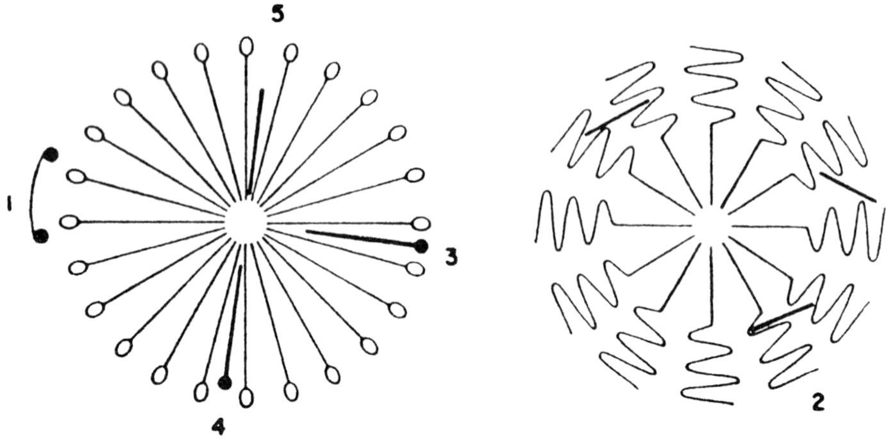

Fig. 2.7. Regions of solubilization of substances in micelles. For explanation of the numbering, see text. Reproduced with permission from Ref. [88].

bilization can be obtained mainly from comparative studies on the solubilizate before and after solubilization. The change in micellar dimensions on solubilization can be followed using X-ray diffraction. Optical absorption and fluorescence spectroscopy, as well as NMR and ESR spectroscopy, can show changes in the microenvironment of the solubilizate following its solubilization. From these types of investigation, the various regions for possible solubilization emerge, as schematically depicted in Fig. 2.7.

Clearly in a dynamic entity like a micelle, a sharp distinction between these regions cannot be expected, and these should therefore be considered at best only at time-averaged loci of the solubilizate. The regions shown in Fig. 2.7 then include the following loci of solubilization:

1. Adsorption at the micellar surface, i.e., at the water/micelle interface.
2. In the case of non-ionic surfactants, between the hydrophilic head groups.
3. In the palisade layer between the hydrophilic head groups.
4. Deeper in the palisade layer.
5. In the inner core of the micelles.

Non-polar molecules such as saturated aliphatic or alicyclic hydrocarbons are solubilized in the inner micellar core. Their UV and NMR spectra indicate an entirely non-polar environment of solubilization. Furthermore, n-alkanes solubilized in sodium dodecanoate micelles have partial molar volumes and compressibilities close to those of liquid alkanes [87]. Determination of the location of aromatic molecules appears to be much more difficult.

A consensus now seems to be emerging concerning the location of arenes in cationic micelles. At least at low concentrations, the arenes seem to reside in the surface layer or the outer palisade layer of these micelles. The polarizability of the π-electron system of the aromatic moiety and the resultant interaction with the positively charged head-group may account for this location close to the micellar interface.

Although it was mentioned that the location of aromatics is rather uncertain, for benzene [55,89], toluene [55] and 2-methylnaphthalene [90], it has been suggested that surface solubilization dominates. Studies of the effect of such solubilizates on the surfactant NMR spectrum indicate that the polar head-groups are much more affected than other groups of the surfactant molecule [89].

In anionic micelles solubilization of arenes may occur in the hydrophobic parts of the aggregates. However, some surface solubilization has been indicated in this case as well [13]. In polyoxyethylenated non-ionic micelles, benzene seems to reside between the polyoxyethylene chains [91].

While interactions of a solubilizate with the micellar aggregate induce, in many cases, pronounced changes in the properties of the host molecule, reciprocity exists in this host–guest relationship. In many cases, marked changes in micellar properties are induced by the solubilizate. Many aromatic molecules induce a sphere to rod transition in HTAB, and it seems that such effects are correlated with solubilization near the head-groups [55,92].

When aliphatic or aromatic hydrocarbons are solubilized in SDS micelles, the aggregate responds by growing until surface area per head-group is the same as in the undisturbed micelle [93]. This, however, does not imply that the solubilized molecules are confined to the centre core of the micelle. The surface area per monomer at the edge of the hydrophobic sphere is about 60 A^2. This is more than twice the cross-section of an hydrophobic chain. In the pure micelle, the excess interfacial area is taken by other monomeric tails. Hydrophobic guest molecules may then replace monomer hydrocarbon chains in this region of the micelle as well.

Large polar molecules such as long-chain alcohols or polar dye molecules are believed to be solubilized with their polar groups near the surface layer and the non-polar portion oriented towards the interior of the micelle. Spectra of solubilizates in these cases indicate that at least part of the molecule resides in a polar microenvironment. The depth of penetration into the palisade layer will depend on the balance between the hydrophobic and electrostatic interactions, which in turn depends on the guest molecule structure. Solubilized alcohols (e.g., butanol, pentanol) increase the volume of SDS micelles as well as the surface area per head-group [93]. It was already mentioned above that added alcohols decrease the surface charge density.

Small polar molecules (e.g., phenols, amines, acids) are generally solubilized close to the micellar surface, in the palisade layer or even at the micelle/water interface. Optical spectra of these molecules indicate that they reside in a relatively polar environment.

Ions are primarily adsorbed at the micellar/water interface, and the interactions involved in this binding are obviously electrostatic in nature. This type of binding has been discussed in Section 2.7. However, when the metal ion is complexed to hydrophobic ligands, the situation may become rather complex. A similar dilemma, regarding the nature of interactions and site of solubilization, may arise for large organic ions (or radical-ions). An example of the former situation is the solubilization of RuL_3^{2+} (L = ligand such as 2,2′-bipyridine (bpy), 1,10-phenanthroline (phen), etc.).

Early results on the effect of SDS micelles on the fluorescence spectrum of $Ru(bpy)_3^{2+}$ indicate a microenvironment which is less polar than n-decanol [95]. Further studies on $Ru(bpy)_3^{2+}$ complexes indicate weak interaction with the non-ionic Triton X-100 micelles. These are probably van der Waals or hydrophobic interactions [96]. Yet the lifetimes of the same photosensitizers in their lowest excited state, when bound to anionic micelles, are sensitive to the isotopic composition of the water (H_2O or D_2O) [97]. This indicates the accessibility of water to the region of the micelle where the complex resides (or vice versa).

Figure 2.8 schematically shows several possible models which try to accommodate the two observations of low polarity and accessibility to water. The model of Fig. 2.8(C) seems to adhere best to most of the experimental observations and to the commonly accepted picture of micellar structure. However, in the dynamic micellar system, different contributions from various structures to a time-averaged observation (such as a fluorescence spectrum or even lifetime of an excited state) have to be considered.

In the case of charged organic molecules, binding is likely to occur at the micelle/water interface when the micelle is oppositely charged. Binding in such cases will increase on going from the neutral molecule to the charged

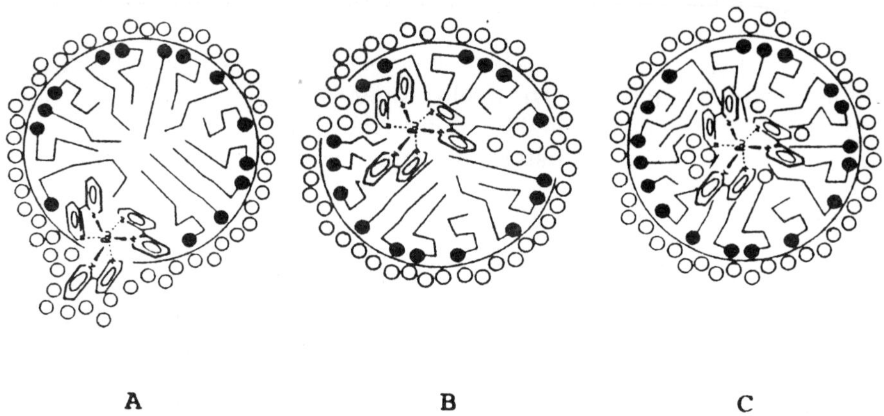

Fig. 2.8. Schematic representation of typical micelle-bound RuL_3^{2+} photo-sensitizers. $Ru(bpy)_3^{2+}$ is used in the figure. Solid circles represent ionic head-groups. Open circles represent water molecules. (A) Modified Dill–Flory model. (B) Simplified Fjord model. (C) Idealized solvent-entrapment model. Taken with permission from Ref. [94].

form. Such is the case, for example, in binding of phenoxide ions, as compared to the corresponding phenol, to HTAB micelles [66].

In the case of a charged amphiphilic substrate interacting with a like-charged micelle, the interaction may be analyzed thermodynamically as a comicellization phenomenon [98]. The incorporation coefficient in such cases is quite sensitive to detergent concentrations and added salts. In the case of amphiphilic substrates and micelles of similar charge, the fine balance between the hydrophobic and electrostatic interactions is of utmost importance in controlling solubilization and thus in affecting the kinetics of a reaction.

A dramatic example of this fine balance was observed in a series of amphiphilic viologens, N-alkyl-N′-methyl-4,4′-bipyridinium dichloride (denoted C_nMV^{2+}, n = 12, 14, 16, 18). The viologen $C_{12}MV^{2+}$ as well as its reduced radical cation form, $C_{12}MV^{+}$, are highly water-soluble and will not partition into positively charged micelles. On the other hand, while the viologen $C_{14}MV^{2+}$ is still water-soluble, its radical cation, $C_{14}MV^{+}$, is very insoluble in water and will strongly bind to positively charged micelles [99].

Finally, we return to the ambiguity involved in the determination of the substrate in the micelle and, at the same time, the utmost importance of such determinations in the analysis of micellar effects on equilibria. Some fundamental difficulties still exist in precisely describing the micellar structure.

The basic Hartley model which depicts the micelle as a well defined sphere sharply separating the apolar inner core from the polar aqueous solution has been challenged by Menger [100,101]. In Menger's model, the micelles are porous clusters into which water can deeply penetrate (the Fjord model). The subject is still a matter of discussion [102,103] and no clear answer is evident.

Added to this basic difficulty, the dynamic processes which continuously vary the aggregate's structure further contribute to the vagueness in determination of substrate locality. Furthermore, different techniques utilized in such determinations may have different sensitivities to particular configurations of the micelle/substrate ensemble and thus different localities for the same system may be reported. In spite of these reservations, it is difficult to overstate the importance of detailed knowledge on the loci of solubilizates in the micellar system.

For example, the naive assumption that electron transfer reactions in micelles occur between solubilizates and acceptors (donors) in the aqueous phase which are spatially isolated from one another, led to the erroneous conclusion that long-range electron tunnelling is involved. In the absence of a definite solution to this problem at present, it should be realized that the term of substrate location is rather loosely used.

2.10. FACTORS INFLUENCING SOLUBILIZATION

The extent to which a substrate can be solubilized into a particular surfactant aggregate depends upon the portion of micelle that can serve as the solubilization site; in the simplest aqueous systems, surfactant and solubilizate structure then play a crucial role.

2.10.1. Effect of the structure of the surfactant

For solubilizates such as hydrocarbons and long-chain polar compounds located in the micellar core or deep in the palisade layer, the amount of solubilized material generally increases with increasing size of the aggregate. Therefore, any factor that causes an increase in the aggregation number of the micelle (see Chapter 1) is expected to increase the solubilization capacity. In particular, the effect of the alkyl-chain length is noticeable.

Examples are the solubilization of ethylbenzene in a series of potassium fatty acid soaps [104], of hydrocarbons in bivalent metal alkyl sulfates [105], and of trans-azobenzene in hexadecylpyridinium with different counterions [106].

A similar increase in solubilizing capacity with increasing alkyl-chain length has been observed for non-ionic surfactants when the solubilizate is located in the hydrophobic core and, conversely, with the decrease of the length of the polyoxyethylene chain (both factors contribute to the increase of the aggregation number). On the other hand, the solubilization of ethylbenzene, which is probably located in both regions of non-ionic micelles, is reported to increase with the number of ethoxy units [107].

In general, the solubilizing capacity toward solutes located in the micellar interior follows the order: non-ionics > cationics > anionics for surfactants having the same hydrocarbon-chain length [108,109].

For polar compounds very few generalizations can be made since solubilization can occur in both the inner and outer regions of the micellar aggregate.

2.10.2. Effect of the structure of solubilizates

Several factors can affect the amount of substance which can be solubilized. The polarity (and polarizability) are a crucial criterion of classification of solubilizates. Other parameters affecting the solubilization extent are chain length and branching, molecular shape and size.

For aliphatic and alkylaryl hydrocarbons MAC decreases with increase of chain length and increases with cyclization and unsaturation (e.g., n-hexane < cyclohexane < benzene), whereas branched chains appear to have little effect. For polycyclic hydrocarbons the extent of solubilization decreases with increase in molecular size [5,110].

For polar substrates, the different possible location in the micelle changing the structure of the molecules makes it difficult to establish rules. Polar compounds solubilized close to the micelle/water interface should have a MAC higher than non-polar substrates located in the core region [111].

Finally, it must be stressed that the selection of the appropriate surfactant to obtain maximum solubilization of a given compound involves the correct matching of all the above considerations.

2.10.3. Effect of added electrolytes

As previously discussed in Chapter 1, the addition of electrolytes to ionic surfactant aggregates decreases the cmc and increases the micellar size. The decrease in mutual repulsion of the ionic head-groups causes a close packing of the amphiphile molecules in the palisade layer, decreasing the volume available for solubilization.

Thus, the addition of small amounts of electrolytes to ionic surfactants (present in concentrations slightly above the cmc) results in an increase of the solubilization for solutes such as hydrocarbons [112], decanol [112,113] and nitrophenylamine [114]. However, for solubilizates located in the palisade region, a decrease is observed as was reported in studies concerning medium-chain alcohols (butanol to heptanol) [115].

When the surfactant concentration is well above the cmc, the electrolyte effect is difficult to predict. Whereas, as reported above, decanol solubilization increases at low surfactant concentrations with NaCl addition, a reverse trend is observed at higher amphiphile concentrations (see Fig. 2.9). The effect of added Na^+ is not only to reduce the cmc, but also to affect the properties of the mixed micelles which are formed.

Similar considerations apply to non-ionic surfactants. Where electrolyte addition leads to an increase in micellar size, the extent of solubilization increases for hydrocarbons. However, until now, the effect of electrolytes on solubilization by non-ionic surfactants is unclear.

2.10.4. Effect of pH

For ionizing substrates, such as acids, amines and phenols, the pH of the aqueous solution may affect the extent of solubilization. Where only the unionized form undergoes solubilization (e.g. phenol or benzoic acid with anionic aggregates) the amount of solubilizate depends on the pH. Potentiometric titrations allow us to obtain information on the extent of solubilization [116].

An interesting case is represented by scarcely soluble organic acids, which exhibit solubilization that is dependent on solution pH. The total solubility (the sum of dissociated and undissociated monomeric acid) increases linearly as a function of $1/a_H^+$ of the solution up to a pH range in which micelles appear. In this domain the solubility is virtually equal to the concentration of NaOH added to the solution.

Above this pH the total solubility increases with decreasing pH of the solution and becomes higher than the added NaOH concentration. This

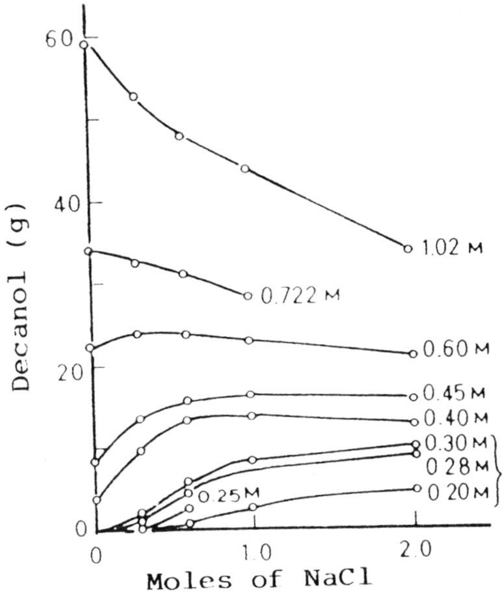

Fig. 2.9. Effect of added NaCl on decanol solubility in aqueous solutions of sodium octanoate, at 20°C. The amounts of NaCl and octanol are given per 1000 g of solution. Concentration range of octanoate: 0.2–1 mol/1000 g of decanol-free solution. Taken with permission from Ref. [113].

extra solubility results from the solubilization of undissociated acid into the micelles of the respective sodium salt [117].

2.10.5. Effect of addition of non-electrolytes

Non-electrolyte additives can have a pronounced effect on the solubilizing properties of surfactants.

A synergistic effect of polar and non-polar compounds has often been observed [118,119]. For example, the presence of solubilized hydrocarbons increases the solubility of polar compounds as the swelling of the micelle allows the incorporation of more polar materials in the palisade layer. On the other hand, the solubilization of polar compounds as long-chain alcohols, amines and fatty acids, appears to increase the solubilization of hydrocarbons.

The addition of some simple non-electrolytes, such as urea, also changes the tendency of micelles to solubilize moderately polar com-

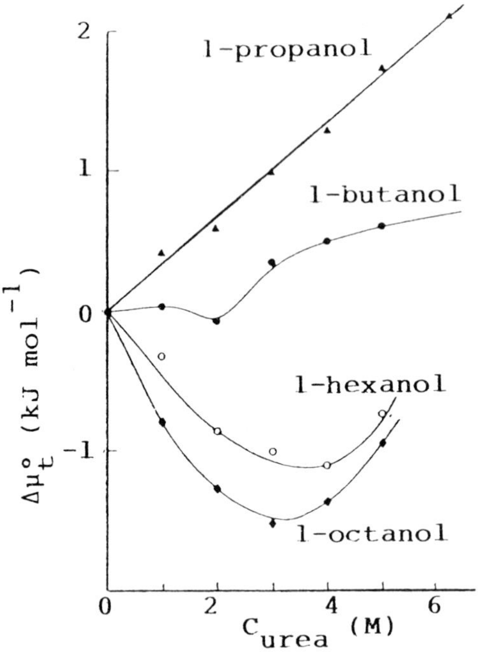

Fig. 2.10. Transfer free energies of primary alcohols from aqueous SDS micelles to SDS aggregates present in urea-containing solutions. T: 25°C. Reproduced with permission from Ref. [120].

pounds. For example, the transfer free energy of alcohols from SDS micelles in water to SDS micelles in aqueous urea solutions changes in a very different way depending on the hydrophobic–hydrophilic balance of the alcohol (see Fig. 2.10) [120].

The observed effects have been qualitatively discussed in terms of parameters such as the extension of the hydrocarbon–water contact region, although further work is still necessary for a better understanding of the phenomena which occur.

Investigations concerning solubilization of highly volatile substances (including hydrocarbons) in surfactant solutions have been also reported. In order to overcome various practical problems related to these measurements, the use of a properly designed suitable apparatus has been proposed [121]. For example, the solubilization of benzene in micelles of n-dodecylammonium perfluoroacetate has been successfully investigated using the above-mentioned device.

2.10.6. Solubilization of mixtures of solubilizate

Despite the effect of the solubilization of mixtures being of relevant practical importance, little attention has been devoted to this aspect. Co-solubilization may cause alterations in the solubilities of compounds if they are located in the same micellar region, e.g. oleic acid and long-chain alcohols [122] or testosterone and ethinylestradiol [123] which are solubilized in the palisade layer.

When solubilizates are preferentially located in different sites, minor effects are observed.

Since the effect of solubilizates on the structural alterations of aggregates described above may enhance the solubilizing capacity, the simultaneous operation of these two processes would explain the occurrence of maxima and minima in the solubility plots of mixtures of solubilizates.

2.10.7. Effect of temperature

The effect of temperature on the micellar properties and on the change in aqueous solubilities of the solubilizates is generally responsible for the increase of solubilization for both polar and non-polar compounds in ionic surfactant aggregates.

Few detailed isotherms of solubilization have been reported in the literature. Solubilization of trichloroethylene by HPC micelles provide an accurate example of a study of thermal effects on solubilization by ionic aggregates [124]. In this case, solubilization exhibits a maximum at approximately 30°C, at which temperature the solubility of the organic solvent in water is a minimum.

For non-ionic polyoxyethylenated surfactants the effect of temperature depends mainly on the nature of solubilizate. Non-polar compounds located in the micellar core appear to show an increase in solubility with increasing temperature, whereas polar materials solubilized in the palisade layer generally exhibit a behaviour going through a maximum.

2.10.8. Solubilization in non-aqueous solvents

Surfactant-mediated solubilization in non-aqueous solvents has been investigated mainly in connection with water solubilization in these media. The addition of water to a non-aqueous solution of surfactant may give rise to aggregation, generating the so-called reverse micelles (see Chapter 1).

Since the polar groups are directed toward the inner core and the hydrophobic groups are in contact with the solvent, small polar molecules can be solubilized within this aqueous pool. The maximum amount of polar solubilizate depends on its nature and on the amount of water present.

In turn, the maximum amount of water solubilized by ionic surfactants in non-aqueous solvents depends on surfactant properties, surfactant concentration, length and structure of the hydrophobic tail and nature of the counterions [125]. In the case of non-ionics, the amount of solubilized water appears to increase with concentration and length of the polyoxyethylene chain.

Further aspects of solubilization in these non-polar solvent systems are examined in other chapters of this book, where the problems related to extractions in reverse micelles and to the use of such structures as chromatographic eluents are discussed in detail.

REFERENCES

1 M.E.L. McBain and E. Hutchinson, *Solubilization and Related Phenomena*. Academic Press, New York, 1955.
2 D. Attwood and A.T. Florence, *Surfactant Systems. Their Chemistry, Pharmacy and Biology.* Chapman and Hall, London, 1985.
3 M.J. Rosen, *Surfactants and Interfacial Phenomena.* 2nd Ed. Wiley, New York, 1989.
4 L. Sepulveda, E. Lissi and F.H. Quina, *Adv. Colloid Interface Sci.*, 25 (1986) 1.
5 H.B. Klevens. *J. Phys. Colloid Chem.*, 54 (1950) 283.
6 P. Ekwall, L. Sjoblom and J. Olsen, *Acta Chem. Scand.*, 7 (1953) 347.
7 D.W. Ownby, W. Prapaitrakul and A.D. King, Jr., *J. Coll. Interface Sci.*, 125 (1988) 526.
8 B.J. Carrol, *J. Coll. Interface Sci.*, 79 (1981) 126.
9 Y.C. Chiu, in: K.L. Mittal and E.J. Fendler (Eds.), *Solution Behavior of Surfactants. Theoretical and Applied Aspects.* Plenum Press, New York, 1982, Vol. II, pp. 1415–1440.
10 A.F. Chan, D.F. Evans and E.L. Cussler, *AIChemE J.*, 22 (1976) 1006.
11 W.I. Higuchi, *J. Pharm. Sci.*, 56 (1967) 315.
12 B. Jonson and H. Wennestrom, *J. Coll. Surface Sci.*, 80 (1981) 482.
13 M. Almgren, F. Grieser and J.K. Thomas, *J. Am. Chem. Soc.*, 101 (1979) 279.
14 J. Burgess and E. Pelizzetti, *Progr. Reaction Kin.*, 17 (1992) 1–170.
15 A. Ben Naim, *Hydrophobic Interactions.* Plenum Press, New York, 1980.
16 S.J. Dougherty and J.C. Berg, *J. Colloid Interface Sci.*, 48 (1974) 110.

17 A.A. Bhalekar and J.B.F.N. Engberts, *J. Am. Chem. Soc.*, 100 (1978) 594.
18 K. Shinoda and T. Soda, *J. Phys. Chem.*, 67 (1963) 2072.
19 R. Dorshow, J. Briggs, C.A. Bunton and D.F. Nicoli, *J. Phys. Chem.*, 86 (1982) 2388.
20 J.P. Kratohvil, *J. Colloid Interface Sci.*, 75 (1979) 271.
21 K. Ikeda, H. Tomida and T. Yotsuzanagi, *Chem. Pharm. Bull. Jpn.*, 25 (1977) 1067.
22 C. Irose and L. Sepulveda, *J. Phys. Chem.*, 85 (1981) 3689.
23 L.J. Winters and E. Grunwald, *J. Am. Chem. Soc.*, 87 (1965) 4608.
24 J.C. Eriksson and G. Gilberg, *Acta Chem. Scand.*, 20 (1966) 2019.
25 S. Ohnishi, J.R. Cyr-Theodore and H. Fukushima, *Bull. Chem. Soc. Jpn.*, 43 (1970) 673.
26 F.H. Quina and V.G. Toscano, *J. Phys. Chem.*, 81 (1977) 1750.
27 T. Miyaji, K. Kurono, K. Vekama and K. Ikeda, *Chem. Pharm. Bull. Jpn.*, 24 (1976) 115.
28 S. Yamaguchi and T. Tsukamoto, *Nippon Kagaku Kaishi*, (1976) 1856.
29 L.K.J. Tong and M.C. Glesmann, *J. Am. Chem. Soc.*, 79 (1957) 4305.
30 E. Pramauro and E. Pelizzetti, *Anal. Chim. Acta*, 126 (1981) 253.
31 E. Pramauro and E. Pelizzetti, *Ann. Chim. (Rome)*, 72 (1982) 117.
32 M. Almgren, F. Grieser and J.K. Thomas, *J. Phys. Chem.*, 83 (1979) 3232.
33 P. Yarmchuk, R. Weinberger, R.F. Hirsch and L.J. Cline Love, *Anal. Chem.*, 54 (1982) 2233.
34 E. Pramauro and E. Pelizzetti, *Anal. Chim. Acta.*, 154 (1983) 153.
35 E. Pelizzetti and E. Pramauro, *J. Phys. Chem.*, 84 (1984) 990.
36 E. Pelizzetti and E. Pramauro, *Anal. Chim. Acta*, 117 (1980) 403.
37 C.A. Bunton and L. Sepulveda, *J. Phys. Chem.*, 84 (1980) 2611.
38 E. Lissi, E. Abuin and A.M. Rocha, *J. Phys. Chem.*, 84 (1980) 2406.
39 D.G. Herries, W. Bishop and F.M. Richards, *J. Phys. Chem.*, 68 (1964) 1842.
40 D.W. Armstrong, R. Seguin and J.H. Fendler, *J. Mol. Evol.*, 10 (1977) 241.
41 D.W. Armstrong and J.H. Fendler, *Biochim. Biophys. Acta*, 478 (1977) 75.
42 D.W. Armstrong and F. Nome, *Anal. Chem.*, 53 (1981) 1662.
43 C. Tanford, *The Hydrophobic Effect*. Wiley, New York, 1980.
44 F. Franks, in: F.Franks (Ed.), *Water*. Plenum Press. New York, 1975. Vol. 4, Ch. 1.
45 A. Wishnia, *J. Phys. Chem.*, 67 (1963) 2079.
46 K.S. Birdi, in: K.L. Mittal (Ed.), *Micellization, Solubilization and Microemulsions*. Plenum Press, New York, 1977. Vol. I, p. 151.
47 E. Pramauro, G. Saini and E. Pelizzetti, *Anal. Chim. Acta*, 166 (1984) 233.
48 R.F. Rekker, in: W.T. Nauta and R.F. Rekker. (Eds.), *The Hydrophobic Fragment Constant, its Derivation and Application*. Pharmacochemistry Library. Elsevier, Amsterdam, 1977, Vol. 1.
49 E. Pramauro, C. Minero, G. Saini, R. Graglia and E. Pelizzetti, *Anal. Chim. Acta*, 212 (1988) 171.

50 K. Martinek, A.V. Levashov and I.V. Berezin, *Mol. Biol.*, 4 (1970) 517.
51 G.A. Smith, S.D. Christian, E.E. Tucker and J.F. Scamehorn, *Langmuir*, 3 (1987) 598.
52 G.A. Smith, S.D. Christian, E.E. Tucker and J.F. Scamehorn, *J. Colloid Interface Sci.*, 130 (1989) 254.
53 W.S. Higazi, F.Z. Mahmoud, A.A. Taha and S.D. Christian, *J. Solution Chem.*, 17 (1988) 191.
54 C. Nguyen, J.F. Scamehorn and S.D. Christian, *Colloids Surfaces*, 30 (1988) 335.
55 J.C. Eriksson and G. Gilberg, *Acta Chem. Scand.*, 20 (1966) 2019.
56 F. Grieser and C.J. Drummond, *J. Phys. Chem.*, 92 (1988) 5580.
57 S.A. Simon, R.V. McDaniel and T.J. McIntosh, *J. Phys. Chem.*, 86 (1982) 1449.
58 E.E. Tucker and S.D. Christian, *Faraday Symp. Chem. Soc.*, 17 (1982) 11.
59 S.J. Rehfeld, *J. Phys. Chem.*, 75 (1971) 3905.
60 P. Mukerjee and J.R. Cardinal, *J. Phys. Chem.*, 82 (1978) 1620.
61 G.A. Smith, S.D. Christian, E.E. Tucker and J.F. Scamehorn, in: W.L. Hinze and D.W. Armstrong (Eds.), *Ordered Media in Chemical Separations.* ACS Symp. Series 342, American Chemical Society. Washington D.C., 1987, Ch.10, pp. 184–198.
62 S.D. Christian, G.A. Smith, E.E. Tucker and J.F. Scamehorn, *Langmuir*, 1 (1985) 564.
63 G.A. Smith, S.D. Christian, E.E. Tucker and J.F. Scamehorn, *J. Solution Chem.*, 15 (1986) 519.
64 S.N. Bhat, G.A. Smith, E.E. Tucker, S.D. Christian, W. Smith and J.F. Scamehorn, *Ind. Eng. Chem. Prod. Res. Dev.*, 26 (1987) 1217.
65 B.H. Lee, S.D. Christian, E.E. Tucker and J.F. Scamehorn, *Langmuir*, 6 (1990) 230.
66 C.A. Bunton and L. Sepulveda, *J. Phys. Chem.*, 83 (1979) 680.
67 C. Minero, E. Pramauro, E. Pelizzetti and D. Meisel, *J. Phys. Chem.*, 87 (1983) 399.
68 D.W. Armstrong and G.Y. Stine, *J. Am. Chem. Soc.*, 105 (1983) 6220.
69 J.T. Davis, *Surface Phenomena in Chemistry and Biology.* Pergamon Press, London, 1958.
70 C. Minero, E. Pramauro and E. Pelizzetti, *J. Phys. Chem.*, 92 (1988) 4670.
71 P. Linse, G. Gunnarson and B. Jonson, *J. Phys. Chem.*, 86 (1982) 413.
72 J Frahm and S. Diekmann, *J. Colloid Interface Sci.*, 70 (1979) 440.
73 E. Pramauro, E. Pelizzetti, S. Diekmann and J. Frahm, *Inorg. Chem.*, 21 (1982) 2432.
74 M. Vincenti, E. Pramauro, E. Pelizzetti, S. Diekmann and J. Frahm, *Inorg. Chem.*, 24 (1985) 4533.
75 P.P. Infelta, M. Gratzel and J.K. Thomas, *J. Phys. Chem.*, 78 (1974) 190.
76 M. Tachiya, *J. Phys. Chem.*, 78 (1983) 5282.
77 Y. Moroi, *J. Phys. Chem.*, 84 (1980) 2186.

78 K. Kalyanasundaram, *Photochemistry in Microheterogeneous Systems.* Academic Press, Orlando, 1987.
79 A.F. Olea, M.V. Encina and E.A. Lissi, *Macromolecules*, 15 (1982) 1111.
80 Y. Miyashita and S. Hayano, *J. Colloid Interface Sci.*, 86 (1982) 344.
81 J. Gettlings, D. Hole, P.L. Jobling, J.E. Russing and E. Wyn-Jones, *J. Chem. Soc. Faraday II*, 74 (1978) 1957.
82 M. Takiya, *Chem. Phys. Lett.*, 33 (1975) 289.
83 J.C. Scaiano and E.B. Abuin, *Chem. Phys. Lett.*, 8 (1981) 209.
84 N.J. Turro, M.T. Zimmt and I.R. Gould, *J. Am. Chem. Soc.*, 105 (1983) 6347.
85 F.H. Quina and V.G. Toscano, *J. Phys. Chem.*, 81 (1977) 1750.
86 I.B. Matheson and R. Massoudi, *J. Am. Chem. Soc.*, 102 (1980) 1942.
87 E. Vikingstad and H. Hoiland, *J. Colloid Interface Sci.*, 64 (1978) 510.
88 P.H. Elworthy, A.T. Florence and C.B. MacFarlane, *Solubilization by Surface-Active Agents.* Chapman and Hall, London, 1968.
89 P. Mukerjee, J.R. Cardinal and N.R. Desai, in: K.L. Mittal (Ed.), *Micellization, Solubilization and Microemulsions.* Plenum Press, New York, 1977, Vol. 1, p. 241.
90 J. Ulmius, B. Lindman, G. Lindblom and T. Drakenberg, *J. Colloid Interface Sci.*, 65 (1978) 88.
91 T. Nakagawa, in: M.J. Schick (Ed.), *Non Ionic Surfactants.* Dekker. New York, 1967, Ch. 17.
92 G. Lindblom, B. Lindman and L. Mandell, *J. Colloid Interface Sci.*, 42 (1973) 400.
93 M. Almgren and S. Swarup, *J. Phys. Chem.*, 86 (1982) 4212.
94 B.L. Hauerstein, W.J. Dressick, S.L. Buell, J.N. Demas and B.A. De Graaf, *J. Am. Chem. Soc.*, 105 (1983) 4251.
95 D. Meisel, M.S. Matheson and J. Rabani, *J. Am. Chem. Soc.*, 100 (1978) 117.
96 K. Mandal, B.L. Hauerstein, J.N. Demas and B.A. De Graaf, *J. Phys. Chem.*, 87 (1983) 328.
97 R. Sriram and M.Z. Hoffman, *Chem. Phys. Lett.*, 85 (1982) 572.
98 L. Miola, R. Blotta Abakerli, M.F. Ginani, P. Berci, V.G. Toscano and F.H. Quina, *J. Phys. Chem.*, 87 (1983) 4417.
99 P.A. Brugger, P.P. Infelta, A.M. Braun and M. Gratzel, *J. Am. Chem. Soc.*, 103 (1981) 320.
100 F.M. Menger, *Acc. Chem. Res.*, 12 (1979) 111.
101 F.M. Menger and B.J. Boyer, *J. Am. Chem. Soc.*, 102 (1980) 5936.
102 H. Wennestrom and B. Lindman, *J. Phys. Chem.*, 83 (1979) 2931.
103 F. Martens and J.W. Verhoeven, *J. Phys. Chem.*, 85 (1981) 1773.
104 H.B. Klevens, *Chem. Rev.*, 47 (1950) 1.
105 I. Satake and R. Matsuura, *Bull. Chem. Soc. Jpn.*, 36 (1963) 813.
106 G.S. Hartley, *J. Chem. Soc.*, (1938) 1968.

107 J. Xia and Z. Hu, in: K.L. Mittal and P. Bhotorel (Eds.), *Surfactants in Solution*. Plenum Press, New York, 1986. Vol. 5, pp. 1055–1065.
108 S. Saito, *J. Colloid Interface Sci.*, 24 (1967) 227.
109 T. Tokiwa, *J. Phys. Chem.*, 72 (1968) 1214.
110 M.J. Schwuger, *Kolloid-Z.Z.Polym.*, 250 (1972) 703.
111 T. Nakagawa and K. Tori, *Koll.-Z.*, 168 (1960) 132.
112 H.B. Klevens, *J. Am. Chem. Soc.*, 72 (1950) 3780.
113 P. Eckwall, L. Mandell and K. Fontell, *J. Colloid Interface Sci.*, 61 (1977) 519.
114 R.S. Stearns, H. Oppenheimer, E. Simon and D.W. Harkins, *J. Chem. Phys.*, 15 (1947) 496.
115 A.M. Blokhus, H. Hoiland, E. Gilje and S. Backlund, *J. Colloid Interface Sci.*, 124 (1988) 125.
116 E. Azaz and M. Donbrow, *J. Colloid Interface Sci.*, 57 (1976) 11.
117 Y. Moroi and Y. Sakamoto, *J. Phys. Chem.*, 92 (1988) 5189.
118 H.B. Klevens, *J. Chem. Phys.*, 17 (1949) 1004.
119 K. Shinoda and H. Akamatu, *Bull. Chem. Soc. Jpn.*, 31 (1958) 497.
120 M. Abu-Hamdiyyah and K. Kumari, *J. Phys. Chem.*, 94 (1990) 6445.
121 Y. Moroi and T. Morisue, *J. Phys. Chem.*, 97 (1993) 12668.
122 R. Matsuura, H. Furudate, H. Tsutsumi and S. Miida, *Bull. Chem. Soc. Jpn.*, 34 (1961) 395.
123 B. Lundberg, T. Lowgren and B. Heikius, *J. Pharm. Sci.*, 68 (1979) 542.
124 H. Uchiyama, E.E. Tucker, S.D. Christian and J.F. Scamehorn, *J. Phys. Chem.*, 98 (1994) 1714.
125 K. Kon-no and A. Kitahara, *J. Colloid Interface Sci.*, 37 (1971) 469.

Chapter 3

Effect of surfactant aggregates on chemical kinetics and equilibria

The important role of reaction media on the kinetics and mechanism of chemical reactions has long been recognized. In particular, the peculiar properties of the environment offered by micelles and other organized assemblies have received increasing attention. The presence of surface active agents usually either accelerates or decelerates reactions, can modify reaction pathways and stereochemistry and allow separations on the basis of products and charge.

Kinetic investigations of the reactions occurring in the micellar environment offer detailed information on the so-called micellar "catalysis" and constitute the physicochemical basis for the analysis of such processes.

As shown in Chapter 1, the dynamic equilibrium in micellar solution is controlled by monomer–micelle exchange in the time scale of microseconds, by micelle dissolution (millisecond range) and, for ionic species, by counterion exchange (microsecond range). Processes occurring at comparable rates have been treated in terms of substrate entry to or exit from the aggregates.

3.1. CLASSICAL KINETIC MODELS

For processes which occur on time scales much longer than those involved in micellar equilibria and substrate-micelle redistribution (usually longer than ca. 10^{-5}–10^{-4} s), simple statistical stochastic models may apply.

Under such long time spans, equilibria involved in reactants/micelles or products/micelles interactions will relax much faster than the reaction itself. This chapter will discuss such processes in which the law of mass action is applicable to the reactants (products)–micelle interactions. Much of this aspect of micellar effects on kinetics has been covered in detail in several reviews and books [1–9]. In the present account, we attempt to provide an updated overview of the various models and compare them with experimental observations.

As a consequence of the underlying assumption of time-averaged reactant/micellar interactions prior to the chemical reaction, most kinetic treatments of micellar effects consider the reactivities to be a sum of two terms attributable to two separate "pseudophases". From this point of view, the micellar effect on unimolecular reaction was treated in a manner analogous to the enzymatic catalysis [10] (see Scheme 3.1).

$$\begin{array}{ccc} M + S_w & \overset{K_M}{\rightleftarrows} & MS \\ \downarrow k_w & & \downarrow k_m \\ \text{Products} & & \text{Products} \end{array} \qquad \text{Scheme 3.1}$$

Such a model assumes association of the reactant, S, to the micellar pseudophase, M, and two different reactivities in the two phases, represented by k_w and k_m, respectively. K_B in this scheme denotes the binding constant of the substrate to the micelle (also often reported as K_M or $K_{(S)}$ for a given solute S) and is defined as:

$$K_M = \frac{[S_m]}{[M][S_w]} \tag{1}$$

$[S_m]$ and $[S_w]$ are the solute concentrations in the micellar and aqueous phase, respectively, whereas [M] represents the micelle concentration which can be calculated from the total surfactant concentrations, C_S, once the cmc and the aggregation number, N_{AG}, are known (see Chapter 2).

From Scheme 3.1, the experimentally observed rate constant, k_{1exp}, can be expressed in terms of the individual rate constants in the two pseudophases and the constant, K_M.

$$k_{1exp} = \frac{k_w + k_m K_M [M]}{1 + K_M [M]} \tag{2}$$

The similarity of Eqn. (2) to the Michaelis–Menten equation is evident and perhaps even more so after rearrangement to Eqn. (3).

$$\frac{1}{k_w - k_{1exp}} = \frac{1}{k_w - k_m} + \frac{1}{k_w - k_m}\frac{1}{K_M[M]} \tag{3}$$

At least phenomenologically, this resemblance provides the rationale for the much sought model system for enzymatic reactions. This analogy between enzyme and micelle catalyzed reactions has been further developed and investigated by Piszkiewicz [11].

A slightly different approach assumes partitioning, rather than stoichiometric association, of the substrate between the micellar and aqueous subphases as is shown in Scheme 3.2.

$$\begin{array}{ccc} S_w & \overset{P}{\rightleftarrows} & S_m \\ \downarrow k_w & & \downarrow k_m \\ \text{Products} & & \text{Products} \end{array}$$

Scheme 3.2

This scheme leads to the following kinetic equation:

$$k_{1exp} = \frac{k_w + k_m P\phi_m}{1 + P\phi_m} \tag{4}$$

where ϕ_m is the volume fraction of the micellar phase ($\phi_m << 1$). When the surfactant concentration greatly exceeds the reactant concentration, the two approaches lead to formally indistinguishable kinetic equations. Since $\phi_m = M\ N_{AG}\bar{V}$, where $\bar{V}$ is the partial molar volume of the surfactant, one obtains $K_B = P\ N_{AG}\bar{V}$. It must be noted that the molar volume is also indicated as V in several references.

Equations (3) and (4) adequately describe unimolecular reactions in nonionic micelles and unimolecular reactions of uncharged substrates in ionic micelles. It thus suggests that the basic assumptions and the pseudophase model are a useful starting point to analyze micellar effects on unimolecular reactions.

From a physical point of view, the micellar effect is related to the polarity of the Stern layer at the micellar–water interface. The polarity at the interface can be estimated spectroscopically by comparison with known solvent effects on charge-transfer or fluorescence spectra (see Chapter 4). Solvent polarities thus obtained are similar to those of protein–water interfaces and slightly larger than that of ethanol. The decrease in the rate of S_N1 reactions observed in micellar solution, for example, may be simply regarded as a local solvent effect. A decrease in solvent polarity often inhibits S_N1 reactions in homogeneous solutions. For ionic reactants, interactions with the charged surface can stabilize intermediate states.

For reactions of higher order, partitioning of both reactants between the two pseudophases has to be considered [1,12].

$$\begin{array}{llll} A_w + B_w & \xrightarrow{k_w} & \text{Products} \\ \uparrow\downarrow P_A \quad \uparrow\downarrow P_B & & \\ A_m + B_m & \xrightarrow{k_m} & \text{Products} \end{array}$$

Scheme 3.3

Assuming that the equilibrium distribution is not altered by the chemical reaction and that the reagents have no effect on the micellar properties (including the cmc), the overall reaction rate can be expressed by:

$$v = k_{2,exp}[A]_t[B]_t = v_m C\bar{V} + v_w(1 - C\bar{V}) \tag{5}$$

where $v_m = k_m [A_m][B_m]$ and $v_w = k_w [A_w][B_w]$, respectively. The relationship between the total concentration of reactants and their true concentration in the two phases is in turn given by:

$$[A]_t = [A_m]\, C\bar{V} + [A_w]\,(1 - C\bar{V}) \tag{6}$$

$$[B]_t = [B_m]\, C\bar{V} + [B_w]\,(1 - C\bar{V}) \tag{7}$$

From the above assumptions and equations, the observed second-order rate constant can be expressed by the following equation:

$$k_{2exp} = \frac{k_m P_A P_B C\bar{V} + k_w(1 - C\bar{V})}{\{1 + (P_A - 1)\, C\bar{V}\}\,\{1 + P_B - 1)\, C\bar{V}\}} \tag{8}$$

In the often encountered situation of dilute surfactant solutions ($C\bar{V} << 1$), and taking into account that $K_B = (P - 1)\bar{V}$, Eqn. (8) can be rewritten as follows:

$$K_{2exp} = \frac{(k_m/\bar{V})\, K_{(A)} K_{(B)} C + k_w}{(1 + K_{(A)}C)\,(1 + K_{(B)}C)} \tag{9}$$

It is important to note that, in Eqn. (9), the binding constants, $K_{B(A)}$ and $K_{B(B)}$, have been written in the simplified form $K_{(A)}$ and $K_{(B)}$. These constants refer to the molar concentration of surfactant monomer rather than to micelle concentration since the latter requires the knowledge of the aggregation number, N_{AG}. The procedure based on Eqn. (9) is preferred since changes in experimental conditions, such as ionic strength, temperature, etc., may alter the aggregation number.

Equation (9) considers only reactions between reactants in the same pseudophase. In a more general case, reactions between a micelle-associated reactant and a non-associated reactant should also be considered. This leads to Eqns. (10) and (11):

$$v = k_m[A_m][B_m]C\overline{V} + k'_m[A_m][B_w]C\overline{V} + k''_m[A_w][B_m] + k_w[A_w][B_w] \tag{10}$$

$$k_{2exp} = \frac{(k_m P_A P_B + k'_m P_A + k''_m P_B)C\overline{V} + k_w(1 - C\overline{V})}{(1 + K_{(A)}C)(1 + K_{(B)}C)} \tag{11}$$

where k'_m is the rate constant for the reaction due to encounters between reactant A in the micellar phase and reactant B in the aqueous phase, while k''_m refers to the reverse situation. The simplified Eqn. (9) was particularly successful in describing reactions involving either uncharged substrates or nonionic surfactants. Thus, it correctly predicts the dependence of k_{2exp} on the surfactant concentration, C. Figure 3.1 shows the dependence of $k_{2exp.}$ on C for several values of P_A, P_B and k_m/k_w [12].

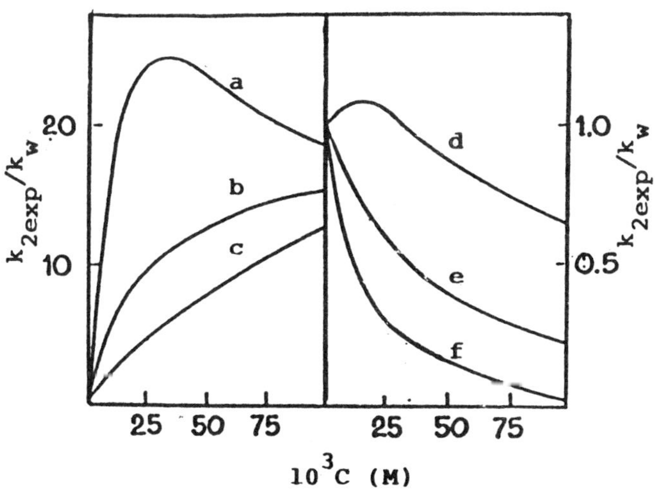

Fig. 3.1. Theoretical plots of apparent second-order rate constant (k_{2exp}) vs. surfactant concentration (C), drawn according to Eqn. (8), using $\overline{V}$ = 0.3 M^{-1}. Reproduced with permission from Ref. [12].

In cases where the aqueous bulk reaction can be neglected (i.e., when $k_w << k_m K_{(A)} K_{(B)} C/\overline{V}$) the concentration-dependent profiles have a maximum at $C_{max} = (K_{(A)} K_{(B)})^{-1/2}$ and, at that concentration, the maximal enhancement ratio is equal to:

$$\left(\frac{k_{2,exp}}{k_w}\right)_{max} = \frac{k_m}{k_w} \frac{K_{(A)} K_{(B)}}{V(K_{(A)}^{1/2} + K_{(B)}^{1/2})^2} \qquad (12)$$

The ratio k_m/k_w characterizes the change in reactivity upon transfer of the reactants from the aqueous to the micellar pseudophases. The second term in the right-hand side of Eqn. (12) is related to the concentration changes in the micellar phase (the so-called "concentration effect").

The dependence of k_{2exp} on C provides a convenient method to experimentally determine the binding constants $K_{(A)}$ and $K_{(B)}$, and the value of $k_m/\overline{V}$. For this purpose, Eqn. (12) is rearranged to:

$$\frac{C}{k_{2exp} - k_w} = X + Y \frac{k_{2exp}}{k_{2exp} - k_w} C + Z \frac{k_{2exp}}{k_{2exp} - k_w} C^2 \qquad (13)$$

where the terms X, Y and Z are given by

$$X = \overline{V}(k_m K_{(A)} K_{(B)})^{-1} \qquad (14)$$

$$Y = X(K_{(A)} + K_{(B)}) \qquad (15)$$

$$Z = X\, K_{(A)} K_{(B)} = \overline{V}/k_m \qquad (16)$$

In the common and important situation where $k_w << k_m K_{(A)} K_{(B)} C/\overline{V}$, Eqn. (13) reduces to:

$$\frac{C}{K_{2exp}} = X + YC + ZC^2 \qquad (17)$$

Plots of C/k_{2exp} against C, while nonlinear, provide a reliable intercept which equals X. With the value of X thus obtained, one may plot $\{(C/k_{2exp}) - X\}/C$ against C to obtain the slope Z and the intercept Y, for the case described by Eqn. (17). For the more general situation of Eqn. (13), the linearity is obtained according to the equation:

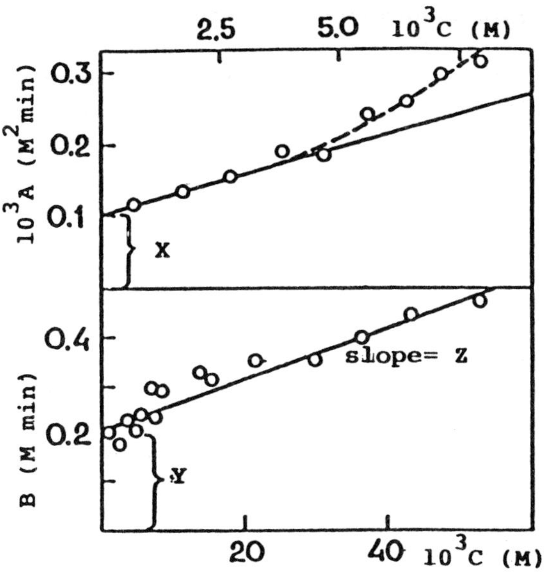

Fig. 3.2. Determination of binding and rate constants according to Eqns. (13) and (17). Plots of A vs. C and B vs. C. $A = C/k_{2exp}$; $B = (C/k_{2exp} - X)/C$. Reproduced from Ref. [12], with permission.

$$\frac{C/k_{2exp} - X}{C}(1 - k_w/k_{2exp}) = Y + ZC \qquad (18)$$

An example of this procedure is given in Fig. 3.2, which plots the experimental data reported in Ref. [13].

The assumptions behind Eqns. (8) and (9) and those derived from them are essentially correct, as substantiated by comparison of the kinetically derived binding constants with those obtained by independent methods. This important point is further discussed later.

In order to calculate the true rate constant, k_m, of the bimolecular reaction in the micellar phase, it is necessary to estimate the volume in which the reaction occurs, V. Since this model makes no distinction between the different regions of the micelle, the choice of V is somewhat arbitrary.

If it is assumed that the reaction occurs within the Stern layer, the value of V for SDS and HTAB ranges from 0.14 to 0.32 M^{-1}, depending on the model used to calculate the thickness of the Stern layer. The value of V would be somewhat larger if the total micellar volume is taken as the reaction volume.

Once the true reaction rate constant and the binding constant are determined, one can evaluate the factors responsible for the micellar effect on chemical reactivity. As already mentioned, the ratio k_m/k_w (e.g., Eqn. (12)) reflects the effects of the microenvironment on the reactivity.

These effects may include effects of polarity or dielectric constant on the rate as well as changes in transition state geometry or orientational effects on the reactants. One example of such an analysis is provided by the studies of micellar effect on the rate of acylation of imidazole derivatives by p-nitrophenyl heptanoate [14]. As can be seen in Fig. 3.3 for neutral nucleophiles, the true rate constant is some two orders of magnitude smaller in the micellar microphase than in water. This may be attributed to the lower dielectric constant and the weaker solvating ability of the micellar phase which reduces the probability of formation of the polar transition state as compared to the situation in water. On the other hand, the micellar medium increases the reaction rate of imidazole anions probably due to desolvation of the anion at the micellar interface [14].

Another important factor can be the location of the reactant in the micellar subphase. This effect is shown in Fig. 3.4, where the rate of

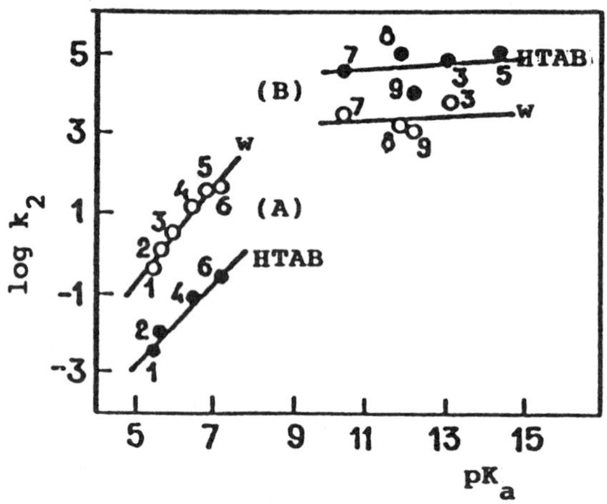

Fig. 3.3. True second-order rate constants (in M^{-1} min^{-1}) for acylation of a series of imidazole derivatives by p-nitrophenyl-heptanoate vs. pK_a values for the nucleophiles. (A) Neutral compounds; (B): Anions. w: bulk aqueous solution. (1) N-methylbenzimidazole; (2) N-phenylimidazole; (3) 4(5)-phenylimidazole; (4) N-benzylimidazole; (5) N-benzoyl-L-histidine; (6) N-heptylimidazole; (7) 5(6)-nitrobenzimidazole; (8) 4(5)-bromoimidazole; (9) benzimidazole. Taken with permission from Ref. [14].

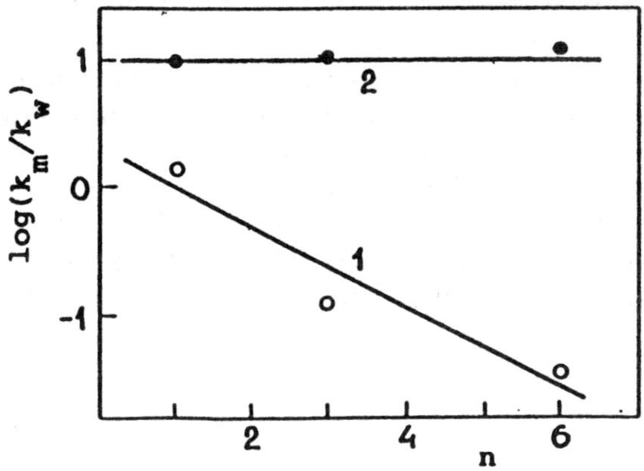

Fig. 3.4. Logarithmic ratios of the true second-order rate constants for acylation of m-bromobenzaldoxime (curve 1) and benzimidazole (curve 2) anions by p-nitrophenylcarboxylates in micelles (k_m) and in water (k_w) as a function of the number (n) of methylene groups in the molecule of the acylating agent, $H(CH_2)_nCOOC_6H_4NO_2$. Reproduced with permission from Ref. [12].

acylation of m-bromobenzaldoxime by p-nitrophenylcarboxylates in micellar solutions is plotted vs. the chain length of the acylating agents which consist of the series $H(CH_2)_nCOOC_6H_4NO_2$. This effect may be attributed to localization of the ester group deeper inside the micelle with increasing the chain length. For this reaction, it was found that addition of one methylene group increases the free energy of association by ca. 400 cal/mole. On the other hand, when the acylated reactant is benzimidazole anion, the rate constant is higher in the micellar solution than that in water and is independent of the chain length of the acylating reagent. This observation has been interpreted to indicate a reaction which takes place in a hydrophobic region of the micelle rather than at its surface [14,15].

The most important factor in the kinetic effects observed for many reactions in micellar systems is the local change in concentration in the micellar pseudophase. Quite often, this is the only reason for the catalytic micellar effect. In other words, the true rate constant in the micellar subphase is generally very similar to that measured in water and the only reason for substantial changes in the observed rates is the increase (or decrease) in local concentrations of the reactants in the micelle subphase.

When one or more of the reactants is a weak acid or base, the apparent shift in its pK_a in the presence of micelles also contributes to the overall kinetic effect. In these cases, the extent of the kinetic effect is determined by the ratio of the binding constants of the two acid/base species to the micelles. Basically, this is another form of concentration effect, yet intrinsic changes in pK_a may occur due to the different microenvironment. These effects are further discussed below.

The Berezin model can be successfully applied also to the effect of a micellar system on an equilibrium involving nonionic species. In these cases, the equilibrium constant for a bimolecular reaction of the type: $A + B \rightleftarrows AB$ in the presence of micelles can be expressed by the equation:

$$K_{exp} = K_w \frac{1 + K_{(AB)}C}{(1 + K_{(A)}C)(1 + K_{(B)}C)} \tag{19}$$

and the true equilibrium constant in the micellar pseudophase, K_m, is given by:

$$K_m = K_w \frac{P_{AB}}{P_A P_B} \tag{20}$$

An example of the effect of SDS on the equilibrium constant is shown in Fig. 3.5. for the reaction:

$$C_6H_5NH_2 + C_6H_5CH{=}O \rightleftarrows C_6H_5N{=}CHC_6H_5 + H_2O \tag{21}$$

Independent measurements of K_{exp} and the rate constants for the forward and back reaction yield practically the same values. When gel filtration is used to determine the binding constants, the mechanism leading to Eqns. (19) and (20) has been confirmed. Furthermore, it was observed that despite the increase in K_{exp}, the true equilibrium constant, K_m, is roughly equal to K_w. Then, the micellar effect is again mainly a concentration effect.

While the Berezin model has proven very useful for nonionic reactants and for reactions in nonionic surfactants, it has been found inadequate for reactions involving ionic species in charged micellar solutions. Although in principle it should be applicable to such systems, the effect of surface potentials on the partition coefficients of the ions should be considered (see Chapter 2, Eqn. (27)).

Even when the electrostatic contribution described by that equation is considered in further developing the model, it should be realized that

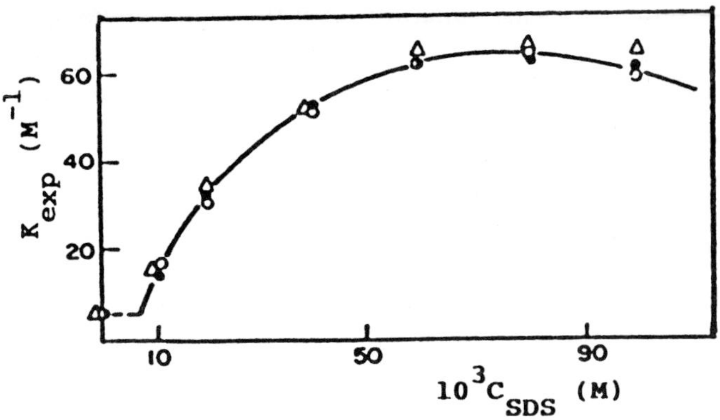

Fig. 3.5. Effect of SDS concentration on an experimental equilibrium constant for the formation of benzylideneaniline: (Δ) values calculated from equilibrium concentrations of benzylideneaniline; (O) from ratios of rate constants for forward and reverse reactions at pH7; (●) ratios at pH 10. Reproduced with permission from Ref. [16].

any specific ion–micelle interaction (which may depend on the nature of the ions) is neglected and, moreover, the dependence of the partition coefficient (P_I) on surfactant concentration is also neglected.

In order to extend the understanding of micellar catalysis to reactions of ionic species in the presence of ionic surfactants, Romsted proposed an ion exchange model for such systems [17]. In his model, Romsted retains the fundamental assumptions of the Berezin model, but the partitioning of the hydrophilic counterions is treated differently. The Stern layer is considered to be saturated by counterions. The degree of ionization, α, is assumed to be independent of surfactant concentration and ionic strength at constant temperature. The hydrophilic ionic reagent, X, and the nonreactive micellar counterions, Y, are assumed to partition between the two subphases according to the ion exchange equilibrium:

$$X_m + Y_m \rightleftarrows X_w + Y_m \qquad (22)$$

This equilibrium is assumed to be achieved rapidly as compared to the rate of the reaction of interest. The selectivity coefficient for the equilibrium is given by:

$$K_{X/Y} = \frac{[X_w][Y_m]}{[X_m][Y_w]} \qquad (23)$$

The observed second-order rate constant for a reaction between the ionic reactant X and a neutral reactant, A, in the presence of counterions Y, assumes the form:

$$k_{2exp} = \frac{k_m \beta S K_{(A)} C}{(1 + K_{(A)}C)([X]_t + K_{X/Y}[Y]_t)} + \frac{k_w}{(1 + K_{(A)}C)} \tag{24}$$

where β is the degree of counterion binding to the Stern layer (β = 1 − α) and S = 1/V. The subscript t indicates total concentrations.

An important consequence of the ion-exchange approach is that it is quite possible to obtain high X_m even when the total concentration of Y exceeds substantially that of X. A high value of $K_{X/Y}$ provides therefore the rationale for specific rate effects in micellar solutions. In order to simplify the mathematical treatment, it is often assumed that $[X]_t >> [X_m]CV$. This assumption limits the applicability of Eqn. (25) to low micelle concentrations and to situations in which the concentration of the reactive ion may be expressed by the total ion concentration. With this assumption, the model cannot therefore be applied to buffered solutions. Nevertheless, Eqn. (25) has successfully predicted the dependence of the observed second-order rate constant on the reactive ion concentration in various systems.

For example, Eqn. (24) predicts an inverse dependence of k_{2exp} on the concentration of the ions (see Refs. [18,19]). This dependence was, indeed, observed for the acid catalyzed hydrolysis of p-nitrobenzaldehyde diethylacetal in SDS solutions (see Fig. 3.6) [19]. When the surfactant concentration is high enough to bind all of the organic substrate, k_{2exp} (calculated in terms of H^+ activity rather than total H^+ concentration and measured by the hydrogen electrode) becomes constant at $k_{2exp} = k_{obs}/a_{H+}$. Equation (24) also predicts that if the ratio $[X]_t/[Y]_t$ is maintained constant, the increase in surfactant concentration will lead to a plateau in $k_{2exp.}$ This prediction has been experimentally verified [20,21].

The effect of added hydrophilic ions in light of the ion-exchange model can be studied also at high surfactant concentrations. Under these experimental conditions, it is assumed that the organic substrate, A, is quantitatively incorporated into the micellar subphase ($K_{(A)}C >> 1$) and the reaction rate constant in the water phase is negligibly small. Then Eqn. (24) simplifies and may be rearranged to:

$$\frac{1}{k_{2exp}} = \frac{[X]_t}{k_m \beta S} + \frac{K_{X/Y}}{k_m \beta S}[Y]_t \tag{25}$$

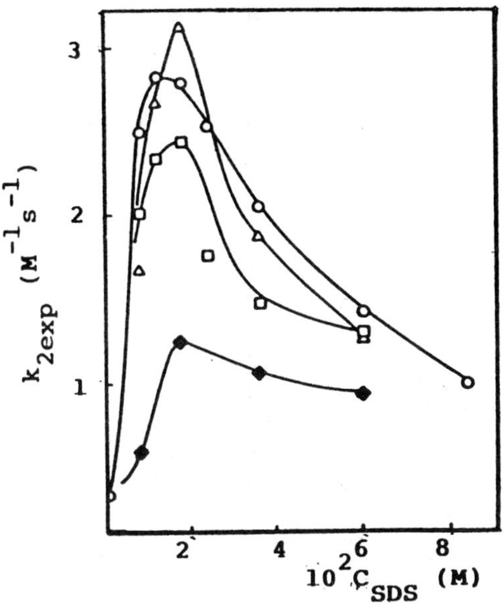

Fig. 3.6. Dependence of k_{2exp} on SDS for the acid-catalyzed hydrolysis of p-nitrobenzaldehyde diethylacetal, in dilute HCl at 25°C. (Δ) 1×10^{-3} M HCl; (O) 3.16×10^{-3} M HCl; (□) 1×10^{-2} M HCl; (♦) 3×10^{-2} M HCl. Reproduced with permission from Ref. [19].

If the values of $[X]_t$, β and S are known or assumed, then k_m and $K_{X/Y}$ can be experimentally determined [21]. An interesting example is the effect of HTAC or TTAC (tetradecyltrimethylammonium chloride) upon the reaction of OH^- with betaine esters [22]. In order to compare the behaviour of substrates having different degrees of hydrophobicity, the following esters were investigated:

$R{-}N^+(CH_3)_2CH_2COOMe$	$R = CH_3$	(Ia)
	$R = C_{12}H_{25}$	(Ib)
	$R = C_{16}H_{33}$	(Ic)

The ion-exchange equilibrium in these cases is:

$$OH^-_m + Y^-_w \rightleftarrows OH^-_w + Y^-_m \quad (26)$$

and from the pseudophase model one may express the first-order rate constant for the overall reaction as:

$$k_{1exp} = \frac{k_w[OH_w^-] + k_m K_{(A)} m_{OH}^S C}{1 + K_{(A)} C} \quad (27)$$

where $m_{OH}^S = [OH_m^-]/C$. (Note that k_{1exp} is often designated k.) Assuming that the ion-exchange constant, $K_{OH/Cl}$, and β are independent of the counterion concentration, Eqn. (23) combined with mass balance considerations yields

$$K_{OH/Cl} = \frac{([OH^-]_t - m_{OH}^S\, C)(\beta - m_{OH}^S)}{m_{OH}^S \{[Cl^-]_t - (\beta - m_{OH}^S)\}} \quad (28)$$

Equations 27 and 28 may be combined to simulate the dependence of k_{1exp} on surfactant or salt concentration. Whenever available experimental values, e.g. for k_w and $K_{(A)}$, were utilized, β was assumed to be in the range 0.75–0.80 and $K_{OH/Cl}$ in the range 3.5–5.0. The fit for the reaction of compound (Ia) is shown in Fig. 3.7(A). Since the reaction occurs primarily in the aqueous pseudophase, the inhibition observed in this figure results from binding of OH^- to the micelles. Addition of NaCl causes an increase in the rate due to the displacement of OH^- from the micellar surface. The agreement between this model and experimental results is further shown in Fig. 3.7(B) for ester (Ib).

The effect of HTAC or NaCl on the rate of the self-micellized ester (Ic) is shown in Fig. 3.7(C). Both HTAC and NaCl reduce the reaction rate, being NaCl more effective. OH^- and Cl^- compete for sites on the surface of the micellized substrate. Figure 3.7(C) also shows that the calculations are relatively insensitive to the exact values of the parameters k_m, $K_{OH/Cl}$ and β. This uncertainty in the values of the above parameters, therefore, limits the usefulness of the model. A further complication may be introduced by neglect of the role of monomeric free substrates.

The assumption that β is independent of ionic concentrations implies that the micelles are effectively saturated with counterions. Measurements of the micellar charge using quasielastic light scattering suggest that the charge decreases upon addition of salts. Furthermore, counterions differ in their affinities to the micellar surface and as a consequence, β may be expected to depend on the nature of the counterions. The situation is further complicated by the fact that different experimental techniques provide varying values for β.

The problem of determining the distribution of small hydrophilic ions between the aqueous and micellar subphases is simplified when a

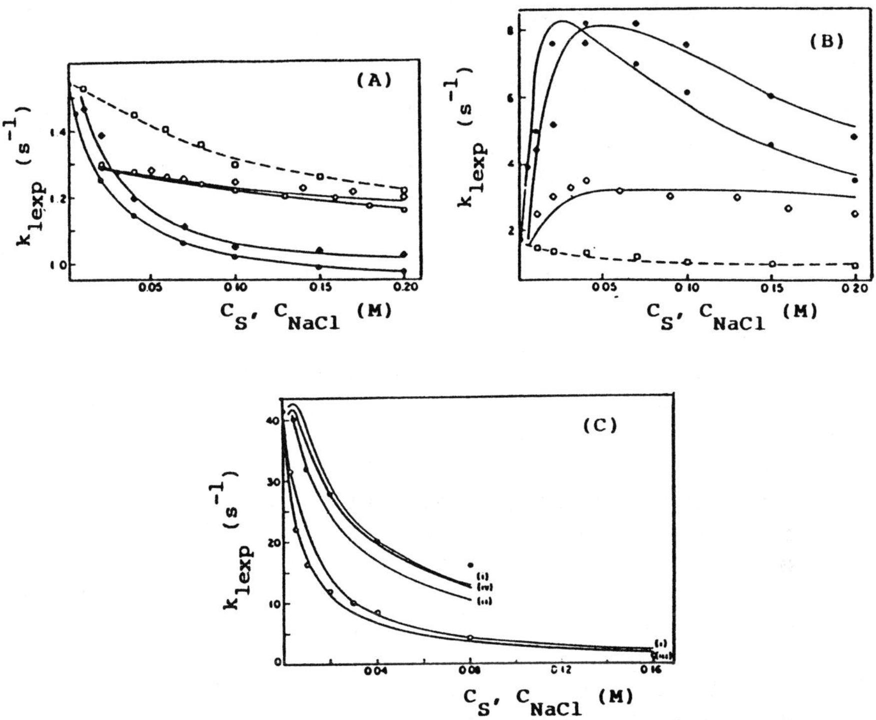

Fig. 3.7. Rate of OH^- reaction with betaine esters in 0.02 M NaOH and 0.003 M ester; solid lines are predicted. (A) ester (Ia): (♦) in TTAC; (□) in TTAC + 0.1 M NaCl; (●) in HTAC; (O) in HTAC + 0.1 M NaCl; (□, dotted line) in NaCl. (B): ester (Ib), symbols as in (A). (C) ester (Ic): (●) in HTAC; (O) in NaCl. Lines are predicted using for (i), (ii) and (iii) rate constants for the reaction of the self micellized ester: 105, 95 and 85 s^{-1} respectively, $K_{OH/Cl} = 4$ and $\beta = 0.15$; (iv): as (iii) with $K_{OH/Cl} = 3.5$ and $\beta = 0.8$. Reproduced with permission from Ref. [22].

reactive counterion surfactant is used (in the absence of inert counterions). In this case, since $m_X^S = \beta$ and $[X_w^-] = \alpha C$, the equation corresponding to Eqn. (27) is:

$$K_{1exp} = \frac{k_w[X_w^-] + k_m \beta K_{(A)} C}{1 + K_{(A)} C} \tag{29}$$

Even simpler is the situation when $K_{(A)}C >> 1$ and the reaction in the aqueous phase is negligibly slow since then:

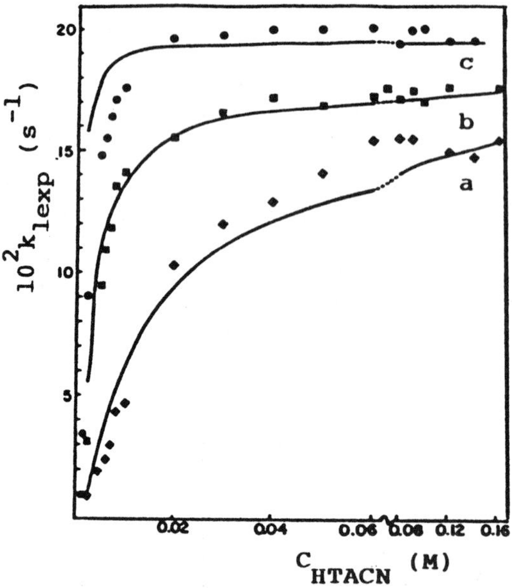

Fig. 3.8. Dependence of k_{1exp} on HTACN for the addition of CN^- to N-alkylpyridinium. Curves (a), (b) and (c) for R = $C_{12}H_{25}$, $C_{14}H_{29}$ and $C_{16}H_{33}$, respectively. Reproduced with permission from Ref. [23].

$$k_{1exp} = \beta k_m \tag{30}$$

The reaction rate constants approach asymptotic values. This prediction is supported by kinetic studies of reactions such as addition of cyanide to N-alkylpyridinium ions in HTACN (hexadecyltrimethylammonium cyanide) micelles [23], nucleophilic attack of bromide on picryl derivatives in HTAB micelles [24], and hydrolysis of acetals in micellized sulphonic acids [25].

An example of such observations is given in Fig. 3.8 for the cyanide reaction. The solid curves in the figure were calculated using Eqn. (30) for which the binding constants were experimentally determined by ultrafiltration. The values of βk_m were found to be virtually independent of the chain length of the alkylpyridinium ion.

With the usual assumptions concerning β and the reaction volume, the second-order rate constant, expressed in terms of moles of reactants per litre (of Stern layer) was essentially the same as the rate constant for the reaction of short chain alkyl pyridinium with NaCN in water. The kinetic micellar effect in this case is, once again, primarily a concentration effect.

While the model discussed above adequately describes the kinetic micellar effect for micelles functionalized with some counterions, it fails when the functional counterion is OH^- or F^- [26]. In this case, the first-order rate constant continues to increase with surfactant concentration even when the substrate is completely bound to the micelles. This behaviour has been observed for the reaction of OH^- with 2,4-dinitrochlorobenzene (DNCB), 2,4-dinitrochloronaphthalene (DNCN) and p-nitrophenyldiphenyl phosphate (pNPDPP) micellized in CTAOH [27]. Figure 3.9 shows the dependence of k_{1exp} on HTAOH with and without added excess of OH^-. Since $K_{M(A)}$ for the organic substrate DNCN is ca. 1600 M^{-1}, the latter should be extensively bound to the micelle at [CTAOH] > 10^{-2} M. Yet k_{1exp} monotonically increases when [HTAOH] increases. The addition of excess OH^- further increases the reaction rate.

Several explanations were suggested for the above described discrepancies, but none of them is completely satisfactory. One addresses the assumption that the reactive counterion is distributed between the aqueous phase and the micellar Stern layer. This assumption makes no distinction between ions in the aqueous pseudophase and ions in the diffuse Gouy–Chapman layer. While in the latter region, the concentration of counterions is higher than in the bulk phase, the distribution of ions in this layer is governed by Coulombic nonspecific interactions. For ions such as OH^- or F^-, which bind only weakly to the Stern layer, neglecting the diffuse layer may introduce a serious inaccuracy. At high surfactant concentrations, this interpretation may be expressed by Eqn. (31), where k_{mw} accounts for the reaction pathway involving ions in the diffuse layer and substrate in the micelles:

$$k_{1exp} = \alpha k_{mw} C + k_m(1 - \alpha) \qquad (31)$$

The first evidence for an additional pathway was observed for the OH^- catalyzed dehydrochlorination of 1,1,1-trichloro-2,2-bis(p-chlorophenyl) ethane and some of its derivatives in HTAB [28]. The same approach was, however, also applied to the reaction of DNCB and DNCN in micelles of p-C_8H_{17}O-$C_6H_4CH_2N^+(CH_3)_3OH^-$ [24]. In spite of its successful application to several systems, it is not clear why it is limited only to OH^- and F^- ions.

In fact, in a realistic system, one would expect a continuous distribution of rate constants (concentrations, if the effect is purely a concentration effect) on moving away from the micellar surface to the bulk aqueous

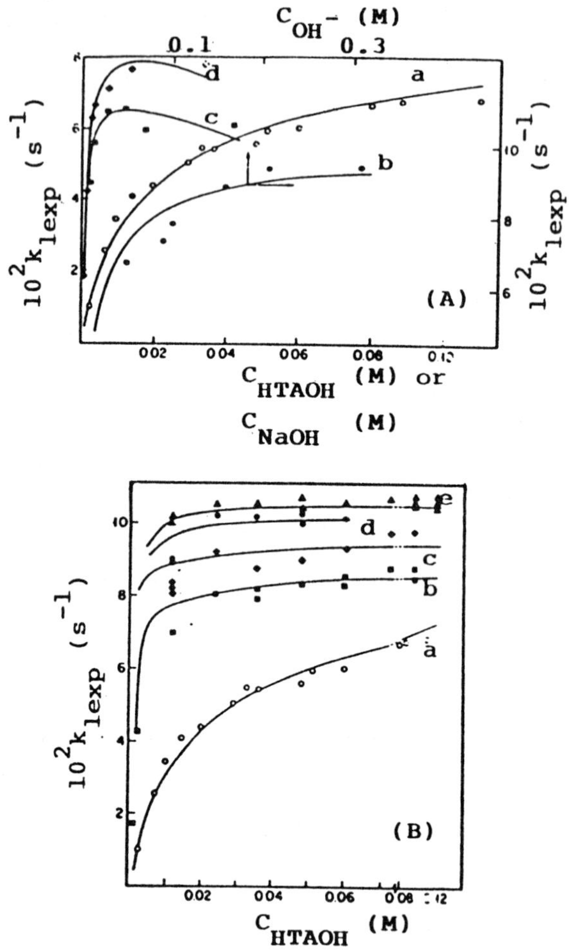

Fig. 3.9. Rate of the reaction of 2,4-dinitrochloronaphthalene with OH^- in HTAOH. (A) with added NaOH. (a) no excess NaOH; (b) 10^{-2} M HTAOH and added NaOH; (c), (d): at 0.05 M and 0.1 M $[OH^-]_t$, respectively. (B) with added KOH. (a) no excess KOH; (b) 0.05 M KOH; (c) 0.1 M KOH; (d) 0.25 M KOH; (e) 0.5 M KOH. Taken from Ref. [27], reproduced with permission.

subphase. The model discussed above assumes such three-rate constants (k_m, k_{mw}, k_w).

Another approach is to assume that β, and thus concentration of bound counterions, increases with increasing surfactant and counterion total concentrations. This is, in fact, a "site model" where the concentration of bound counterions increases with the concentration of free sites. The binding constant for counterions may then be expressed as:

$$K_{(X)} = [X_m^-]\left\{[X_w^-](C - [X_m^-])\right\}^{-1} \qquad (32)$$

Combined with mass balance consideration, this leads to a quadratic equation in $[X_m^-]$:

$$K_{(X)}[X_m^-]^2 - (K_{(X)}C + K_{(X)}[X^-]_t + 1)[X_m^-] + K_{(X)}C[X^-]_t = 0 \qquad (33)$$

Equation (33) can be used to calculate the concentrations of the reactive counterions in the aqueous and micellar subphases from the known C and $[X^-]_t$ for various assumed values of the ionic binding constant $K_{(X)}$. The solid curves of Fig. 3.9 were calculated taking $K_{(OH)}$ = 55 M^{-1}. The fit to the experimental results seems quite satisfactory, although its physical implications are less obvious. This approach actually disagrees with the basic assumption of the ion-exchange model which requires constant charge neutralization. Equation (32) puts no restrictions on the value of β within the formal limits of 0 to 1.

Figure 3.10 predicts the value of β as a function of surfactant concentration for various values of $K_{(X)}$. For a strongly bound anion with $K_{(X)}$ = 2000 M^{-1}, β would be practically constant at ca. 0.85 over a wide concentration range. Since $K_{OH/X} = K_{(X)}/K_{(OH)}$ one could estimate for Br^- that $K_{(Br)} \approx 10^3$ M^{-1} (using $K_{(OH)}$ = 55 M^{-1} and the published values of $K_{OH/Br}$ = 10 to 40). Thus, for HTAB in the presence or absence of NaBr, β would be expected to be nearly constant. The nucleophilic attack of Cl^- in HTAC or Br^- in HTAB on methylbenzenesulphate can be interpreted in terms of constant β. The experimental data for this reaction in HTAC + NaCl solution, however, could not be fitted by the simple model.

Instead, Eqn. (33) and $K_{(Cl)}$ = 220 M^{-1} had to be invoked. This value seems reasonable in comparison with $K_{(OH)}$ = 55 M^{-1}, $K_{(F)}$ = 40 M^{-1}, $K_{(X)}$ = 80 M^{-1} for acetate and formate (A20) and $K_{(Br)}$ = 2300 M^{-1}. That the degree of ionization, α, changes on increasing $[OH^-]$ has also been verified for a reaction which occurs exclusively in the aqueous pseudophase [29].

Another possible explanation for the increase in β for OH^- or F^- functionalized micelles may arise from an increase in the micelle size from n-mers to the normal micelle size. This change would occur over a wide $[OH^-]$ or $[F^-]$ range, but over a much smaller range of concentrations of stronger bound anions such as Br^- or CN^-. Structural changes, reflected in the unusual behaviour of hydroxide surfactants have been reported [30].

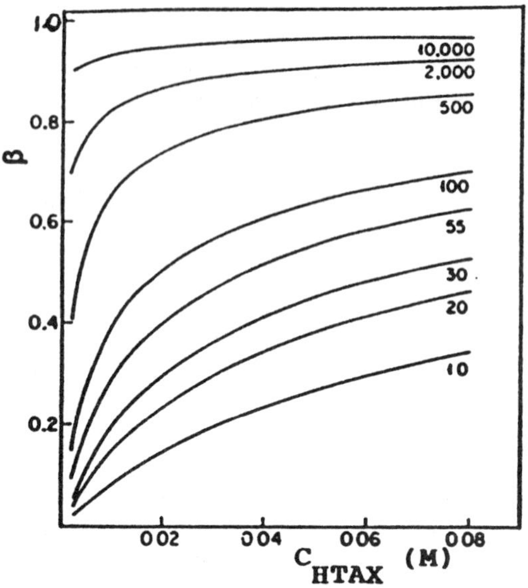

Fig. 3.10. Predicted variation of charge neutralization of head groups, with surfactant concentration and counterion binding constant $K_{(X)}$. Reproduced with permission from Ref. [27].

As noted above, Romsted's model can be applied to situations in which the reactive ion concentration may be expressed in terms of the total ion concentration. This restriction excludes reactivity studies in buffered micellar solutions. Modified ion-exchange models have been developed in which the relevant equations are expressed in terms of the concentration of the hydrophilic ions where the latter are either independently determined or calculated from micellar surface potentials.

The ion-exchange model developed by Quina and Chaimovich [31] is based on the following assumptions:

(i) The distribution of micellar aggregate sizes can be presented in terms of the most probable aggregation number, N_{AG}.

(ii) Ion–ion and ion–head group interactions are noncooperative.

(iii) The degree of ionization, α, of the individual micellar species are all equal to one another.

(iv) Ion–ion exchange rates are rapid relative to the reaction lifetime.

(v) Activities of micellar and ionic species can be expressed as concentrations.

All of the above assumptions refer to any given set of conditions. They therefore allow for changes in $K_{X/Y}$ as experimental conditions vary.

With these assumptions and the appropriate mass balance equations, for a micellar solution containing analytical concentrations of surfactant, ($[D^+Y^-]_t$) and a reactive counterion ($[B^+X^-]_t$), with or without added common salt ($[B^+Y^-]_t$), $K_{X/Y}$ assumes the form:

$$K_{X/Y} = \frac{[X_m^-]}{[X^-]_t - [X_m^-]} \; \frac{\alpha C + \text{cmc} + [X_m^-] + [B^+X^-]_t}{(1-\alpha)C - [X_m^-]} \tag{34}$$

At high surfactant concentration:

$$\lim_{C \to \infty} K_{X/Y} = \frac{[X_m^-]}{[X_w^-]} \; \frac{\alpha}{(1-\alpha)} \tag{35}$$

Equation (35) predicts that $[X_m^-]/[X_w^-]$ will approach a limiting value and provides a way to estimate $K_{X/Y}$. Equation (35) may be solved for $[X_m^-]$ to give:

$$[X_m^-] = \frac{\left\{-(A_1 + [B^+Y^-]_t) + [(A_1 + [B^+Y^-]_t)^2 + 4(1 - K_{X/Y})[X^-]_t K_{X/Y}(1-\alpha)C]^{1/2}\right\}}{2(1 - K_{X/Y})} \tag{36}$$

where:

$$A_1 = \alpha C + \text{cmc} + [X^-]_t K_{X/Y} + (1-\alpha)K_{X/Y} C \tag{37}$$

If the system is buffered towards the reactive counterion, X^-, $[X_w^-]$ remains constant. Combined with the assumption of non-interacting components of the buffer, this leads to:

$$[X_m^-] = \{-(A_2 + [B^+Y^-]_t) + [(A_2 + [B^+Y^-]_t)^2 + 4[X_w^-]K_{X/Y}(1-\alpha)C]^{1/2}\}/2 \tag{38}$$

where:

$$A_2 = \alpha C + \text{cmc} + [X_w^- w]K_{X/Y} \tag{39}$$

The dependence of $[X_m^-]$ on C, as revealed by Eqns. (38) and (39), shows that buffering of the intermicellar aqueous phase does not imply buffering of the exchangeable reactive species at the micellar surface.

According to this model, for a reaction of a univalent ionic substrate S^- in an oppositely charged micelle D^+Y^-, the rate constant is given by:

$$k_{1exp} = \frac{k_w + k_m K_{S/Y}([X_m^-]/[X_w^-])}{K_{X/Y}([X_m^-]/X_w^-])} \quad (40)$$

Comparing Eqns. (2) and (40), it is clear that the substrate partitioning dependence in the classical model is replaced by the selectivity of the micelle-bound ions in this model. Similarly, for a bimolecular reaction between an uncharged substrate A and a reactive univalent counterion X^- in an unbuffered system, the rate constant follows the equation:

$$k_{2exp} = \frac{[Y^-]_t \{k_w + (k_m/V)K_{(A)}K_{X/Y}([Y_m^-] / [Y_w^-])\}}{(1 + K_{(A)}C)\{1 + K_{X/Y}([Y_m^-] / [Y_w^-])\}} \quad (41)$$

and for the corresponding buffered system:

$$k_{2exp} = \frac{[Y_w^-] \{k_w + (k_m/\overline{V})K_{(A)}K_{X/Y}([Y_m^-] / [Y_w^-])\}}{(1 + K_{(A)}C)} \quad (42)$$

A first test of the validity of these equations was the kinetic study of the alkaline hydrolysis of N-methyl-4-cyanopyridinium ions (4-MCP) in cationic HTAB and TTAC micelles [32]. Since the substrate in these cases resides exclusively in the intermicellar aqueous phase, the pseudo first-order rate constant is proportional to $[OH_w^-]$. The experimental results can be fitted using Eqns. (37) and (38) ($[B^+Y^-]_t = 0$), as shown in Fig. 3.11 with the values specified in the captions to this figure. The value obtained for $K_{OH/Cl}$ was successfully used in the analysis of the alkaline hydrolysis of 3,5-dinitroanisole in TTAC micelles in the presence of excess Cl^- [33].

The validity of the same approach to buffered solutions has been verified by studies [34] of: (i) micellar excluded substrates such as 4-MCP and 2-MCP; (ii) strongly bound substrates such as p-nitrophenyloctanoate (NPO); (iii) partitioned substrates such as p-nitrophenylacetate (NPA). All these solutes were dissolved in borate buffered HTAB micelles. Figure 3.12(A) shows that the rate constant for the reaction described in item (i) above is independent of [HTAB]. One may conclude, therefore, that the intermicellar bulk pH in the aqueous phase is maintained constant by 0.020 M borate buffer at least up to 0.10 M HTAB.

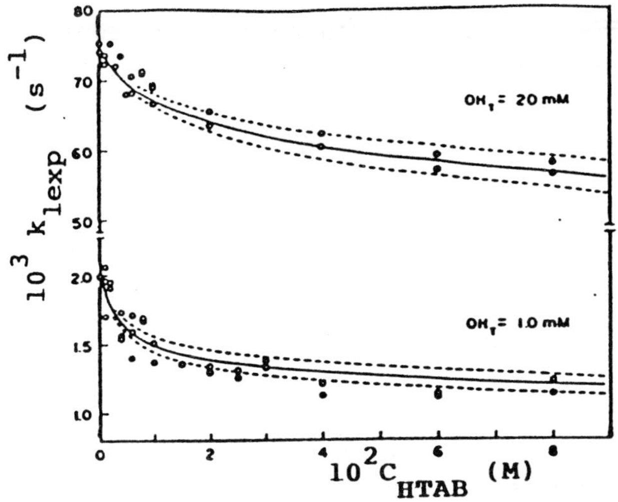

Fig. 3.11. Effect of [HTAB] on the rate of the alkaline hydrolysis of N-methyl-4-cyanopyridinium ion. Curves were calculated using: (a) $[OH^-]_t$ = 0.02 M: k_w^0 = 0.0985 s^{-1}, α = 0.20, cmc = 6×10^{-4} M, B = 1.50; the solid line is obtained using $K_{OH/Br}$ = 0.080; the upper and lower dashed curves represent $K_{OH/Br}$ = 0.060 and 0.10, respectively. (b) $[OH^-]_t$ = 0.001 M: k_w^0 = 2.15×10^{-3} s^{-1}; B = 1.50, α = 0.20, cmc = 7.5×10^{-3} M. The curves were calculated using the $K_{OH/Br}$ values as above. Reproduced with permission from Ref. [32]. $k_w^0 = k_w\ 10^{[1.02\,\mu^{1/2}/(1+B\mu^{1/2})]}$.

Situation (ii) is described in Fig. 3.12(B). When the concentration of HTAB exceeds ca. three times the cmc, more than 95% NPO is incorporated into the micelles. The observed rate constant under such conditions is given by:

$$k_{1exp} = k_m\ [OH_m^-] \tag{43}$$

and thus should reflect $[OH_m^-]$ at [HTAB] 2×10^{-3} M. As can be seen in Fig. 3.12(B), k_{1exp} decreases upon increasing surfactant concentration; a direct indication that $[OH_m^-]$ decreases in a similar manner, even though the borate buffer maintains $[OH_w^-]$ constant. Quantitatively, however, Eqns. (38), (39) and (43) could not satisfactorily describe the data of Figs. 3.12(B,C). This failure, though, can be amended using modified mass balance ion-exchange equations which take into account binding of borate ions to the micelle.

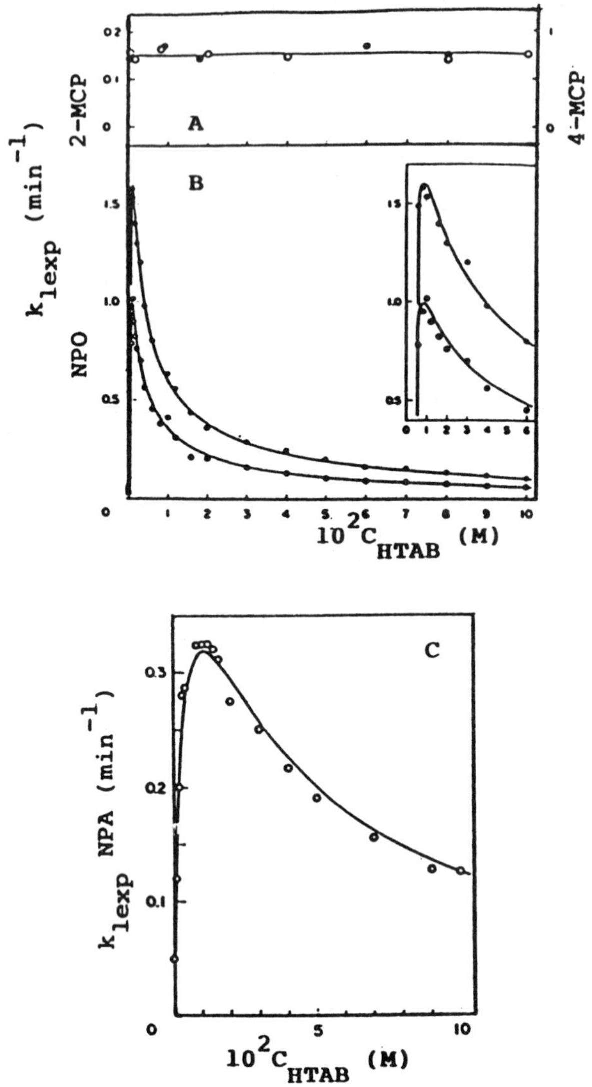

Fig. 3.12: Effect of [HTAB] on observed rate constant for alkaline hydrolysis in the presence of 0.020 M borate buffer: (A) 2-MCP (O) at pH 9.50 and 4-MCP (●) at pH 9.80 (the reported k_{1exp} values are multiplied by 10^2); (B) NPO at pH 9.5 (O) and pH 9.80 (●). The corresponding data for low [HTAB] are shown in the inset (the concentration values are in $Mx10^3$); (C) NPA at pH 9.50. The curve is calculated. Taken from Ref. [34], reproduced with permission.

Introducing ion-exchange selectivity coefficients for borate (denoted B), $K_{OH/B}$ and $K_{B/Br}$, which are interrelated through $K_{B/Br} = K_{OH/Br}/K_{OH/B}$, one obtains the mass balance equation:

$$[OH_m^-]^3 + L[PH_m^-]^2 + n[OH_m^-][OH_w^-] - P[OH_w^-]^2 = 0 \quad (44)$$

where the terms L, N and P correspond to:

$$L = A_2 + K_{OH/B}[OH_w^-] + [B]_t \quad (45)$$

$$N = A_2K_{OH/B} + K_{OH/Br}\{[B]t - (1 - \alpha)C\} \quad (46)$$

$$P = K_{OH/Br}(1 - \alpha)CK_{OH/B} \quad (47)$$

The fit to the experimental values (see Fig. 3.12B,C) was obtained with $k_m/k_w = 0.37$ for NPA and 0.29 for NPO, and $K_{OH/B} = 2.3$. From this last value $k_{B/Br} \approx 0.035$ is estimated. This small value indicates an inefficient binding of borate to HTAB, as was also predicted from other considerations [35].

The applicability of the ion-exchange model to polyvalent ions has been considered for the following reaction occurring in positively charged micelles (HTAB or hexadecyltrimethylammonium thiosulphate, HTAT [36]):

$$S_2O_3^= + CH_3(CH_2)_3I \rightarrow CH_3(CH_2)_3S_2O_3^- + I^- \quad (48)$$

Bromide ions can displace thiosulphate ions from the micellar surface, and this exchange reaction has been verified spectroscopically. The following equilibrium can then be considered:

$$S_2O_{3m}^= + 2\,Br_w^- \rightleftarrows S_2O_{3w}^= + 2\,Br_m^- \quad (49)$$

From Eqn. (49), the corresponding ion-exchange equation holds:

$$K_{T/Br} = \frac{[T_m][Br_w^-]^?}{[T_w][Br_m^-]^2} \quad (50)$$

where T represents thiosulphate.

The micellar effect on the spectral properties allowed determination of $K_{T/Br}$. The pseudo first-order rate constant in HTAT can be described

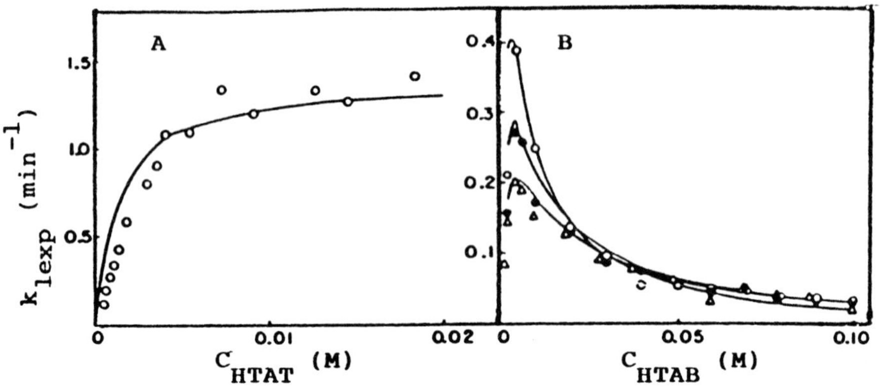

Fig. 3.13. Effect of HTAT and HTAB on reaction (48). [n-butyl iodide] = 10^{-4} M. Lines are calculated using $k_w^0 = 0.146\ M^{-1}\ min^{-1}$ and $K_{(A)} = 450\ M^{-1}$. (A) Effect of HTAT. (B) Effect of HTAB, $[Na_2S_2O_3] = 1.57\times10^{-3}$ M; (O): [NaBr] = 0 M, $K_{T/Br} = 0.025$; (●): [NaBr] = 4.9×10^{-3} M, $K_{T/Br} = 0.008$; (Δ): [NaBr] = 9.75×10^{-3} M, $K_{T/Br} = 0.011$. Reproduced with permission from Ref. [36].

according to Eqn. (31). A value for $\alpha = 0.15$ was estimated conductometrically, and $K_{(A)} = 450\ M^{-1}$ was obtained spectrophotometrically. The effect of HTAB can be described by the ion-exchange model, taking into account the equilibrium Eqn. (50), as follows:

$$k_{1exp} = [T]_t \frac{k_w + (k_m/V)K_{(A)}K_{T/Br}([Br_m^-] / [Br_w^-])^2 / C\overline{V}}{(1 + K_{(A)}C)\{1 + K_{T/Br}([Br_m^-] / [Br_w^-])^2 / C\overline{V}\}} \tag{51}$$

Figure 3.13 shows the fit of the experimental results to Eqns. (29) and (51). The same value of k_m fits both systems, indicating that the reactivity of thiosulphate is identical in HTAT and HTAB. The value of k_m is approximately eight times larger than k_w, probably due to the change in the dielectric constant of the micellar medium.

Despite its success [7,37–39], the pseudophase model has major weakness in some of the assumptions listed above. The failure of the applicability in solutions containing moderately concentrated electrolytes [27,40–44] led to the abandoning of the concept of a constant β and to describing the ionic distribution in terms of equations with the form of Langmuir or Volmer isotherms.

Nonspecific, coulombic and specific ion–micelle interactions have been considered, and the Poisson–Boltzmann equation has been solved with inclusion of a term accounting for specific interactions. The frac-

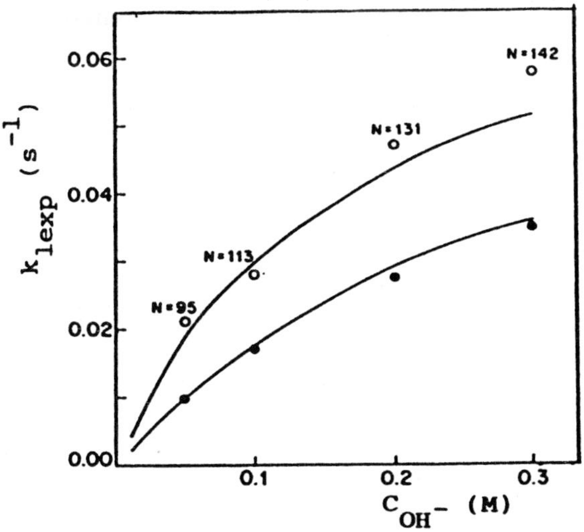

Fig. 3.14. Reaction of DDT with OH^- in 0.02 (O) and 0.06 (●) M HTAB, respectively. The fit was performed by assuming: $\Delta = 2.4$ Å, $\sigma = 0.4$ Å, $a = 24$ Å, and $N_{AG} = 173 + 60 \log [OH^-]$. Reproduced with permission from Ref. [42].

tional coverage, f, (1 – f describes the neutralization of micellar head-groups) is derived in terms of the Volmer isotherm as:

$$f = \frac{\delta \exp\{-f/(1-f)\}[X_w]}{1 + \delta \exp\{-f/(1-f)\}[X_w]} \tag{52}$$

where δ is a specific parameter related to the noncoulombic affinity of X for the micellized cationic surfactant. Values of δ will be low for hydrophilic ions having high charge density, whose interactions with the micelle are largely coulombic, and will be larger for polarizable, low charge density ions. Assuming $\delta = 0$ for the very hydrophilic ion OH^-, and $\delta = 15$ and 120 M^{-1} for Cl^- and Br^-, respectively, satisfactory agreement has been observed in different studies performed in HTAC and HTAB micellar media [42,45–47].

For example, in the reaction of OH^- with p-nitrophenyldiphenylphosphate (dephosphorilation), with 2,4-dinitrochloronaphthalene (aromatic nucleophilic substitution) and with 1,1,1-trichloro-2,2-bis(p-chlorophenyl)ethane (DDT) (elimination, see Fig. 3.14) the experimental data showed a good fit with the model. The fits deteriorate when the reactive ion, e.g. OH^-, is at concentrations higher than 0.1 M. Improvements are however obtained introducing a roughness concept of the micellar surface [42].

The combination of Eqn. (52) with the Poisson–Boltzmann equation (see Chapter 1, Eqn. (29)), with the appropriate boundary conditions at the micellar surface and the cell wall (for the cell model, see the next section) leads to:

$$d\phi/dr = (1 - f)\, N_{AG} e^2/\varepsilon a^2 kT \tag{53}$$

where ϕ is the reduced micellar potential, $\phi = e\psi/kT$, ε is the dielectric constant of water and a is the radius of the charged micellar surface.

The first-order rate constant is expressed as:

$$k_{1exp} = k_{\psi} = \frac{k_w[OH_w^-] + k_m K_{(S)}([HTAX] - cmc)[OH_{\Delta}^-]}{1 + K_{(S)}([HTAX] - cmc)} \tag{54}$$

where the subscript Δ indicates the thickness of the shell in which the reaction occurs.

The above derived equations are valid assuming that the roughness of the micellar surface is given by a Gaussian distribution, with a coefficient σ, and that N_{AG} is a function of the concentration of reactive ions. Computer simulations gave satisfactory agreement, as can be seen in Fig. 3.14.

In this last version of the pseudophase ion-exchange model, the inclusion of the Poisson–Boltzmann approach involves the assumption that the increase in electrolyte concentration will decrease micellar head-groups repulsion and therefore increase micellar surface charge density. Correspondingly, the increase in N_{AG} upon addition of electrolytes is taken into account.

Moreover, a distinction between coulombic nonspecific and specific micelle–ion interactions is introduced and the comparison between k_w and k_m depends upon the thickness of the reactive shell, Δ. Some of the above introduced concepts will be also present in the models described in the following sections.

3.2. ELECTROSTATIC MODELS

Following the principles reported by Berezin [1] and Romsted [17] it was shown to be possible to calculate the rate of the reaction at the micellar surface, with the introduction of the surface potential. The pseudophase ion-exchange (PIE) model was originally developed for the quantitative prediction of micellar effects upon the reaction rates and to overcome the limited predictability of the pseudophase model when charged reactive species are involved.

Almgren and Rydholm [48] defined the ion binding coefficients, β_i, by:

$$\beta_i = \theta_i /\{[i_w](1 - \theta)\} \tag{55}$$

where $\theta = \Sigma_i \theta_i$, θ_i being the ratio of bound ions of type i to the number of head groups of the micelle. It was shown that β_i is not a true constant, but that the ratio $\theta_i/([i_m]\ (1 - \theta)$ should be constant. Through an expression relating the concentration outside the Stern layer with that in bulk solution:

$$\beta_i = \beta_i^o \exp\ (e\psi/kT) \tag{56}$$

it follows:

$$\theta_i = \beta_i^o(1 - \theta)[i_w] \exp\ (e\psi/kT) \tag{57}$$

Funasaki [49] used the similar expression:

$$[Y_m] = [Y_w]\ PP_Y^o \exp(-Z_Y F\psi/RT) \tag{58}$$

for the calculation of concentration at the micellar surface, P_Y^o being the distribution coefficient at surface potential $\psi = 0$.

It was shown by Chaimovich et al. [50] that Eqns. (57) and (58) are formally identical to an equation derived from an expression of θ_i written as

$$\theta_i = \frac{N_i \exp\ (\phi_i /RT - Z_i F_\psi /RT)}{1 + \Sigma_k N_k \exp\ (\phi_k /RT - Z_k F_\psi /RT)} \tag{59}$$

where N_i is the mole fraction of counterion i in the aqueous phase; ϕ_i is the nonelectrostatic component of the Stern adsorption potential for counterion i, and k are all the counterion species present.

Thus for any pair i and j of univalent counterions:

$$\frac{\theta_i}{\theta_j} = \frac{N_i}{N_j} \exp\ (\Delta\phi_j /RT) \tag{60}$$

an expression which is equivalent to:

$$K_{i/j} = \exp(\Delta\phi_{ij}/RT) = \frac{[i_m][j_w]}{[j_w][j_m]} \tag{61}$$

where $\Delta\phi_{ij} = \theta_i - \theta_j$. Furthermore, using Eqns. (60) and (61) it can readily be shown that:

$$\theta_i = (1-\theta)\, N_i \exp(\phi_i/RT - Z_iF/RT) \tag{62}$$

which is identical to Eqns. (57) and (58).

Since Eqns. (57) and (58) relate the micellar surface concentration with the surface potential, the consistency of these expressions in the change of this parameter can be checked. Since a major factor determining ψ is the ionic strength, the Gouy–Chapman theory allows us to relate ψ with the free counterion concentration as:

$$\psi = 2.303\, RT\, (-\delta \log \Sigma_k [k_w] + \log\gamma)/F \tag{63}$$

where δ and γ are constants.

Almgren and Rydholm found that β values increase with increasing ionic strength, as can be expected, but β varied also when the ionic strength in the bulk phase is constant. The results were compatible with the assumptions that the variations of ψ are entirely due to the variation of θ (Romsted assumes that θ is constant) or that the variations in ψ are due to other factors and θ is constant (different properties of counterions induce change in the micelle size, surface, charge density and water penetration, which are factors determining ψ). It was suggested that the real situation is probably something between these extremes.

It was also shown that Eqn. (61) describes the effect of added common salt on the pK_{app} of totally bound acid–base indicators in the presence of HTAB [51,52]. However in the presence of added foreign salts (Cl^-, F^-) the values of ψ were found to be dependent on the nature of the added salt.

Starting from Eqn. (62) it can in fact be shown [50] that:

$$\frac{ZF_\psi}{2.303\, RT} = \log \frac{(1-\theta)}{\theta} + \log \Sigma_k N_k \exp(\phi_k/RT) \tag{64}$$

This equation reduces to the same form as Eqn. (63) only when: (i) there is but one counterionic species bound to the micelle, or (ii) the ϕ_k are equal for all counterions k. Condition (i) corresponds to the Romsted limit in the presence of added common salt. When $\phi_k = 0$ for all k, the description is that of Gouy–Chapman. An alternate form of condition (ii), i.e. equal ϕ_k values but nonzero, implies via Eqn. (62) that all the selectivity coefficients are unity. Various experimental evidence has been reported showing that these coefficients are in general not unity.

It can be concluded that these electrostatic models and the ion-exchange model, since they were developed under similar assumptions and conditions, are quite similar and the descriptions of kinetic data concerning basic hydrolysis of p-nitrophenylacetate [48] or p-nitrophenylbutyrate [52], as well as acid dissociation constants [52] are comparable.

The characterization of the electrostatic environment of a charged organized structure and the related ion distributions in the vicinity of the charged aggregate were analyzed using a potential approach based on a generalization of the Poisson–Boltzmann theory [53,54]. The basic ideas of the Poisson–Boltzman theory are properly generalized to take into account a spatially-varying permittivity, writing:

$$\varepsilon = \varepsilon(r) = \varepsilon_0 D_0 \, f(r) \tag{65}$$

where ε_0 is the permittivity in free space, D_0 is the bulk dielectric constant and f(r) is a short-range function which specifies the dependence of the function $\varepsilon(r)$ on the radial coordinate r.

The electrostatic potential is considered to be generated by a charged organizate of a given net charge. For example in the simulation for the reaction of hydrated electron with pyrene solubilized in HTAB [55] two different values for HTAB charge were used, namely 15 and 30; the results obtained were essentially the same [56] and this was considered as a proof for the essential correctness of an approach based on the full nonlinear Poisson–Boltzmann equation rather than its linearization. Also the choice of different models of dielectic saturation was found to be of less importance than retaining the full functionality between the electrostatic potential and the potential of average force as linked in the nonlinear Poisson–Boltzmann (PB) equation.

Regardless of the model so far described, the phenomenon referred to as micellar catalysis seems to be a consequence of an electrostatic effect (for ionic reactions) which influences a relatively large volume of solution and sweeps oppositely charged species to the surface of the aggregate. Obviously, for organized assemblies and reacting species of like sign, the rate is depressed as consequence of an exclusion volume.

An electrostatic description of micellar systems based on a modified nonlinearized PB equation was proposed by Frahm and Diekmann [57]. A numerical method was described with two modifications: (i) the influence of finite ion sizes on the charge distribution and (ii) the dependence of the dielectric constant on the electric field strength. The nonlinearized PB equation is then solved for spherical geometry, becoming:

$$\frac{1}{r^2}\frac{\partial}{\partial r}\left\{r^2\,\varepsilon\,\frac{\partial}{\partial r}\,\psi\right\} = -\,\rho/\varepsilon_o \tag{66}$$

A quantitative measure for the influence of the two modifications upon the solution of the nonlinearized PB equation was obtained by performing calculations on surface concentration of 2-2 electrolyte.

Good agreement is observed for small surface potential ($\psi \leq 100$ mV) and small concentrations of 2–2 electrolyte ($\leq 10^{-3}$ M). However, if one of these values is increased the deviation between the two models becomes considerable. The unmodified nonlinearized PB equation always provides the greater value for the number of attracted ions as a consequence of its neglect of the ion sizes (see Fig. 3.15).

In addition, in the calculations according to this model the micellar parameters (aggregation number, surface potential, aggregate radius, cmc) are independently measured as well as their dependence on external parameters (temperature, addition of salt, surfactant concentration) and are not free to change. Application of this model to complex formation and electron-transfer will be examined in the following sections.

3.3. GENERALIZED PSEUDOPHASE MODEL

The generalized pseudophase (GP) thermodynamic approach has been developed to treat micellar kinetics (and micellar equilibria) in the presence of different counterions. When applied to the micellar systems the GP model assumes the two pseudophase approximation (as in the previously discussed models) and the "dressed" micelle approach [58, 59] by means of which the bound species are considered as parts of the aggregate, without any contribution to the bulk ionic strength. This model, which was first applied to the analysis of the redox reaction between $IrCl_6^{2-}$ and Fe^{2+} in cationic micelles [60], has been successfully employed to describe other chemical equilibria and kinetics [9,61].

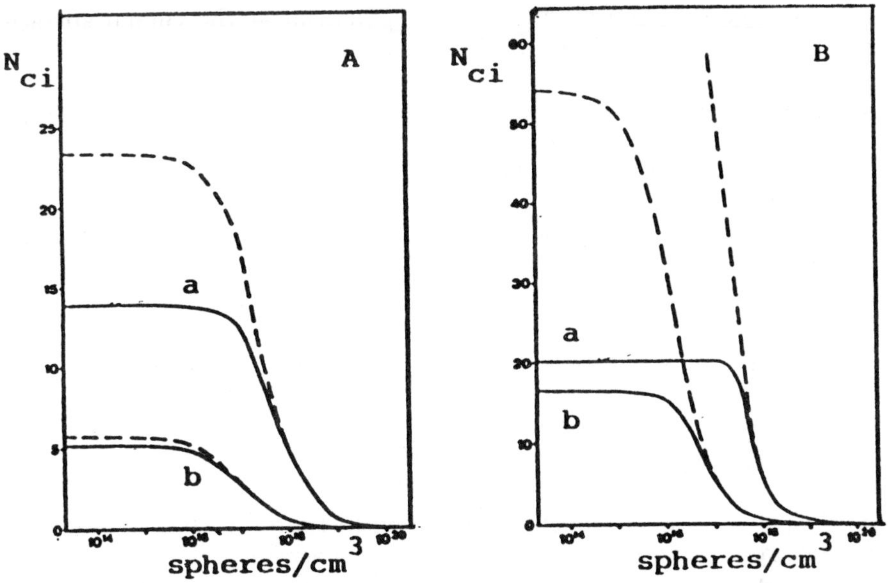

Fig. 3.15. Application of the cell model to the calculation of the number of 2+ counterions (N_{ci}) as a function of charged sphere concentrations at two different surface potentials (A): $\psi_o = -100$ mV; (B): $\psi_o = -150$ mV. Two 2–2 electrolyte concentrations considered: (a): 1×10^{-2} M; (b): 1×10^{-3} M. Curves: (—) modified nonlinearized PBE; (- - - -) unmodified nonlinearized PBE (Gouy–Chapman approximation). Reproduced with permission from Ref. [57].

The development of the corresponding theory is based on the analysis of the free energy of transfer of the reactive species between the two pseudophases, which is related to the difference in their coupling work in the corresponding environments. If two adjacent volume elements of the solution are considered and taking into account that, at constant pressure, the electrochemical potentials of both elements are equal, the following expression can be written:

$$\mu^{o}_{i,r1} + RT\ln(\chi_{i,r1}) + z_i F\phi_{i,r1} = \mu^{o}_{i,r2} + RT\ln(\chi_{i,r2}) + z_i F\phi_{i,r2} \qquad (67)$$

where $\chi_{i,r}$ is the volume fraction of solute i at position r, z_i is the absolute electric charge, $\phi_{i,r}$ is the mean electric potential in the volume element, F is the Faraday constant, and the product $z_i F\phi_i$ represents the coulombic work of coupling with the local environment.

Equation (67) is valid for all the adjacent volume elements, thus, in the presence of an interface the colloidal solution is characterized by the equality between the electrochemical potential at the colloid surface and that of any other point in the solution:

$$n_b^i = n_f^i K_i^o K_i^e w_i \tag{68}$$

where n_b^i and n_f^i are the number of molecules of the solute i in the bulk and at the surface, respectively, $K_i^o = \exp[(\mu_{i,f}^o - \mu_{i,b}^o)/RT]$ and $K_i^e = \exp(z_i F\Delta\phi/RT)$. The two states describing the whole system are usually referred to as pseudophases and the colloidal system is considered as composed of two domains in which the homogeneity of physical and chemical conditions is assumed. The concentration of the species bound to the micellar aggregate is, in turn, given by the equation:

$$[I_b] = \frac{K_i^o K_i^e C_i w_i}{1 + K_i^o K_i^e w_i} \tag{69}$$

$[I_f] = C_i - [I_b]$, where the subscripts b and f represent bound and free species, respectively. The term w_i takes into account the correction for the volume fraction of the pseudophase ($w_i = f_i/(1 - f_c)$, where f_i is the volume fraction of solution that bound ions can occupy and f_c is the volume fraction of the colloidal aggregates.

The hydrophobic free energy of transfer ($\Delta G_i^o = RT \ln K_i^o$) includes all the contributions to the coupling work, in each environment, which are not strictly coulombic (dipole–dipole, higher electrostatic interactions, specific solute–medium interactions, solvation, etc.). Attempts to introduce specific interactions between ions and micelles by assuming the behaviour described by the Volmer isotherms have also been reported [45,62].

The hydrophobic transfer constant K_i^o is here treated as a true constant, although a more refined treatment could be useful.

The electrostatic free energy of transfer ($\Delta G_i^e = RT \ln K_i^e$) is, in turn, a mean field approximation that relates the existing differences between the electrostatic coupling work of the investigated ions in the two pseudophases. It must be emphasized that the contribution of the coupling work with the free pseudophase is generally neglected in the previously examined models. The term K_i^e depends on the difference in the electrical coupling work of the ion i with the bulk at a well defined ionic strength (the effect of ionic atmosphere due to other ions is taken into account) and with the micellar surface environment, which is extremely sensitive to the electric field when ionic aggregates are considered. Under these assumptions, the following equation holds:

$$-\ln K_i^e = \frac{x_i e \phi_s}{K_{HB} T} - \ln \Gamma_{i,f} \quad (70)$$

where K_{HB} represents the hydrophobic transfer constant, $\Gamma_{i,f}$ is the activity coefficient of the free species in the bulk solution. After introduction of the linearized Poisson–Boltzmann equation for bulk ions we obtain:

$$-\ln K_i^e = \frac{z_i e \phi_s}{K_{HB} T} + \frac{z_i^2 B_1 \kappa}{2(1 + \kappa r_{di})} \quad (71)$$

where $B_l = e^2/4\pi\varepsilon\varepsilon_0 K_{HB} T = 7.192$ Å (at 25°C), κ is the Debye length ($\kappa = 0.329\sqrt{I}$ at 25°C), r_{di} is the mean approach distance between the examined ion and the other ions (a r_{di} value in the range 3.8–4.3 Å is appropriate for most experimental conditions [63])

The surface potential of the micelle depends on the number of "condensed" ions and, in the case of spherical symmetry and under the Debye–Hückel approximation, it is related to the net surface charge through the equation:

$$\phi_s = \frac{Q B_l K_{HB} T}{e R_{mic}} \frac{1}{1 + \kappa R_{mic}} \quad (72)$$

where Q is the net micellar surface charge and R_{mic} is the micellar radius.

The possible positions of the ions at the micellar surface have been examined in detail by Stigter [64]. If the "dressed" micelle model is considered and assuming that the counterions are located in the plane of the surfactant head-groups, the bound ion potential is approximately equal to the micellar surface potential given by the Q charge. Considering the micelle as a pseudophase and taking into account the ion mass balance ($C_i = [I_b] - [I_f]$):

$$Q = V_{mic} \sum_{i=1} z_i \frac{K_i^o K_i^e C_i f_i^v}{1 + f_c(K_i^o K_i^e f_i^v - 1)} \quad (73)$$

where C_i are the analytical concentrations of the species i, f_i^v is the fraction of micellar volume, V_{mic}, that bound ions can occupy, and f_c is the volume fraction of the micellar pseudophase. As the contribution of micelles and bound ions to the ionic strength is negligible [65], the following expression for the ionic strength holds:

$$I = \frac{1}{2} \sum_{i=1} \frac{z_i^2 C_i}{1 + f_c (K_i^o K_i^e f_i^v - 1)} \tag{74}$$

Equations (73) and (74) (which can only be solved numerically), together with the definitions of K_i^o and K_i^e, define well the micellar system at constant micellar radius. The deviations from ideality due to electrostatic and hydrophobic interactions are taken into account, together with the influence of surfactant concentration and added salt, if present. The distribution of the species between the pseudophases can be evaluated and the dependence of the concentrations of bound and free species on the hydrophobic transfer constants, salt and surfactant concentrations can also be estimated.

The GP model exhibits some advantages compared to other reported approaches. For example, when the PIE model fails (as in the case of asymmetric electrolytes), the GP model can quantitatively interpret the experimental data [60]. Furthermore, although with double-charged species the validity of the Poisson–Boltzmann equation and its linearization is doubtful, the GP model allows a better treatment of the data through a more refined evaluation of the K_i^e terms.

3.4. ACID–BASE EQUILIBRIA

The changes in pK_{app} of a weakly acidic substrate HA in the presence of a micelle forming surfactant was interpreted in terms of substrate partitioning by Tong and Glesmann [66] in their analysis of the effect of nonionic micelles on the dissociation of naphthols and phenols. The dissociation of the reactant according to:

$$HA \overset{K_a}{\rightleftarrows} H^+ + A^- \tag{75}$$

and assuming partitioning of the species HA and A^- lead to the following expression for the apparent dissociation constant:

$$K_{a(app)} = K_{a(w)} \frac{1 + K_{(A^-)}C}{1 + K_{(HA)}C} \tag{76}$$

An equation identical to (76) was proposed by Berezin [1] and can also be derived from the treatment of Mukerjee and Banerjee [67]. Equation (76) predicts an approach to a constant value of $K_{a(app)}$ at high surfactant concentrations. It does not account directly for a salt effect on $pK_{a(app)}$, although such an effect may be introduced through a salt effect on $K_{(A^-)}$.

The ion-exchange formalism for the dissociation of HA in the presence of a positively charged amphiphile, D^+Y^-, leads to the following equation for $K_{a(app)}$:

$$K_{a(app)} = K_{a(w)} \frac{1 + K_{A/Y}(Y_m / Y_w)}{1 + K_{(HA)} C} \tag{77}$$

where:

$$Y_m = a - \alpha)C - [OH^-_m] - [A^-_m] \tag{78}$$

$$Y_w = \alpha C + cmc + [OH^-_m] + [A^-_m] + [BY]_t \tag{79}$$

Equation (77) accounts directly for a salt effect through the Y_m/Y_w term. For a bimolecular reaction between a neutral substrate B and a nucleophile A^-, in the presence of the D^+Y^- surfactant, the observed second-order rate can be derived:

$$k_{2exp} = [HA]_t\, K_{a(w)} \frac{(k_m / \overline{V}) K_{(B)} K_{A/Y}(Y_m / Y_w) + k_w}{(1 + K_{(HA)} C)\,([H^+] + K_{a(app)})\,(1 + K_{(B)} C)} \tag{80}$$

Reactions involving a nucleophile, A^-, can thus be analyzed in terms of Eqn. (80). Examples include thiolysis and oximolysis of p-nitrophenyl octanoate, reported in Ref. [50]. The effect of HTAB on the dissociation of weak acids, such as phenols, mercaptans and oximes [50,68] was analyzed according to the procedure outlined above. Figure 3.16 shows plots of $[A^-]/[HA]$ as a function of [HTAB] for phenol.

From the ratio $[A^-]/[HA]$ one can readily obtain $K_{a(app)}$ at a known pH. The solid lines in this figure were calculated according to Eqns. (77–79) using $\alpha = 0.2$, $K_{OH/Br} = 0.08$ and $pK_w = 13.833$ (30°C) [50]. In the absence of added salts, a minimum in the $pK_{a(app)}$ vs. C is observed and furthermore the addition of a common salt produces a dramatic change in the shape of these curves.

A comparison between the deprotonation of a weak acid (5-nitroindole) in HTAOH and in HTAB + NaOH has also been reported [69]. In HTAB + NaOH a plot of the fraction of the deprotonated form as a function of C passes through a maximum as is predicted. In HTAOH micelles, the deprotonated form increases continuously with [HTAOH]. The same latter behaviour is observed when OH^- is added even under conditions where the 5-nitroindole is completely bound to the micelle.

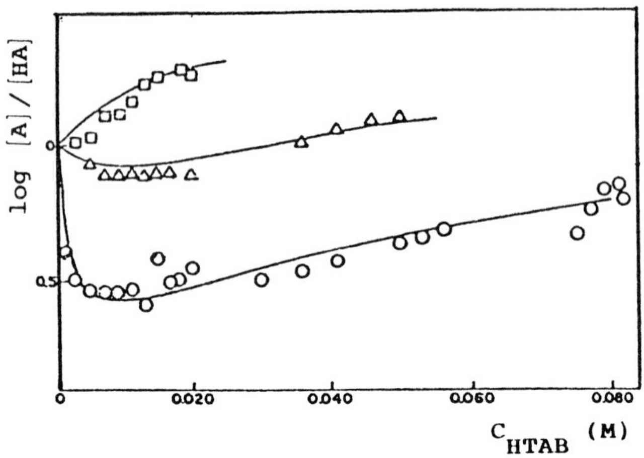

Fig. 3.16. Effect of HTAB on the ratio $[A^-]/[HA]$ of phenol at 30°C, 0.02 M borate buffer pH 9.9 in the absence (O) and presence of added NaBr 0.01 M (Δ) and 0.1 M (□). Solid lines are calculated according to Eqn. (77) using the following values of the parameters $K_{(HA)}$ and $K_{A/Y}$: 34 M^{-1}, 1.05 (no NaBr); 60 M^{-1}, 3.9 (0.01 M NaBr); 123 M^{-1}, 5.5 (0.10 M NaBr); $K_{OH/Borate}$ = 2.25, pK_a = 9.90. Reproduced with permission from Ref. [50].

This observation suggests that the Stern layer is not saturated with OH^-, except at very high $[OH^-]$.

We may recall here that similar conclusions could be drawn from the kinetic effects of HTAOH micelles. This hypothesis seems to better rationalize the above-mentioned observations than the assumption that the reaction occurs between OH^- in the aqueous phase and a substrate in the micellar pseudophase. The latter assumption would predict the kinetic effect, but not an effect on the equilibrium constant. The approximations involved in the model, the influence of the degree of dissociation, α, the role of buffers and the significance of the basicity constant in the micellar phase are extensively discussed in Ref. [68].

More recently, quantitative treatments of acid–base equilibria in micellar systems have been performed on the basis of the GP model [61]. The pK_a shifts have been attributed to the binding of the acid–base couple to the micellar aggregate, due to both electrostatic and hydrophobic interactions.

Starting from the classical treatment of acid–base equilibria in micellar media, the ratio $K_{a(app)}/K_{a(w)}$ for a monoprotic weak acid, HA, depends on the free energy difference due to partitioning of the acid between the micelle and the bulk through the expression:

$$K_{a(app)} = K_{a(w)} \frac{F_A - F_H^+}{F_{HA}(1 - f_c)} \tag{81}$$

$K_{a(w)}$ describes, as usual, the acid dissociation in the bulk pseudophase. The terms F_i are in turn given by $F_i = \{1 - f_c(P_i - 1)\}$, where P_i is the partition coefficient of the species i and f_c has the meaning indicated in the previous section.

By applying the generalized pseudophase approach, it can be demonstrated that the following equation also holds:

$$pH + \frac{\beta}{1 - \beta} = pK^o_{a(w)} + \log \frac{F_{HA}\,\Gamma_{A^-}}{F_A - \Gamma_{HA}} \tag{82}$$

where $K^o_{a(w)}$ is the thermodynamic dissociation constant in the free pseudophase (composed of water plus the added salt), β is the degree of protonation of the weak acid HA, and Γ_i are the activity coefficients of the species i in the free pseudophase. The factors F_i take into account the dependence of the ionic strength on the degree of dissociation of the micelle and the surfactant concentration, according to the equation:

$$F_i = 1 + V_m C_D(K_i^o\, K_i^e\, f_i^v - 1) \tag{83}$$

Equations 81 and 83 describe the behaviour expected for uncharged solutes ($K_i^e = 1$), for charged solutes as the salt is increased ($K_i^e \rightarrow 1$), and for dilute surfactant concentrations where $pK_{a(app)} \rightarrow pK_{a(w)}$. When the hydrophobic transfer constant, K_i^o, is known, the corresponding pK_a shift can be calculated.

The generalized pseudophase model has been applied to the treatment of other acid–base systems of analytical concern, including the case of highly hydrophobic weak acids and the investigation of dissociation equilibria in the presence of large excess of inert salts in the micellar solution [61].

The evaluation of a weak acid titration curve in the presence of ionic micellar solutions is also interesting from the analytical point of view. By applying the electroneutrality condition to the surfactant solution and expressing the fraction of the free dissociated species using Eqn. (81) and the mass balance, the following equation can be derived [61]:

$$[H_f^+] - \frac{K_w}{[H_f^+]} + C_{HAtot}\left(\Phi - \frac{K_{a(w)}}{K_{a(w)}F_{A^-} + [H_f^+]\,F_{HA}}\right) = 0 \tag{84}$$

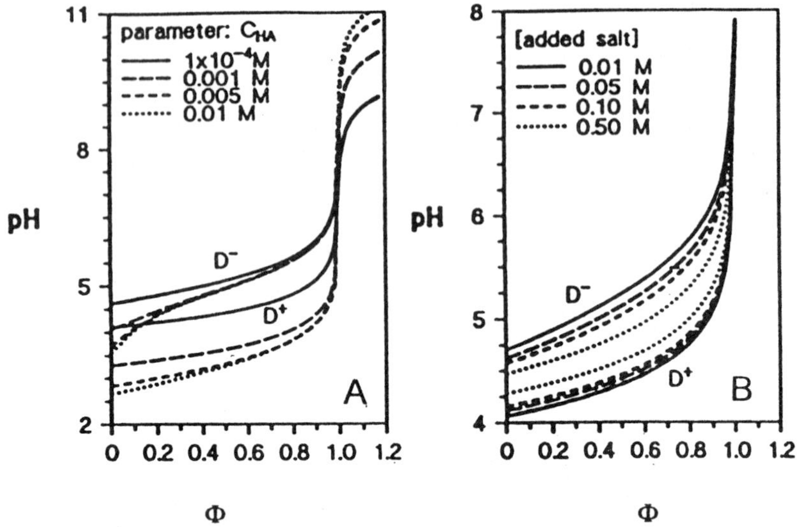

Fig. 3.17. (A) Calculated pH vs.Φ as a function of the analytical concentration of a weak acid HA using the GP model. Surfactant: D^-M^+ and D^+X^-(C_s = 0.05 M); added salt: M^+X^- 0.05 M. Micellar parameters: N_{AG} = 100, R_{mic} = 25 Å, r_{di} = 2.5 Å; K^0_{OH} = 1, K^0_M = 3; K^0_X = 7; K^0_{HA} = 1000, K^0_A = 1000 (the charges of ionic species are omitted); $pK_{a(w)}$ = 4. Titrant concentration (MOH) = C_{HAtot}. (B) Dependence of pH from Φ at different MX concentrations. Other conditions as in (A). Reproduced with permission from Ref. [61].

Figure 3.17(A) shows the changes of pH predicted by the model as a function of the titrated fraction, Φ, for a weak acid (concentration in the range 1×10^{-4}–1×10^{-2} M). Remarkable variations of titration curves are predicted for Φ < 1. Moreover, strong pH shifts before the equivalence point are calculated in the presence of increasing concentrations of added 1–1 salt (see Fig.3.17(B)).

The analysis of ΔpK_a variation as a function of added surfactant concentration (at fixed salt concentration) and as a function of added salt (at constant surfactant concentration, above the cmc), using the GP approach, can also be found in Ref. [61].

3.5. COMPLEX FORMATION IN MICELLAR SYSTEMS

The effect of micellar systems on inorganic reactions has received increasing attention recently. Studies on complex formation or ligand-exchange are of interest not only from the point of view of inorganic reaction mechanisms, but are also useful models for reactions occurring at the surfaces of biomembranes or proteins. A list of the observed effects of aqueous micellar systems on different complex formation reactions is compiled in Table 3.1.

As far as the complexation of metal ions is concerned, two main groups of ligands have been investigated: the first one includes usual (more or less hydrophobic) chelating compounds, whereas the second one is composed of specifically designed ligands containing a tuned hydrophobic moiety, which is often an alkyl chain of proper length. Some of the compounds pertaining to the second group can form micelles and most of them form mixed chelating aggregates with common surfactants. The successful applications of such mixed micelles in recently developed separation methods are described in detail in Chapter 6.

TABLE 3.1

Examples of micellar effects on complex formation reactions

Investigated reaction	Surfactant media	Observed effect	Ref.
Cu(II) + protoporphyrin dimethyl ester	SDS, HTAB	The reaction rate is accelerated by a factor 2×10^4 in SDS relative to HTAB.	70,71
Hemin + CN^-	SDS, HTAB, Triton X100	The dissociation of hemin dimer is favoured. Strong effects on equilibrium and kinetics of the complex.	72,73
Vitamin B-12 + glycine, imidazole, azide	SDS, HTAB	Modest effect on formation rates. A 19-fold decrease in the aquation rate of glycine complex is observed.	74
Vitamin B-12 + cysteine and N-alkanoylcysteine	SDS, HTAB	Modest effect on aquation with cysteine.	75
Cu(I) + porphyrins and ferrihemes	SDS	Formation of complexes between micellar solubilized ligands and the metal ion. In some cases the mixed complex undergoes electron transfer.	76
$Co(NH_3)_5Cl_2^+$ + H	SDS, STS[a], SHS[b]	Strong acceleration up to 105-fold.	77

continued

TABLE 3.1 *continuation*

Investigated reaction	Surfactant media	Observed effect	Ref.
$Fe(phen)_3^{2+}$ dissociation and racemization	ATAB[c]	Increase of racemization rate (up to 3-fold). Decrease of dissociation rate (ca. one half).Effect proportional to the surfactant alkyl chain length.	78
$Cu(phen)X_2^+ + phen \rightarrow Cu(phen)_2^+ + 2\ X$	SDS	5-fold and 100-fold inhibition of forward and reverse reaction rate, respectively.	79
Aquation of $Fe(phen)_3^{2+}$	Dodecyl-pyrazinium iodide	10-fold increase in reaction rate.	80
Reaction of O_2 with Co(II) complexes with the ligands: 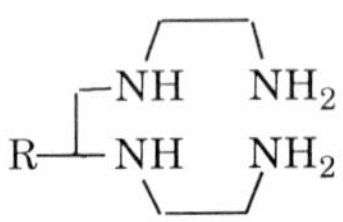 $R = C_{12}H_{25}$		Functionalized micelles formed by the Co(II)-ligand complexes greatly enhance the stability of the O_2-Co(II)-ligand complexes.	81
Ni^{2+} (or Mn^{2+}) + PADA	SDS	Enhancement of complex formation rate (up to ca. 10^3).	82–84
Ni^{2+} + PADA	SAS[d]	Rate enhancement decreases with decreasing alkyl chain length.	85
Ni^{2+} + PADA	HTAN, Triton X-100	Complex formation is markedly inhibited.	86
Ni^{2+} + bpy and dmbpy	SDS	Rate enhancement up to ca. 100-fold.	87
Ni^{2+} (Zn^{2+}) + Murexide	SDS	Reduction of the complexation rate. Decrease in the equilibrium constant.	88,89
Cu^{2+} + benzoylacetone	DTAC, SDS	The stability of the 1:1 complex is decreased in DTAC. Enhanced reaction rate in SDS.	90
Cu^{2+} + ethanedioic acid-bis hydrazides	SDS, HTAB	Strong ground-state complexation in SDS (from photophysical data). Chelation in HTAB dependent on ligand structure.	91

Investigated reaction	Surfactant media	Observed effect	Ref.
Fe^{3+} + PAS-Cn[e]	Brij 35	Reduction of the complexation rate. Different reaction sites depending on the alkyl chain length.	92
Ni^{2+} + oxine[f] and C_{11}-oxine[g]	HTAB	Drastic reduction of k_{obs} at neutral pH for oxine (bulk reaction) and slight changes of k_{obs} for C_{11}-oxine (surface reaction).	93
	water/ dodecane/ SDS/1-pentanol microemulsion	Kinetics of complexation slowed by the presence of C_{11}-substituent, with intermediate formation (interfacial reaction).	94
Ni^{2+} + oxine	HTAN, SDS and Triton X100	Higher reaction rate in SDS (in respect to water). Evaluation of reaction rates of oxine and oxinate species (faster reaction in nonionic micellar media).	95
Aquation and hydrolysis of cis-$[Co(en)_2BCl]_h^{2+}$	SDS	Reduction of aquation rate proportional to the alkylamine hydrophobicity. Strong retardation of basic hydrolysis.	96
Cu^{2+} + Cn-NHMePyr[i]	HTAB, SDS, $C_{12}E_6$	Reaction rate decrease (in HTAN and $C_{12}E_6$) proportional to n. Reaction scarcely sensitive to HTAB for less hydrophobic terms (C_1-C_8).	97

a: Sodium tetradecylsulphase; *b*: sodium hexadecylsulphate; *c*: alkyltrimethylammonium bromide; *d*: sodium alkylsulphates; *e*: 4-alkylamidosalicylic acids; *f*: 8-hydroxyquonoline; *g*: 7-(4-ethyl-1-methyloctyl)-8-hydroxyquinoline (Kelex 100); *h*: en=1,2-diaminoethane, B: alkylamines, imidazole, N-methylimidazole; *i*: 6-(alkylamino)methyl-2-(hydroxymethyl) pyridines.

3.5.1. Reactions between metal ions and current ligands

Initially, kinetic analysis was carried out by adopting the Berezin model [1]. More detailed kinetic and thermodynamic investigations of two types of reactions provide a comparison between the different

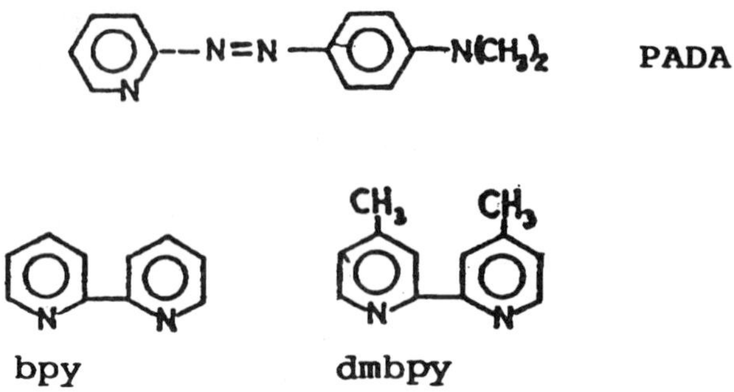

Fig. 3.18. Molecular structures of some ligands used in complexation studies in micellar media. PADA: pyridine-2-azo-p-dimethylaniline; bpy: 2,2′-bipyridine; dmbpy: 4,4′-dimethyl-2,2′-bipyridine.

models discussed in previous sections. These two types of reactions include:

(i) complex formation occurring at the micelle/water interface, e.g., reaction between Ni^{2+}_{aq} and hydrophobic ligands, such as PADA, bpy or dmbpy (see formulas in Fig. 3.18) in anionic micelles.

(ii) complex formation in the bulk aqueous phase between a metal ion (e.g., Ni^{2+}_{aq}, Zn^{2+}_{aq}) and an hydrophilic ligand (e.g., murexide anion) in the presence of anionic micelles.

The reaction indicated by Eqn. (85), where L represents PADA, bpy or dmbpy, describes the complex formation. In the following we analyze the different kinetic models reported in previous sections as applied to this reaction.

$$Ni(H_2O)_6^{2+} + L \underset{k_{-1}}{\overset{k_1}{\rightleftarrows}} Ni(H_2O)_4L^{2+} + 2\,H_2O \tag{85}$$

The corresponding kinetic equation of rate of approach to equilibrium is given by:

$$k_{obs} = k_1[Ni^{2+}]_t + k_{-1} \tag{86}$$

when $[Ni^{2+}]_t >> [L]$ and the concentration of the outer-sphere intermediate complex Ni^{2+}–L is << $[NiL^{2+}]$.

The presence of SDS strongly enhances the observed rate constants, as shown in Fig. 3. 19. The maximum of the rate enhancement is located around the cmc value. Below the cmc, only a slight increase of the observed rate is apparent which could be attributed to the presence of premicellar aggregates. Above the cmc the maximum rate is followed by a steady decline due to the dilution of metal ions (at constant overall concentration) at the surface of an increasing number of micelles.

The increase at the cmc is determined by the binding of the hydrophobic ligands and of the aquo-ions to the micelles. In a series of studies [82,84,87], the analysis of the experimental data (when all the Ni^{2+}_{aq} ions are assumed to bind to the micelle–water interface) led to:

$$k_{obs} = k_{1m}([Ni^{2+}]_t/C_D) + k_{-1m} \tag{87}$$

Plots of k_{obs} as a function of $[Ni^{2+}]_t/C_D$ gave reasonably good straight lines for PADA, whereas for bpy and dmbpy partitioning of the ligands between the micellar and aqueous pseudophases had to be taken into account. In the case of ligands which can participate in protonation equilibria, an additional term is added to account for the acid/base equilibrium. The similarity of the values of k_{1m} for the three different ligands (with different hydrophobicities, being the binding constants with SDS: 2800 M^{-1}, 250 M^{-1} and 40 M^{-1} for PADA, dmbpy and bpy, respectively) indicates that ligand exchange between the surface region of the micelles and its interior does not occur to any significant extent.

The experimental data of Fig. 3.19 are in accordance with the Berezin treatment as well. The equation for the maximum catalytic factor becomes:

$$\frac{k_{obs(max)}}{k_w} = \frac{k_m K_{(Ni)} K_{(L)}}{k_w \cdot 0.3\,(K_{Ni}^{1/2} + K_L^{1/2})^2} \tag{88}$$

As discussed below, $k_m = k_w$ and, assuming that $K_{(Ni)} \approx 10^3$ M^{-1} (from Ni^{2+}/Mu^- complexation, see below), a maximum increase of ca. 100 can be calculated. This is to be compared with the experimental value of 130.

The optimal surfactant concentration, C_{max}, is in turn given by:

$$C_{max} = cmc + (K_{(Ni)} K_{(L)})^{-1/2} \tag{89}$$

For PADA in SDS, $(K_{(Ni)} K_{(L)})^{-1/2} \cong 6 \times 10^{-4}$ M, so that the maximum rate is observed very close to the cmc.

The analysis of the kinetic data and activation parameters [82,83],

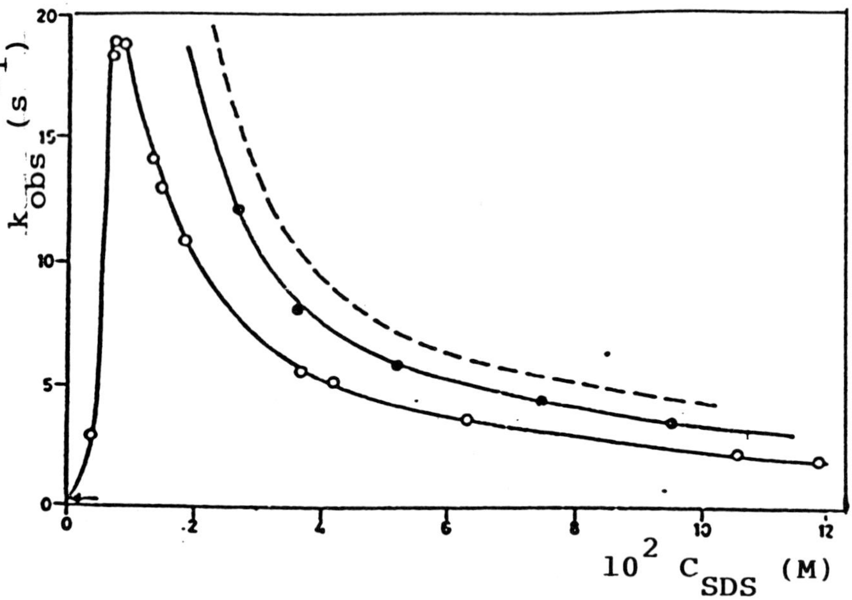

Fig. 3.19. Plots of the observed rate constant, k_{obs}, vs. [SDS] for reaction (85). (- - -) L = PADA; (O) L = dmbpy; (O) L = bpy. The arrow indicates k_{obs} at [SDS] = 0; $[Ni^{2+}]_t = 1\times10^{-4}$ M. Reproduced with permission from Ref. [87].

led to the conclusion that these reactions follow the Eigen–Wilkins mechanism also at the micellar surface. If k_1 is expressed as $K_{os}k_{ex}$, where K_{os} in the equilibrium constant for outer-sphere complex formation and k_{ex}, the first-order rate constant for water loss, it can be shown that k_{ex} in bulk water and at the micellar surface are very similar. Thus, the rate constants for these processes remain essentially unaffected by changing the environment from water to micelles.

The concentration of Ni^{2+}_{aq} ions at the micellar surface can be calculated using the numerical method of Frahm and Diekmann [57], which was already outlined in Chapter 1, Section 1.10.2.

The concentration of the doubly charged counterions near the surface is then determined in competition with other counterions of different charges. In the case of decylsulphate micelles, a spherical geometry is assumed, r_{mic} = 15 Å and the diameter of a hydrated ion is ca. 5 Å. The possible reaction environment is assumed to be the shell of a sphere of thickness r_s. The concentration of Ni^{2+}_{aq} is calculated by integration of the appropriate ion distribution over a shell from 15 to (15 + r_s) Å.

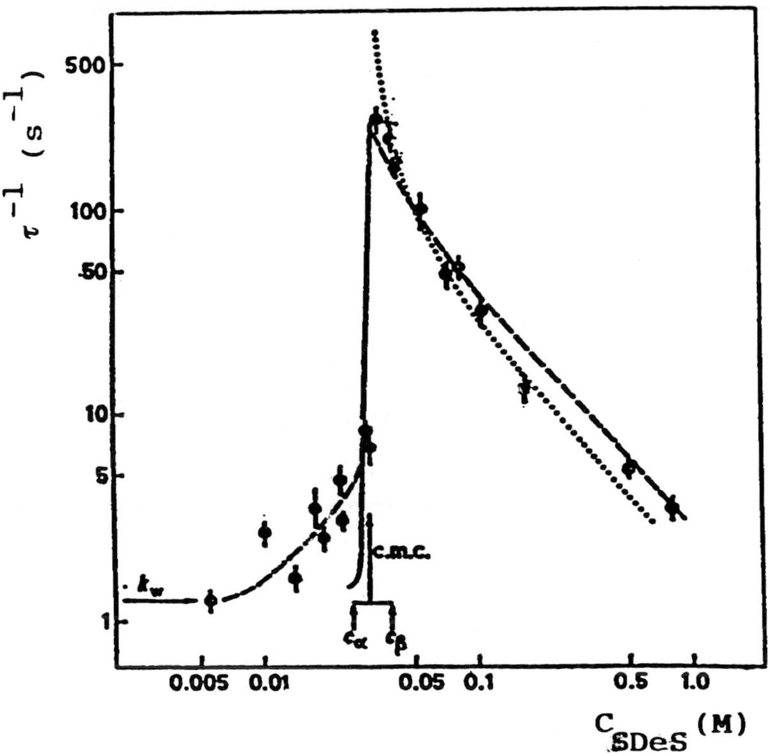

Fig. 3.20. Complexation rate of Ni^{2+} with PADA in SDeS solutions. Curves calculated with the modified nonlinearized PB equation using d_{ion} = 5 Å (- - -) cell model (r_s = 8 Å); (——) infinite dilution model (r_s = 8 Å); (····) from Ref. [82]. (r_s = 10 Å), (both using r_{mic} = 15 Å), below the cmc: (- - -) infinite dilution model with tetramers. Reproduced with permission from Ref. [98].

The experimental data (rate constants vs. [SDeS] for the Ni^{2+}_{aq} + PADA reaction in SDeS is shown in Fig. 3.20, together with the fit to different models. The dotted line represents the previously reported simple method of James and Robinson [82]. The latter assumes that the total amount of Ni^{2+} is bound at the micellar surface. This can be considered as a limiting case at higher surfactant concentrations, but it fails for surfactant concentration slightly above cmc, where Ni^{2+} concentration is in excess over micelle concentration.

This last situation may be represented by another limiting case referred to as the infinite dilution method: the number of Ni^{2+} ions attracted to the surface of a micelle in the latter model remains constant when the micelle concentration is further decreased. Thus, the total number of Ni^{2+} at the surface decreases too, so that the bulk Ni^{2+} concentration eventually reaches the total Ni^{2+} concentration. Therefore, the condition for the infinite dilution model is $[Ni^{2+}]_t = [Ni^{2+}]_{bulk}$. The results are shown as a solid line in Fig. 3.20.

With the cell model of Frahm and Diekmann, choosing $\psi = -102$ mV for SDeS (measured with the aid of coumarin dyes [51,99]), a dielectric constant at the surface of 35, and r_s = 8Å, the calculations lead to the dashed line in Fig. 3.20. As can be seen in this figure, the cell model can adequately explain the kinetic results for complexation reaction occurring at the micellar surface over the entire concentration range of surfactant.

Another interesting case is the reaction between Ni^{2+}_{aq} and the hydrophilic negatively charged murexide ligand, in the presence of anionic micelles. It can be represented by the following scheme:

$$\begin{array}{ccccc} Ni^{2+}_w + Mu^-_w & \rightleftarrows & (NiMu)^+ & \rightleftarrows & (NiMu)_w + H^+ \\ \downarrow\uparrow + M & & \downarrow\uparrow + M & & \downarrow\uparrow + M \\ Ni^{2+}_m & & (NiMu)^+_m & & (NiMu)_m \end{array}$$

Scheme 3.4

With the reasonable assumption that the binding of $NiMu^+$ and NiMu to the micellar surface can be ignored, since these species are hydrophilic, the equilibrium and rate constants in the presence of micelles assume the form:

$$K_{1(app)} = K_1 \frac{[Ni^{2+}_w]}{[Ni^{2+}_w] + [Ni^{2+}_m]} \tag{90}$$

and:

$$\tau^{-1} = k_1[Ni^{2+}_w] + k_{-1m}(1 + K_{(H)}[H^+_w]^{-1})^{-1} \tag{91}$$

where K_1 and $K_{(H)}$ are known from spectral measurements in the absence of surfactants.

From Eqn. (90) it is evident that $K_{1(app)}/K_1$ is equal to the fraction of metal ions present in the bulk phase. Figure 3.21(A) gives this ratio as a function of surfactant concentration and Fig. 3.21(B) shows a plot of the relaxation times observed, where the lines are calculated according to Eqn. (90), using $[Ni^{2+}_w]$ values obtained from the equilibrium measurements of Fig. 3.21(A).

The numerical treatment employing the diffuse double-layer theory presented by Frahm and Diekmann can also provide the fraction of ion

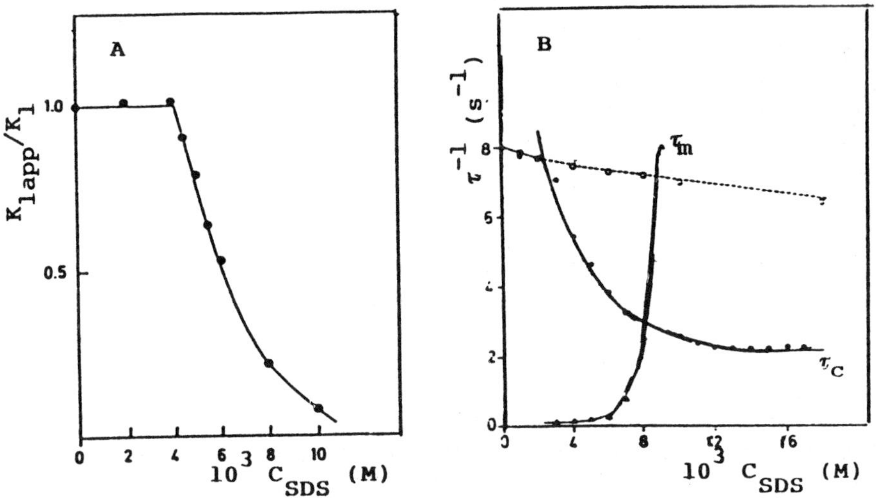

Fig. 3.21. (A): Plot of $K_{1(app)}/K_1$ against [SDS]. (B): Reciprocal relaxation time, τ^{-1}, for complex formation against [SDS] at 25°C. Dotted curve shows ionic strength effect on the rate (NaCl instead of SDS). Figures taken from Refs. [88,89]. Reproduced with permission.

concentration in the bulk of the solution at the cell boundary.

The solid line in Fig. 3.22 represents this calculated concentration compared to the experimental points of Fig. 3.21(A). For the calculations, a value of cmc = 4.3×10^{-3} M (below the value 8×10^{-3} M for pure, aqueous SDS) was taken. The other parameters used were r_m = 18 Å, N_{AG} = 71 and ψ = −125 mV. The latter value was interpolated from values obtained for aqueous solution of pure SDS (−130 mV) and for SDS containing $[Ni^{2+}] = 10^{-3}$ M (−106 mV). The general shape of Fig. 3.22 could be determined without any adjustable parameter.

This analysis and the resultant good agreement between theory and experiments leads to the conclusion that the rate enhancement (reaction between M^{n+} and hydrophobic ligands) or decrease in forward formation rate and equilibrium constant (reaction between M^{n+} and hydrophilic ligands) is due to the local change in the concentration of the metal ion at the micellar surface. The concentration of the reactive species (metal ions and ligands) in the different regions (micellar "surface" and bulk solution) can thus be accurately predicted. The mechanism of these reactions follows Eigen's model and is not changed in the micellar system. Moreover, the rate limiting step of complex formation is the water exchange from the inner coordination sphere of the metal ion.

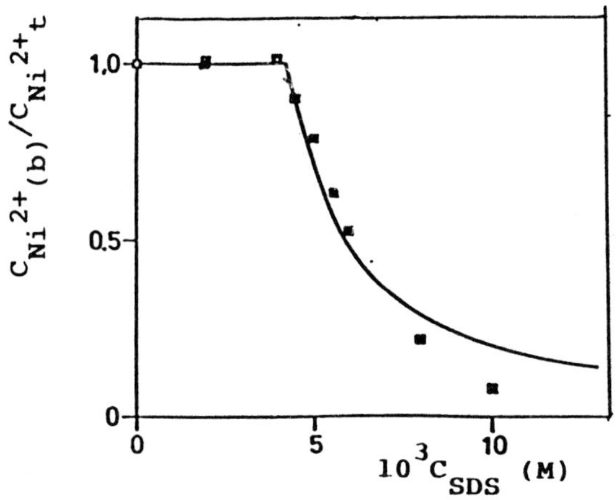

Fig. 3.22. $[Ni^{2+}]_{bulk}/[Ni^{2+}]_t$ as a function of [SDS]. $[Ni^{2+}]_t = 2\times10^{-4}$ M. Experimental points taken from Ref. [88]. Reproduced with permission from Ref. [100].

3.5.2. Reactions between metal ions and amphiphilic ligands

In the case of functionalized aggregates, formed by amphiphilic ligands or by mixtures of these ligands and usual unreactive surfactants, the complex formation may occur in different aggregate sites or in the bulk depending on the ligand structure. For example, the kinetic analysis of the reaction between Fe^{3+} and PAS-Cn ligands (n = 2,4) in nonionic micellar solutions of Brij 35 (acid media) indicates that the complex formation takes place in the bulk or at the micelle–water interface [92], whereas the micellar phase is the preferred site for the more hydrophobic terms of the series.

The evaluation of apparent equilibrium constants and partition data from kinetic analysis is another important aspect of these studies. For the above model system, the following reaction scheme can be proposed:

$$Fe^{3+} + H_2L_w \underset{k_{-1}}{\overset{k_1}{\rightleftarrows}} FeL_w^+ + 2H^+$$ Scheme 3.5

$$Fe^{3+} + HL_w^- \underset{k_{-2}}{\overset{k_2}{\rightleftarrows}} FeL_w^+ + H^+$$

$$FeOH^+ + H_2L_w \underset{k_{-3}}{\overset{k_3}{\rightleftarrows}} FeL_w^+ + H^+$$

$$FeOH^+ + HL^-_w \underset{k_{-4}}{\overset{k_4}{\rightleftarrows}} FeL^+_w$$

The reaction scheme takes into account the partitioning of the two forms of the ligand and of the 1:1 complex between the aqueous and micellar phase by introducing the corresponding binding constants $K_{(H2L)}$, $K_{(HL^-)}$ and $K_{(FeL^+)}$, respectively. The following kinetic equation is thus obtained:

$$\frac{d[FeL^+]}{dt} = \frac{k_f C_{Fe}(C_L - [FeL^+])}{(1 + K_{OH}[H^+]^{-1})} - k_r[FeL^+] \tag{92}$$

from which the observed pseudo first-order rate constant, k_{obs}, is derived:

$$k_{obs} = k_f C_{Fe}/(1 + K_{OH}/[H^+]) + k_r \tag{93}$$

where k_f and k_r are the forward and reverse rate constants, respectively. C_{Fe} is the analytical concentration of the cation and KOH is here the hydrolysis constant of Fe^{3+}. In Eqn. (92), C_L is the analytical concentration of the ligand. The terms k_f and k_r are, in turn, given by:

$$k_f = \frac{k_{fw}\, K_{a(app)}}{(1 + K_{(HL^-)}C)\,(K_{a(app)} + [H^+])} \tag{94}$$

$$k_r = \frac{k_{rw}}{(1 + K_{(FeL^+)}C)} \tag{95}$$

Knowing the forward and reverse rate constants, the equilibrium constant can be easily evaluated. $K_{a(app)}$ is the first dissociation constant of salicylic acid, which behaves as a monoprotic acid. The terms k_{fw} and k_{rw} include the pH dependence of the reaction, according to:

$$k_{fw} = k1[H^+]/K_{aw} + (k_2 + k_3K_{OH}/K_{aw}) + k_4K_{OH}/[H^+] \tag{97}$$

$$k_{rw} = k_{-1}[H^+]^2 + (k_{-2} + k_{-3})[H^+] + k_{-4} \tag{98}$$

Equations (93–98) thoroughly describe the reaction system. The apparent complex formation constant (calculated from kinetic data) is dependent on the surfactant concentration and agrees well with the values determined via spectrophotometric analysis.

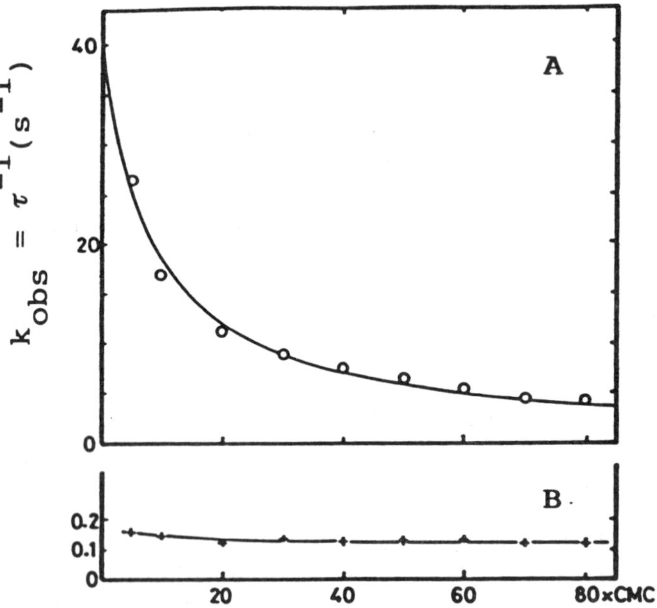

Fig. 3.23. Variation of k_{obs} with surfactant concentration (expressed as number of times the cmc). $C_M = 5\times10^{-3}$ M; $C_L = 1\times10^{-4}$ M; pH 8 (buffer 0.1 M triethanolamine/HCl). (A) HOx (curve calculated from Eqn. (99)). (B) C_{11}-HOx (experimental points). Reproduced with permission from Ref. [93].

The variation of reaction site with the amphiphilic ligand hydrophobicity has also been clearly shown by Kim and Tondre [93] in their studies on Ni^{2+} complexation with 8-hydroxyquinoline derivatives in HTAB micellar solution. The observed rate constant (or reciprocal relaxation time) is here expressed by the equation:

$$\tau^{-1} = k_{fapp}C_M\{P_L\phi_m/(1-\phi_m)+1\} + k_{dapp} \quad (99)$$

where C_M is the total metal concentration, P_L is the partition coefficient of the ligand (mainly present as uncharged HOx under the examined conditions), ϕ_m is the volume fraction of the micellar pseudophase, k_{fapp} and k_{dapp} are the apparent rate constants for complex formation and dissociation, respectively.

The satisfactory fit of experimental data for HOx at pH 8 (assuming a P_L value of ca. 380 and $k_{fapp} = k_{fw} = 7.95\times10^3$ $M^{-1}s^{-1}$) with Eqn. (99) is shown in Fig. 3.23(A). The observed behaviour is consistent with a reaction occurring in the bulk aqueous phase, with electrostatic repulsion of Ni^{2+} from the micellar surface and fast ligand exchange between the two phases.

A very different behaviour is observed in the presence of the hydrophobic derivative C_{11}-HOx (see Fig.3.23(B)). The absence of effect of the surfactant concentration (at the pH examined) indicates that the ligand is here fully located in the aggregates and, thus, the reaction occurs only when Ni^{2+} ions are able to penetrate the electrostatic barrier of the HTAB aggregates. The very low values of k_{obs} are related to the difficulty of the metal ion–ligand encounter.

Another detailed kinetic analysis of complex formation between amphiphilic ligands (forming mixed aggregates with usual surfactants) and metal ions can be found in Ref. [97], where the reaction of a series of 6-(alkylamino)methyl-2-(hydroxymethyl)pyridines with Cu^{2+} in different micellar systems is examined. The role of hydrophobic and electrostatic interactions is clearly shown and the evaluation of activation parameters led to the conclusion that cationic and nonionic aggregates act in these systems mainly through the modification of the local concentration of the reagents.

The peculiar properties of hydrophobic and amphiphilic ligands have also been evidenced in other cases, where the complex formation concerns nonmetallic species. For example, it has been shown that the micelle formation of cobaltous complexes with long-chain polyamines can lead to a dramatic enhancement of the dioxygen stability constant. The kinetic measurements were analyzed in order to assess the respective contribution of the dimensionality of the reaction, the orientation of the reactants upon the micellar effect and the local concentration of the reactants. It appears that the magnitude of the latter factor is sufficient to explain the micellar effect, since calculations of the mean distance of the reactant molecules in the presence and in the absence of micellar aggregates agree with the observed rates [81].

3.6. ELECTRON-TRANSFER REACTIONS

Electron transfer processes are the simplest and most common reactions, and thus are of great importance in naturally occurring biological cycles and in various industrial processes. Organized assemblies can resemble metallo-proteins and enzymes in that they possess distinct regions of hydrophobic and hydrophilic character. Investigations of coupled systems composed of electron-transfer reactions and micellar aggregates may contribute to the understanding of the electron-transfer processes at biological interfaces and to the possibility of modifying reaction rates and mechanisms.

By choosing reactants with different properties (positively or negatively charged, uncharged, hydrophobic or hydrophilic) in different micelle-forming surfactants (anionic, cationic, nonionic), it is possible to select different locations in the micellar system for the reaction to occur. Different kinetic models have been adopted to rationalize the variety of observations in such systems.

When one of the reactants bears a charge like to that of the micelles, three limiting cases are observed:

(i) The second reactant is oppositely charged and is very strongly bound to the micellar aggregate. The reaction then takes place at the micelle-water interface. This is the case of ion complexes with hydrophobic ligands such as FeL_3^{3+} reacting with negatively-charged complexes, in anionic micelles [101–104]. In these cases, the reaction rate experiences a sudden decrease at the cmc, then remains constant (a slight increase can be due to kinetic salt effects). An important case is represented by the effect of SDS on the reaction of a series of FeL_3^{3+} complexes with $Mo(CN)_8^{4-}$. Whereas the reaction rates in the absence of surfactant are diffusion controlled, independent of the redox potentials of the $FeL_3^{3+/2+}$ couples, in the presence of micellar aggregates the electron-transfer rates become dependent on redox potentials [102], as can be seen in Fig. 3.24.

(ii) The reactants are partitioned between the micellar and aqueous pseudophases and the reaction progresses in both pseudophases. This case has been observed in electron transfer between partially solubilized quinones in SDS micelles [105].

(iii) One reactant is bound to the micelles, but the reaction proceeds in the bulk solution. This is the case of the reaction between $IrCl_6^{2-}$ and Fe^{2+} in the presence of SDS [106]. The addition of SDS causes a decrease of the reaction rate as a result of the attraction of Fe^{2+} to the anionic micelles. The bulk concentration of Fe^{2+} and $IrCl_6^{2-}$ has been calculated using the numerical method of Frahm and Diekmann [98]. The results obtained at various experimental conditions and different ionic strengths have been already presented in Chapter 2, Section 2.7 (Fig. 2.5).

If the "true" second-order rate constants are calculated using these bulk concentrations, they are approximately independent of SDS concentration and are equal to the rate constants determined in the absence of SDS, thus confirming the suggestion that the reaction proceeds in the bulk of the solution.

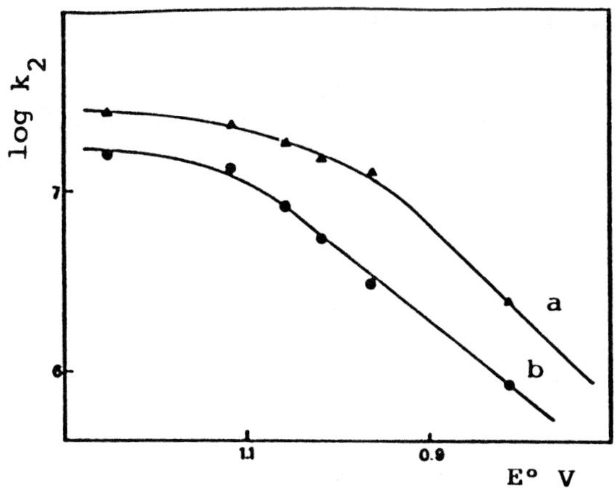

Fig. 3.24. Plots of log k as a function of the reduction potentials of $FeL_3^{3+/2+}$. Curve (a): $Fe(CN)_6^{4-}$; curve (b): $Mo(CN)_8^{4-}$. [SDS] = 0.01 M; T = 8°C. Reproduced with permission from Ref. [102].

$$
\begin{array}{ccc}
MPTZ_w + Fe_w^{3+} & \underset{k_{-1w}}{\overset{k_{1w}}{\rightleftarrows}} & MPTZ_w^{+} + Fe_w^{2+} \\
\uparrow\downarrow K_{(MPTZ)} & & \uparrow\downarrow K_{(MPTZ^{+})} \\
MPTZ_m + Fe_w^{3+} & \underset{k_{-1m}}{\overset{k_{1m}}{\rightleftarrows}} & MPTZ_w^{+} + Fe_w^{2+}
\end{array}
$$

Scheme 3.6

For reactions involving an uncharged reactant, this case (iii) is a very common one. For example, for the equilibrium between N-methylphenothiazine with Fe^{3+} in HTAN, Scheme 3.6 leads to Eqn. (100) for the equilibrium constant.

$$K_{1exp} = \frac{K_{1w}}{1 + K_{(MPTZ)}C} \tag{100}$$

where K_{1exp} and K_{1w} are the equilibrium constants in the presence and absence of surfactant, respectively, and $K_{(MPTZ)}$ is the binding constant of MPTZ to HTAN. For the forward reaction:

$$k_{1exp} = \frac{k_{1w} + k_{1m}K_{(MPTZ)}C}{1 + K_{(MPTZ)C}} \tag{101}$$

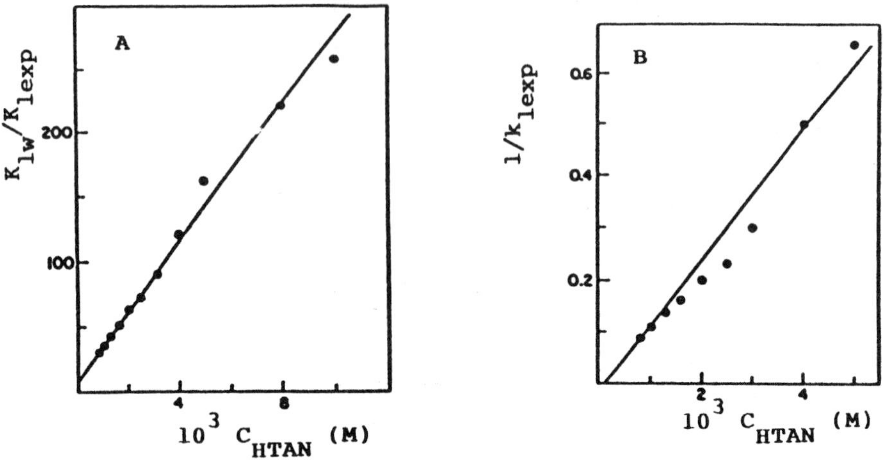

Fig. 3.25. Effect of [HTAN] on equilibrium and rate constants for the reduction of Fe^{3+} by MPTZ. (A): Linear dependence of K_{1w}/K_{1exp} on [HTAN] according to Eqn. (101) ($K_{(MPTZ)}C >> 1$). (B): Plots of $1/k_{1exp}$ vs. [HTAN], according to Eqn. (102). Reproduced with permission from Ref. [107].

where k_{1w} and k_{1m} represent the rate constants in the aqueous and in the cationic micellar pseudophase, respectively.

Assuming that k_{1m} is much smaller than k_{1w}, Eqn. (101) assumes the form:

$$\frac{1}{k_{1exp}} = \frac{1}{k_{1w}} + \frac{K_{(MPTZ)}C}{k_{1w}} \tag{102}$$

or when $K_{(MPTZ)}C >> 1$:

$$k_{1exp} = k_{1m} + k_{1w}/(K_{(MPTZ)}C) \tag{103}$$

Plots according to Eqns. (100) and (102) are shown in Fig. 3.25.

Equations 100–103 also describe other reported systems, such as ferrocene oxidation by positively charged complexes in HTAB [108], quinols and catechols oxidation in micellar systems [109,110], phenothiazine oxidation [111], and oxidation of $Cu(phen)_2^+$ by $Co(EDTA)^-$ in SDS [112].

From Eqn. (102) it is evident that, in cases in which k_{1w} cannot be directly measured (very fast reaction, poor solubility of the reactant), plots of $1/k_{1exp}$ as a function of C can be used to determine k_{1w} from the intercept. Where the intercept is too small, k_{1w} can be estimated from the slope, provided the binding constant of substrate, $K_{(S)}$, is independently determined. This procedure has been successfully applied to the investigation of the reaction of ferrocene with $Fe(CN)_6^{3-}$ [113] and of MPTZ with $Mo(CN)_8^{3-}$ [107].

The latter reaction in SDS is, in addition, an interesting case of micellar effect on equilibrium, as shown in Scheme 3.7.

$$\begin{array}{ccc} MPTZ_w + Mo(CN)_8^{3-} & \underset{k_{-2w}}{\overset{k_{2w}}{\rightleftarrows}} & MPTZ_w^+ + Mo(CN)_8^{4-} \\ \uparrow\downarrow K_{(MPTZ)} & & \uparrow\downarrow K_{(MPTZ^+)} \\ MPTZ_m + Mo(CN)_8^{3-} & \underset{k_{-2m}}{\overset{k_{2m}}{\rightleftarrows}} & MPTZ_m^+ + Mo(CN)_8^{4-} \end{array}$$

Scheme 3.7

In this system, while $Mo(CN)_8^{3-/4-}$ will be excluded from the micellar phase, MPTZ and $MPTZ^+$ are expected to strongly bind to the anionic SDS aggregates. Under these assumptions, the experimentally observed equilibrium constants K_{2exp} is related to the equilibrium constant in aqueous solution K_{2w}, through:

$$K_{2exp} = K_{2w} \frac{1 + K_{(MPTZ)}C}{1 + K_{(MPTZ^+)}C} \tag{104}$$

where $K_{(MPTZ)}$ and $K_{(MPTZ^+)}$ represent the binding constants of MPTZ and $MPTZ^+$ to SDS, respectively. K_{2exp} is experimentally found to be independent of SDS concentration, indicating that both products $K_{(MPTZ)}C$ and $K_{(MPTZ^+)}$ are >> 1. Under such conditions, Eqn. (104) can be rearranged to give:

$$K_{2exp}/K_{2w} = K_{(MPTZ)}/K_{(MPTZ^+)} \tag{105}$$

Since the complex $Mo(CN)_8^{3-/4-}$ does not interact with the micellar system, these ratios provide a method to determine the change in redox potential of the $MPTZ^+$/MPTZ couple upon micellization.

When both reactants interact strongly with the micelles, plots of the observed rate constants as a function of surfactant concentration can show the usual behaviour with a maximum around the cmc. However, in some cases, the reaction rates decrease monotonically relative to the

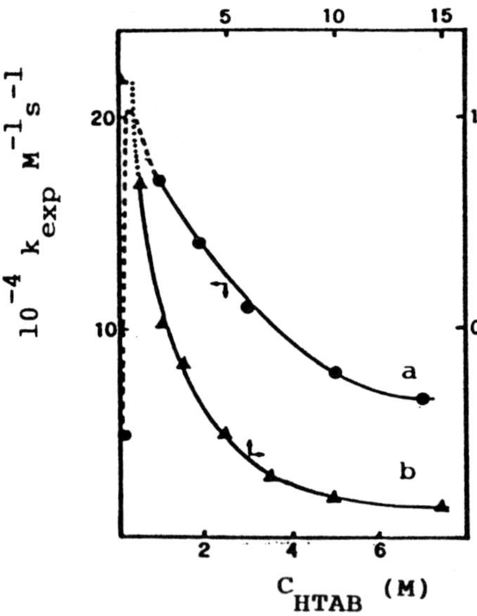

Fig. 3.26. Effect of HTAB on k_{exp} for the following reactions: (a) tert-butylbenzene-1,4-diol/$Mo(CN)_8^{3-}$; (b) tert-butylbenzene-1,2-diol/$IrCl_6^{2-}$. Reproduced with permission from Ref. [109].

rate in bulk solution. These different behaviours are shown in Fig. 3.26.

Such an observation suggests that, in the case of electron-transfer reactions, other factors in addition to the concentration effect at the micellar surface may be operating. The pseudophase approach suggested in some cases that the specific rate constant in the micellar pseudophase (k_m) is very close to that in the bulk [113,114]. The change in the observed rate constant was essentially attributed to the concentration effect at the micellar interface. However, for other investigated systems [108,115,116], the estimated k_m values differ significantly from the bulk values.

As described in Chapter 5, micellar systems affect redox potentials, thus studies of the effect of micelles on the kinetics and equilibria of properly chosen redox reactions could shed some light on these effects. For example, the equilibrium constants between benzenediols and $Mo(CN)_8^{3-}$, in HTAB were found to shift by 3–4 orders of magnitude relative to homogeneous solution [109]. In the same systems, k_w/k_m ratios are in the range 10^2–10^4. On the basis of a pure concentration effect, the latter ratio would have been 1, thus several microenvironmental effects may contribute to this pronounced effect on the k_w/k_m ratio. Reduced frequency of encounters might change the diffusion-controlled limit, as has been mentioned before.

However, it is quite conceivable that reorganization energies, which accompany electron transfer, may change due to either electrostatic or hydrophobic interactions, and these in turn may substantially change k_m in such systems. Location of the reactants within the micelle may also affect these rate constants. Thus, the ratio of k_w/k_m is ca. 50 times higher for 1,2-benzenediols than for the 1,4-derivatives. Such effects are also reflected in the binding constants to the micelle.

Since the numerical computation of concentrations of ionic species, discussed above for complex formation, was found to explain successfully the "catalytic" effect of micelles [57], the same approach was adopted in some selected redox systems, e.g., reactions (106) and (107):

$$Os(bpy)_3^{3+} + Fe^{2+} \rightleftarrows Os(bpy)_3^{2+} + fe^{3+} \quad (106)$$

$$(Os(dmbpy)_3^{3+} + Fe^{2+} \rightleftarrows Os(dmbpy)_3^{2+} + Fe^{3+} \quad (107)$$

Due to the high charge and the hydrophobicity of the ligands, $OsL_3^{3+/2+}$ can be assumed to entirely reside at the micellar interphase (SDS). For the other reactants, Fe^{2+} and Fe^{3+}, the concentration at the surface was calculated and Fig. 3.27 displays the electrostatic surface catalytic factor for Fe^{2+} This factor represents the increase in the rate due to electrostatic interactions, and is calculated as a ratio between the estimated concentration at the micellar surface and the concentration of the same species dissolved in the actual volume of the aqueous phase.

In the case of reaction (106), the rate enhancement could not entirely be explained by the catalytic factors of Fig. 3.27. The equilibrium described by Eqn. (107) offered an opportunity to investigate the influence of micelles on both the forward and backward reactions since the micellar effect on the equilibrium constant can be measured.

From Fig. 3.28(A), it is evident that the increase in the reaction rate is greater for the backward than for the forward reaction. This finding is totally unexpected on purely electrostatic considerations, since the attraction of Fe^{3+} ions to the micellar surface is higher than that of the Fe^{2+} ion. The ratio of the non-electrostatic catalytic factors shown in Fig. 3.28(B) should be related to the change in redox properties of $OsL_3^{2+/2+}$ on association with micelles. Since the rate constant of electron-transfer

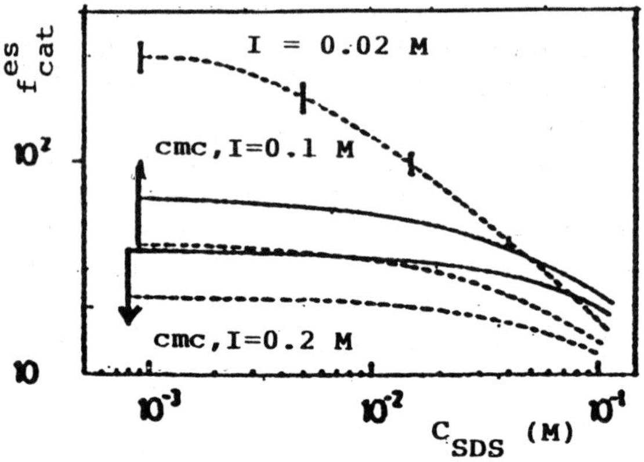

Fig. 3.27. Calculated electrostatic surface enrichment factors, f_{cat}^{es}, for Fe^{2+} ions at the surface of SDS micelles as a function of [SDS]. The values are calculated for three ionic strengths and two different thicknesses of the reaction shell at the surface: Solid curves 5 Å; dashed curves 8 Å. Error bars correspond to uncertainty in the surface potential (±5 mV) drawn only for I = 0.02 M. Reproduced with permission from Ref. [106].

reactions depends on the free energy change in the reaction, this effect will also influence the reaction kinetics. In the case of the equilibrium between MPTZ and $Fe^{3+/2+}$ in SDS, Scheme 3.8 is applicable:

$$\begin{array}{ccccccc} MPTZ_w & + & Fe_w^{3+} & \underset{k_{-1w}}{\overset{k_{1w}}{\rightleftarrows}} & MPTZ_w^+ & + & Fe_w^{2+} \\ \uparrow\downarrow K_{(MPTZ)} & & \uparrow\downarrow K_{(Fe^{3+})} & & \uparrow\downarrow K_{(MPTZ^+)} & & \uparrow\downarrow K_{(Fe^{2+})} \\ MPTZ_m & + & Fe_m^{3+} & \underset{k_{-1m}}{\overset{k_{1m}}{\rightleftarrows}} & MPTZ_m^+ & + & Fe_m^{2+} \end{array}$$

Scheme 3.8

and a conventional analysis of the observed equilibrium constant leads to:

$$K_{1exp} = K_{1w} \frac{(1 + K_{(MPTZ^+)}C)(1 + K_{(Fe^{2+})}C)}{(1 + K_{(MPTZ)}C)(1 + K_{(Fe^{3+})}C)} \quad (108)$$

Since for $C > 10^{-2}$ M the value of K_{1exp} does not change significantly with increasing C, it can be assumed that the K_BC terms are much greater than 1. Then Eqn. (108) simplifies to:

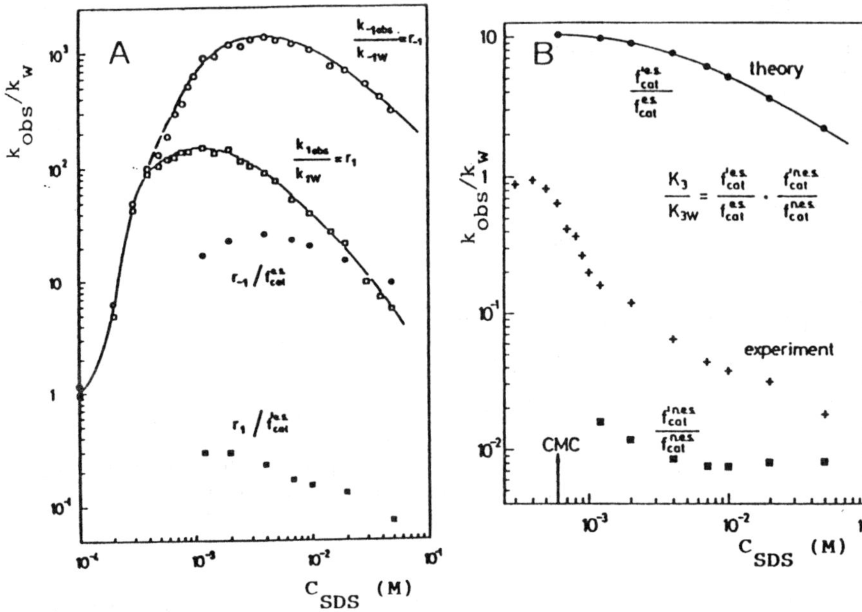

Fig. 3.28. Separation of electrostatic and nonelectrostatic effects on rates and equilibrium constants for the reaction $Os(dmpby)_3^{2+/3+}$–$Fe^{3+/2+}$ in SDS (I = 0.1 M). (A): k_{obs}/k_w for the forward and back reactions vs. [SDS] at I = 0.1 M. Solid points represent the non-electrostatic catalytic factors for the forward and backward reactions, f'^{nes}_{cat} and f^{nes}_{cat}, respectively. These were calculated from the total rate enhancement divided by the theoretical values for the electrostatic enrichment factors. (B): Experimental equilibrium constants, K_{obs}/K_w, vs. [SDS]. Electrostatic and non-electrostatic contributions were calculated as indicated above. Taken from Ref. [106], reproduced with permission.

$$K_{1exp}/K_{1w} = (K_{(MPTZ^+)}/K_{(MPTZ)})(K_{(Fe^{2+})}/K_{(Fe^{2+})}) \tag{109}$$

From the values determined and reported for $K_{(MPTZ^+)}/K_{(MPTZ)}$ in SDS, the ratios of the binding constants of Fe^{2+} and Fe^{3+} to the same micelles were obtained. The value thus obtained could be explained in terms of electrostatic contribution to micellar binding. For this system, the non-electrostatic catalytic factors calculated as previously described are larger than for both the forward and reverse rate constants. A summary of reactions for which the effect of SDS micelles on equilibrium constants has been measured is given in Table 3.2.

TABLE 3.2

Equilibrium constants for some electron-transfer reactions in the absence of surfactant and in 0.1 M SDS[a]

Redox reaction		K_{1w}	K_{1SDS}	Ref.
(1) $OsL_3^{2+} + Fe^{3+}$	$OsL_3^{3+} + Fe^{2+}$	3.8[b]	0.7[b]	106
(2) MPTZ + Fe^{3+}	$MPTZ^{+} + Fe^{2+}$	3.2[b,e]	0.6[b]	107
		1.9[f]	0.6[f]	
(3) $Mo(CN)_8^{3-}$ + MPTZ	$Mo(CN)_8^{4-} + MPTZ^{+}$	3.0[c,e]	15.0[c]	107
(4) OsL_3^{3+} + MPTZ	$OsL_3^{2+} + MPTZ^{+}$	0.4[d,e]	11.0[d]	117
(5) $Mo(CN)_8^{3-} + OsL_3^{2+}$	$Mo(CN)_8^{4-} + OsL_3^{3+}$	75[d]	3.8[d]	117

a: 25C, I = 0.1 M ($NaClO_4$), spectrophotometrically evaluated; *b*: $HClO_4$ 0.01 M; *c*: $HClO_4$ 0.1 M; *d*: $HClO_4$ 0.02 M; *e*: 10% ethanol; *f*: $HClO_4$ 0.01 M, I = 0.02 M ($NaClO_4$).

In addition to the classical pseudophase models, the generalized pseudophase approach can also be applied to treat redox reactions. The reaction between $IrCl_6^{2-}$ and Fe^{2+} has, in particular, been reexamined in order to investigate the influence of hydrophobic counterions of hexadecyltrimethylammonium aggregates on reactivity and to evaluate the influence of surfactant composition when mixed cationic–nonionic aggregates are present [9]. Without entering into the details concerning such specific investigations, it must be emphasized that the ability of the GP model to describe quite complex systems has been again confirmed.

It can be seen from this table that exclusion of one reactant from one of the subphases could divert the reaction primarily to the other phase. Whether the reaction occurs at the interface or the bulk will be determined by the relative contribution of these interactions between the reactants and the two phases, as previously discussed. In systems where both reactants associate with the micelles, the various computational methods provide estimates of both the electrostatic and non-electrostatic contributions.

However, the electrostatic interactions affect the hydrophobic ones, thus making the distinction and separation of the two interactions somewhat artificial, and their individual effect unpredictable (compare for example the reactions of MPTZ with iron(III) and with octacyano-

TABLE 3.3

Summary of micellar effects on electron-transfer rates and equilibria

Reactants		Products		Micelle[a]	Reaction site	Catalytic effect[b]		Ref.
A	B	C	D			F.r	B.r	
MPTZ	Fe^{3+}	$MPTZ^{+}$	Fe^{2+}	(+)	bulk	I	NE	107
MPTZ	Fe^{3+}	$MPTZ^{+}$	Fe^{2+}	n.i.	bulk[c]	I	NE	
MPTZ	Fe^{3+}	$MPTZ^{+}$	Fe^{2+}	(–)	interface	Ha+Ea	Ha+Ea	
MPTZ	$Mo(CN)_8^{3-}$	MPTZ+	$Mo(CN)_8^{4-}$	(–)	bulk	I	I	
RPTZ[d]	Fe^{3+}	$RPTZ^{+}$	Fe^{2+}	(+)	bulk	I	NE	118
				(–)	interface	Ha+Ea	Ha+Ea	
				n.i.	bulk	I	NE	
OsL_3^{2+}	Fe^{3+}	OsL_3^{3+}	Fe^{2+}	(–)	interface	Ea+Hi	Ha+Ea	106
DQ[e]	AQS^{-}[e]	DQ^{-}	AQS	(–)	bulk and interface	hardly any	hardly any	119
RuL_3^{2+}[f]	$Mo(CN)_8^{4-}$	RuL_3^{3+}	$Mo(CN)_8^{3-}$	(–)	interface	I	—	101
$IrCl_6^{2-}$	Fe^{2+}	$IrCl_6^{3-}$	Fe^{3+}	(+)	bulk	Ei	—	60
				(–)	bulk	Ei	—	106
$IrCl_6^{2-}$	Diols[g]	$IrCl_6^{3-}$	Quinones	(+)	interface	Ei or Ea[h]		109
				(–)	interface	Ei		
$Mo(CN)_8^{3-}$	Diols	$Mo(CN)_8^{4-}$	Quinones	(+)	interface	Ei or Ea[h]		109
				(–)	interface	Ei		

a: (+) cationic, n.i: nonionic, (–) anionic; *b*: catalytic effect on forward reaction (F.r) and backward reaction (B.r). I: inhibition, N.E.: no effect, Ha: hydrophobic acceleration, Ea: electrostatic acceleration, Ha: hydrophobic acceleration, Hi: hydrophobic inhibition; *c*: a little contribution arises from reaction at micelle–bulk interface; *d*: R = H, CH_3, C_2H_5, C_4H_9, $C_{12}H_{25}$; *e*: DQ = duroquinone, AQS^{-} = anthraquinonesulphonate; *f*: L = bpy; *g*: derivatives of catechol and quinol. Changes of the electron-transfer equilibria attributed to the microenvironment of micellized diols (variation of E° values). *h*: depending on the diol hydrophobicity.

molybdate(V) in anionic micelles). Clearly, a better understanding of the factors controlling electron-transfer rates at the interface would allow a more definitive prediction of these changes in rate.

3.7. EFFECT OF AMPHIPHILES AND ORGANIZED ASSEMBLIES ON PRECIPITATION/DISSOLUTION PROCESSES

The investigation of the effect of surfactants on the formation of precipitates has been carried out for assessing the ability of amphiphiles in preventing (or promoting) the rate of precipitation (dissolution) [120, 121], in controlling the crystalline phase [122] and the dimensions and shapes of the precipitate particles [123].

The effect of organized structures, such as micelles, w/o microemulsions and vesicles, on precipitation has recently received increasing attention. All these investigations originate from the interest in several fields including scale prevention or dissolution [124], uniform particle formation [125], biomineralization processes [126,127], catalyst and photocatalyst preparation in controlled conditions [128,129]. As an example, the effect of surfactants on precipitation of calcium carbonate [130] is illustrated.

Anhydrous calcium carbonate is known to exist in three crystalline phases: calcite (the stable form at ordinary temperature and pressure), aragonite and vaterite. By rapid mixing, at 25°C, of equimolar solutions of $CaCl_2$ and Na_2CO_3, crystals of calcite are formed. The addition of SDS (below the cmc) generates a mixture of calcite and vaterite, whereas the increase of SDS concentration up to the cmc value provides amorphous $CaCO_3$ precipitates.

The adsorption of SDS at low concentrations onto calcium carbonate particles is appreciable, while for higher concentrations, the equilibrium:

$$Ca^{2+} + 2\,DS \rightleftarrows Ca(DS)_2 \qquad (110)$$

affects significantly the formation of calcium carbonate.

The presence of negatively charged (in particular polycharged) hydrophobic species may strongly inhibit crystal growth since these compounds not only adsorb onto surfaces but also may disrupt the zone of diffusion around the crystal surface, further enhancing the exclusion of incoming lattice ions.

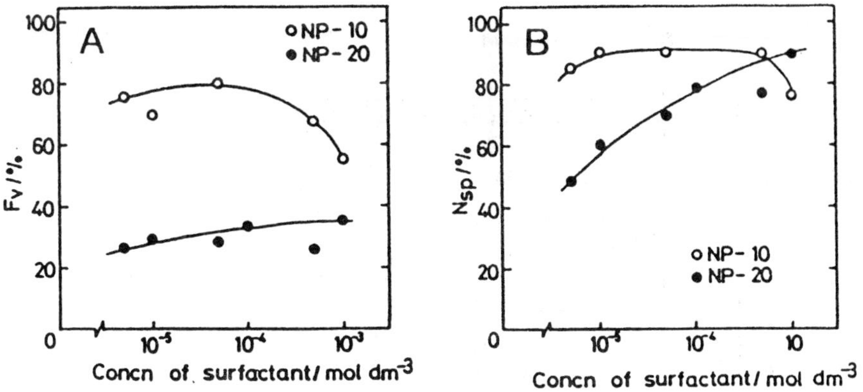

Fig. 3.29 Effects of surfactant concentration on $CaCO_3$ precipitation.(A). Variation of vaterite ratio (F_v). (B) Dependence of spherical particle ratio (N_{sp}). Reproduced with permission from Ref. [131].

The presence of HTAB below the cmc generates again a mixture of vaterite and calcite, but increasing HTAB around and above the cmc the morphology becomes that characteristic of calcite. The decrease in the vaterite crystal ratio might be caused by the slow diffusion of water owing to the thick layer of HTAB adsorbed around the calcium carbonate particles.

Since nonionic surfactants have no charged groups, the formation of calcium carbonate in the presence of such surfactants proceeds without taking into consideration the charge effect. The effect of nonylphenolethoxylated surfactants (NP-10 and NP-20) on the vaterite ratio and on spherical particle ratio is depicted in Fig. 3.29.

The film thickness of the surfactant adsorbed on the particles increases with decreasing the length of oxyethylene chain and with increasing the surfactant concentration. The film prevents the surface of the particles from contacting the water, thus stabilizing the vaterite crystal. A high percentage of spherical particles may be obtained.

The confinement in organized systems such as w/o microemulsions or vesicles can provide fundamental insights in the biomineralization processes and offer a new approach for obtaining controlled-size monodispersed particles [127,129,130].

The controlled nucleation and growth of particles involves the temporal and spatial regulation of the chemical processes and the essential property to be controlled is the level of supersaturation within the

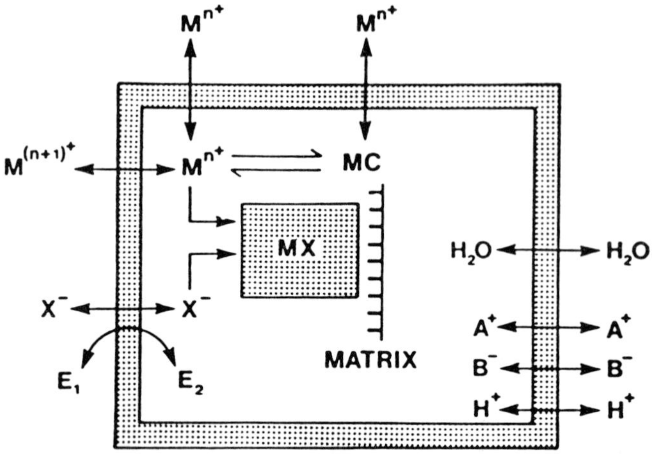

Fig. 3.30. Proposed mechanisms for controlling the local supersaturation of MX precipitate in biomineralization processes. MC indicates a soluble cation complex; E_1, E_2 represent enzymes. Reproduced with permission from Ref. [133].

mineralization zone. The precision of replication of precipitates requires a high degree of control of the reaction volumes and of the structural, compositional and electrochemical nature of organic boundaries. Possible mechanisms for controlling localized supersaturation in biomineralization are reported in Fig. 3.30.

Extensive work has appeared in the last decade on the formation of particles in organized systems, ranging from silver halides to carbonates, magnetic oxides, noble metals (Pt,Pd,Rh,Ir), Ni, Co, Fe borides [132] and semiconductor chalcogenides. Figure 3.31 shows a model for the preparation of monodisperse colloidal bimetallic boride particles which applies w/o microemulsions to regulate the nucleation process. Control of size, stabilization and localization of the particles can be achieved.

There are extensive references [129,130] to the main recent advances in this field.

REFERENCES

1 I.V. Berezin, K. Martinek and A.K. Yatsimirskii, *Russ. Chem. Rev. (Engl. Transl.)*, 42 (1973) 487.

2 C.A. Bunton, in: E.H. Cordes (Ed.), *Reaction Kinetics in Micelles*. Plenum

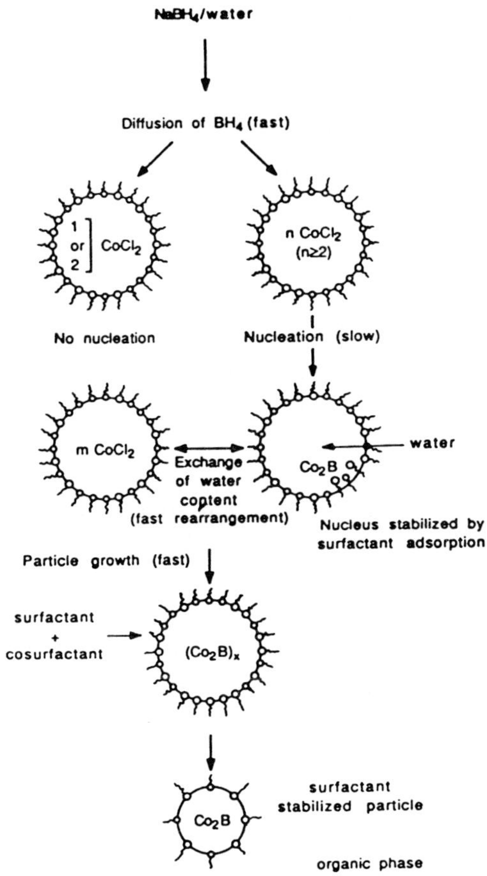

Fig. 3.31. Schematic representation of cobalt boride particles formation within microemulsion aggregates. Reproduced with permission from Ref. [132].

Press, New York, 1973.
3 J.H. Fendler and E.J. Fendler, *Catalysis in Micellar and Macromolecular Systems.* Academic Press, New York, 1975.
4 J.H. Fendler, *Membrane Mimetic Chemistry*. Wiley, New York, 1982.
5 C.A. Bunton, *Catal. Rev. Sci. Eng.*, 20 (1979) 1.
6 C.A. Bunton and G. Savelli, *Adv. Phys. Org. Chem.*, 22 (1986) 213.
7 C.A. Bunton, F. Nome, F.H. Quina and L.S. Romsted, *Acc. Chem. Res.*, 24 (1991) 357.

8 R.A. Mackay, *Adv. Coll. Interface Sci.*, 15 (1981) 131.
9 C. Minero and E. Pelizzetti, in: S.E. Friberg and B. Lindman (Eds.), *Organized Solutions*. Dekker, New York, 1992, p. 307.
10 F.M. Menger and C.E. Portnoy, *J. Am. Chem. Soc.*, 89 (1967) 4698.
11 D.J. Piszkiewicz, *J. Am. Chem. Soc.*, 99 (1977) 7695.
12 R. Martinek, A.K. Yatsimirskii, A.V. Levashov and I.V. Berezin, in: K.L. Mittal (Ed.), *Micellization, Solubilization and Microemulsions*. Plenum Press, New York, 1977, Vol. 2, p. 489.
13 A.K. Yatsimirskii, Z.A. Strelsova, K. Martinek and I.V. Berezin, *Khim. Catal.*, 15 (1974) 354.
14 K. Martinek, A.P. Osipov, A.K. Yatsimirskii and I.V. Berezin, *Tetrahedron Lett.* (1975) 1279.
15 A.K. Yatsimirskii, K. Martinek and I.V. Berezin, *Tetrahedron*, 27 (1971) 2855.
16 K. Martinek, A.K. Yatsimirskii, A.P. Osipov and I.V. Berezin, *Tetrahedron*, 29 (1973) 963.
17 L.S. Romsted, in: K.L. Mittal (Ed.), *Micellization, Solubilization and Microemulsions*. Plenum Press, New York, 1977, Vol. 2, p. 509.
18 C.A. Bunton and L.G. Ionescu, *J. Am. Chem. Soc.*, 95 (1973) 2912.
19 C.A. Bunton and B. Wolfe, *J. Am. Chem. Soc.*, 95 (1973) 3742.
20 C.A. Bunton, L.H. Gan and G. Savelli, *J. Phys. Chem.*, 87 (1983) 336.
21 L.S. Romsted and E.H. Cordes, *J. Am. Chem. Soc.*, 90 (1968) 4404.
22 H. Al-Lohedan, C.A. Bunton and L.S. Romsted, *J. Phys. Chem.*, 85 (1981) 2123.
23 C.A. Bunton, L.S. Romsted and C. Thamavit, *J. Am. Chem. Soc.*, 102 (1980) 3900.
24 C.A. Bunton, L.S. Romsted and G. Savelli, *J. Am. Chem. Soc.*, 101 (1979) 1253.
25 C.A. Bunton and L.S. Romsted, in: K.L. Mittal and E.J. Fendler (Eds.), *Solution Behavior of Surfactants: Theoretical and Applied Aspects*. Plenum Press, New York, 1982, Vol. 1, p. 975.
26 C.A. Bunton, J. Frankson and L.S. Romsted, *J. Phys. Chem.*, 84 (1980) 2607.
27 C.A. Bunton, L.H. Gan, J.R. Moffat, L.S. Romsted and G. Savelli, *J. Phys. Chem.*, 85 (1981) 4118.
28 F. Nome, A.F. Rubira, C. Franco and L.G. Ionescu, *J. Phys. Chem.*, 86 (1982) 1881.
29 H. Chaimovich, I.M. Cuccovia, C.A. Bunton and J.R. Moffatt, *J. Phys. Chem.*, 87 (1983) 3584.
30 S. Hashimoto, J.K. Thomas, D.F. Evans, S. Mukherjee and B.W. Ninham, *J. Coll. Interface Sci.*, 95 (1983) 594.
31 F.H. Quina and H. Chaimovich, *J. Phys. Chem.*, 83 (1979) 1844.
32 H. Chaimovich, J.B.S. Bonilha, M.J. Politi and F.H. Quina, *J. Phys. Chem.*, 83 (1979) 1851.

33 J.B.S. Bonilha, H. Chaimovich, V.G. Toscano and F.H. Quina, *J. Phys. Chem.,* 83 (1979) 2463.
34 F.H. Quina, M.J. Politi, I.M. Cuccovia, E. Baumgarten, S.M. Martins-Franchetti and H. Chaimovich, *J. Phys. Chem.,* 84 (1980) 361.
35 T. Sasaki, M. Hattori, J. Sasaki and K. Nukina, *Bull. Chem. Soc. Jpn.*, 48 (1975) 1397.
36 I.M. Cuccovia, R.M.V. Aleixo, N.E. Erismann, N.T.E. van der Zee, S. Schreirer and H. Chaimovich, *J. Am. Chem. Soc.*, 104 (1982) 4544.
37 K.A. Wilk, *J. Phys. Chem.,* 96 (1992) 901.
38 M. da Graça Nascimento, M.A. Lezcano and F. Nome, *J. Phys. Chem.,* 96 (1992) 5537.
39 A.A. Ruzza, M.R.K. Walter, F. Nome and D. Zanette, *J. Phys. Chem.,* 96 (1992) 1463.
40 R. Germani, P. Ponti, G. Savelli, N. Spreti, C.A. Bunton and J.R. Moffatt, *J. Chem. Soc. Perkin Trans.*, 2 (1989) 40.
41 M.F.S. Neves, D. Zanette, F.H. Quina, M.T. Morelli and F. Nome, *J. Phys. Chem.,* 93 (1989) 1502.
42 C.A. Bunton and J.R. Moffatt, *Langmuir*, 8 (1992) 2130.
43 R. Germani, G. Savelli, T. Romeo, N. Spreti, G. Cerichelli and C. A. Bunton, *Langmuir*, 9 (1993) 55.
44 R. Germani, G. Savelli, N. Spreti, G. Cerichelli, C. Mancini and C. A. Bunton, *Langmuir*, 9 (1993) 61.
45 C.A. Bunton and J.R. Moffatt, *J. Phys. Chem.,* 90 (1986) 538.
46 C.A. Bunton and J.R. Moffatt, *J. Phys. Chem.,* 92 (1988) 2896.
47 A. Blasko', C.A. Bunton, Y.S. Hong, M.M. Mhala, J.R. Moffatt and S. Wright, *J. Phys. Org. Chem.*, 4 (1991) 618.
48 M. Almgren and R. Rydholm, *J. Phys. Chem.,* 83 (1979) 360.
49 F. Funasaki, *J. Phys. Chem.,* 83 (1979) 237.
50 H. Chaimovich, R.M.V. Aleixo, I.M. Cuccovia, D. Zanette and F. H. Quina, in: K.L. Mittal and E.J. Fendler (Eds.), *Solution Behavior of Surfactants. Theoretical and Applied Aspects*. Plenum Press, New York, 1982, Vol. 1, p. 949.
51 M.S. Fernandez and P. Fromherz, *J. Phys. Chem.,* 81 (1977) 1755.
52 N. Funasaki, *J. Phys. Chem.,* 83 (1979) 360.
53 S. Diekmann and J. Frahm, *J. Chem. Soc. Faraday I*, 76 (1980) 446.
54 W.J. Gettings and E. Wyn-Jones (Eds.), *Techniques and Applications of Fast Reactions in Solutions*. Reidel, Dordrecht, 1979.
55 M. Gratzel, J.J. Kozak and J.K. Thomas, *J. Phys. Chem.,* 62 (1975) 1632.
56 M.H. Dung and J.J. Kozak, *J. Chem. Phys.*, 76 (1982) 984.
57 J. Frahm and S. Diekmann, *J. Coll. Interface Sci.*, 70 (1979) 440.
58 D.F. Evans, D.J. Mitchell and B.W. Ninham, *J. Phys. Chem.,* 88 (1984) 6344.
59 Y.S. Chao, E.Y. Shen and S.H. Chen, *J. Phys. Chem.,* 89 (1985) 4862.

60 C. Minero, E. Pramauro and E. Pelizzetti, *J. Phys. Chem.,* 92 (1988) 4670.
61 C. Minero and E. Pelizzetti, *Adv. Coll. Interface Sci.,* 37 (1992) 319.
62 E. Rodenas, C. Dolcet and M. Valiente, *J. Phys. Chem.,* 94 (1990) 1472.
63 E.A. Guggenheim and H. Stokes, *Equilibrium Properties of Aqueous Solutions of Single Strong Electrolytes.* Pergamon Press, Oxford, 1969.
64 D. Stigter, *J. Phys. Chem.,* 79 (1975) 1015.
65 R.M. Pashely and B.W. Ninham, *J. Phys. Chem.,* 91 (1987) 2902.
66 L.K.J. Tong and M.C. Glesmann, *J. Am. Chem. Soc.*, 79 (1957) 4305.
67 P. Mukerjee and K. Banerjee, *J. Phys. Chem.,* 68 (1964) 3567.
68 C.A. Bunton, L.S. Romsted and L. Sepulveda, *J. Phys. Chem.,* 84 (1980) 2611.
69 A. Cipiciani, G. Savelli and C.A. Bunton, *J. Phys. Chem.,* 87 (1983) 5259.
70 M.B. Loewe and J.N. Phillips. *Nature (London),* 190 (1961) 262.
71 M.B. Loewe and J.N. Phillips. *Nature (London),* 194 (1962) 1058.
72 J. Simplicio, *Biochemistry,* 11 (1972) 2525.
73 J. Simplicio and K. Schuenzer, *Biochemistry,* 12 (1973) 1923.
74 J.H. Fendler, F. Nome and H.C. Van Woert, *J. Am. Chem. Soc.*, 96 (1974) 6745.
75 F. Nome and J.H. Fendler, *J. Am. Chem. Soc.*, 99 (1977) 1558.
76 E.A. Deardorff, R.A. Cour and J.K. Hurst, *J. Am. Chem. Soc.*, 103 (1981) 6611.
77 J.R. Cho and H. Morawetz, *J. Am. Chem. Soc.*, 96 (1974) 375.
78 S. Tchiyashiki and H. Yamatera. *Bull. Chem. Soc. Jpn.*, 55 (1982) 759.
79 L.L. Hodges and M.A. Araujo, *Inorg. Chem.*, 21 (1982) 3236.
80 H.D. Burrows, J. Ige and S.A. Umoh, *J. Chem. Soc. Faraday I,* 79 (1982) 947.
81 J. Simon, J. LeMoigne, D. Markovitsi and J. Dayantis, *J. Am. Chem. Soc.*, 102 (1980) 7247.
82 A.D. James and B.H. Robinson, *J. Chem. Soc. Faraday I,* 74 (1978) 10.
83 B.H. Robinson and N.C. White. *J. Chem. Soc. Faraday I,* 74 (1978) 2625.
84 J. Holzwarth, W. Knoche and B.H. Robinson, *Ber. Bunsenges. Phys. Chem.,* 82 (1978) 1001.
85 P.D.I. Fletcher and V.C. Reinsborough, *Can. J. Chem.*, 59 (1981) 1361.
86 A.I. Carbone, F.P. Cavasino, E. Di Dio and C. Sbriziolo, *Int. J. Chem. Kinet.*, 18 (1986) 609.
87 V.C. Reinsborough and B.H. Robinson, *J. Chem. Soc. Faraday I,* 75 (1979) 2395.
88 M. Fischer, W. Knoche, B.H. Robinson and J.H. MacLagan, *J. Chem. Soc. Faraday I,* 75 (1979) 119.
89 M. Fischer, W. Knoche, P.D.I. Fletcher, B.H. Robinson and N.C. White, *Coll. Polym. Sci.*, 258 (1980) 733.
90 Y. Miyake, M. Shigeto and M. Teramoto, *J. Chem. Soc. Faraday I,* 82 (1986) 1515.
91 E. Blatt, *Coll. Surfaces,* 34 (1988) 205.

92 F.P. Cavasino, C. Sbriziolo, E. Pelizzetti and E. Pramauro, *J. Phys. Chem.*, 93 (1989) 469.
93 H.S. Kim and C. Tondre, *Langmuir*, 5 (1989) 395.
94 C. Tondre and M. Boumezioud, *J. Phys. Chem.*, 93 (1989) 846.
95 S. Muralidharan, W. Yu, S. Tagashira and H. Freiser, *Langmuir*, 6 (1990) 1190.
96 A.C. Dash, J. Prusti, J. Pradhan and P.K. Das, *J. Chem. Soc. Faraday I*, 86 (1990) 507.
97 S.G. Son, M. Hebrant, P. Tecilla, P. Scrimin and C. Tondre, *J. Phys. Chem.*, 96 (1992) 11072.
98 S. Diekmann and J. Frahm, *J. Chem. Soc. Faraday I*, 75 (1979) 2199.
99 P. Fromherz, *Chem. Phys. Lett.*, 26 (1974) 221.
100 S. Diekmann and J. Frahm, *J. Chem. Soc. Faraday I*, 76 (1980) 446.
101 E. Pelizzetti and E. Pramauro, *Inorg. Chem.*, 18 (1979) 882.
102 E. Pelizzetti and E. Pramauro, *Inorg. Chim. Acta*, 46 (1980) L29.
103 Bruhn and J.F. Holzwarth, *Ber. Bunsenges. Phys. Chem.*, 82 (1978) 1006.
104 V. Eck, M. Marcus, G. Stange, J. Westerhousen and J.F. Holzwarth, *Ber. Bunsenges. Phys. Chem.*, 112 (1982) 261.
105 M. Almgren, F. Grieser and J.K. Thomas, *J. Phys. Chem.*, 83 (1979) 3232.
106 E. Pramauro, E. Pelizzetti, S. Diekmann and J. Frahm, *Inorg. Chem.*, 21 (1982) 2432.
107 C. Minero, E. Pramauro, E. Pelizzetti and D. Meisel, *J. Phys. Chem.*, 87 (1983) 399.
108 F.P. Cavasino, C. Sbriziolo and E. Pelizzetti, *Ber. Bunsenges. Phys. Chem.*, 87 (1983) 843.
109 E. Pelizzetti and E. Pramauro, *J. Phys. Chem.*, 88 (1984) 990.
110 E. Pelizzetti, E. Pramauro and D. Croce, *Ber. Bunsenges. Phys. Chem.*, 84 (1980) 265.
111 K.V. Ponganis, M.A. De Araujo and H.L. Hodges, *Inorg. Chem.*, 19 (1980) 2704.
112 M.A. De Araujo and H.L. Hodges, *Inorg. Chem.*, 21 (1982) 3167.
113 C.A. Bunton and G. Cerichelli, *Int. J. Chem. Kinet.*, 11 (1980) 519.
114 E. Pramauro and E. Pelizzetti, *Inorg. Chem.*, 19 (1980) 1407.
115 A.A. Bhalekar and J.B.F.N. Engberts, *J. Am. Chem. Soc.*, 100 (1978) 5914.
116 E. Pelizzetti and E. Pramauro, *Ber. Bunsenges. Phys. Chem.*, 83 (1979) 996.
117 E. Pelizzetti, E. Pramauro, D. Meisel and E. Borgarello, in: K.L. Mittal and B. Lindman (Eds.), *Surfactants in Solution*. Plenum Press, New York, 1984, Vol. 2, p. 1159.
118 E. Pelizzetti, E. Fisicaro, C. Minero, A. Sassi and H. Hidaka, *J. Phys. Chem.*, 95 (1991) 761.
119 M. Almgren, F. Grieser and J.K. Thomas, *J. Phys. Chem.*, 83 (1979) 3232.
120 G.H. Nancollas (Ed.), *Biological Mineralization and Demineralization*. Springer Verlag, Berlin, 1982.

121 D. Attwood and A.T. Florence, *Surfactant Systems*. Chapman and Hall, London, 1983.
122 E. Staab, L. Addadi, L. Lieserowitz and M. Lahav, *Adv. Mater.*, 2 (1990) 40.
123 C.S. Sikes and A.P. Wheeler, *Chem. Tech.*, 18 (1988) 620.
124 M.F. Hague, N. Kallay, V. Primvan and E. Matijevic, *J. Coll. Surface Sci.*, 137 (1990) 36.
125 E. Matijevic, *Pure Appl. Chem.*, 64 (1992) 1703.
126 E. Matijevic, *Chem. Mater.*, 5 (1993) 412.
127 S. Mann, J. Webb and R.J.P. Williams, *Biomineralization*. VCH, Weinheim, 1989.
128 L. Addadi and S. Weiner, *Angew. Chem. Int. Ed. Engl.*, 31 (1992) 153.
129 J.H. Fendler, *Chem. Rev.*, 87 (1987) 877.
130 J.H. Fendler, *Membrane–Mimetic Approach to Advanced Materials*. Springer Verlag, Berlin, 1994.
131 T. Suhara, K. Esumi and K. Meguro, *Bull. Chem. Soc. Jpn.*, 56 (1983) 2932.
132 J.B. Nagy, *Coll. Surfaces.*, 35 (1989) 201.
133 S. Mann, *J. Inorg. Biochem.*, 28 (1986) 363.

Chapter 4

Spectral analysis in organized amphiphilic media

The use of surfactants to improve determination schemes based on spectral methods was one of the first applications of organized systems in analytical chemistry. During the past 20 years, the unique features of surfactant aggregates have been recognized as being very helpful to overcome a lot of problems associated with the use of organic solvents in molecular absorption/luminescence spectroscopy.

The analytical applications concern the usual molecular spectroscopic techniques, in particular: UV-VIS spectrophotometry, fluorimetry, phosphorimetry and chemiluminescence methods. Since 1979, some reviews (or parts of reviews) have been published dealing with these applications [1–6].

Surfactants and their assemblies, in particular emulsions and microemulsions, have also been introduced in atomic spectroscopy, where they play an important role as dispersants in several metal determinations [7].

In most cases surfactant-modified procedures allow an improvement in the sensitivity and/or the selectivity of spectroscopic determinations, commonly performed in homogeneous solutions, whereas certain analytical methods (such as micelle-stabilized room-temperature phosphorescence) can be exclusively conducted in organized media. The features of these different methods will be separately examined.

4.1. SURFACTANT-MEDIATED UV-VIS SPECTROSCOPY

The current experimental problems encountered in UV-VIS spectral analysis are mainly concerned with at least one of the following points:

- scarce solubility of analytes and/or their coloured derivatives in water, which often leads to the use of extraction in organic solvents;
- lack of stability of the species determined;
- low selectivity due to spectral overlap between the absorbing products and the reagents;
- interference effects from other components of the sample;
- slow reaction rates between the species to be determined and the reagents.

Most of the above-mentioned unfavourable processes can be adequately overcome by working with proper surfactant molecules and with their assemblies.

Ionic amphiphiles are usually employed as ion-pair forming reagents, whereas micellar solutions formed from either ionic and nonionic surfactants are used as suitable solvents or as reaction media. In the first case, the formation of hydrophobic ion pairs favours the analyte extraction into organic solvents, the advantage being related to the increased solubility of the extracted species. Thus, the UV-VIS determination becomes more sensitive.

Another crucial point is that the solubilization in micellar solutions allows the location of the analyte in different regions of the aggregates, altering the solute spectral properties through electrostatic, hydrophobic and medium effects.

4.1.1. Modification of UV-VIS spectra of organic analytes upon binding to surfactant aggregates

Changes in the ultraviolet absorption maxima as well as in vibrational fine structure of several organic molecules have been observed upon solubilization in aqueous micellar solutions. The similarities between absorption spectra obtained in micellar solutions and in low-polarity solvents was generally interpreted as indicating a low-polar environment of the analyte in the micelle. Conversely, the resemblance of the absorption spectra of solutes in micellar solutions and in polar solvents was assumed to indicate a probe location near or at the micelle-water interface. Using this simple qualitative approach, a comparison of the UV spectra of ethylbenzene obtained in heptane, water and 0.1 M SDS (see Fig. 4.1), suggests a non-polar solubilization site, probably in the core region.

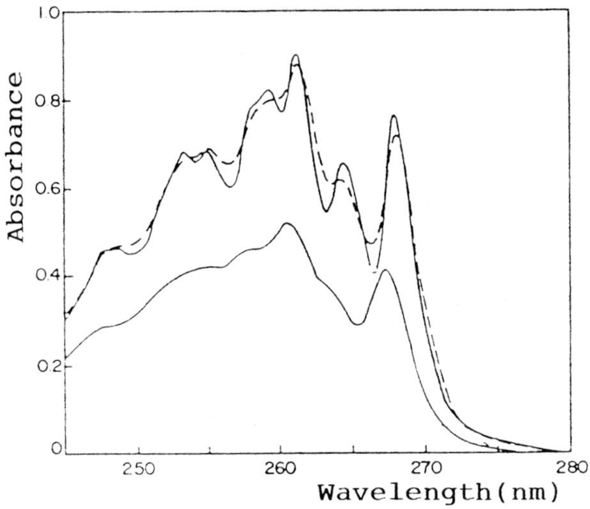

Fig. 4.1. UV spectra of ethylbenzene in different solvent systems. Upper continuous line: heptane; dashed line: 0.1 M SDS; lower continuous line: water. Reproduced from Ref. [8], with permission.

However, considerable caution is necessary when spectral data in micellar media are analyzed. For example, the UV spectra of naphthalene in potassium dodecanoate, dodecylammonium chloride and Brij 35 are very similar to that in water if the region between 240 and 300 nm is considered, whereas the absorption bands at 220 and 310 nm resemble those observed in non-polar solvents. In order to explain these contradictory observations, a peculiar orientation of the naphthalene molecule (in part imbedded in the hydrocarbon region of the micelle and in part placed near the polar head groups) was invoked [9].

On the contrary, an aqueous solubilization site was assumed for chloroxylenol in Cetomacrogol 1000 (polyoxyethylene(24)hexadecanol) after examination of the corresponding absorption spectra in micellar solution, in water and in cyclohexane [10]. In fact, the absence of fine structure in water and in Cetomacrogol 1000 in the 270–285 nm region was attributed to hydrogen-bond formation, possible only in hydrated micellar regions.

The solute concentration is also a very important factor. Qualitative comparisons of UV spectra led some investigators to indicate a preferential location in the less polar region of ionic micelles for ethylbenzene and benzene [11,12], whereas from the spectral shifts of the same molecules in solutions containing a very low mole fraction of solute (high concentration of surfactants), other authors [13] concluded that the probes were imbedded in a moderately polar environment.

The above contradictory results have been interpreted assuming that the solutes are partitioned between a "dissolved" state (where the hydrophobic interactions with the hydrocarbon-like core play a dominant role) and an "adsorbed" state where the π-electron system of aromatic molecules is able to interact with water dipoles. The surface activity of the investigated aromatic species controls the adsorbed fraction, which is higher for solutes having lower surface tension (γ) values and for solutes present at very low concentrations [13].

Although a general scheme for the interpretation of the spectral effects observed upon micellization is not available, practical applications based on either the wavelength shifts and the absorbtivity coefficient variations have been abundantly reported. Chromogenic derivatization is another interesting approach [138]. In some cases, as discussed in the next section, the effective polarity of the microscopic environment can be correlated with the spectroscopic properties of probe solutes.

4.1.2. Changes of UV-VIS spectra of micellized organic molecules as indicators of the microaggregate polarity

Some UV spectroscopic characteristics of probe solutes, sensitive to the polarity of the medium, have been used to determine the average polarities of the micellar microenvironments. In many instances, linear correlations between the position of the absorption maximum (λ_{max}) and solvent polarity parameters (such as Reichardt and Dimroth's $E_T(30)$, Kosower's Z or Grunwald–Winstein's γ) have been found.

In particular, the solvatochromic parameter $E_T(30)$ is one of the most widely used [14,15]. It is based on the $\pi \rightarrow \pi^*$ absorption band of a pyridinium-N-phenoxide betaine dye, which is soluble in many organic solvents and solvent mixtures and shows one of the largest peak shifts upon changing the solvent polarity. In fact, its longest-wavelength intramolecular charge-transfer (CT) absorption band is hypsocromically shifted by 375 nm (9730 cm^{-1}) on going from diphenyl ether (λ_{max} = 810 nm) to water (λ_{max} = 453 nm).

$E_T(30)$ is defined as the excitation energy (expressed in kcal mol^{-1} or kJ mol^{-1}) of the dye, which is in turn related to the maximum of the spectrum according to the equation:

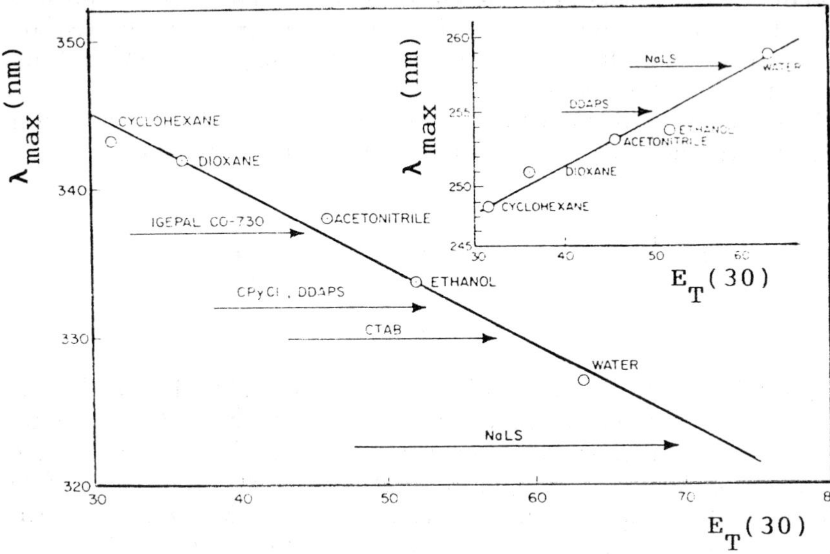

Fig. 4.2. Absorption maxima of benzophenone as a function of the solvent polarity parameter $E_T(30)$. Surfactant concentration: 0.1 M. CPyCL = HPC; NaLS = SDS; DDAPS: 3-(dimethyldodecylammonio)propane-1-sulphonate; Igepal CO-730: polyoxyethylene(15)-nonylphenol. Reproduced from Ref. [10] with permission.

$$E_T(30) = hc\nu_{max}N_A = 2.8591 \times 10^{-3}\ \nu/\text{cm}^{-1} \tag{1}$$

where ν_{max} is the wave number (in cm^{-1}) corresponding to the absorption maximum.

The $E_T(30)$ values for more than 200 solvents have been determined and other indicator molecules have been also used. Figure 4.2 shows the behaviour of the $n \rightarrow \pi^*$ transition of the probe solute benzophenone as a function of $E_T(30)$ for different solvent systems.

A good linear correlation exists between the wavelength maximum and the microscopic solvent polarity of the reported pure solvents. If the same solute is dissolved in aqueous micellar solutions, its absorption maximum reflects different local micropolarities experienced by the chromophore, which in turn depend on the nature of the surfactant used.

Other organic solutes, such as pyrene-3-carbaldehyde [16] or surface-active merocyanine dyes [17], have been employed as suitable probes of micellar environments. In particular, merocyanine dyes exhibit large shifts in UV-visible absorption maxima towards shorter wavelengths as the medium polarity increases and this effect has been attributed to a different stabilization energy of charge separation in the ground and first excited electronic states [18].

The structure of the indicator **1** (4-[2-(1-hexadecyl-1,4-dihydropyridin-4-ylidene)ethylidene]cyclohexa-2,5-dien-1-one) is shown below:

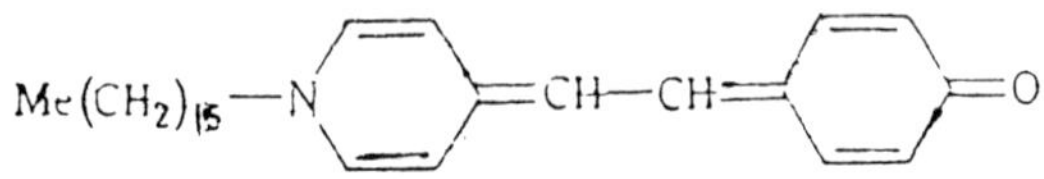

(**1**)

Figure 4.3 shows the absorption spectra of compound **1** at pH 6, at three different concentrations of the cationic surfactant dodecylammonium chloride (DAC): below (**a**), near (**b**), and above the cmc (**c**), respectively. The marked spectral change is due to the incorporation of the probe molecules within the aggregates. The absorption at 390 nm was attributed to the protonated dye and that at 475 nm to the neutral form.

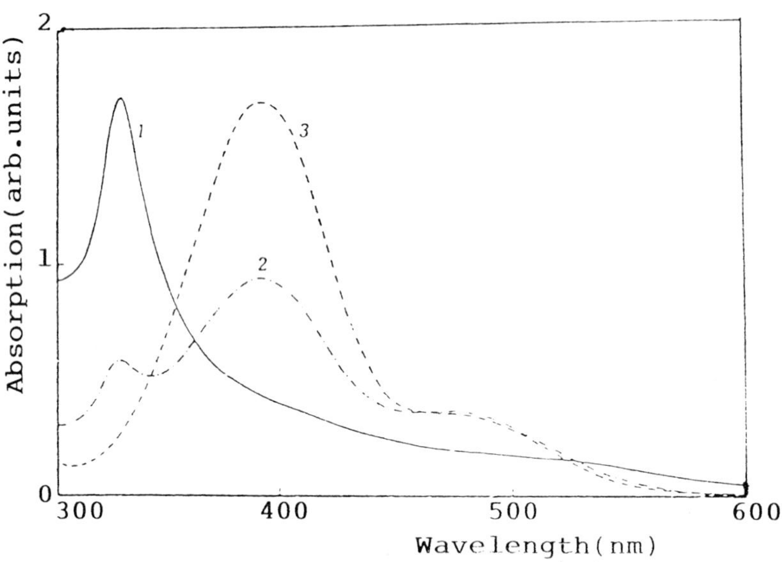

Fig.4. 3. UV-VIS absorption spectra of the merocyanine dye **1** in DAC micellar solutions. [dye]: 1×10^{-5} M. [DAC]: 9×10^{-3} M (1); 1.6×10^{-2} M (2); 2.2×10^{-2} (3). Reproduced from Ref. [17] with permission.

A comparison of the λ_{max} values determined in different organic solvents indicates an E_T value of 235 kJ mol^{-1} for the absorption at 475 nm, which means that the micellized dye is imbedded in a polar environment (the dielectric constant, D, corresponds to that of methanol). The relative intensities of the two observed maxima in the micellar solution reflect the hydrogen ion concentration in the Stern layer and, thus, can give useful information about the local acidity in different cationic micelles.

Mukerjee and Cardinal [13] introduced other UV-spectroscopic parameters that are sensitive to the polarity of the medium, applied to aromatic molecules used as probe solutes. The following terms were defined:

– A_s/A_p: ratio of the absorbances at the position of a solvent-induced band and the adjacent vibronic peak (at the lower wavelength).

– R_C: the ratio of the absorbances at two fixed wavelengths, one near a peak and other near the adjacent valley. $R_{C(m)}$ refers to a compound solubilized into the micelles, whereas $R_{C(w)}$ corresponds to the aqueous spectra.

– R_{pv}: the ratio of the absorbances at the peak and at the valley obtained from the complete spectrum.

The wavelengths (λ_{max}) of the absorption peaks in reference solvents and in the investigated micellar solution were compared. In general, the refractive index of the environment is the main controlling factor for λ_{max} (with the exception of methanol) and a shift towards the longer wavelength is observed as the solvent becomes less polar.

The $R_{C(m)}$ value leads to an estimate of the average polarity of solubilized compounds using the calibration curves of R_C against the dielectric constants of reference solvents. The analyte mass balance is given by:

$$C_T = C_m\phi_m + C_w\phi_w \tag{2}$$

where C_m and C_w are the solute concentrations in the micelles and in water, respectively; ϕ_m is the volume fraction of micelles and ϕ_w is the volume fraction of water.

$$R_C = \frac{A(\lambda_1)}{A(\lambda_2)} = \frac{\varepsilon_{1(m)}C_m\phi_m + \varepsilon_{1(w)}C_w\phi_w}{\varepsilon_{2(m)}C_m\phi_m + \varepsilon_{2(w)}C_w\phi_w} \tag{3}$$

Taking into account the definition of the partition coefficient of solutes between micelles and water ($P_{MW} = C_m/C_w$), the following equation can be derived:

$$R_C = R_{C(m)} - \frac{\varepsilon_{2(w)}(R_C - R_{C(w)})\phi_w}{\varepsilon_{2(m)}P_{MW}\phi_m} \qquad (4)$$

According to Eqn. (3), from plots of R_C vs. $(R_C - R_{C(w)})\phi_w/\phi_m$, both $R_{C(m)}$ and P_{MW} can be determined (see Fig. 4.4). For example, the estimated P_{MW} value for benzene in SDS is in the range 65–85, in good agreement with partition data from other experimental techniques. The $R_{C(m)}$ value gives an estimate of the average polarity of the dissolved probe, provided that calibration curves of R_C against the dielectric constants of pure solvents are available.

Table 4.1 lists the dielectric constants of some organic solvents and those attributed to different regions of SDS from the comparison of the R_C values of alkylbenzene solutes. It can be seen that even the more hydrophobic aromatic solutes are imbedded in a partially polar environment. In fact, for a probe dissolved in a fully hydrocarbon medium, a lower D value should be expected (D ≈ 2 for heptane).

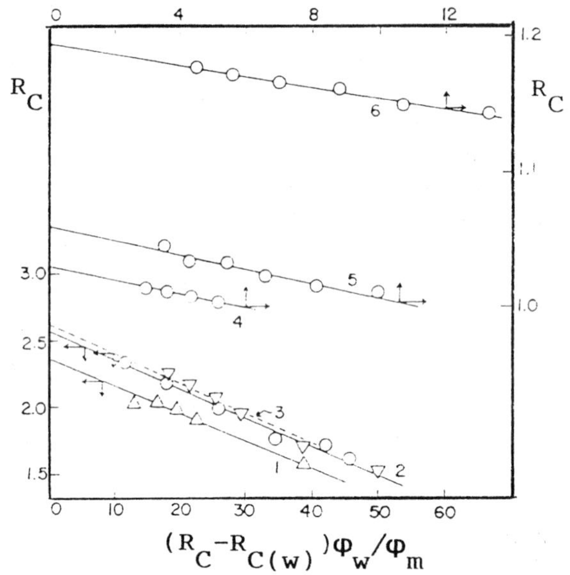

Fig. 4.4. Plots of R_C vs. $(R_C - R_{C(w)})\phi_w/\phi_m$ for different aromatic solutes. 1: benzene in HTAC (R_C: 260.4/259.0); 2: benzene in SDS;(R_C: 260.4/259.0); 3: benzene in HTAC (R_C: 260.7/259.0); 4: ethylbenzene in SDS (R_C: 264.4/263.3); 5: β-phenethyl alcohol in SDS (R_C: 264.2/263.0); 6: p-xylene in SDS (R_C: 268.4/267.1). The wavelengths at which the absorbances were measured are indicated. Reproduced from Ref. [13] with permission.

TABLE 4.1

Different polarities experienced by alkylbenzenes in SDS micelles[a]

Organic solvent	D	Probe solute (in SDS)	D
50%Methanol–water	55	benzene	49
70%Methanol–water	45	p-xylene	40
		ethylbenzene	41
Methanol	33	n-butylbenzene	36
95%Ethanol	26	hexamethylbenzene	23
2-Propanol	18	p-di-tert-butyl-	16

a: Data taken from Refs. [8] and [13].

A correction for the spectral shifts occurring on solubilization is necessary, and this is usually performed by estimating the R_C change in a reference solvent having a similar dielectric constant. For example, the benzene spectrum in HTAC is red-shifted about 0.3 nm in respect to that obtained in reference solvents having similar D values. The corrected R_C measurements are those of curve 3 (Fig. 4.4). However, these corrections do not change very much the estimated dielectric constant of the local microenvironment, which is lowered by ca. 6%.

4.1.3. UV-VIS Spectroscopy of metal–chelate complexes

One of the more popular instrumental methods for the analysis of metal ions is based on their complexation reaction with suitable ligands, usually termed chelometric indicators. If the experimental conditions are carefully chosen, the absorbance of the metal chelate is proportional to the metal ion concentration.

Typical problems connected with these spectral methods are the lack of specificity and sensitivity and the poor solubility in water of some of the chelates formed. The positive effects obtained by adding surfactants, below and above their cmc, are related to the following factors:

- increased solubility of hydrophobic chelates in the micellar pseudophase;
- spectral effects including the increase of the molar absorptivity of the complexes and a more or less relevant batochromic shift of the absorption maximum;
- modification of the complexation equilibria, with the formation of

ternary species including the metal ion, the ligand and the surfactant molecules;
- changes in the acid-base equilibria of the ligands;
- favourable kinetic effects.

Although in most cases sensitization is due to a combination of the above-mentioned effects, each factor will be separately analyzed.

4.1.3.1. Solubilization of neutral chelates by micellar aggregates

The organic dyes used as metallochromic reagents contain polar groups with donor atoms capable of dissociating protons and/or coordinating the metal ion (see Fig. 4.5). The solubility in water of the formed chelates depends on the presence of free ionizable groups, otherwise the solubility of the complexes is usually lower than that of the corresponding free ligands.

The addition of surfactants (generally non-ionics) allows the solubilization within or at the interface of the micelles. In these cases sensitization suddenly appears just above the actual surfactant cmc and the spectral changes become more pronounced as the surfactant concentration increases, up to a limit corresponding to the complete micellar solubilization of the complex. Triton X-100 and other polyoxyethylene-type surfactants are largely used in these methods.

Table 4.2 lists a series of improved spectrophotometric determinations of metal chelates using the above simple approach.

4.1.3.2. Spectral effects due to the micellar environment

The differences in the absorption spectra between the coloured metal-chelates and the free ligands are due to the different electron arrangement within the molecular orbitals. In fact, the coordination with a metal cation implies the delocalization of the π-electrons in the dye and the corresponding appearance of the complex absorption bands at longer wavelengths.

A great number of published papers on UV-VIS surfactant-sensitized metal complexes determinations have been concerned with the use of a variety of organic metallochromic reagents (in particular triphenylmethane derivatives) in the presence of cationic surfactants.

Comparison of spectral properties of the complexes in the absence and in the presence of such amphiphiles usually shows an up to ten-fold increase in the molar absorptivity coefficient (ε can attain values in the range of 10^5 M^{-1} cm^{-1}) and batochromic shifts of the maximum wavelength in the range 20–100 nm.

A

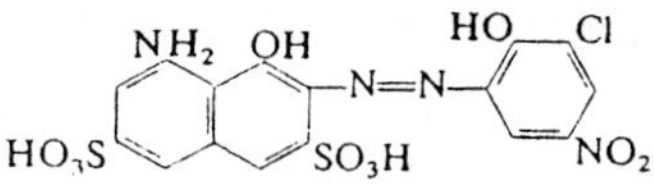

Gallion

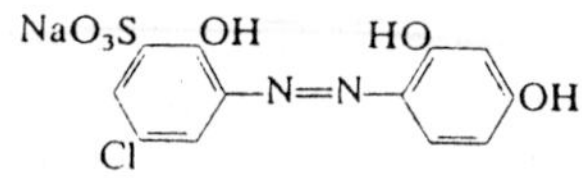

Lumogallion

B

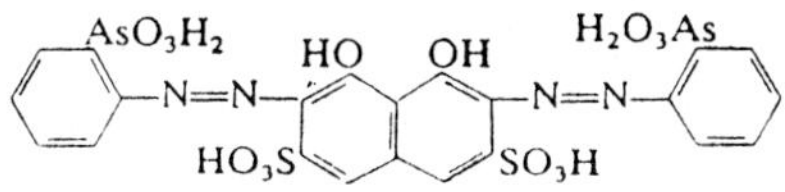

Arsenazo III

C

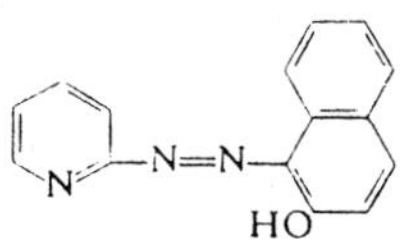

PAN

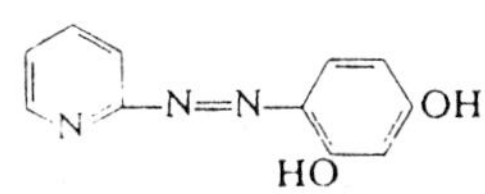

PAR

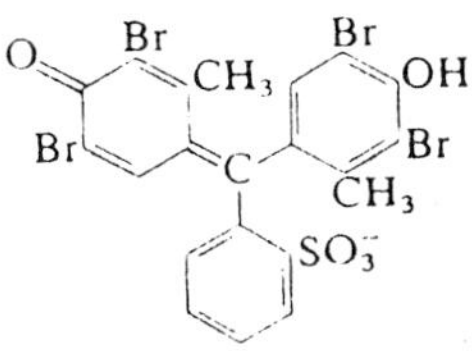

Bromcresol green

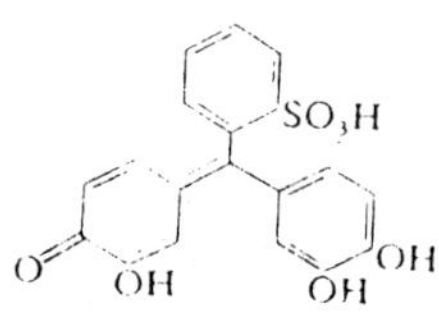

Pyrocatechol violet

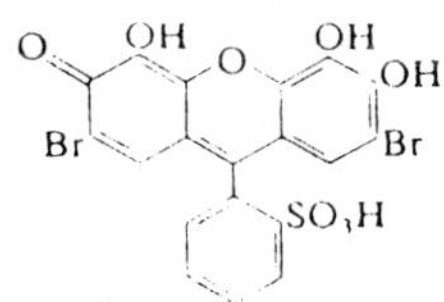

Brompyrogallol red

D

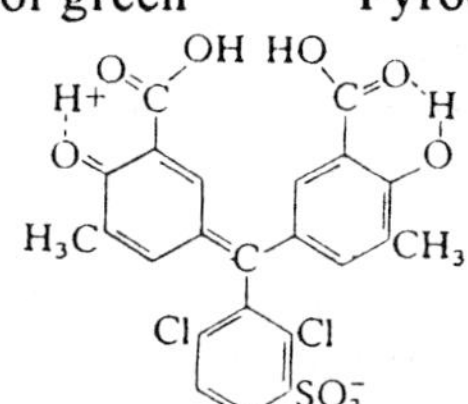

Chrome Azurol S

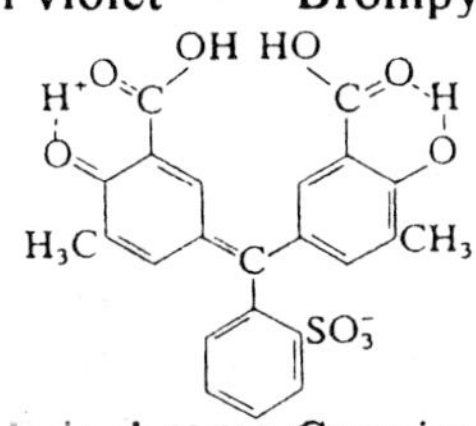

Eriochrome Cyanine R

Fig. 4.5. Molecular structures of some metallochromic ligands used in spectrophotometric determinations of metals. Typical organic groups: aryl monoazo (A); aryl bisazo (B); heterocyclic monoazo (C); triphenylmethane derivatives (D).

TABLE 4.2

Spectrophotometric determination of metal chelates in non-ionic surfactant media

Metal	Ligand	Surfactant	λ_{max}(nm)	εC^a	pH	Ref.
Cd(II)	Cadion[b]	Triton X100	477	1.7	12.5	19
Cd(II)	Cadion	Triton X100	480	11.9	[c]	20
Ag(I)	Cadion 2B	Triton X100	565	10.0	9.2	21
Cd(II)	PAN	Nemol K1030	555	4.9	9.0	22
Ti(IV)	PAN/H_2O_2	Nemol K130	529	1.1	4.0	24
Bi(III)	QNPH	Brij 58	520	9.0	1.8	25
Fe(II)	5-BrPADAP	PONPE 10	558	7.1	4.7	26
Fe(III)			595	7.4	4.4	
Cu(II)			559	9.5	9.8	
Zn(II)			558	13.0	9.8	
Cd(II)			561	13.0	9.8	
Ni(II)			563	11.0	9.8	
Co(III)			590	8.7	4.4	

a: molar absorptivity coefficient (in M^{-1} $cm^{-1} \times 10^{-4}$); *b*: dual-wavelength method; *c*: KOH 0.4 M. QNPH: o-hydroxyhydroquinonephthalein; 5-Br-PADAP: 2-(5-bromo-2-pyridylazo)-5-diethylaminophenol; PONPE 10: polyoxyethylene(10)-4-nonylphenyl ether; Nemol K1030: polyoxyethylenenonylphenol (HLB: 15.1).

If the structures of the ligands commonly used in these spectral methods are considered, it appears that a common feature of all these molecules is the presence of some characteristic polar groups, such as $-SO_3H$, –OH, –COOH. The quinoid system is usually the chromophore of these compounds and the free ligands are characterized by the corresponding absorption bands, which are in turn dependent on the degree of ionization of the polar groups.

The existence of ion-association complexes formed below the cmc between ionic surfactant molecules and ligands with opposite charge has been reported [2,27–29]. These species are electrically neutral and usually poorly soluble in water, but readily soluble in the micelles. Similar interactions are also observed between ionic metal chelates and oppositely charged amphiphiles.

However, these electrostatic interactions cannot be considered as the sole cause of the spectral effects. In fact, hydrophobic bulky ions (such as tetraethylammonium or diphenylguanidinium) which are able to give ion pairs but do not form micelles, are inactive. A more realistic approach considers the incorporation of the anionic complexes (or ligands) in the micelle Stern layer, where hydration of the ionic groups, attractive interactions between surfactant cationic heads and aromatic π-electrons, and hydrophobic interactions between surfactant chains and ligand organic moieties, can simultaneously occur. The corresponding drastic changes of the chromophore environment may be considered responsible for the observed spectral effects [4].

4.1.3.3. Formation of new complex species

The addition of cationic surfactants to a negatively charged metal–chelate complex may originate in the formation of new analytical species, usually indicated as "ternary complexes", in which one or more molecules of surfactant bind to the anion, mainly through electrostatic interactions.

These complexes are electrically neutral and are scarcely soluble in water. They exhibit stoichiometric dye/surfactant ratios.

Figure 4.6 shows that the spectrum of the ternary complex Nb-bromopyrogallol red (BPR)-hexadecylpyridinium, obtained in the presence of HPB micelles, is very different from that of the binary Nb-BPR complex [30]. In the absence of surfactant two binary complexes were observed in 1 M HCL, corresponding to 1:1 and 1:3 (Nb:BPR) species. In the presence of surfactant (even below the cmc) sensitization occurs with formation of

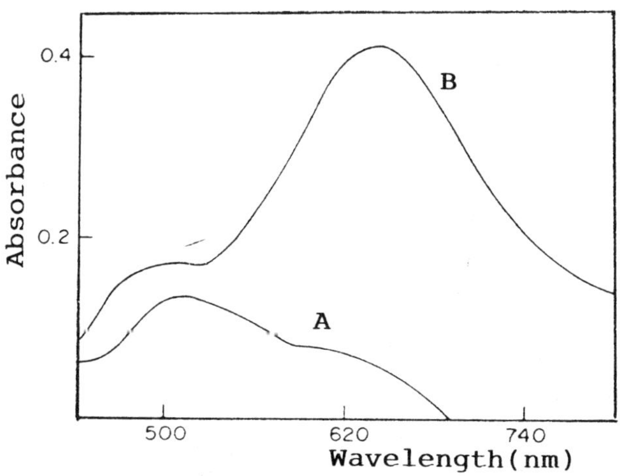

Fig. 4.6. Absorption spectra of Nb-BCR complexes in 1 M HCl. (A): binary complex; (B): ternary complex in the presence of HPB. Reproduced with permission from Ref. [30].

ternary complexes having different stoichiometric compositions. When both ligand and surfactant are in large excess, the continuous variation method indicated the formation of $NbL_2^{2-}(HP^+)_2$ species.

Since electrostatic interactions are at the basis of the ternary complex formation, significant ionic strength effects are expected in these determinations. These effects, however, seem to operate in opposite directions depending upon the system considered.

For example, the interaction of the anionic ligand BPR with HPB micelles at pH 4 was examined and, upon increasing the ionic strength, a released dye into the aqueous bulk was reported [4]. The saturation of the Stern layer by strong electrolytes explains the dissociation of the dye–micelle pairs.

However, for Be-CAS complexes dissolved in HTAB micellar solutions,

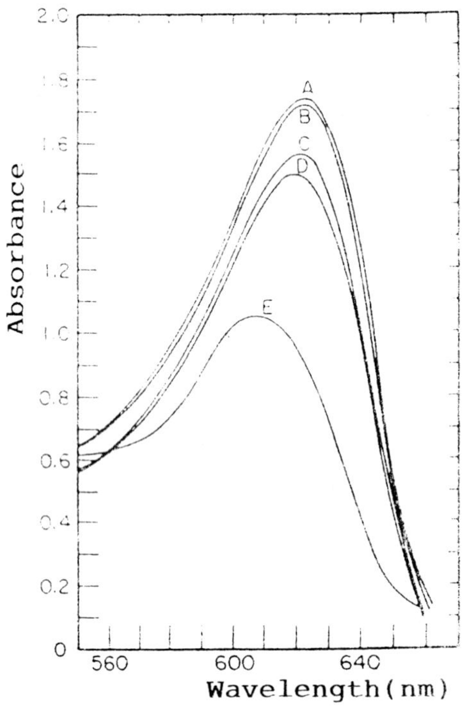

Fig. 4.7. Effect of different added salts on Be-CAS visible spectra. Be(II): 2.2×10^{-5} M, CAS: 1.0×10^{-4} M, HTAB: 1.0×10^{-3} M, pH 6.7. A: = 0.1 M $NaNO_3$; B: 0.1 M NaBr; C: 0.1 M NaCl; D: 0.1 M Na_2SO_4; E: no added salt. Reproduced from Ref. [31] with permission.

molar absorptivity increases have been observed upon addition of different sodium salts, in the order: no salt < Na_2SO_4 < NaCl < NaBr < $NaNO_3$ [31]. These absorbance increases levelled off at salt concentrations greater than ca. 0.2 M. In the absence of HTAB, the effect of added salts (up to 0.4 M) on the absorbance of aqueous Be-CAS solutions was negligible.

The observed salt effects, which also include red shifts of the absorption maxima (see Fig. 4.7) cannot be explained on the basis of competitive ion-exchange between the salt anions and the negatively charged complexes for the cationic micelle surface since the complex release must be accompanied by an absorbance decrease. In this case, other changes due to electrolyte addition should be considered (such as the increased aggregation number and micellar size), which lead to a modified microenvironment around the micelle-bound complex.

Table 4.3 lists some selected examples of surfactant-sensitized determinations in which the formation of ternary complexes has been clearly recognized.

4.1.3.4. Changes in the acid–base equilibria of the ligands

Surfactant aggregates can affect not only the electronic structure of the ligands but also their pK_a values, since stabilization (or destabilization) of charged forms occur depending on the type of ionic micelle used. It was shown in Chapter 3 that micellar effects on acid–base equilibria can be rationalized using different models, provided the surface potential and other medium parameters (such as the ionic strength) are known. The typical effects observed on some common metallochromic indicators will be examined in this light.

Most of the complexing agents using in spectrophotometry are weak organic acids, thus the proton release (which favours the chelate formation) is facilitated in cationic micellar solutions. The interactions between the anionic dyes and cationic surfactants induce spectral changes comparable to those observed by increasing the pH of the dye solution.

The pK_a changes reflect the reduction of the free energy of the anionic form, which decreases more than that of the undissociated one, when interacting with the positive surfactant head groups. Hydrophobic interactions also play a fundamental role.

Several quantitative treatments have been developed for explaining micellar effects on indicator equilibria (see Chapter 3). For example, a detailed study concerning the indicator bromocresol green (BCG) in the presence of Septonex (1-ethoxycarbonylpentadecyltrimethylammonium bromide) is described in Ref. [41]. In that work, the pK_a changes with

TABLE 4.3

Examples of surfactant-sensitized UV-VIS determinations of metal chelates through the formation of ternary species

Metal/Ligand/Surfactant	λ (nm)	ε (M^{-1} cm^{-1})$\times 10^{-4}$	pH	Ref.
La(III)/XO/HPB(1:2:4)	625	9.20	7.5	32
Al(III)/PCV/HTAB(1:2:5)	670	0.53	10.5	33
Pb(II)/PCV/HPC	645	5.06	7.6	34
Pb(II)/PCV/Zeph	665	4.20	8.0	
Pb(II)/PCV/HDMAC	663	4.67	8.3	
Fe(III)/CAS/HTAB	628	13.20	2.5	35
Fe(III)/ECR/HTAB	605	10.00	3.1	
Fe(III)/PCV/HPC	680	6.40	7.2	
Fe(III)/CBG/HTAC	693	14.30	5.5	36
Ga(III)/CBG/HTAC	662	14.40	6.5	37
Nb(V)/APAC/HPB (1:1:4)	560	1.87	HCL 1 M	38
Nb(V)/APAC/HPB (1:1:2)	650	0.74	HCl 2–4 M	
Be(II)/CAS/HTAB (1:2:2)	608	5.50	6.7	39
	622	8.37^{a}	6.7	
Cu(II)/CAS/HTAB (1:2:2)	620	4.80		
	620	11.10^{a}		
Al(III)/CAS/HTAB (1:3:2)	650	3.20		
	650	7.70^{a}		
U(VI)/CAS/ST	610	9.20	5.0	40

a: 0.1 M NaBr added. HDMAC: hexadecyldimethylbenzylammonium chloride. Zeph: tetradecyldimethylbenzylammonium chloride. ST: Sterinol (benzyldimethyldodecylammonium chloride). APAC: p-arsonophenylazochromotropic acid. XO: Xylenol Orange.

micellized surfactant concentration have been explained in terms of ion pair formation between the dye anion and the cationic surfactant molecules.

More recently, the pseudophase ion-exchange (PIE) model has been successfully applied by Romsted and Zanette [42] to describe the pK_a changes of the ligand PADA (pyridine-2-azo-p-dimethylaniline) in anionic micelles. The derived equations are of general interest since they can also be applied to anionic ligands interacting with cationic micellar aggregates.

For hydrophobic ligands (as PADA) which are slightly soluble in water, both the forms L and HL^+ are assumed to be completely bound to the anionic aggregates. If ionization occurs in the micellar pseudophase, the intrinsic acidic constant is given by:

$$K_{a(m)} = \frac{[L]_m \, C_{H^+m}}{[HL^+]_m} \tag{5}$$

where the subscript m indicates the solute location in the micellar pseudophase and C_{H^+m} is defined as:

$$C_{H^+m} = [H^+]_m / C_D \tag{6}$$

When hydrogen ion concentration in the micellar phase is considered, a location in the Stern layer is assumed, whereas no restrictions are considered for the ligand location in different micellar sites. Taking into account the ion-exchange equilibrium at the micellar surface:

$$H^+_m + Na^+_w \rightleftarrows H^+_w + Na^+_m \tag{7}$$

The micellar selectivity constant can be defined as follows:

$$K_H^{Na} = \frac{[H^+]_w \, C_{Na^+m}}{[Na^+]_w \, C_{H^+m}} \tag{8}$$

where $C_{Na^+m} = [Na^+]_m / C_D$. Sodium ions are also considered to be located in the Stern layer. Defining β (assumed as constant) as the fraction of micelle surface covered by the counterions:

$$\beta = C_{H^+m} + C_{Na^+m} \tag{9}$$

and, taking into account that for pH values higher than ca. 4 and surfactant concentrations higher than ca. 0.05 M, $C_{Na^+m} >> C_{H^+m}$, the following approximation is valid under these experimental conditions:

$$\beta = \sim C_{Na^+m} \tag{10}$$

By combining Eqns. (8) and (10) and rearranging:

$$C_{H^+m} = \frac{\beta [H^+]_w}{K_H^{Na} \, [Na^+]_w} \tag{11}$$

After concentrations conversion to activities, combination of Eqns. (5) and (11) leads to:

$$\frac{[L]}{[HL^+]} = \frac{[L]_m}{[HL^+]_m} = \frac{K_{a(m)} K_H^{Na} \, a_{Na^+}}{\beta \, a_{H^+}} \tag{12}$$

The ratio between the two forms of the indicator can be experimentally determined from spectrophotometric measurements, according to the equation:

$$\frac{[HL^+]}{[L]} = \frac{A - A_L}{A_{HL^+} - A} \tag{13}$$

where A is the measured absorbance of the indicator, A_L is the absorbance of the fully deprotonated form, and A_{HL^+} is the absorbance at low pH, when the ligand is completely protonated. Starting from literature data, both K_H^{Na} and β are available for SDS [43,44], whereas a_{Na^+} can be estimated from:

$$a_{Na^+} = \gamma^{\pm}[Na^+]_w = \gamma^{\pm} \left\{[Na^+]_{int} + cmc + (1 - \beta)C_D\right\} \tag{14}$$

where $[Na^+]_{int}$ represents the sum of concentrations of the introduced Na^+ ions (from buffers, inert salt, etc.).

Plots of $[L]/[HL^+]$ vs. a_{Na^+}/a_{H^+} satisfactorily fit a simple linear equation with zero intercept, from which $K_{a(m)}$ can be calculated (see Fig. 4.8). For PADA a pK_a increase of ca. 1.6 units (in respect to the corresponding value in water) has been measured. This value is within the pK_a shift range determined for other acid–base indicators dissolved in ionic micelles. The effect of ionic strength on surfactant-sensitized pK_a changes can be also satisfactorily predicted using the PIE model.

4.1.3.5. Kinetic effects on chelate formation reactions

The complex formation is usually the rate-determining step in several liquid–liquid extractions of metal chelates, performed prior to the spectrophotometric analysis. Since the reaction rates can be drastically varied in the presence of surfactant aggregates, these effects are usually exploited in surfactant-modified spectrophotometric methods.

Accumulation of reactive cations and ligands in the low-volume micellar pseudophase favours the chelate formation and, thus, speeds up the analysis. For example, anionic SDS micelles increase the reaction rate between Ni^{2+} and PADA (pyridine-2-azo-p-dimethylaniline) due to the

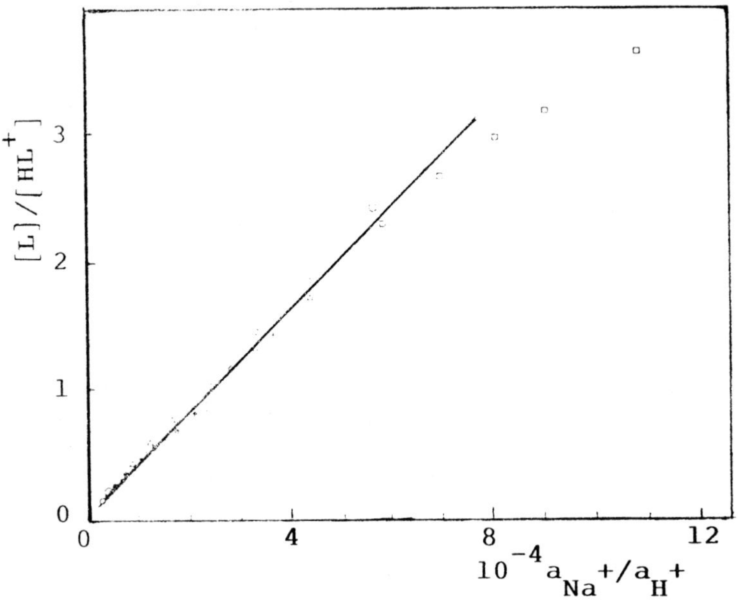

Fig. 4.8. Plot according to Eqn. (12) for the ligand PADA in 0.02 M SDS and 0.02 M succinate buffer, at different initial pH values: (□) 5.876; (Δ) 5.496; (●) 5.185; (O) 4.820. Taken from Ref. [42], reproduced with permission.

relevant micelle–ligand binding and to the electrostatic attraction of the metal ions, as demonstrated by James and Robinson [45].

On the contrary, cationic micelles reduce the reaction rate between hydrophobic ligands and metals since electrostatic repulsion between aggregates and analytes occurs.

Non-ionic micelles can also inhibit the reaction if the ligand is partitioned between the aggregates and the bulk because the compartmentalization of the reagents reduces the active collisions in the aqueous phase. However, the above-described effects can be exploited to suppress undesired interferent reactions.

Some examples of positive and negative effects of surfactant aggregates on chelate formation kinetics are listed in Table 4.4, where the charge of the complexes is also indicated. The ligands were present in their neutral form.

The main features of micellar catalysis on reaction-rate methods and their useful applications in flow-injection analysis will be further examined in Chapter 7.

TABLE 4.4

Observed micellar effects on complex formation kinetics

Metal	Ligand	Surfactant	Charge	Reaction rate	Ref.
Ni(II)	PAP[a]	SDS	+	Increased	46
Co(II)	PADA	SDS	++	Increased	
Ni(II)	PAN	SDS	+	Increased	
Ni(II)	PADA	HTAN	++	Decreased	47
Ni(II)	PADA	Triton X100	++	Decreased	
Cu(II)	BZA[b]	SDS	+	Increased	48
Cu(II)	BZA	DTAC	+	Slightly decreased	
Ni(II)	8-HQ[c]	HTAB	+	Decreased[d]	49
Ni(II)	C_{11}-HQ[e]	HTAB	+	Strongly decreased	
Ni(II)	8-HQ	SDS	+	Increased	50
Fe(III)	PAS-C_n[f]	Brij 35	+	Decreased	51

a pyridine-2-azo-p-phenol; *b*: 1-phenylbutane-1,3-dione (benzoylacetone); *c*: 8-hydroxyquinoline;*d*: reaction rate compared with that measured in methanol; e: 7-(4-ethyl-1-methyloctyl)-8-hydroxyquinoline; *f*: 4-((alkylcarbonyl)amino)-2-hydroxybenzoic acids.

4.2. MICELLAR EFFECTS ON FLUORESCENCE

Studies on unimolecular and bimolecular processes involving electronically excited molecules in organized amphiphilic media are very important from two points of view:

(i) knowledge of the photophysics and the photochemistry of probe substrates allows information to be gained concerning the structure and properties of the surfactant host aggregates;

(ii) the peculiar properties of the excited molecules, when present in such microheterogeneous environments, can be exploited for analytical purposes.

The medium effects provided by microaggregates can be better understood by considering the excitation and deactivation processes that a given organic molecule can undergo upon absorption of light of proper wavelength. The classical Jablonski diagrams represent the electronically excited singlet and triplet states, their possible interconversion and the internal deactivation pathways (see Fig. 4.9).

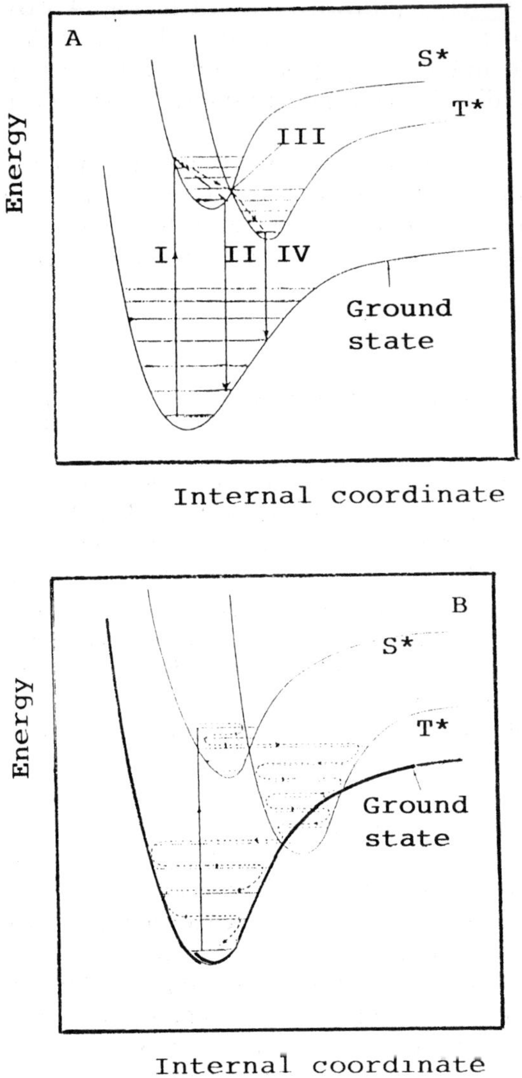

Fig. 4.9. (A) Radiation emission from molecular excited states. I: excitation process; II: fluorescence emission from S*; III: intersystem crossing between singlet and triplet states; IV: phosphorescence emission from T*. (B) Radiationless energy dissipation from singlet excited states by vibrational deactivation and internal conversion.

In addition to the internal decay pathways shown in Fig. 4.9(B), other competitive deactivation processes can occur, usually by collision between the excited molecule and other species present in the system (quenching). For example, the dissolved oxygen is a typical fluorescence quencher.

When the quencher is another analyte molecule, the deactivation is termed self-quenching. Quenching deactivation can proceed, in some cases, through the formation of transient luminescent complexes (exciplexes, excimers). The possible quenching processes are summarized below:

$S^* + Q \rightarrow S + Q$	Quenching	(15)
$S^* + S \rightarrow S + S$	Self-quenching	(16)
$S^* + Q \rightarrow [SQ]^* \rightarrow S + Q + h\nu$	Exciplex formation	(17)
$S^* + S \rightarrow [SS]^* \rightarrow S + S + h\nu'$	Excimer formation	(18)

In condensed homogeneous media, the solvent molecules can intervene in some of the excited-state evolution pathways. Usually solvents interact more strongly with excited states that have higher polar character than the ground state, affecting the relative populations of the excited-state vibrational levels as well as the relative positions of the curves. The radiative emissions from these solvent-modified states are therefore different. A detailed, rigorous and comprehensive analysis of the basic physico-chemical aspects of photochemical processes occurring in the presence of microheterogeneous surfactant systems can be found in Ref. [52].

4.2.1. Medium effects on fluorescence spectra

In organized media, changes in the nature of the environment experienced by a given solute on transfer from the bulk aqueous medium to the host aggregate (or migration from one region to another within the aggregate) are strongly reflected in the fluorescence emission. In particular, the observed emission intensities are usually many times greater than in the corresponding homogeneous media because the emitting excited state is protected from quenching by the microaggregate environment.

Different medium-dependent properties of probe solutes can be monitored to investigate the solute location within the aggregate. For example, aromatic molecules such as benzene or pyrene show the Ham effect, with a marked intensity enhancement of symmetry forbidden (low-intensity) $S_0 \rightarrow S_1$ transition in polar environments, whereas the

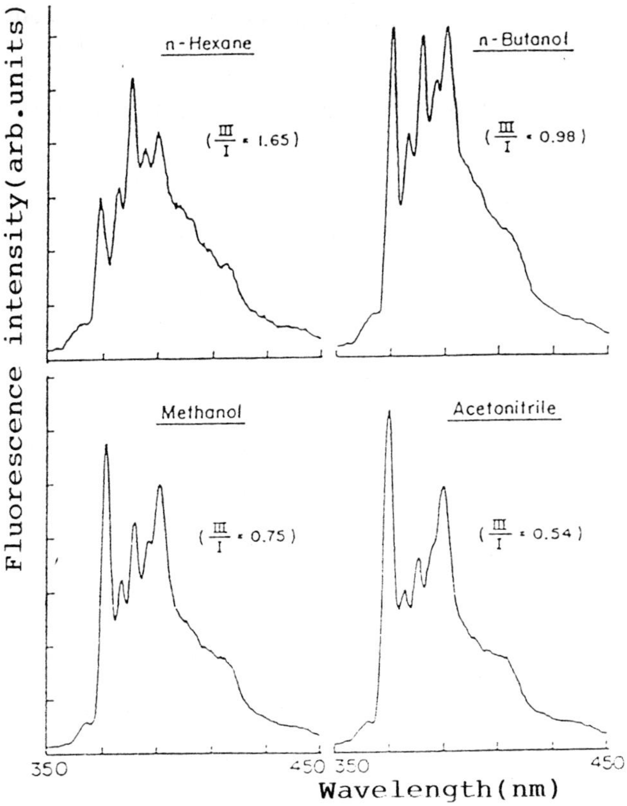

Fig. 4.10. Solvent dependent intensity ratio between the vibronic bands I and III of pyrene at room temperature. [Pyrene] = 2×10^{-6} M; λ_{ex} = 310 nm. Reproduced from Ref. [53] with permission.

strong (allowed) vibronic bands are practically insensitive to medium polarity. The peak intensity ratio between the fluorescence bands (I_{III}/I_I or I_I/I_{III}) of pyrene depends on the medium polarity (see Fig. 4.10) and has been used to investigate the solubilization site in micellar aggregates. In fact, this ratio varies from 1.87 in pure water to ca. 0.60 in pure aliphatic hydrocarbons, with intermediate values in micellar media.

Other lumiphor properties have been also used to investigate micellar systems [52]. For compounds having polar groups, such as pyrenecarboxaldehyde, the position of the fluorescence maximum (λ^F_{max}) and the fluorescence quantum yields have been related to the medium polarity.

In this molecule, fluorescence was assigned to an n-π* electronic transition in non-polar solvents (peaks around 400–420 nm in heptane), whereas the fluorescent emission in more polar media (D ≥, 10) was attributed to a π–π* state. A linear variation of λ^F_{max} (towards longer wavelengths) with the medium dielectric constant has been found [54].

4.2.2. Micellar effects on acid-based properties of the excited states

Most of the excited molecules (present in singlet, triplet or other spin states) show pK_a values which differ greatly from those measured in their ground states. During the lifetime of the excited states, protonation and deprotonation can occur, since they are very fast processes (see Scheme 1).

	pK_a^*		
HA^*	b	$A^{-*} + H^+$	
hυ↑↓fluor.		hυ↑↓fluor.	Scheme 1
	pK_a^*		
HA	b	$A^- + H^+$	

For example, the pK_a of 2-naphthol in water is about 9.5 [55], whereas for the excited molecule in the same solvent a pK_a* value of 2.8 was reported [56]. Differences up to 8 pH units have been observed in some cases.

As for the ground state, the acid–base equilibria in the excited state are drastically altered by surfactant aggregates. The extent of protonation (or deprotonation) of the excited states of micellized lumiphors can be evaluated from luminescence measurements, and the analytical sensitivity can be improved by properly choosing the acid (or basic) emitting form.

In particular, ionic micelles induce relevant pK_a* shifts on solutes having functional groups (such as –OH, –COOH, $-NH_2$, –N=NH, etc.) which are capable of protonation/deprotonation and interaction with either π- or n-type electrons in the excited state.

Some conditions must be satisfied in order to observe such interactions: the fluorescence or absorbance spectrum must be different for the ground and excited state, the fluorescence lifetime should be long enough to permit the protonation (or deprotonation) of the excited state; the analytes must (at least in part) bind to the micelles. Only a limited number of compounds satisfy all the above-mentioned conditions.

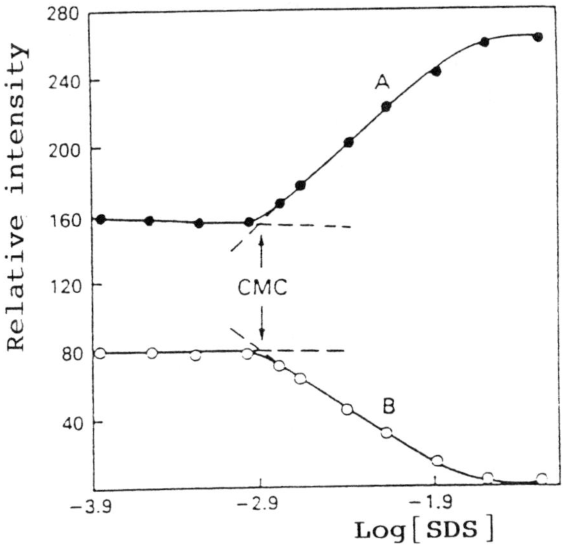

Fig. 4.11. Effect of SDS micelles on fluorescence emission of 2-naphthol. $C_{solute} = 3\times10^{-6}$ M; 0.1 M $NaClO_4$ M; pH: 6–7. (A): HA*, $\lambda_{em} = 355$ nm. (B): A^{-*}, $\lambda_{em} = 420$ nm. Reproduced from Ref. [57] with permission.

Figure 4.11 shows the influence of SDS concentration on the relative fluorescence intensities of neutral and anionic forms of 2-naphthol. The maximum effect is observed when complete solubilization of the acid in the micelles occurs. The observed fluorescence intensities are pH-dependent. In particular, the emission from the excited form HA* decreases below pH ca. 2 due to quenching by H^+.

4.2.3. Micellar-enhanced fluorescence of organic molecules

The effect of different micellar aggregates (SDS, HDTC, Triton X100) on the spectrofluorimetric determination of pyrene was reported by Singh and Hinze [58]. In that work, a 3–16 fold increase in sensitivity (compared to ethanol) was observed for the investigated analyte, which was attributed to the decreased polarity, increased microviscosity and shielding against deactivation processes for emitting molecules imbedded in a hydrocarbon-like region. Figure 4.12 shows the typical calibration curves of pyrene, obtained working in different organized and homogeneous media.

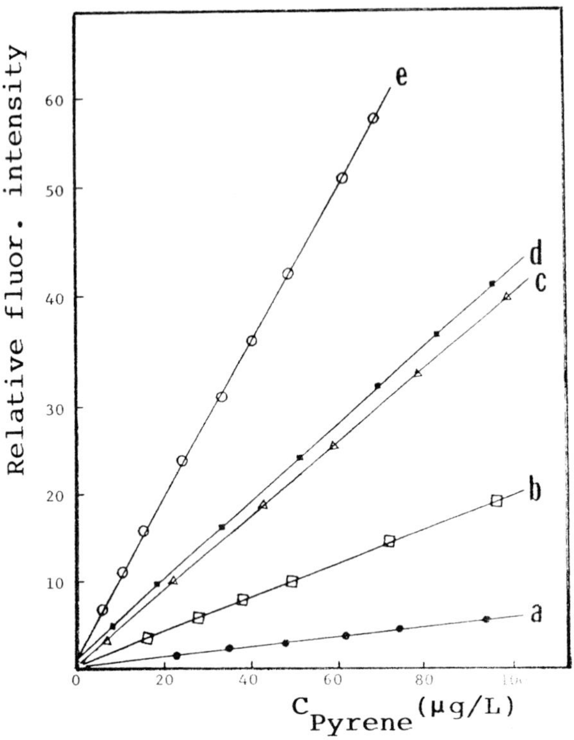

Fig. 4.12. Fluorescence of pyrene as a function of its concentration in different solvent media. λ_{ex} = 241 nm; λ_{em} = 390 nm. **a**: ethanol; **b**: water; **c**: SDS 3.5×10^{-2} M; **d**: HTAC 6.25×10^{-3} M; **e**: Triton X100 4×10^{-3} M. Reproduced from Ref. [58] with permission.

These working curves usually exhibit a longer linearity range in micelles if compared to that observed in homogeneous solvents. For example, the linear dynamic range (LDR) of pyrene in SDS (10 g l^{-1}) and HTAC (2 g l^{-1}) is 0.25–500 μ, g l^{-1}, whereas the corresponding values in water and methanol are 2.0–100 μ, g l^{-1} and 1.8–250 μ, g l^{-1}, respectively [58].

The gain in sensitivity is usually measured by the micellar enhancement factor (MEF), defined as the ratio between the fluorescence intensity in micellar solution and that observed in homogeneous solvents (at constant lumiphor concentration). Selected examples of such determinations are given in Table 4.5.

The determination of non-fluorescent analytes through their derivatization with suitable fluorescent reagents is also possible, extending

TABLE 4.5

Micellar-enhanced fluorescence of some organic analytes

Analytes	λex,λem (nm)	Micellar solution	DL (μg l^{-1}	MEF	Ref.
Aromatic hydrocarbons[a]:					59
Naphthalene	282,336	SDS 0.024 M	0.30	4	
Fluoranthene	288,458		2.50	1.5	
1,2-Benzanthracene	290,390		0.50	4	
Biphenyl	260,310		0.18	4	
Anthracene	250,390		0.20,		
Pyrene	254,390		0.25	10	
Benzo[e]pyrene	291,394		0.20	2.5	
Benzodiazepines[b]:					60
Diazepam	360,500	SDS 1%		6.2	
Prazepam	360,490			5.7	
Oxazepam	360,490			3.3	
Lorazepam	370,495			2.2	
Other compounds:					
Fenproporex[c]	260,560	SDS 0.5%			61
	pH 6	0.80	2.6		
	pH 1 (HCl)		2.7		
	pH 13 (NaOH)		1.9		
1-Naphthol[d]	312,449 (pH 7)	HTAB 0.4%	0.20	2.0	62
	341,449 (pH 12)	HTAB 0.1%	0.20	5.0	

a: MLC coupled with fluorescent detection; *b*: DL not reported, lower analyte concentration (from calibration graphs): 1 μg l^{-1}; *c*: (±)-3-(α-methylphenethylamine)propionitrile; *d*: 1-naphthol determined alone (neutral pH) and in mixtures with carbaryl (1-naphthyl-N-methylcarbamate) in basic media.

the number of compounds which can be analyzed using micellar-enhanced fluorescence.

Non-fluorescent amino acids can be converted into fluorescent analytes after derivatization with dansyl chloride (1-N,N-dimethylaminonaphthalene-5-sulphonyl chloride) and OPT (o-phthalaldehyde-2-mercaptoethanol) [63]. The reported detection limits of dansylglycine in cationic micellar media (HTAC, DTAC) are one order of magnitude lower than those measured in water, whereas a similar increase in sensitivity can be obtained using the OPT fluorescent derivative in the presence of Brij 35, Triton X100 or SDS micelles.

Another example of use of fluorescent derivatives concerns the determination of fluorescein-labelled phenobarbital in different cationic micelles and/or cyclodextrin media [64].

More recently [65,66], bile salt media has attracted attention as an alternative to usual micellar systems, since these peculiar aggregates allow the solubilization of the emitting compounds in a highly hydrophobic and viscous microenvironment. The fluorescence lifetimes of polycyclic aromatic hydrocarbons in sodium taurocholate and sodium taurodeoxycholate media reflect the higher medium rigidity, being greater than those obtained in micellar SDS.

4.2.4. Micellar-enhanced fluorescence of metal chelates

The effect of micellar aggregates on the luminescence of organic compounds can be reasonably explained, whereas the introduction of a third component (a metal ion) in these microheterogeneous systems significantly increases the difficulty of rationalizing the observed effects. In fact, the mechanisms which are operating in micellar-enhanced fluorescence of metal complexes have not been completely elucidated.

Sanz-Medel and coworkers [67] recognized two main effects as responsible for enhanced fluorescent emission of metal chelates and for shifts (usually towards longer wavelengths) of their excitation maxima: (i) a simple medium effect due to the solubilization of neutral chelates and ligands within the micellar aggregates; (ii) interaction of ionic complexes with surfactant molecules and aggregates having opposite charge.

The fluorescence emission of several water-insoluble metal chelates originating from neutral ligands, such as 8-hydroxyquinoline (and its haloderivatives) or flavonoids (morin, quercetin, etc.) is enhanced through a solubilization mechanism (i). For such (neutral) complexes the emitting species appear to be protected against deactivation pathways by the surrounding micellar environment.

A comparison between the measured fluorescence intensities of the Al(III)–oxine complex in different micellar solutions indicates that sensitization effects are similar and this can be simply attributed to a complete solubilization of both the chelate and the ligand within the aggregates (see Fig. 4.13).

Charge-dependent sensitization effects (ii) have been observed between ionic chelates (usually anionic)and oppositely charged micelles.

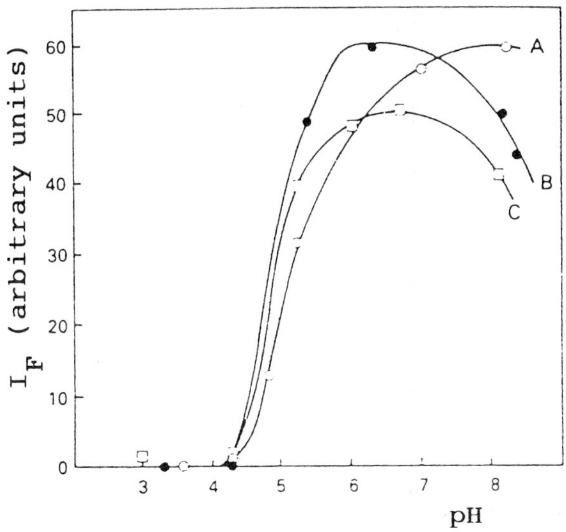

Fig. 4.13. Fluorescence intensity of Al(III)-8-hydroxyquinoline complex in different surfactant solutions. A: SDS 0.6% (w/v); B: HTAB 0.6% (w/v); C: Brij 35 3% (w/v). $C_{Al(III)} = 3.7\times10^{-5}$ M; $C_{oxine} = 1.86\times10^{-4}$ M. Reproduced from Ref. [67], with permission.

For example, ligands bearing charged sulphonate groups, such as 8-hydroxyquinoline-5-sulphonic acid (HQS), give rise to the formation of several anionic complexes, which can strongly interact with cationic surfactants and their aggregates. Figure 4.14 shows that the increase in fluorescence intensity of Al(III)–HQS complexes in HTAB solutions is relevant (about 5-fold), and is higher than that obtained when working with the neutral ligand oxine under the same conditions.

Several metal ions have been determined as metal chelates in different micellar media (see Table 4.6) and, moreover, the fluorescence of such species can be exploited for the determination of organic molecules. For example, tetracycline antibiotics show enhanced emission of the two most intense bands (at 590 nm and 613 nm) in the presence of an excess of europium(III) [82], due to an intra-chelate energy transfer process. The ligand absorbs the excitation radiation and transfers its energy to the emitting ion, thus increasing the fluorescence intensity of the complexed metal.

The study of fluorescence quenching effects can also be useful for metal ion determinations. For example, the zinc–morin complex shows intense fluorescence emission in Triton X100 micellar media, which is

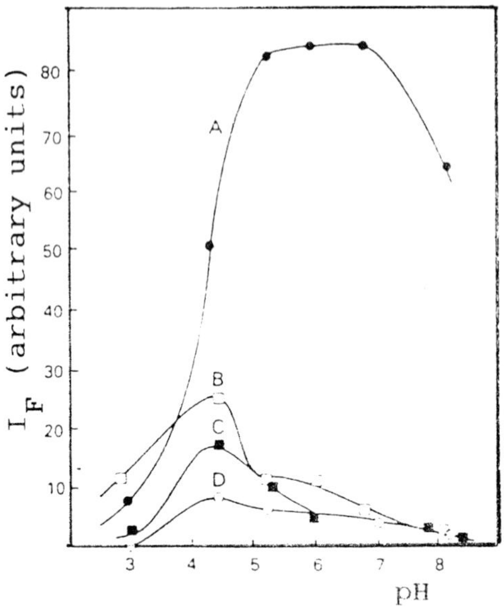

Fig. 4.14. Variation of fluorescence intensity of Al(III)–HQS chelate with pH in different micellar systems. A: HTAB; B: Brij 35; C: homogeneous solution; D: SDS. Other conditions as reported in Fig. 4.13. Reproduced from Ref. [67] with permission.

quenched by iron(III). Since the intensity reduction is proportional to the quencher concentration, this indirect method has been used to determine iron(III) with a detection limit of 5 ng ml^{-1} [83].

4.2.5. Synchronous fluorescence in micellar media

The analysis of multicomponent mixtures present in micellar media can be performed by introducing the synchronous wavelength scanning luminescence [84,85]. The experiments are carried out by scanning the excitation and emission monochromators at a fixed rate, with a constant wavelength offset maintained between them. The instrument plots the emission intensity vs. the wavelength position of the excitation monochromator; thus, the position of the emission monochromator is given by:

$$\lambda_{em} = \lambda_{ex} + \Delta\lambda \quad (19)$$

$\Delta\lambda$ is the chosen wavelength offset, often taken as the wavelength difference between the most intense peaks of excitation and emission, respectively.

TABLE 4.6

Micellar-enhanced fluorescence analysis of some chelate complexes at room temperature

Metal	Ligand	Surfactant	pH	$\lambda_{ex}/\lambda_{em}$ [a]	DL[b]	Ref.
Tb	PTA/TOPO	0.5% BL-9EX	3.4	309/542	0.1	68
Al	Morin	3% Genapol PF-20[c]	3.8	430/495	0.2	69
Al	Lumogallion	0.5% T X100	4.7	485/590	0.02	70
Al	HQS	1.2×10^{-3} M HTAB[d]	5.5	390/497	1	71
Nb	Morin	0.5% HTAB	1.0[e]	433/503	1	72
Nb	Quercetin	0.5% HTAB	0.3[e]	435/497	3	73
Ta	Quercetin	0.3% HTAB	0.05[e]	430/485	50	
Ta	Morin	0.2% HTAB	0.2[e]	422/496	30	
Nb	Lumogallion	0.5% T X100 + tartrate	1.0	495/610	7	74
Pb	Morin	2% Genapol PF-20	3.3	420/495	0.06	75
Zn	Morin	1.5% Genapol PF-20	4.7	433/503	3	76
Eu	o-Phen/TTA	0.05% T X100	6–8	371/612	[f]	77
Dy	EDTA/Tiron	3×10^{-3} M HTAB	11.0	316/577	16	78
Zn	DCHQ	4×10^{-3} M Brij 35	6–7	397/534	3	79
V	QNPH	0.1% PVA	6.0	400/540		80
Ga	PAN	0.3% SDS	4.0	546/573	0.8	81

a: nm. *b*: detection limit expressed in $\mu g\ l^{-1}$. *c*: ethylene oxide–propylene oxide condensate (1.6:1 w/w). *d*: determined using first and second derivative synchronous fluorimetry. *e*: Sulphuric acid concentration (M). *f*: DL < 0.1 l^{-1} ng in the presence of Tb(III). PTA: pivaloyltrifluoroacetone. TOPO: tri-n-octylphosphine oxide. BL-9EX: nona-oxyethylene dodecyl ether. o-Phen: 1,10-phenanthroline. TTA: thenoyltrifluoroacetone. DCHQ: 5,7-di-chloro-2-methyl-8-hydroxyquinoline. QNPH: o-hydroxyhydroquinonephthalein. Percent of surfactants in w/v.

Using this experimental approach the analyte spectral profile is significantly simplified and it is possible to evaluate the contribution of each component to the fluorescence of mixtures, even when the excitation and emission spectra of the analytes show overlap. By changing the applied $\Delta\lambda$ offset, the position and band width of the synchronous fluorescence peaks can be adequately varied. For example, benzo[a]pyrene and perylene can be simultaneously determined in Triton X100 micellar solution [86].

Figure 4.15 shows the conventional and synchronous fluorescence spectra of perylene, and the synchronous fluorescence spectra of an

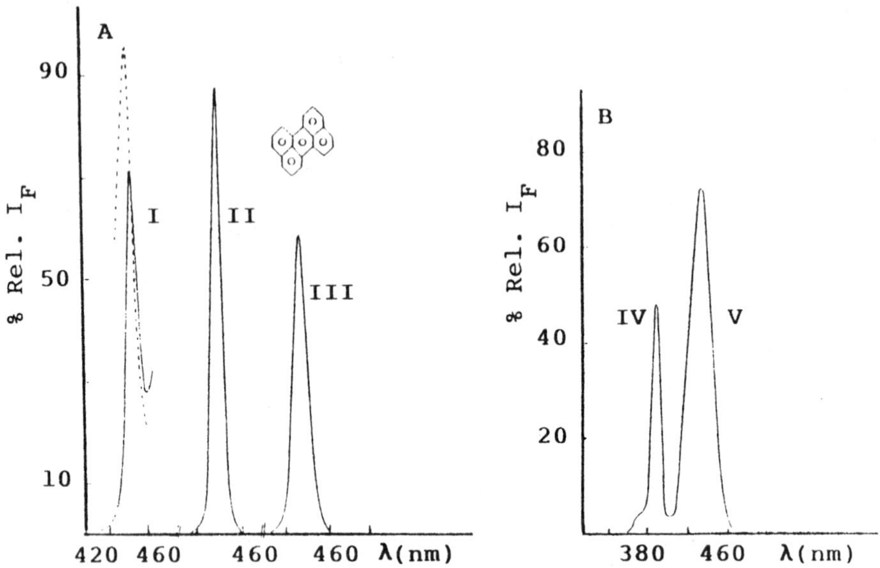

Fig. 4.15. (A) Fluorescence spectra of perylene. I normal; II synchronous ($\Delta\lambda$ = 3 nm); III synchronous ($\Delta\lambda$ = 18 nm). [Perylene] = 2×10^{-6} M; $C_{T\,X100} = 2\times10^{-3}$ M. (B) Synchronous fluorescence spectra of 1×10^{-8} M benzo[a]pyrene (IV) and 1×10^{-8} M perylene (V). $C_{T\,X100}$ = 2×10^{-3} M. $\Delta\lambda$ = 18 nm. Reproduced from Ref. [86] with permission.

equimolar mixture of benzo[a]pyrene and perylene. The dashed lines correspond to the excitation spectra of perylene.

The most intense synchronous fluorescence peaks are obtained (for a given component) when $\Delta\lambda = \lambda_{max(em)} - \lambda_{max(ex)}$. Increasing $\Delta\lambda$ the synchronous spectra show bands having lower intensities and larger half-widths. However, the applied wavelength offset must also originate a distinct (and possibly intense) peak for the other component. For the above reported binary mixture, a $\Delta\lambda$ value of 18 nm allows the simultaneous determination of both analytes, their respective concentrations being proportional to the corresponding peak area (or peak height).

The analysis of other binary and ternary mixtures of PAHs in seawater has been recently reported [87]. Different micellar media have been used, with detection limits in the range 0.13–0.22 ng ml^{-1} depending on the analyte considered.

The sensitive determination of aluminium with HQS in the presence of HTAB (see Table 4.6) is an example of application of synchronous fluorimetry to the analysis of metal chelates.

4.2.6. Fluorescence analysis in microemulsions and reversed micellar solutions

Fluorescence, together with other spectroscopic and electroanalytical techniques, has been used to investigate the physicochemical properties of microemulsions and reversed micellar systems. As mentioned previously (see Section 4.2.1), pyrene, which is located in the micellar interphase, is one of the most popular fluorescent probes for sensing local polarity in normal micelles. The ratio of the intensities of the first and third vibronic bands (I_I/I_{III}) of pyrene is a very sensitive parameter which has been also used to investigate the polarity of oil–water microemulsions [88], where the continuous phase is (as for normal micelles) aqueous. Red shifts in the emission spectrum can also be expected upon increase of local polarity.

In SDS/1-pentanol/heptane/water microemulsions, different medium polarity was revealed by pyrene depending upon the structural configuration of the quaternary system. For an L1 structure (aqueous continuous phase) water penetration into the oil droplets is scarce, even in the absence of added hydrocarbon, and this is reflected in the I_I/I_{III} value, which is almost constant. Only a fraction of 1-pentanol (about one third of the total) is present in the palisade layer of the microaggregate, whereas most of the alcohol is in the droplet interior.

The fluorescence quantum yields of the non-polar probes decrease continuously passing from micellar solutions to alcohol/water/surfactant ternary mixtures and to oil-rich microemulsions. The corresponding I_{II}/I_{III} values vary from ca. 0.85 (SDS) to ca. 0.6 (heptane-rich microemulsion).

Studies on reversed micelles, in which the continuous phase is hydrophobic, require other suitable probes, such as 1-anilino-8-naphthalene sulphonic acid (ANS). These compounds are much more soluble than pyrene in the polar microdroplets and can adequately sense the polarity changes occurring between different regions of the aggregate. The emission peak shifts to longer wavelengths as polarity increases and a decreased fluorescence quantum yield is often observed

Only a few analytical determinations have been reported in the literature. They include the determination of lipase with 4-methyl-umbelliferyl heptanoate [89] and of hydrophobic amines with o-phthalaldehyde [90] (both performed in reversed micelles). The analysis of Eu(III) and Sm(III) chelates, using β-diketones and trioctylphosphine oxide as ligands in a SDS/water/1-butanol/heptane oil–water microemulsion has also been recently reported [91].

4.3. ROOM TEMPERATURE PHOSPHORESCENCE IN MICELLAR SOLUTIONS

Phosphorimetry is potentially a very useful analytical technique due to its high sensitivity, with limits of detection in the nanomole range, and often relevant selectivity. However, the widespread use of this technique is limited by the experimental difficulties related to the need of operating at very low temperatures. Since the phosphorescent excited state of organic molecules easily undergo radiationless decay by molecular collision processes, supercooling of analyte-containing solutions (usually at 77 K in liquid nitrogen) has been largely used in order to observe strong phosphorescent emission.

In the past decade, some peculiar analytical methods which allow the phosphorescence of a great number of organic compounds at room temperature have been developed. The adsorption of the emitting analytes onto suitable solid supports [92,93] and the solute incorporation in micellar aggregates were at the basis of these successful applications.

In particular, micelle-stabilized room-temperature phosphorescence (MS-RTP) exerts its beneficial effect through the compartmentalization of solubilized lumiphores in a micellar region surrounded by a relevant concentration of heavy-atom counterions. The collisions between lumiphores and external quenchers are greatly reduced, whereas the heavy atoms induce enhanced spin–orbit coupling. This latter process favours the mixing of a higher energy S_1 singlet (fluorescence emitter) with a lower energy T_1 triplet (intersystem crossing), thus increasing the phosphorescence yield (see Fig. 4.9).

Appreciable stabilization of the triplet state does not occur at concentrations below the cmc, confirming the essential role played by the micelles.

The ratio between the concentrations of host micelles and guest lumiphors is also an important factor which can influence the deactivation pathways. For example, for aromatic hydrocarbons the main deactivation processes at room temperature are triplet–triplet annihilation (TTA) and excimer formation, which can be prevented only if the probes are not closely positioned within the same aggregate.

Taking into account that the probe molecules are distributed among the micelles according to Poisson–Boltzmann statistics (see Chapter 2) and that the micelle concentration can be estimated from the cmc and the aggregation number, the probability of finding more than one probe in a given micelle and the percent of micelles in this situation can be calculated. The decrease in phosphorescence intensity due to intramicellar deactivation processes can, thus, be predicted. Generally speaking, a large excess of micelles is necessary in order to minimize these undesirable effects.

The presence in the working solution of classical phosphorescence quenchers, like dissolved oxygen, must also be avoided. Careful degassing of the samples prior to the analysis or the use of chemical oxygen-scavengers, such as sodium sulphite [94], is essential in MS-RTP analysis.

4.3.1. MS-RTP of organic compounds

The first analytical application of this technique was reported by Cline Love et al. [95] and concerned the very sensitive determination of naphthalene, pyrene and biphenyl in the presence of mixed-counterion dodecylsulphate micelles.

In Fig. 4.16 the phosphorescence spectrum of naphthalene in supercooled ethanol is compared with that obtained in 0.15 M mixed micellar system composed by 70% SDS and 30% Tl dodecyl sulphate at room temperature.

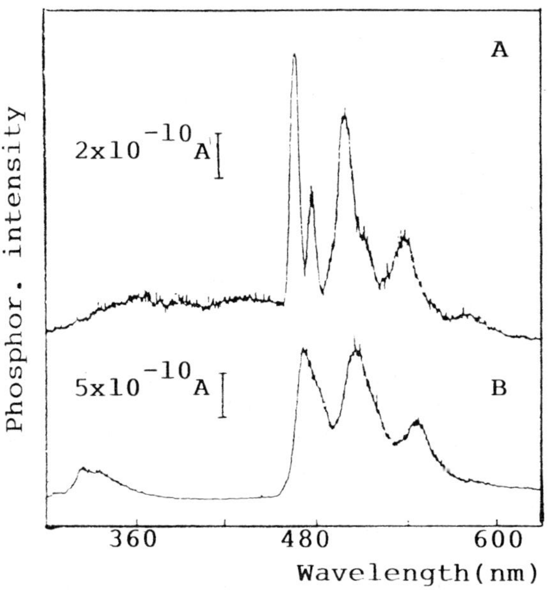

Fig. 4.16. Luminescence spectra of 6×10^{-5} M naphthalene in: (A) ethanol at 77 K; (B) 0.15 M Tl/Na dodecylsulphate micelles (see text). Taken from Ref. [95], reproduced by permission.

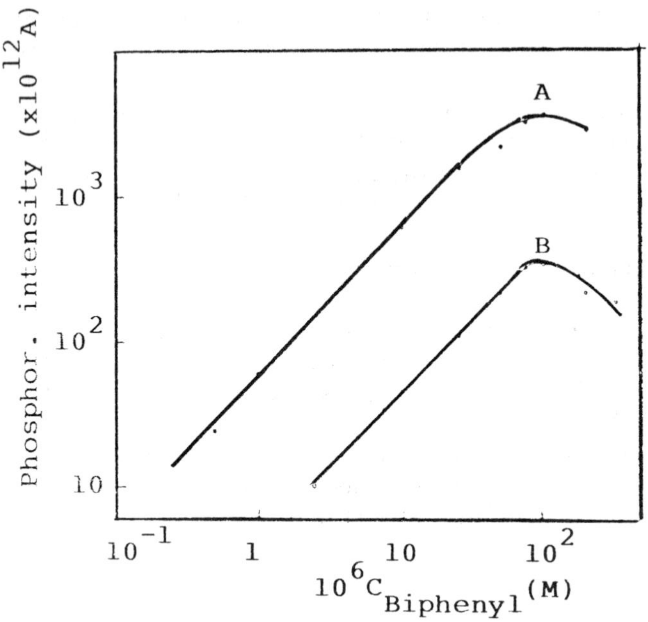

Fig. 4.17. Variation of RTP intensity (measured in pA) vs. biphenyl concentration (in μ, mol l^{-1}) in different micellar systems.(A): Surfactant concentration: 0.15 M; T = 23°C; TlDS/NaDS = 0.3; λ_{em} = 470 nm. (B): Surfactant concentration 0.08 M; T = 20°C; AgDS/NaDS = 0.5; λ_{em} = 490 nm. Excitation wavelength = 266 nm for both systems. Reproduced from Ref. [95] with permission.

An efficient lumiphore protection against non-radiative decay processes is observed. It must be noted that the effective concentration of counterions on the micellar assembly for a 0.1 M solution of mixed Tl/Na dodecyl sulphate is ca. 4 M.

Typical analytical working curves, reporting the phosphorescence intensity vs. the probe concentration, are shown in Fig. 4.17. for biphenyl. The phosphorescence yield is strongly dependent upon the nature of the metal ion counterion used (Tl^+ or Ag^+, instead of Na^+) and on the heavy metal/sodium ratio. Since the sulphonate head-group interacts differently with these different counterions (see Chapter 1), cmc values and aggregation numbers of the corresponding mixed aggregates are expected to be different.

The RTP intensities show a maximum for Tl^+ (or Ag^+) of ca. 30%. Above this limit, the favourable spin–orbit interaction provided by the heavy atom is probably counterbalanced by relevant changes in the aggregate geometry and, moreover, the precipitation of the corresponding dodecylsulphate salts also occurs. The above inconveniences can be drastically reduced by using mixed surfactant systems including SDS and short chain alkylsulphates, such as octylsulphate [96].

Curvatures observed in the analytical curves at higher analyte concentrations have been attributed in part to the increase of intramicellar deactivation processes (such as TTA) and mainly to the so-called inner filter effect, related to the geometrical configuration of the apparatus.

4.3.1.1. Dynamic aspects of MS-RTP

In the absence of secondary deactivation processes, the radiative decay of the triplet state follows a first-order kinetics, described by the equation:

$$[P^*]_t = [P^*]_0 \exp.(-t/\tau) \tag{20}$$

The phosphorescence lifetime, τ, of the emitter is given by [97]:

$$\frac{1}{\tau} = k_- + k_{MP} + k_{qIN} [Q]_{IN} \frac{k_- k_+ C_M}{k_p + k_{qEX} [Q]_{EX} + k_+ C_M} \tag{21}$$

where k_- is the rate constant for lumiphor exit from micelles, k_{MP} is the deactivation rate within micelles, k_p is the deactivation rate outside the aggregates; k_{qIN} and k_{qEX} are the quenching rates corresponding to internal and external quenchers, respectively; k_+ is the reentry rate of lumiphors; C_M is the concentration of micelles.

When the quenching processes are negligible, Eqn. (21) reduces to:

$$\frac{1}{\tau} = k_- + k_{MP} \frac{k_- k_+ C_M}{k_p + k_+ C_M} \tag{22}$$

Taking into account that $k_+ >> k_-$ and that the concentration of micelles is usually about 1×10^{-3} M, it can be seen that, at constant surfactant and lumiphor concentration, the phosphorescence lifetime depends mainly on k_- and k_{MP}. The lifetimes are generally proportional to the average residence time t_r (given by $1/k_-$), which is in turn affected by factors such as internal micelle polarity, lumiphore size, lumiphore solubility in the bulk, etc. For example, arenes present in Tl/Na (or Ag/Na) dodecylsulphate micellar solutions show MS-RTP lifetimes

TABLE 4.7

Characteristics of some phosphorescent aromatic hydrocarbons in micellar media at 295 K[a]

Analyte	Sw (M)	t_r (μs)	Lifetime (ms)	
		NaDS	Ag/NaDS	Tl/NaDS
Naphthalene	2×10^{-4}	4	0.31	0.45
Biphenyl	4×10^{-5}	10	0.33	0.43
Pyrene	6×10^{-7}	244	0.60	0.93
1-Bromonaphthalene	5×10^{-5}	40		1.15
1-Methylnaphthalene	2×10^{-4}	10		1.00

a: Data taken from Refs. [97] and [98]. Heavy metal/sodium ratio:0.3/0.7.

which are inversely proportional to their corresponding solubilities in water (see Table 4.7).

The MS-RTP lifetimes are longer when Tl heavy atoms are used, according to the expected increase in the spin–orbit coupling constants with increasing atomic numbers. However, other factors, such as the specific interactions occurring between the delocalized aromatic electrons and the heavy metal ions, must be also taken into account in order to explain the observed effects.

If two components are present in the sample, their exit rates k_- and micellar phosphorescence deactivation rates k_{MP} remain practically unchanged and this allows their individual characterization, through the measurement of the experimental MS-RTP lifetime. Figure 4.18 shows the decay curve obtained for a mixture of naphthalene and pyrene in the presence of NaDS/TlDS 70:30, which indicates the presence of two emitting species.

4.3.1.2. Sensitized RTP through energy transfer between donors and acceptors in organized media

The utility of micellar media for phosphorescence enhancement arises not only from the protection of the emitting triplet state against deactivation, but also from the organization of energy donors and acceptors in a reduced volume, thus favouring the triplet–triplet energy transfer between them. The transfer can occur only if the triplet state of the donor molecules possess a greater energy than the triplet state of the acceptor.

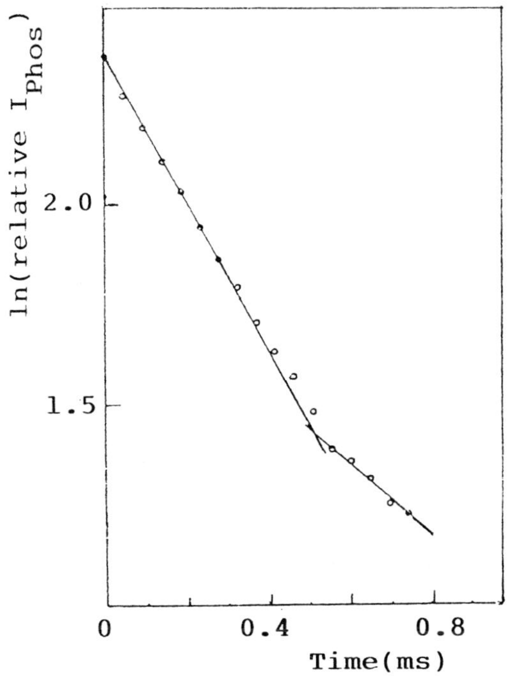

Fig. 4.18. Phosphorescence decay of 1.7×10^{-5} M pyrene and 3.2×10^{-5} M naphthalene. Total surfactant concentration: 0.15 M. Reproduced from Ref. [97] with permission.

For example, excitation of biacetyl at 280 nm in cationic micellar media do not produce measurable phosphorescence, whereas this luminophor can act as triplet energy acceptor if naphthoate anions replace the usual halogen counterions of the surfactant [99]. Excitation at 280 nm in micellar media containing both naphthoate and biacetyl show the typical phosphorescence spectrum of the analyte.

As biacetyl emission can only be observed in the wavelength range corresponding to the naphthoate absorption spectrum and considering the relative energy levels of these molecules (see Fig. 4.19), sensitization is attributed to the transfer of triplet energy from naphthoate to biacetyl.

A similar diacetyl behaviour can be observed in reversed micelles of AOT containing cetyltrimethylammonium naphthoate as bound energy donor. Both diacetyl and 2-naphthoate molecules are localized at the interfacial region of the aggregates and the efficiency of energy transfer is related to the water-pool size. In fact, sensitized phosphorescence

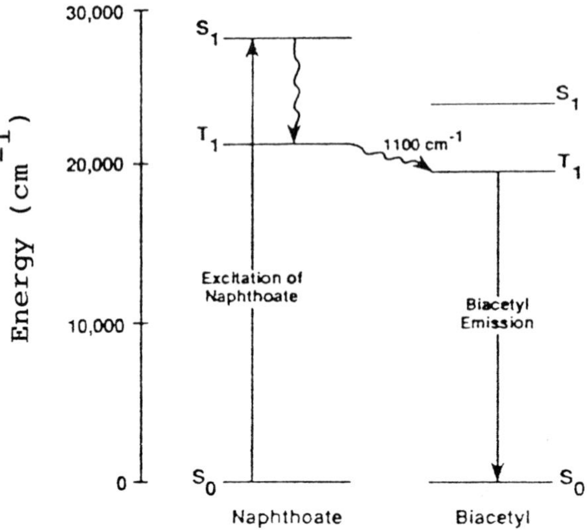

Fig. 4.19. Relative energy levels of the excited states of biacetyl (acceptor) and naphthoate (donor). Taken from Ref. [99], reproduced with permission.

decreases when the distance between donor and acceptor molecules becomes larger than about 15 Å.

Enhanced RTP of different organic compounds can also be observed in AOT reversed micelles containing the lanthanide counterions Tb(III) or Eu(III) [100]. This effect is due to the donor behaviour of the analytes, which are able to transfer their triplet state energy to the 4f energy level of the lanthanide atoms, increasing the acceptor emission intensity.

Both the lanthanide ions, strongly attracted by the surfactant sulphonate groups, and the polar groups of the organic donor (2-naphthoate) are preferentially located at the aggregate interface (see Fig. 4.20). The sensitized emission increases if sodium ions of AOT are fully replaced by lanthanide counterions, as in surfactants $Tb(III)(EHS)_3$ and $Eu(III)(EHS)_3$ (EHS: bis((2-ethylhexyl)sulphosuccinate) [101].

In the presence of a lanthanide excess, the measured phosphorescence intensity is proportional to the concentration of the organic donors, thus allowing their determination. A comparison between the luminescence of the excited $Tb(EHS)_3$ alone and in the presence of equivalent amounts of different organic donors permits the evaluation of the corresponding enhancement factors.

Compounds possessing polar groups, which are preferentially oriented towards the water pool of the aggregate, show very large enhancement

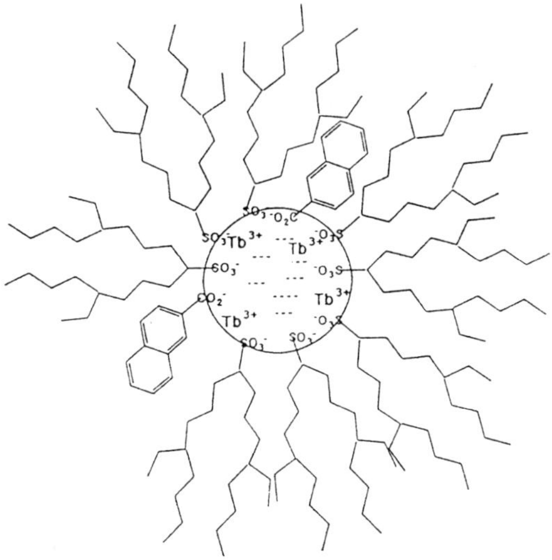

Fig. 4.20. Schematic representation of the distribution of donors and acceptors within the reversed micelles. Reproduced from Ref. [100] with permission.

factors (EF). For example, caffeine, acetylsalicylic acid and naphthaleneacetic acid show EF values in the range 100–150, whereas for theophylline EF is ca. 500 [100]. Less polar compounds such as biphenyl and its monochloro- and dichloro- derivatives exhibit lower EF values (in the range 3–10), indicating a solute location in a more hydrophobic environment (probably in the hydrocarbon region), at a greater distance from the water pool.

The low detection limits of the above polar analytes, which are in the range 10^{-7}–10^{-9} M, together with the extended dynamic range (over two orders of magnitude) indicate that lanthanide-sensitized RTP may be considered a promising new method for the determination of such organic compounds.

4.3.1.3. Use of synchronous scanning to increase MS-RTP selectivity

The technique of synchronous wavelength scanning luminescence can also be applied to MS-RTP measurements. The chosen monochromator offset is usually the difference between the longest wavelength peak of the excitation spectrum and the shortest wavelength peak of the phosphorescence profile. Fluorene, which shows an excitation maximum at 303 nm and an emission maximum at 439 nm, originates a synchronous peak at 303 nm after the application of a wavelength offset of 136 nm.

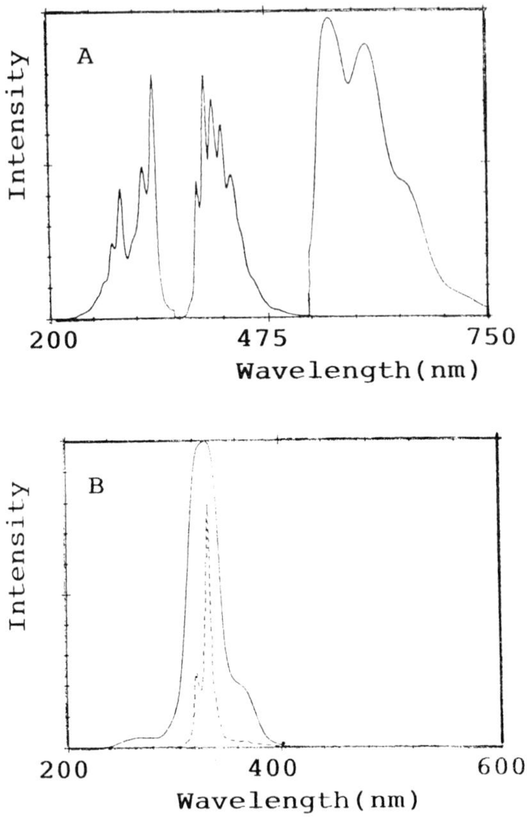

Fig. 4.21. (A) Luminescence profile for benzo[e]pyrene. λ_{ex} = 332 nm. (B) Synchronous MS-RTP and fluorescence (dashed profile) spectra for 1.8×10^{-4} M benzo[e]pyrene. λ_{fluor} = 63 nm; $\lambda_{MS\text{-}RTP}$ = 223 nm. Peak maxima at 332 nm (excitation wavelength). Taken from Ref. [102], reproduced with permission.

In Fig. 4.21 the typical luminescence profile of benzo[e]pyrene is compared with the synchronous MS-RTP spectrum. The synchronous fluorescence profile is also reported.

Phosphorescent analytes having different structures show different wavelength offsets. As indicated in Table 4.8, these offset values vary significantly from one compound to another, even within the same chemical family. If a mixture of lumiphors is scanned at a given wavelength offset, corresponding to a specific analyte, it may be possible to identify such a compound even in the presence of other emitting impurities.

TABLE 4.8

Synchronous MS-RTP parameters for some PNA

Hydrocarbon	λ_{ex} (nm)	λ_{ex} (nm)
Biphenyl	272	180
Phenanthrene	298	185
Triphenylene	287	152
Fluorene	303	136
Chrysene	305	215
Pyrene	335	268
Benzo[e]pyrene	332	223

Data taken from Ref. [101].

Figure 4.22 shows the MS-RTP identification of fluorene and pyrene, present in a four components mixture composed by fluorene (7.7×10^{-6} M) phenanthrene (7.2×10^{-6} M), pyrene (2.5×10^{-6} M), and triphenylene (7.7×10^{-6} M). The identification of the other two components has been also reported in Ref. [102].

Synchronous fluorescence usually shows smaller peak widths than MS-RTP, thus allowing better resolution. On the other hand, fluorescence measurements require narrow slit widths, with corresponding decrease of the exciting radiation intensity and weaker emission.

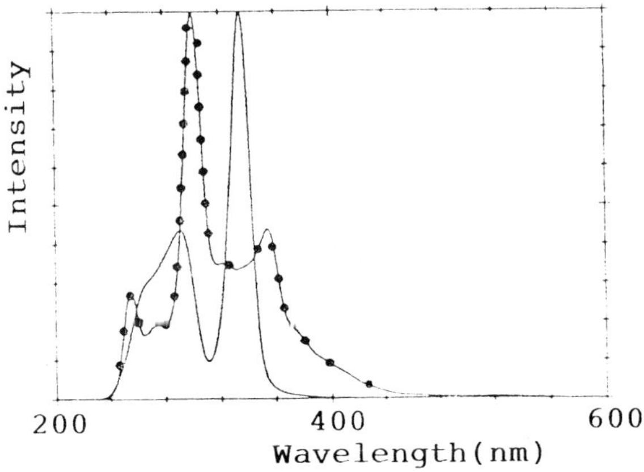

Fig. 4.22. Synchronous wavelength MS-RTP spectra of a PNA mixture. (●) fluorene (offset 136 nm; peak at 303 nm). (——) pyrene (offset 268 nm; peak at 335 nm). Taken from Ref. [102], reproduced with permission.

In synchronous wavelength scanning MS-RTP wide slit widths are allowed, but problems can arise from spectral interferences of grating second-order radiation due to the large wavelength offsets employed. Moreover, triplet state energy transfer can also occur during the analysis of mixtures. Details concerning the removal of these interferences can be found in Ref. [102].

4.3.2. MS-RTP of metal chelates

The application of micelle-stabilized room-temperature liquid phosphorimetry to the determination of trace metals was first proposed by Sanz Medel et al. [103], who reported the analysis of niobium as chelate complex with ferron (7-iodo-8-quinolinol-5-sulphonic acid). Although the detailed mechanism of chemical and photochemical processes involving the phosphorescent metal chelates is still open to discussion, the influence of several experimental parameters upon the phosphorescence yield has already been recognized and few analytical determinations have been reported in the literature.

In addition to the classical chemical parameters which regulate the complex formation (pH, metal to ligand ratio, etc.) and to temperature (usually kept in the range 20–25°C), the following factors must be considered in MS-RTP:

(i) The surfactant concentration, which is usually well above the cmc, is crucial since the mean occupation number of the micelles by the emitting chelate must tend to one. The presence of more than one solute molecule within the aggregate drastically increases the non-radiative deactivation of the lumiphor.

(ii) The charge-type of the micellar aggregate employed is essential in order to maximize the binding interaction and, thus, the phosphorescence yield.

(iii) The choice of suitable external heavy atoms to increase the intersystem-crossing is also necessary. Particular attention is devoted to the nature of these heavy atoms since they must be inactive towards the complex components and, in particular, towards the metal ion.

As far as the surfactant concentration is concerned, the observed constancy of the phosphorescent emission with C_D (keeping the complex concentration constant) seems to confirm that micelles exert their beneficial effects by protecting the triplet excited state from quenchers present in the aqueous solution and by separating the complex molecules from each other, thereby reducing some typical deactivation

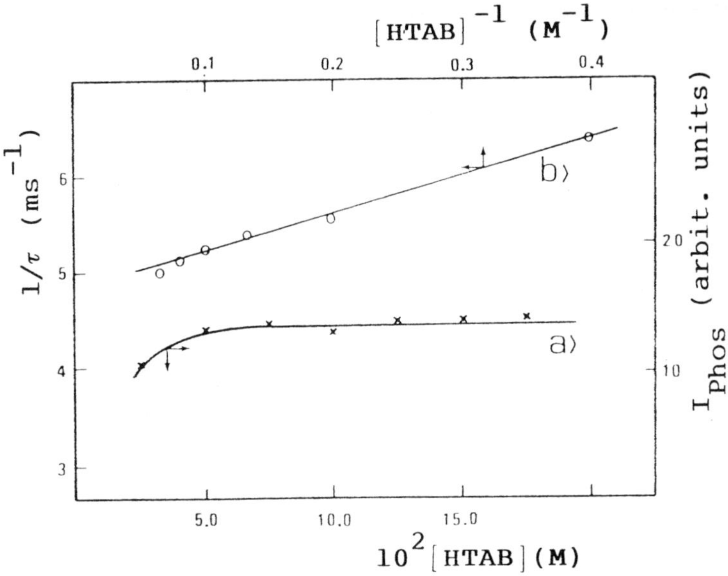

Fig. 4.23. Effect of surfactant concentration on phosphorescence of Al(III)-ferron complex in HTAB micellar solution. (**a**): phosphorescence intensity against [HTAB]. (**b**): inverse triplet lifetimes against $[HTAB]^{-1}$. [Al(III)] = 1.85×10^{-5} M; [ferron] = 7.1×10^{-5} M; λ_{ex} = 386 nm, λ_{em} = 586 nm; delay time = 0.06 ms, gate time = 2 ms. 0.01 M sodium sulphite was used as oxygen scavenger. Reproduced from Ref. [104] with permission.

pathways such as self-quenching or T–T annihilation. It was also found that the triplet lifetimes of the complexes decrease (moderately) with C_D, indicating a reduced mobility of the lumiphors due to the increased microviscosity associated with the formation of packed amphiphile aggregates. The above effects can be seen in Fig. 4.23, where the phosphorescent intensity and the triplet lifetimes of the Al(III)–ferron complex [104] are plotted against the surfactant concentration.

The charge and type of surfactant employed is another essential point. For example, aluminium reacts with ferron to form the anionic complex AlL_3^{3-}, which shows phosphorescence in cationic and zwitterionic micellar media but not in the anionic SDS. In the latter case, the electrostatic repulsion prevents the effective binding of the complex to the micelles and radiationless deactivation occurs in the bulk. Phosphorescence of anionic complexes has also been observed in non-ionic micellar media, where dipole–dipole interactions would account for solubilization and stabilization of the emitting analytes. However, the emission intensity is usually lower in these systems.

The effect of heavy atoms, which facilitate the population of the complex triplet state, is largely dependent upon the system considered. The Nb–ferron–HTAB system does not produce significant phosphorescent emission unless an additional heavy atom (bromoform) is added [103], whereas the Al–ferron–HTAB system exhibits phosphorescence without any addition [104]. Although a tentative explanation of this different behaviour has been proposed [105], the lack of information due to the low number of examined systems does not allow generalizations. At present, the search for suitable heavy atoms appears to be mostly empirical.

4.4. CHEMILUMINESCENCE IN ORGANIZED MEDIA

Chemiluminescence (CL) is a phenomenon that originates from a chemical reaction (typically a redox process) able to produce an electronically excited compound. This product can deactivate through a non-CL pathway, can return to its ground state by emitting light or can transfer the excess energy to another molecule (sensitizer), which in turn exhibits luminescence (see Scheme 2).

$$\begin{array}{lllllll} & & & \rightarrow P & & & \\ A + B & \rightarrow & P^* & \rightarrow P + h\nu & & & \\ & & + S & \rightarrow P + S^* & \rightarrow & S + h\nu & \end{array} \qquad \text{Scheme 2}$$

The maximum CL intensity and the duration of CL are altered by the presence of micelles or other organized media, such as cyclodextrins. The basic requirements for the production and observation of CL emission, the beneficial effects of amphiphilic organized assemblies upon the chemiluminescent reactions, and the practical applications of these systems in CL-based analytical methods, have been extensively examined by Hinze et al. in a recently published book [5].

Among the different types of CL reactions exploited for analytical purposes, those based on lucigenin have been largely investigated. The influence of organized assemblies on such reactions will be examined with more detail.

4.4.1. Lucigenin-based CL systems

4.4.1.1. Lucigenin-hydrogen peroxide CL reaction

Lucigenin (10,10′-dimethyl-9,9′-biacridinium dinitrate) reacts with H_2O_2 in basic media to form excited N-methylacridone (NMA*), which emits light to return to its ground state, according to the following scheme:

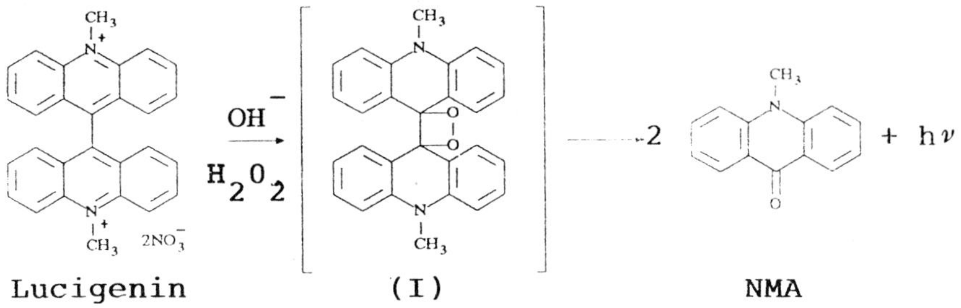

Scheme 3

At surfactant concentrations below the cmc, the net CL intensities are about the same as in water, regardless of the charge type of the surfactant employed. Above the cmc, the presence of suitable micelles (or vesicles) enhances the CL quantum yield from 4 to ca. 100 times by stabilizing the reaction intermediate (**I**) in the less polar aggregate environment [106–109]. Moreover, the energy-transfer processes responsible for competitive dark reactions are inhibited in these organized media [107,109] and protection against metal ion interferences can also be obtained using polyoxyethylene-type non-ionic aggregates [110].

The high binding constants of lucigenin and NMA to different micellar and vesicular systems (which are in the range 10^3–10^4 M^{-1}) ensure the preferential accumulation of these species in the micellar phase [111].

A less favourable partition of hydrogen peroxide occurs, being the corresponding binding constants to different micelles in the range 100–200 M^{-1} [112]. However, in strong alkaline media the anionic HO_2^- form exists, which is more strongly attracted by cationic aggregates and, conversely, repelled by anionic micelles.

The effect of different micellar systems on the lucigenin-hydrogen peroxide CL reaction is shown in Fig. 4.24, where the integrated CL emission intensity is reported as a function of surfactant concentration. The net CL intensity, integrated over a defined time interval, depends on the CL quantum efficiency Φ_{CL} and on the number of molecules reacting per unit time, dC/dt, according to the equation [111]:

$$I_{CL} = \int \Phi_{CL}(t)(dC/dt)dt \qquad (23)$$

where Φ_{CL} is in turn given by the product $\Phi_{ex}\Phi_l$, being Φ_{ex} the excitation

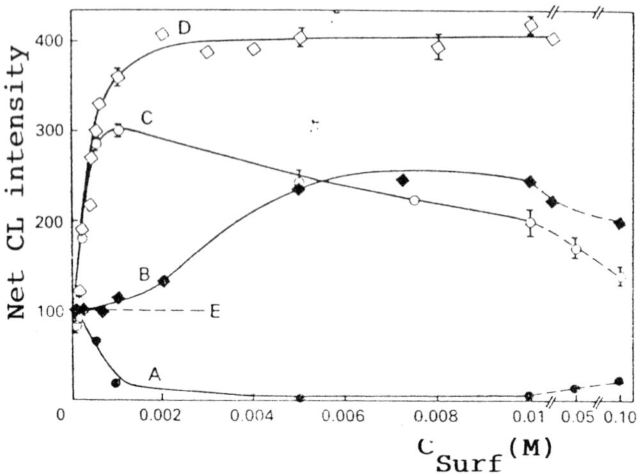

Fig. 4.24. Effect of surfactant concentration on the integrated CL emission for the lucigenin–H_2O_2 CL reaction at pH 12.5. Lucigenin 3.3×10^{-4} M; H_2O_2 2.9×10^{-4} M. Instrument delay and run time: 10 s. A, SDS; B, SB-12 (3-(N-dodecyl-N,N-dimethylammonio)propane-1-sulphonate; C, HTAC (pH 12.8); D, Brij 35; E, no surfactant added. Taken from Ref. [112], reproduced with permission.

yield of the excited state and Φ_l the luminescence yield of the primary emitter, respectively.

Working with the non-ionic surfactant Brij 35, the CL intensity is linearly proportional to H_2O_2 concentration in the range 3×10^{-7}–1×10^{-3} M [110], thus, the sensitive determination of this analyte can be performed.

In addition to lucigenin, other acridinium compounds can also react with hydrogen peroxide or oxygen to give rise to CL emission. Although the reaction mechanism can differ, the product emitter is usually NMA* or some NMA* derivative. Selected examples of micelle-sensitized CL determinations of H_2O_2 based on the use of lucigenin and other acridinium compounds are listed in Table 4.9.

4.4.1.2. Determination of analytes producing H_2O_2 by coupling with lucigenin CL reaction

The micellar-enhanced CL reaction can also be combined with certain enzymatic reactions to determine analytes which give rise to the formation of H_2O_2. For example, cholesterol can be oxidized by molecular oxygen in the presence of the enzyme cholesterol oxidase to Δ^5-cholesten-3-one and H_2O_2, which reacts with lucigenin in basic media to give CL emission.

TABLE 4.9

Micelle-sensitized CL determinations using lucigenin and related compounds

CL reagent (analyte)	Micellar system	pH	EF[a]	DL[b]	Ref.
Lucigenin[c]	HTAC 1×10^{-3} M	12	1.9	9.5	110
(H_2O_2)	SB-12 6×10^{-3} M	12	2.4	7.5	
	Brij 35 2×10^{-3} M	11.5	3.3	3.3	
H_2O_2[d]	HTAC 1×10^{-3} M	12	2.6		110
(Lucigenin)	Brij 35 2×10^{-3} M	12	4.0	2.9	
H_2O_2[e]	SB-12 1×10^{-2} M	6	4.3		113
(Acridinium ester)[f,g]		9	1.6		
	HTAC	6	67		
		9	3		

a: Enhancement factors obtained from $I_{CL(mic)}/I_{CL(water)}$. *b*: Detection limits: 10^7 M. *c*: Reagent concentration: 3.3×10^{-4}–1.0×10^{-3} M. *d*: Reagent concentration: 2.9×10^{-4}–3.3×10^{-4}M. *e*: Reagent concentration: 5.0×10^{-4} M. *f*: 2,4,6-Trichloromethylacridinium carboxylate fluorosulphonate 3.8×10^{-5} M. g: Reported detection limit (other experimental conditions): 0.5 fmol [113].

The CL intensity is related to the amount of H_2O_2 produced, which is in turn dependent on the cholesterol concentration.

Working in the presence of Triton X-100 or reduced Triton X-100 micelles, the CL output is enhanced up to ca. 8 times. This beneficial effect was attributed to the increased rate of hydrogen peroxide formation (enzymatic reaction) in micellar media and to the protection of the emitter against deactivation in the subsequent CL step [5].

4.4.1.3. CL reactions between lucigenin and organic reductants

A certain number of biological reductants (such as ascorbic acid, uric acid, glucose and fructose) can be determined by measuring the light produced from their corresponding CL reactions with lucigenin in basic media. The use of this analytical procedure is, however, limited by the insolubility of NMA in water.

The addition of SDS (under the cmc) or SDS and other surfactants (above the cmc) can help to overcome this problem. Moreover, the sensitivity is enhanced by a factor of ca. 2–4 in the presence of HTAC or SB-12 micelles, compared with that in water.

For example, the lucigenin–glucose CL reaction, reported in Scheme 4, shows an increased light output in the presence of HTAC micelles which was mainly attributed to favourable kinetic effects on the base-catalyzed keto-enol tautomerization of the sugar (assumed as the rate-determining step). Since it is known that micellar catalysis of bimolecular reactions involving the participation of OH^- (such as the tautomerization reaction mentioned) can be further enhanced using the corresponding HTA^+OH^- surfactants, the increase of CL intensity observed using these micelles confirms the above hypothesis [114].

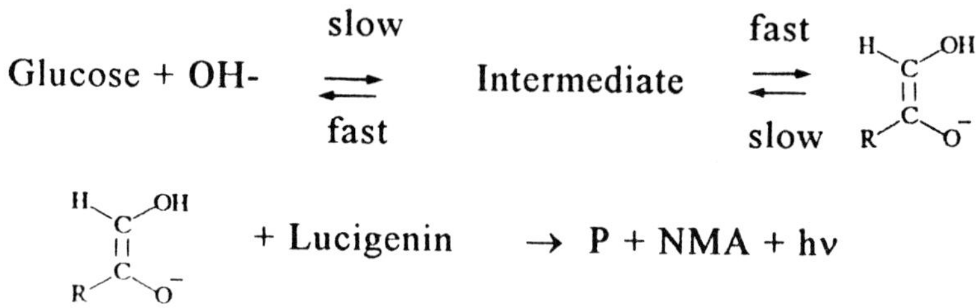

Scheme 4

The following integrated CL intensities, relative to water, have been measured for some biological reductants in HTAC (and HTAOH) micellar media: uric acid, 3.5(15.7); glucose, 1.7(2.8); fructose, 1.4(2.1); ascorbic acid, 3.9(9.3) [114].

Catecholamines (adrenaline, noradrenaline, dopamine) have also been determined through their CL reaction with lucigenin in cationic micellar media [115]. The negative effect of the halide counterions, which act as reaction quenchers, can be overcome by using CTAOH aggregates [116].

4.4.2. Other CL systems of analytical interest

4.4.2.1. Luminols

Luminol (5-amino-2,3-dihydro-1,4-phthalazinedione) is another popular CL reagent which, upon reaction with H_2O_2 in alkaline media, can give rise to the formation of the primary emitter 3-aminophthalate (see Scheme 5). Molecules containing peroxy or endoperoxide groups are postulated as possible reaction intermediates.

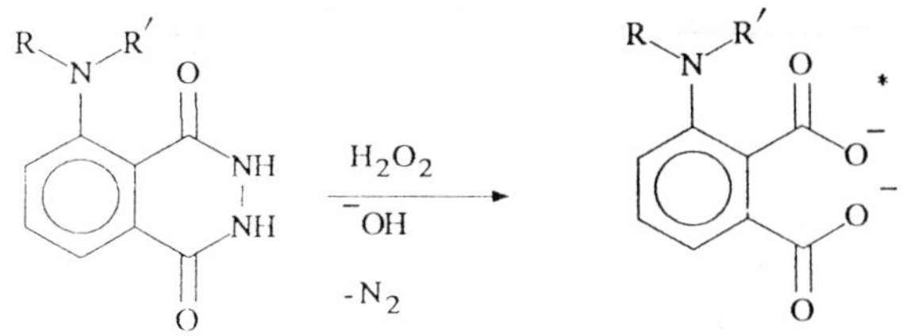

Scheme 5

Luminol and luminol derivatives can be exploited for the determination of the oxidizing agent (hydrogen peroxide) or for the analysis of certain metal ions (such as Cu^{2+} or Co^{2+}) which act as catalysts in these reactions. The presence of cationic surfactant micelles (HTAB or Septonex) improves the sensitivity of the luminol–H_2O_2 CL reaction through the enhancement of the light output by a factor of ca. 8 [117].

The addition of organic sensitizers further facilitates the efficient energy-transfer step from the donor (primary emitter) to the acceptor. For example, the observed enhancement factor of the luminol–H_2O_2 CL emission becomes ca. 35 in the presence of fluorescein.

Several basic investigations concerning the effect of different ordered media on luminol CL reactions are reported in Ref. [5]. Analytical applications include the determination of some organic substrates (glucose, glucose oxidase and amino acids) in HTAC (or HTAB) reversed micelles [118,119] and the use of such aggregates for the determination of gold through the reaction between chloroauric acid and luminol [120,121].

4.4.2.2. Peroxyoxalate-based CL

Micelle-sensitized CL reactions based on the association of peroxyoxalate compounds with fluorescer molecules are also of analytical interest. The first step in scheme 6 is the reaction between oxalates and H_2O_2(I), which through the formation of hydroperoyoxalate intermediates gives rise to dioxetanedione. The last compound reacts with a fluorescer reagent (F) according to steps (II) and (III), originating the light emission with a relevant quantum yield (ca. 0.3).

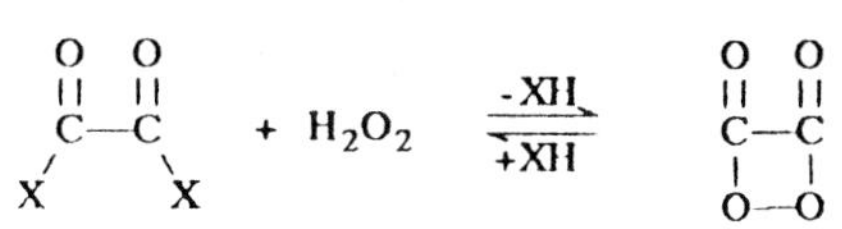

(I)

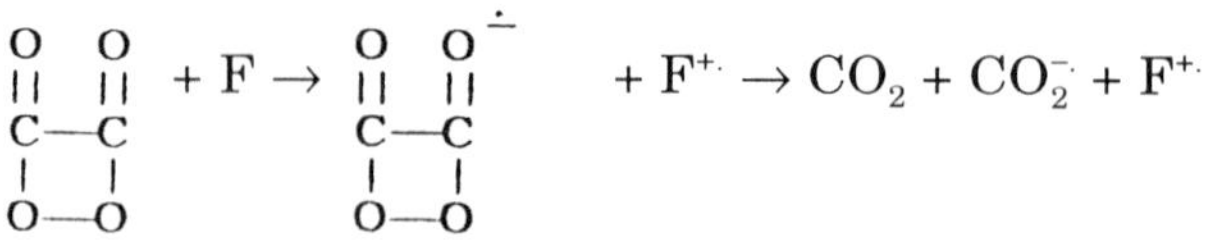

(II)

$$CO_2^{-\cdot} + F^{\cdot} \rightarrow CO_2 + F^{*} \rightarrow F + h\nu \qquad \text{(III)}$$

Scheme 6

An important application of this reaction as a chemical illuminating source has been found and relevant enhancement of light intensity and emission time have been obtained using different ordered media. Most basic studies on surfactant-modified CL of peroxyoxalates concern this practical field and, in particular, the characterization of the CL emission of the system METQ-H_2O_2-rubrene [5]. METQ is the oxalate ester 4,4′-[oxalylbis] (trifluoromethyl)sulphony]limino[ethylene]bis(4-methyl morpholiniumtrifluoromethane)sulphonate; the fluorescer rubrene is 5,6,11,12-tetraphenylnaphthacene.

From the analytical point of view, the sensitive determination of hydrogen peroxide and/or fluorophors is of interest, although the scarce solubility of most oxalic derivatives in water and their very fast emission kinetics limit their application in this medium. By using surfactant aggregates these difficulties can be overcome.

ANS (8-anilino-1-naphthalenesulphonic acid), rhodamine B (RB), and dansyl chloride (1-(dimethylamino)naphthalene-5-sulphonyl chloride, DANC), which are commonly used as fluorophor labels, can be also used in peryoxalate-based CL reactions [122]. The CL enhancement depends on the efficiency of fluorescence, excitation and deactivation processes, in turn controlled by factors such as pH, fluorophor and surfactant charge, surfactant hydrophobicity, etc. The observed beneficial effects given by micellar systems in some analytical determinations are summarized in Table 4.10.

Fluorophor-labelled proteins, such as albumin-dansyl chloride, can be also determined using this micelle-sensitized CL approach.

4.4.2.3. CL reactions based on luciferins

An interesting example of bioluminescence is that of firefly luciferin and related compounds [123,124]. The *Cypridina* luciferin-oxygen CL reaction, shown in Scheme 7, does not occur in bulk aqueous solution, whereas in cationic HTAB or non-ionic Tween 60, the CL emission is observed.

TABLE 4.10

Effect of surfactant systems on peroxyoxalate (METQ)-H_2O_2-fluorophor CL[a]

Analyte	Surfactant	Analytical parameter improved[b]	EF
ANS	Brij 35	Detection limit	3
DANC	Brij 35		2.5
	HTAC		5
RB, ANS	Brij 35	Dynamic linear ranges	2–3[c]
H_2O_2	Brij 35	Detection limit	3[d]
	HTAC		22[d]
	HTAC		2[e]

a: Data taken from Ref. [122]; *b*: at pH 7, compared to aqueous buffered solution; *c*: orders of magnitude; *d*: using ANS as fluorophor; *e*: using RB as fluorophor.

The ability of the less polar environment to enhance the luminescence quantum yield of the oxyluciferin primary emitter, together with the micellar catalytic effects, are responsible for the observed sensitization. In SDS the reaction does not occur because the luciferins, being anionic, are repelled from the micellar surface.

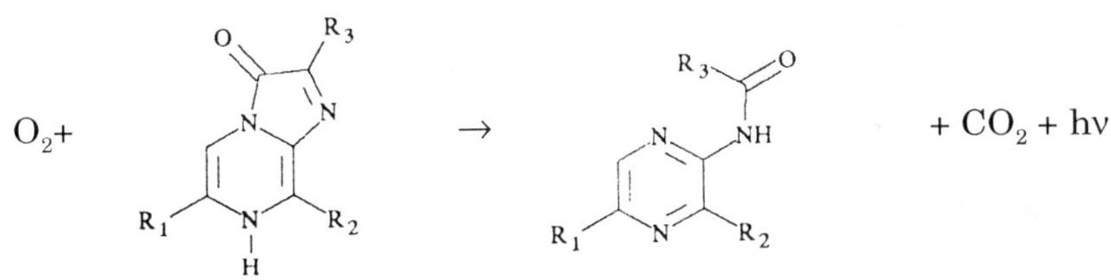

R_1 = –3-indolyl; R_2 = –H; R_3 = $-CH_3$ Scheme 7

Among the analytical applications, the luminescent ATP assay based on the reaction:

$$\text{Luciferin} + \text{ATP} + O_2 \rightarrow \text{oxyluciferin} + \text{AMP} + \text{P} + CO_2 + h\nu \qquad (24)$$

where P represents pyrophosphate. This reaction, which occurs in the presence of the hydrophobic enzyme firefly luciferase and magnesium ions, allows us to determine the very important bioanalyte adenosine triphosphate (ATP) (or species which react with ATP). CL enhancements in the range 2–5 (in respect to water) can be obtained in aqueous micellar media, but the best results were found by entrapping the active enzyme in reversed micelles of Brij 96 in cyclohexane. Working in this medium, the bioluminescent emission of the system luciferin–luciferase–ATP is stable for several minutes, while the corresponding time profile in water shows complete deactivation in few seconds [125,126].

4.5. USE OF SURFACTANT AGGREGATES IN ATOMIC SPECTROSCOPY

The determination of metals present in oily matrices by atomic spectroscopy is usually performed after the dry ashing of the samples, followed by the wet dissolution of the ashes in suitable acids. Since these preliminary treatments are time-consuming and can lead to metal losses, alternative procedures have been proposed.

The direct introduction of organic solutions into the atomization devices has been, in some cases, simply carried out after the dilution of the sample with organic solvents. However, the accurate preparation of organometallic standards (which must be soluble in organic media) is required. Moreover, when organic solutions are introduced in the flame, the measurements are often poorly reproducible.

On the other hand, the addition of organic solvents to aqueous acid solutions of metal ions prior to their flame-absorption analysis has been proven to be useful. In fact, these solvents can enhance the sensitivity of aqueous atomic absorption (AA) determinations by a factor of 2 or more and, moreover, they can favour the aspiration and nebulization steps, mainly through surface tension and viscosity effects. Working under these conditions, the formation of very small droplets in the nebulizer can be obtained.

The use of surfactant aggregates, in particular emulsions and microemulsions, as a way to introduce directly samples and standards into the atomizers has been proposed as a compromise strategy which combines the direct feed of organic solutions with the advantages of ashing methods. The analysis of metals present in lubricating oils, fuels, ointments, and other liposoluble samples, has been successfully performed through the formation of oil-in-water or water-in-oil emulsions, which allow the use of aqueous solutions for the preparation of standards. De la Guardia and Salvador published an extensive review on this topic [7].

TABLE 4.11

Advantages and disadvantages related to different sample preparation steps

Technique	Advantages	Disadvantages
Direct analysis	Higher sensitivity. Scarce sample manipulation.	Standard preparation is cumbersome.
Back extraction	Easy introduction of the samples. Use of aqueous standards allowed.	Time-consuming pre-treatments. Lower sensitivity.
Use of microemulsions and emulsions	Direct introduction of the sample. Use of aqueous standards. Higher stability and sensitivity[a]. Higher aspiration rates (water–oil).	Dilution of the extracts. Slower aspiration rates (water–oil). The flame is less stable.

a: Not always observed.

A comparison of the performances obtained using different sample treatments is given in Table 4.11.

4.5.1. Mechanism of sensitivity enhancement

Although several practical applications of surfactants in atomic spectrometry have been reported, only few studies were centred on sensitization mechanisms. A detailed analysis of surfactant concentration effects on the droplet size distribution of aerosols, performed under constant flame conditions, showed a relevant change in the droplet mean diameter [127]. This parameter is reduced to about half the value in water in the presence of 5×10^{-2} mol l^{-1} SDS or HTAC. Since smaller droplets are more efficiently sampled by the premix burner, a net enrichment results and enhancement of the analytical signal is observed.

In order to explain the enhanced atomic absorption shown by several metal ions (Ca^{2+}, Mg^{2+}, Ni^{2+}, Mn^{2+}, Cr^{3+}, Cu^{2+}) in the presence of anionic surfactants (SDS), the preferential distribution of these cations amongst the smaller aerosol droplets in the premix burner chamber was invoked [128].

Working with ionic surfactants having charge opposite to that of the analyte ion, Kornahrens et al. [129] proposed a model for the surfactant-induced analyte enrichment in the aerosol, based on the theory of aerosol ionic redistribution [128].

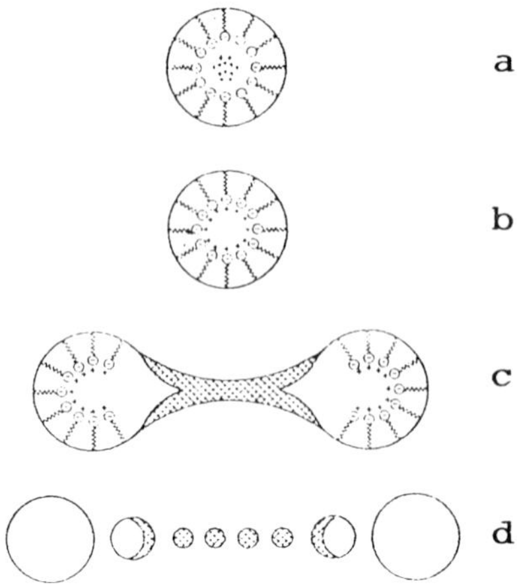

Fig. 4.25. Metal ion redistribution within the aerosol droplets: (a) surfactant migration to liquid–air interface; (b): metal ions diffusion within the droplet; (c) stripping of larger drops; (d) analyte accumulation in smaller droplets. Taken from Ref. [129] with permission.

The model assumes that the ionic surfactant molecules accumulate at the air–water interface and simultaneously increase the analyte concentration in the double layer formed on the surface of large drops. These large drops subdivide during the nebulization step: smaller droplets form from a stripping of the drop surface, while larger droplets arise from the analyte-depleted bulk drop centre (see Fig. 4.25). The analyte enrichment takes place in the smaller droplets and, since these droplets are more efficiently collected by the premix burner, sensitization occurs.

The observed signal enhancement is greatest just below the cmc, because at higher surfactant concentrations micelles form and the analyte counterion is partially attracted from the drop surface to the bulk, thus reducing the enrichment effect. Figure 4.26 shows the variation of copper absorbance as a function of SDS concentration. The maximum sensitivity occurs in the cmc range.

4.5.2. Use of emulsions in atomic spectroscopy

Some basic requirements must be met by emulsions in order to be employed in sample preparation: the emulsion stability is necessary, at least during the time of analysis; the emulsified sample must be physi-

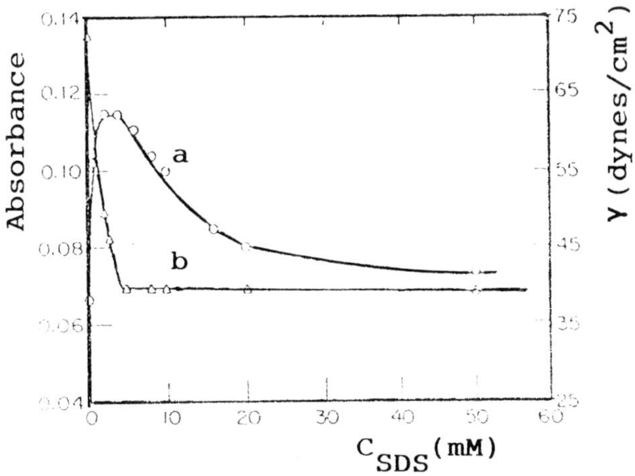

Fig. 4.26. SDS-sensitized AA signal for copper. (a) Cu absorbance; (b) surface tension of the aspirated solution. Taken from Ref. [129] with permission.

cally suitable for aspiration and introduction into the atomization systems; atomization of samples and standards must give identical results for identical analyte concentrations.

It is thus necessary that an adequate selection of the surfactants are used as emulsifying agents and several factors must be considered. They include the nature of the samples to be emulsified, the type of emulsion (oil–water or water–oil) to be prepared, the structure of the hydrophilic group of the amphiphile (which can interact with the analyte), the structure of its non-polar part (which must be similar to the solvents to be emulsified), etc.

Usually, one of the most important parameters considered for the selection of the proper surfactant is the HLB, since according to its value, different oily samples can be emulsified. For example, in order to obtain stable (few hours) oil–water emulsions from lubricating oils, an HLB in the range 8–18 is recommended, whereas other HLB values are required to match adequate emulsions with other samples. The proper HLB range can be obtained also by mixing together different surfactants. After dispersion, the emulsion system must remain stable during a time interval large enough to complete the determination procedure.

Table 4.12 lists some examples of reported applications of oil-in-water emulsions in atomic spectroscopic analysis of metals.

TABLE 4.12

Use of oil-in-water emulsions in atomic absorption/emission analysis

Sample	Analyte	Surfactant system	Method	Ref.
Lubricating oils	Pb	MS-12/Tween 20	AAS	130
	Fe	MIBK/Nemol K39	AAS	131
	Fe,Pb	Brij 35/n-octanol	DCPS	132
	Mo	MIBK/Nemol K39	AAS	133
MIBK	Cu,Fe	Nemol K38/TergitolXD	AAS	134
Gasoline	Pb	Brij 30/Tween 80	AAS	135
Undecanoate ointments	Zn	Brij 30/Tween 20	AAS	136
Toluene	Mn	Non-ionic emulsion	ICPS	137

MIBK: methylisobutylketone; MS-12; Nemol K39: nonylphenol(polyoxyethylene)$_9$glycol ether; Nemol K38 nonylphenol(polyoxyethylene)$_8$glycol ether; Tergitol XD.

One important factor in the choice of emulsions to be directly aspirated into the flame is the stability and homogeneity of such dispersed systems, which are in turn related to the size of the dispersed oil (or water) droplets. The use of highly stable microemulsions, having a mean diameter of the dispersed droplets in the range 20–200 nm, can improve the reproducibility of atomic emission/absorption measurements.

4.5.3. Use of on-line emulsification in flow systems

On-line emulsification of oily samples can be accomplished in an ultrasonic bath using highly stirred chambers where a very effective mixing of the sample and the added surfactant is achieved. The resulting emulsions can be continuously pumped to the premix burner of an AA unit. The analysis of additives present in gasolines has been performed using the manifold reported in Fig. 4.27.

Another interesting and recent application of on-line emulsification is the possibility of treating solid dispersions coming from a microwave oven with a surfactant-rich carrier [138]. The emulsified sample can be continuously introduced into a flame and the time required for the analysis is drastically reduced, allowing the use of emulsions which are stable only for a short time.

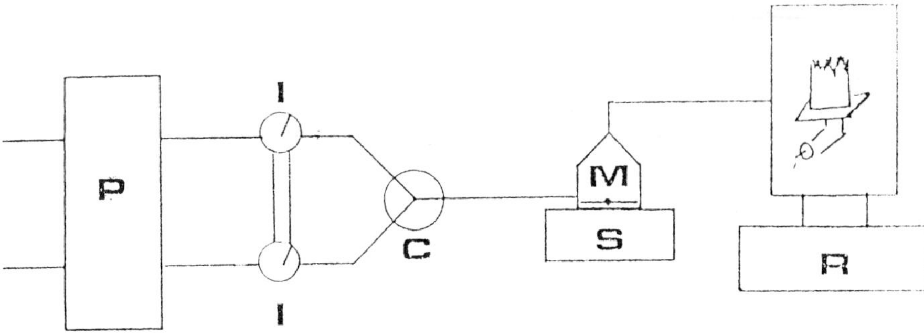

Fig. 4.27. Continuous flow analysis of lead in gasolines. P: peristaltic pump. A: gasoline. B: aqueous surfactant mixture. I: double injector. C: merging zone. M:mixing chamber. S: magnetic stirrer. F: flame. R: recorder unit. Taken from Ref. [7], reproduced with permission.

REFERENCES

1 W.L. Hinze, in: K.L. Mittal (Ed.), *Solution Chemistry of Surfactants*. Plenum Press, New York, 1979, pp. 79–127.
2 L. Cermakova, in: K.L. Mittal and B. Lindman (Eds.), *Surfactants in Solution*. Plenum Press, New York, 1984, Vol. 2, p. 1217.
3 E. Pelizzetti and E. Pramauro, *Anal. Chim. Acta*, 169 (1985) 1.
4 M.E. Diaz Garcia and A. Sanz Medel, *Talanta*, 33 (1986) 255.
5 W.L. Hinze, N. Srinivasan, T.K. Smith, S. Igarashi and H. Hoshino, in: *Advances in Multidimensional Luminescence*. JAI Press, Tokyo, 1991, Vol. 1, pp. 149–206.
6 G.L. McIntire, *CRC Crit. Rev. Anal. Chem.*, 21 (1990) 257.
7 M. de la Guardia and A. Salvador, *Analusis*, 19 (1991) 52.
8 J.R. Cardinal and P. Mukerjee, *J. Phys. Chem.*, 82 (1978) 1614.
9 J.H. Fendler and E.J. Fendler, *Catalysis in Micellar and Macromolecular Systems*. Academic Press, New York, 1975. Ch. 3, p. 51.
10 B.A. Mulley and A.D. Metcalf, *J. Pharm. Pharmacol.*, 8 (1956) 774.
11 S. Riegelman, N.A. Allawala, M.K. Hrenoff and L.A. Strait, *J. Colloid Sci.*, 13 (1958) 208.
12 S.J. Rehfeld, *J. Phys. Chem.*, 75 (1971) 3905.
13 P. Mukerjee and J.R. Cardinal, *J. Phys. Chem.*, 82 (1978) 1620.
14 C. Reichardt, *Solvents and Solvent Effects in Organic Chemistry*. VCH, Weinheim, 2nd edn., 1988.
15 K. Dimroth, C. Reichardt, T. Siepmann and F. Bohlmann, *Liebigs Ann. Chem.*, 661 (1963) 1.
16 K. Kalyanasundaram and J.K. Thomas, *J. Phys. Chem.*, 81 (1977) 2176.

17 K.F. Donchi, G.P. Robert, B. Ternai and P.J. Derrick, *Aust. J. Chem.,* 33 (1980) 2199.
18 H.G. Benson and J.N. Murrell, *J. Chem. Soc., Faraday Trans.,* 68 (1972) 137.
19 H. Watanabe and H. Ohmori, *Talanta*, 26 (1979) 954.
20 H. Chung-Gin, H. Chao-Sheng and J. Ji-Hong, *Talanta*, 27 (1980) 676.
21 W. Fu-Sheng and Y. Fang, *Talanta*, 30 (1983) 190.
22 J. Medina Escriche, M. Llobot Estelles and F. Bosch Reig, *Talanta*, 30 (1983) 915.
23 K. Hayashi, Y. Sasaki, S. Tagashira and E. Kosaka, *Anal. Chem.,* 58 (1986) 1444.
24 M. Llobat Estelles, J. Medina Escriche, A. Tomas Primo and A. Sevillano Cabeza, *Analyst*, 111 (1986) 53.
25 I. Mori, Y. Fujita, K. Fujita, R. Fujita, Y. Nakahashi and K. Kato, *Talanta*, 35 (1988) 879.
26 J. Miura, *Analyst*, 114 (1989) 1323.
27 S.P. Moulik, S. Ghosh and A.R. Das, *Colloid Polym. Sci.,* 257 (1979) 645.
28 J. Rosendorfova and L. Cermakova, *Talanta*, 27 (1980) 705.
29 V. Buresova, L. Kuban and L. Sommer, *Collect. Czech. Chem. Commun.,* 46 (1981) 1090.
30 M.E. Diez Garcia and A. Sanz Medel, *Talanta*, 32 (1985) 189.
31 J.H. Callaham and K.D. Cook, *Anal. Chem*., 54 (1982) 59.
32 V. Svoboda and V. Chromy, *Talanta*, 13 (1966) 237.
33 J.E. Chester, R.M. Dagnall and T.S. West, *Talanta*, 17 (1970) 13.
34 M. Jarosz and A. Swietlow, *Microchem. J*., 37 (1988) 322.
35 M. Jarosz and M. Malat, *Microchem. J.,* 37 (1988) 268.
36 M. Miyawaki and K. Uesugi, *Mikrochim. Acta*, I (1985) 135.
37 K. Uesugi and M. Miyawaki, *Microchem. J.,* 26 (1981) 288.
38 A. Sanz Medel, C. Camara Rica and J.A. Perez Bustamante, *Anal. Chem.,*52 (1980) 1035.
39 J.H. Callahan and K.D. Cook, *Anal. Chem.,*56 (1984) 1632.
40 G. Kwapulinska and B. Janoszka, *Fresenius J. Anal. Chem.,* 338 (1990) 641.
41 J. Havel, I. Buresova Jankarova and V. Kuban, *Collect. Czech. Chem. Commun.,* 48 (1983) 1290.
42 L.S. Romsted and D. Zanette, *J. Phys. Chem.,* 92 (1988) 4690.
43 C.A. Bunton, K. Ohmenzetter and L. Sepulveda, *J. Phys. Chem.,* 81 (1977) 2000.
44 P. Mukerjee, K. Mysels and P.J. Kapauan, *J. Phys. Chem.,* 71 (1967) 4166.
45 A.D. James and B.H. Robinson, *J. Chem. Soc. Faraday Trans. I*, 74 (1978) 10.
46 P.D.I. Fletcher and B.H. Robinson, *J. Chem. Soc. Faraday Trans. I*, 80 (1984) 2417.
47 A.I. Carbone, F.P. Cavasino, E. Di Dio and C. Sbriziolo, *Int. J. Chem. Kinet.,* 18 (1986) 609.

48 Y. Miyake, M. Shigeto and M. Teramoto, *J. Chem. Soc. Faraday Trans. I*, 82 (1986) 1515.
49 H.S. Kim and C. Tondre, *Langmuir*, 5 (1989) 395.
50 S. Muralidharan, W. Yu, S. Tagashira and H. Freiser, *Langmuir*, 6 (1990) 1190.
51 F.P. Cavasino, C. Sbriziolo, E. Pelizzetti and E. Pramauro, *J. Phys. Chem.*, 93 (1989) 469.
52 K. Kalyanasundaram, *Photochemistry in Microheterogeneous Systems*. Academic Press, London, 1987, pp. 388.
53 K. Kalyanasundaram and J.K. Thomas, *J. Am. Chem. Soc.*, 99 (1977) 2039.
54 K. Kalyanasundaram and J.K. Thomas, *J. Phys. Chem.*, 81 (1977) 2176.
55 E.L. Wehry and L.B. Rogers, *J. Am. Chem. Soc.*, 88 (1966) 351.
56 T. Forster, *Z. Electrochem.*, 54 (1950) 531.
57 E. Djoufac-Woumfo, N. Arnaud and J. Georges, *Analyst*, 113 (1988) 447.
58 H.N. Singh and W.L. Hinze, *Anal. Lett.*, 15 (1982) 221.
59 D.W. Armstrong, W.L. Hinze, K.H. Bui and H.N. Singh, *Anal. Lett.*, 14 (1981) 1659.
60 M. de la Guardia Cirugeda and F. Rodilla Soriano, *Analyst*, 114 (1989) 77.
61 M.V. Galdù, M. de la Guardia and L. Braco, *Analyst*, 112 (1987) 1047.
62 J. Sancenòn, J.L. Carriòn and M. de la Guardia, *Fresenius J. Anal. Chem.*, 336 (1990) 389.
63 H.N. Singh and W.L. Hinze, *Analyst*, 107 (1982) 1073.
64 T.L. Keimig and L.B. McGown, *Talanta*, 33 (1986) 653.
65 S.T. Meyerhoffer and L.B. McGown, *Anal. Chem.*, 63 (1991) 2082.
66 P.M. Ritenour Hertz and L.B. McGown, *Anal. Chem.*, 64 (1992) 2920.
67 A. Sanz Medel, R. Fernandez de la Campa and J.I. Garcia Alonso, *Analyst*, 112 (1987) 493.
68 T. Taketatsu, *Talanta*, 29 (1982) 397.
69 J. Medina Escriche, M. de la Guardia Cirugeda and F. Hernandez Hernandez, *Analyst*, 108 (1983) 1386.
70 A.G. Howard, A.J. Coxhead, I.A. Potter and P.A. Watt, *Analyst*, 111 (1986) 1379.
71 F. Salinas, A. Munoz de la Pena and M.S. Duran, *Anal. Lett.*, 21 (1988) 1457.
72 A. Sanz-Medel and J.I. Garcia Alonso, *Anal. Chim. Acta*, 165 (1984) 159.
73 A. Sanz-Medel, J.I. Garcia Alonso and E. Blanco Gonzalez, *Anal. Chem.*, 57 (1985) 1681.
74 A. Sanz-Medel, M. Fernandez Perez, M. de la Guardia Cirugeda and J.L. Carrion Dominguez, *Anal. Chem.*, 58 (1986) 2161.
75 J. Medina, F. Hernandez, R. Marin and F.J. Lopez, *Analyst*, 111 (1986) 235.
76 F. Hernandez Hernandez, J. Medina Escriche and M.T. Gasco Andreu, *Talanta*, 33 (1986) 537.
77 Y. Jing-He, Z. Gui-Yun and W. Bo, *Anal. Chim. Acta*, 198 (1987) 287.

78 Z. Guiyun, Y. Jinghe and S. Zhikun, *Anal. Chim. Acta*, 215 (1988) 331.
79 R. Compano, A. Grima, A. Izquierdo and M.D. Prat, *Anal. Chim. Acta*, 227 (1989) 219.
80 I. Mori, Y. Fujita, K. Fujita, Y. Nakahashi, R. Matsuo and K. Kato, *Fresenius Z. Anal. Chem.*, 333 (1989) 42.
81 L.M. Polo Diez, M.W. Kabbani and S.J. Durand Alegria, *Fresenius Z. Anal. Chem.*, 334 (1989) 41.
82 J. Georges and S. Ghazarian, *Anal. Chim. Acta*, 276 (1993) 401.
83 M. Tarek, M. Zaki, L.F. Esmail and A.Y. El-Sayed, *Fresenius Z. Anal. Chem.*, 331 (1988) 607.
84 J.B.F. Lloyd, *Nature (Lond.)*, 231 (1971) 64.
85 P.H. Baudot and J.C. Andre, *Anal. Lett.*, 15 (1982) 471.
86 J.J. Santana Rodriguez, Z. Sosa Ferrera, A. Afonso Perera and V. Gonzalez Diaz, *Anal. Chim. Acta*, 255 (1991) 107.
87 J.J. Santana Rodrìguez, J. Hernàndez Garcìa, M.M. Bernal Suàrez and A. Bermejo Martìn-Làzaro, *Analyst*, 118 (1993) 917.
88 G. Ramis Ramos, M.C. Garcia Alvarez Coque, A. Berthod and J.D. Winefordner, *Anal. Chim. Acta*, 208 (1988) 1.
89 M.H. Memon and P.J. Worsfold, *Anal. Proc.*, 23 (1986) 418.
90 M.H. Memon and P.J. Worsfold, *Anal. Chim. Acta*, 183 (1986) 179.
91 H. Watarai and K. Ogawa, *Anal. Chim. Acta*, 277 (1993) 73.
92 R.T. Parker, R.S. Freedlander and R. Bruce Dunlap, *Anal. Chim. Acta*, 119 (1980) 189.
93 R.T. Parker, R.S. Freedlander and R. Bruce Dunlap, *Anal. Chim. Acta*, 120 (1980) 1.
94 M.E. Diaz Garcia and A. Sanz-Medel, *Anal. Chem.*, 58 (1986) 1436.
95 L.J. Cline Love, M. Skrilec and J.G. Habarta, *Anal. Chem.*, 52 (1980) 754.
96 N.E. Nugara and A.D. King, Jr., *Anal. Chem.*, 61 (1989) 1431.
97 L.J. Cline Love, J.G. Habarta and M. Skrilec, *Anal. Chem.*, 53 (1981) 437.
98 M. Almgren, F. Grieser and J.K. Thomas, *J. Am. Chem. Soc.*, 101 (1979) 279.
99 L.A. Blyshak and I.M. Warner, *Anal. Chem.*, 62 (1990) 1953.
100 A.G. Mwalupindi, T.T. Ndou and I.M. Warner, *Anal. Chem.*, 64 (1992) 1840.
101 A.G. Mwalupindi, L.A. Blyshak, T.T. Ndou and I.M. Warner, *Anal. Chem.*, 63 (1991) 1328.
102 R.A. Femia and L.J. Cline Love, *Anal. Chem.*, 56 (1984) 327.
103 A. Sanz-Medel, P.L. Martinez Garcia and M.E. Diaz Garcia, *Anal. Chem.*, 59 (1987) 774.
104 M.E. Diaz Garcia, M.R. Fernandez de la Campa, W.L. Hinze and A. Sanz-Medel, *Mikrochim. Acta*, III (1988) 269.
105 M.R. Fernandez de la Campa, M.E. Diaz Garcia and A. Sanz-Medel, *Anal. Chim. Acta*, 212 (1988) 235.

106 C.M. Paleos, G. Vassilopoulos and J. Nikokavouras, in: M.A. DeLuca and W.D. McElroy (Eds.), *Bioluminescence and Chemiluminescence*. Academic Press, New York, 1981, p. 729.
107 C.M. Paleos, G. Vassilopoulos and J. Nikokavouras. *J. Photochem.,* 18 (1982) 327.
108 J. Nikokavouras, G. Vassilopoulos and C.M. Paleos. *J. Chem. Commun.,* (1981) 1082.
109 J. Nikokavouras and G. Vassilopoulos. *Z. Physik Chem.,* 265 (1984) 618.
110 C.L. Malehorn, T.E. Riehl and W.L. Hinze, *Analyst*, 111 (1986) 941.
111 W.L. Hinze, T.E. Riehl, T.E. Singh and Y. Baba. *Anal. Chem.,* 56 (1984) 2180.
112 T.E. Riehl, C.L. Malehorn and W.L. Hinze, *Analyst*, 111 (1986) 931.
113 C.L. Howie and N.L. Grayeski in: J. Scholmerich, R. Andreesen, A. Kapp, M. Ernest and W.G. Woods (Eds.), *Bioluminescence and Chemiluminescence: New Perspectives*. Wiley, New York, 1987, p. 415.
114 A. Ingvarsson, C.L. Flurer, T.E. Riehl, K.N. Thimmaiah, J.M. Williams and W.L. Hinze, *Anal. Chem.,* 60 (1988) 2047.
115 T. Kamidate, K. Yoshida, T. Segawa and H. Watanabe, *Anal. Sci.,* 5 (1989) 359.
116 T. Kamidate, T. Kaneyasu, T. Segawa and H. Watanabe, *Bull. Chem. Soc. Jpn.,* 64 (1991) 1991.
117 J. Lasovsky and F. Grambal, *Bioelectrochem. Bioenerg.,* 15 (1986) 95.
118 S. Igarashi and W.L. Hinze, *Anal. Chem.,* 60 (1988) 446.
119 S. Igarashi and W.L. Hinze, *Anal. Chim. Acta*, 225 (1989) 147.
120 Imdadullah, T. Fujiwara and T. Kumamaru, *Anal. Chem.,* 63 (1991) 2348.
121 Imdadullah, T. Fujiwara and T. Kumamaru, *Anal. Chem.,* 65 (1993) 421.
122 N. Dan, M. Ling Lau and M.L. Grayeski, *Anal. Chem.,* 63 (1991) 1766.
123 T. Goto and H. Fukatsu, *Tetrahedron Lett.,* 49 (1969) 4299.
124 T. Goto, I. Kuboto, N. Suzuki, Y. Kishi, S. Inoue in: J. Cormier, D.M. Hercules and J. Lee (Eds.), *Chemiluminescence and Bioluminescence*. Plenum Press, New York, 1973, p. 325.
125 K. Martinek, I.V. Berezin, Y.L. Khmelnitski, N.L. Klyachko and A.V. Levashov, *Collect. Czech. Chem. Commun.,* 52 (1987) 2589.
126 N.L. Klyachko, M.Y. Rubtsova, A.V. Levashov, E.M. Gavrilova, A.M. Egorov, K. Martinek and I.V. Berezin, *Ann. N. Y. Acad. Sci.,* 501 (1987) 267.
127 M. Kodama and S. Miyagawa, *Anal. Chem.,* 52 (1980) 2358.
128 J.A. Boroweic, A.W. Boorn, J.H. Dillard, M.S. Cresser and R.F. Browner, *Anal. Chem.,* 52 (1980) 537.
129 H. Kornahrens, K.D. Cook and D.W. Armstrong, *Anal. Chem.,* 54 (1982) 1325.
130 J. Hernandez-Mendez, L. Polo-Diez and A. Bernal-Melchor, *Anal. Chim. Acta,* 108 (1979) 39.
131 A. Salvador, M. de la Guardia and A. Berenguer, *Talanta*, 30 (1983) 986.
132 E. Pramauro, E. Pelizzetti and G. Saini, *Ann. Chim.* (Rome), 74 (1984) 673.

133 J.B. Beferull Blasco, M. de la Guardia Cirugeda and A. Salvador Carreno, *Anal. Chim. Acta*, 174 (1985) 353.
134 M. de la Guardia and M.T. Vidal, *Talanta*, 31 (1984) 799.
135 L. Polo-Diez, J. Hernandez-Mendez and F. Pedraz-Penalva, *Analyst*, 105 (1980) 37.
136 L. Polo-Diez, J. Hernandez-Mendez and J.A. Rodriguez-Gonzalez, *Analyst*, 106 (1981) 737.
137 M. de la Guardia, G. Legrand, M. Druon and J.L. Lovrier, *Analusis*, 10 (1982) 476.
138 J.S. Esteve-Romero, E.F. Simò-Alfonso, M.C. Garcia Alvarez-Coque and G. Ramis-Ramos, *TrAC*, 14 (1995) 29.

Chapter 5

Electroanalytical chemistry in organized surfactant systems

The solubilization of non-polar organic compounds in aqueous micellar solutions for electrochemical measurements was the first reported application of surfactant assemblies in electroanalysis [1]. It has long been established that surface-active agents act as supressors in polarography (usually at concentrations below the cmc) and that amphiphile adsorption onto the electrodes is responsible for the observed suppression effect. The relationship between the height of the polarographic wave and the concentration of solubilized electroactive species and amphiphiles was investigated many years ago and reported in the literature [2].

More detailed investigations on the electrochemical processes occurring in the presence of surfactant assemblies as well as on their applications in electroanalytical chemistry have been performed in the last fifteen years. Some reviews [3–7] and a recently published book [8] summarize the main reported applications in this field.

Important advantages are related to the use of surfactant aggregates, which include the stabilization of unstable chemical species (e.g. ion-radicals) produced at electrodes by coulombic and hydrophobic interactions with micelles, the alteration of the product distribution arising from electrodic processes due to partitioning of electroactive reagents and/or products, favourable catalytic effects due to the increase in the electron transfer rate (usually resulting in a sensitivity increase), and the use of reducible and oxidizable solutes to measure diffusion in microaggregate media. In fact, electroactive substances solubilized in micellar aggregates can participate in redox processes taking place at the electrodes and/or in the solution, the current being generally controlled by the diffusion of the host aggregate.

Most of the electrochemical investigations in microaggregates are performed in order to gain information about the microaggregate itself, the guest solubilizate, or the mechanism of the redox processes occurring at the aggregate–bulk interface. Important parameters of the microaggregate, such as the cmc, the degree of dissociation (α), the size, and the properties of the microenvironment, can be evaluated using a proper electroanalytical technique.

Data concerning the solubilizate properties can also be obtained from electrochemical measurements. In particular the partitioning of electroactive compounds, the electrochemical behaviour of water-insoluble compounds in water-rich systems, and the effect of aggregates on redox processes involving the analytes, can be extensively investigated.

Another relevant point is the study of kinetics and mechanisms of electron transfer at interfaces, which includes the electron transfer at aggregate/electrode interfaces, the influence of adsorption processes and the presence of electrocatalytic phenomena. A typical practical application arising from these studies is, for example, the extension of the useful potential range in several analytical determinations.

The electrochemical investigations in organized surfactant systems offer some advantages: the measurements are usually simple, fast and

TABLE 5.1

Summary of the electroanalytical techniques in which surfactant microaggregates have been applied

Basic technique	Type of measurement	Experimental variables
Conductometry	Conductometric titration	Λ vs. $C_{analyte}$ E vs. $C_{analyte}$
Potentiometry	Direct titrations ($i \cong 0$)	E vs. $C_{titrant}$ pH vs. $C_{titrant}$
Voltammetry	Polarography (DME)[a]	i vs. E
	RDE[b]	i vs. E
	Cyclic voltammetry (CV)	i vs. E
	Microelectrode voltammetry	i vs. E
	Chronopotentiometry	E vs. time
	Chronocoulometry	Q vs. time

a: Dropping mercury electrode; *b*: rotating disk electrode.

inexpensive; they are applicable to opaque systems; prior information about physical properties (except viscosity) is unnecessary; the results are, generally, quite easily explained.

The main disadvantages are related to the use of a probe which can modify the aggregate structure. Furthermore, the adsorption phenomena can strongly alter the instrumental response and the kinetics at the electrode may be complex

Table 5.1 summarizes the basic features of the techniques where the application of amphiphilic assemblies has thus far been reported. The main aspects of each surfactant-modified method will be separately examined and discussed in the following sections.

5.1. CONDUCTOMETRIC MEASUREMENTS IN AMPHIPHILIC ORGANIZED MEDIA

5.1.1. Conductometry in surfactant/water systems

According to Onsager, the equivalent conductance, Λ, of an electrolyte solution is related to the equivalent conductance at infinite dilution, Λ_0, and to the electrolyte concentration, C, through the equation

$$\Lambda = \Lambda_0 - (A + B\Lambda_0)\sqrt{C} \qquad (1)$$

If the equivalent conductance of anionic ethoxylated surfactants is plotted against the square root of the amphiphile concentration, diagrams as shown in Fig. 5.1 are obtained. At low surfactant concentration ($C < \text{cmc}$), the equivalent conductance of an alkyl sulfate salt lies close to the Onsager slope for an univalent strong electrolyte. At the cmc, an abrupt descending portion of the curve is typically observed.

Since it was pointed out that the formation of ionic micelles involves the aggregation of a number of ions of like charge, an increase rather than a decrease in electrical conductivity is expected when aggregation occurs. Moreover, micelle formation would give rise to a single entity having a charge and volume n times that of a single amphiphilic ion. From Stokes' law, the viscous resistance to its movement in an electric field is inversely proportional to the radius of the aggregate and will therefore be equal to $(n^{1/3})^{-1}$ of that exerted by a single ion, whereas the force is proportional to the net charge. The micelle should then move more rapidly and lead to a conductivity which is $n^{2/3}$ times that due to n monomers.

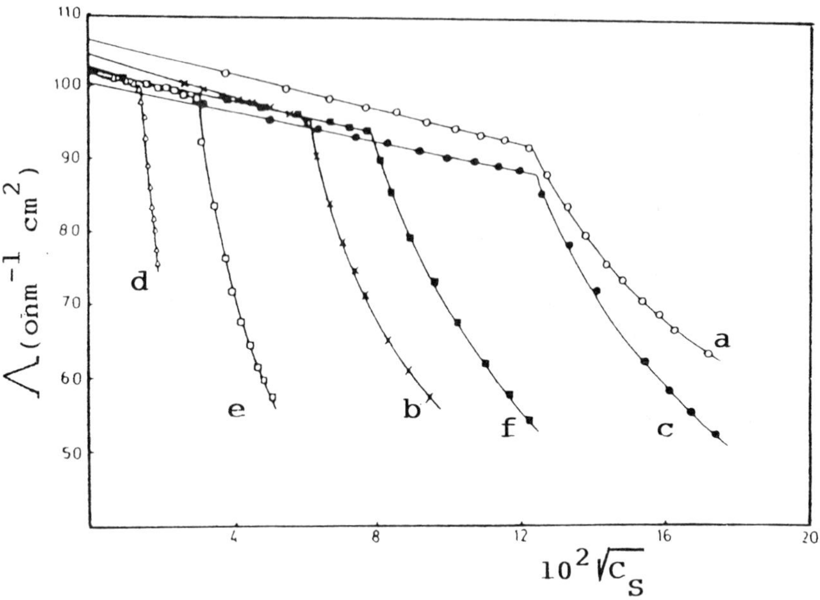

Fig. 5.1. Equivalent conductance of sodium alkylethoxysulfates at 25°C. (a) $C_{10}(EO)_2O$–SO_3Na; (b) SDS; (c) $C_{12}(EO)O$–SO_3Na; (d) $C_{12}(EO)_2O$–SO_3Na; (e) $C_{14}(EO)_2O$–SO_3Na; (f) $C_{16}(EO)_2O$–SO_3Na. Taken from Ref. [9]. Reproduced with permission.

The decrease in conductivity accompanying micelle formation must necessarily be due to the inclusion within the aggregate of a considerable number of counterions. These counterions effectively neutralize an equivalent charge on the micelle and are themselves withdrawn from the conduction process. Both effects lead to a reduction of the equivalent conductance.

If the surface charge density is small, the fraction of the counterions bound to the micelle is correspondingly small and an increase of Λ must be observed when the aggregates form. This behaviour was confirmed for some cationic surfactants (see Fig. 5.2), where the maximum can be attributed either to increased binding to higher surfactant concentrations or to a change in micellar shape [10].

5.1.1.1. Conductometric determination of the cmc

Conductivity measurements of ionic micellar solutions have been largely used to estimate the cmc. The measured electric conductivity as a function of surfactant concentration shows a linear behaviour, with a

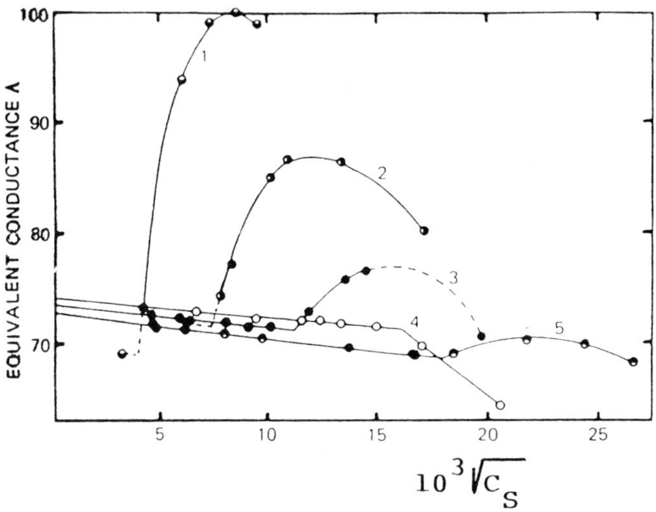

Fig. 5.2. Equivalent conductance of quaternary ammonium bromates. (1) octadecyl triamylammonium; (2) octadecyltributylammonium; (3) octadecyltripropylammonium; (4) octadecyltriethylammonium. (5) hexadecyltrimethylammonium. Reproduced with permission from Ref. [10].

break in the cmc range, both in the presence of pure surfactant and when low concentrations of added solubilizates are present (see Fig. 5.3). The abrupt change in the experimental variable associated with the aggregate formation is evident, and the corresponding cmc can be easily estimated. Figure 5.3 shows an example of these plots [11].

A large amount of the cmc data of ionic amphiphiles reported in the literature has been determined using this experimental approach [12].

5.1.1.2. Determination of the micellar fraction of charge

If interionic interactions are neglected, the degree of dissociation (α) of the micellar aggregates can be approximated by the ratio of the slopes in the conductivity plot, above and below the cmc range [13,14]. In fact, by applying the mass balance and electroneutrality condition to a solution containing micelles, amphiphilic ions and counterions, the following equations are obtained:

$$C_{tot} = C_S + n\,C_m \tag{2}$$

$$C_i = C_S + \alpha n\,C_m \tag{3}$$

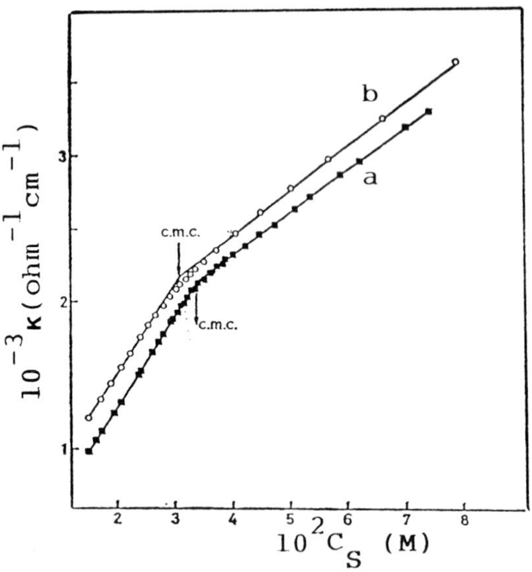

Fig. 5.3 Conductivity of aqueous solutions of sodium decylsulfate (SDeS), at 25°C. (a) Pure surfactant; (b) surfactant containing a solubilized ligand (2×10^{-5} M PADA[a]) and 1×10^{-3} M Ni(II). Reproduced with permission from Ref. [11].

where C_{tot} is the analytical concentration of surfactant, C_S and C_m are the concentrations of monomers and micelles, respectively, $n = N_{AG}$, C_i is the concentration of free counterions. From Eqns. (2) and (3) we obtain:

$$C_i = C_S + \alpha\,(C_{tot} - C_S) \tag{4}$$

The measured specific conductivity, κ, is given by a sum of contributions:

$$\kappa = C_i\Lambda_i + C_S\Lambda_S + C_m\Lambda_m \tag{5}$$

where Λ_i, Λ_S and Λ_m are the equivalent conductivities of counterion, amphiphile monomer and micelles, respectively. Assuming that above the cmc C_S is constant (and equal to the cmc), the concentration of counterion is given by:

$$C_i = \alpha C_{tot} + (1 - \alpha)\mathrm{cmc} \tag{6}$$

Recalling that $C_{tot} - \mathrm{cmc} = C_D$ and indicating with κ_{cmc} the specific conductivity at the cmc, the combination of Eqns. (5) and (6) gives:

$$\kappa = (\Lambda_i + \Lambda_S)\mathrm{cmc} + (\alpha\Lambda_i + \Lambda_m/n)C_D = \kappa_{cmc} + (\alpha\Lambda_i + \Lambda_m/n)C_D \tag{7}$$

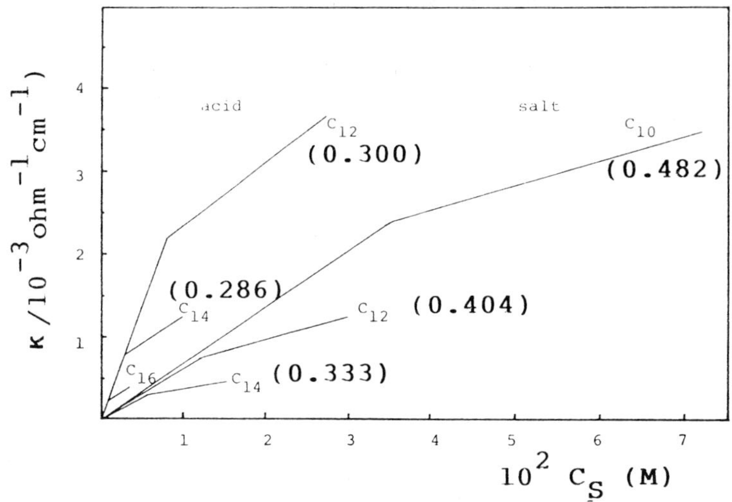

Fig. 5.4. Specific conductance vs. amphiphile concentration for some alkylsulfonic acids and their sodium salts. Reproduced with permission from Ref. [15].

Assuming Λ_i equal to the limiting conductivities of the counterions and introducing the approximation $\alpha\Lambda_i >> \Lambda_m/n$, Eqn. (7) allows the estimation of α.

Several examples of these calculations are reported in the literature. Figure 5.4 illustrates the case of alkylsulfonic acids and their sodium salts [15]. The corresponding α values of some of them are indicated in the figure.

A better description of conductometric measurements in micellar media can be obtained introducing more complex models, based on mass action micellization thermodynamics and on the Debye–Hückel–Onsager conductivity theory. In a recently published work, typical micellar parameters were estimated using the above approach and the interionic interactions were taken into account [16].

5.1.2. Conductometry in emulsions and microemulsions

The measurement of the electrical conductivity in emulsion systems can be exploited to investigate the nature (water or oil) of the continuous phase, since water–oil emulsions and oil–water emulsions yield oil-like and water-like conductivity, respectively.

In emulsion science, the R-ratio measures the tendency for a given amphiphile to disperse into oil, divided by its tendency to dissolve in water. According to Winsor [17], R is qualitatively defined in terms of cohesive energy as follows:

$$R = A_{CO}/A_{CW} \tag{8}$$

where A_{CO} is the cohesive energy of the solute in the oil phase and A_{CW} represents the equivalent term in water. A more complete version of this empirical parameter, which includes different interaction terms between the oil, water and surfactant molecules, can be found in Ref. [18]. From the practical point of view, when one of these tendencies far exceeds the other, the interfacial region tends to take on a definite curvature.

For example, if $R >> 1$, the cohesive energy between the amphiphile and the oil is much larger than that between the amphiphile and the aqueous phase. The interfacial area in contact with oil tends to be maximized and that with water minimized, thus oil tends to become the continuous phase and a type II microemulsion system is obtained. For $R << 1$, the characteristic system is type I because the interface is convex toward the oil and water tends to be the continuous phase. When $R = 1$, a type III system is obtained.

It should be recalled that the differences between these systems depend upon the alcohol/surfactant ratio. As this ratio increases, a I–III–II transition is observed. A path representing a sequence of compositions, in which the amount of surfactant plus alcohol is kept constant while the oil/water ratio is increased, is usually indicated in each phase diagram.

A very important feature of electrical transport in micellar solutions concerns the point at which the inversion from water to oil continuity takes place as the water/oil ratio is systematically changed or as the system undergoes a I–III–II transformation. For type I systems the R-ratio is less than 1 and the interface tends to be convex toward the oil phase. Because of this intrinsic bending tendency, water continuous systems tend to resist the transformation to oil continuity even though the proportion of oil to water may be large. Thus type I systems tend to retain a conductivity characteristic of the aqueous phase even at relatively large oil/water ratios.

Type II systems ($R > 1$) show the electrical conductivity properties of oil, even at rather small oil/water ratios, whereas Type III systems yield an intermediate behaviour.

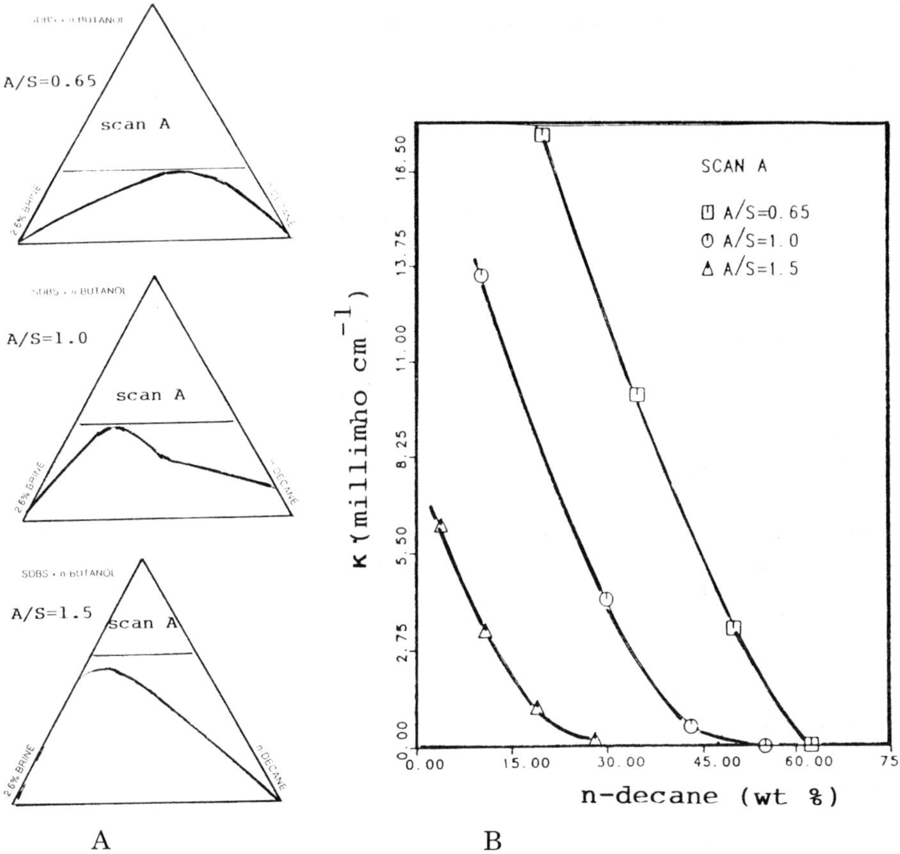

Fig. 5.5. (A) Pseudoternary phase diagrams of a microemulsion system composed of sodium dodecylbenzene sulfonate/n-butanol/n-decane/brine, at three alcohol to surfactant ratios. (B) Variation of the corresponding electrical conductivities at 30°C. Reproduced and adapted with permission from Ref. [19].

The composition and conductivities of three microemulsions are reported in Fig. 5.5A and B, respectively. It can be seen that at the highest alcohol/surfactant ratio (A/S = 1.5) the system is type II; thus R > 1 and the interface is convex toward the aqueous phase. The transition to an oil continuous phase appears complete at a composition of about 30 wt% of n-decane, since at this point the electrical conductivity becomes approximately equal to that of oil. Conversely, for A/S = 0.65, the system is type I and at approximately 60 wt% n-decane the conductivity drops to that of oil.

5.2. POTENTIOMETRIC MEASUREMENTS

5.2.1. Methods using surfactant ion-specific electrodes

The main reason for the use of ion-specific electrodes in the field of surfactant solutions stems from the fact that aggregation taking place in the system results in a change in the activity of both counterions and surfactant ions. Ion-specific electrodes have been available for a long time and, in recent years, a number of surfactant ion-specific electrodes have been developed. As in the case of the classical ion-specific electrode, the potentiometric setups based on these electrodes measure a potential E which is dependent on the activity a_S of the surfactant ion, according to the Nernst equation:

$$E = E_0 + 2.303 \frac{RT}{zF} \log a_S \tag{9}$$

where z is the charge of the surfactant ion (positive or negative for cations and anions, respectively). At 25°C, the term 2.303 RT/F is equal to 59.1 mV. E_0 is a constant depending on the nature of the reference electrode.

Surfactant ion electrodes are used to measure the activity of non-micellized surfactant ions and their applications tend to be developed. Moreover, these electrodes are very useful for the study of the solution chemistry of surfactants, because they allow investigatation of the behaviour of these molecules below and above the cmc and the presence of adsorption processes on particles and surfaces. No added salt is needed in these experiments.

5.2.1.1. Types of surfactant ion-specific electrodes

Some important factors must be considered when an ion-specific electrode is chosen for studies in surfactant solutions: (i) the response of the electrode, intended as the concentration range (below the cmc) where Eqn. (9) is obeyed; (ii) the selectivity, measured by the amount of other ions that can be present in the amphiphilic solution without altering the electrode response; (iii) the lifetime of the electrode, expressed as the number of experimental measurements performed without alteration of the electrode response; (iv) the time necessary (in each measurement) to reach a constant response.

The electrodes fulfilling the above requirements belong to four main types:

(a) ion-exchange membrane electrodes;
(b) solid state electrodes;
(c) liquid ion-exchange membrane electrodes;
(d) surfactant-sensitive polymeric membrane electrodes.

Of these, liquid ion-exchange membrane electrodes are more convenient because their response time is in the range of few minutes [20]. Other electrodes, such as solid membrane electrodes, need longer times (up to ca. 1 hour) before the readings stabilize. A typical scheme of the experimental setup follows:

Ag/AgCl electrode	Electrolyte–agar bridge	Surfactant solution (C_{A^-} or C_{C^+}	Liquid ion-exchange membrane	Surfactant solution (C_0 < cmc) + salt	Ag/AgCl electrode

The liquid ion-exchange membrane is usually composed of a water-insoluble complex containing two ionic surfactants, a cation C^+ and an anion A^-, dissolved in a non-polar solvent (decanol, nitrobenzene, dichlorobenzene, etc.) immiscible in water. A similar arrangement is described for ion-exchange membrane electrodes using solid ion-exchangers instead of liquids. Further details concerning the manufacture and performances of other electrodes can be found in Ref. [21].

For the liquid membrane systems described above, the potential difference measured between the two Ag/AgCl reference electrodes is given (if a cationic surfactant is considered) by:

$$E = 2.303 \frac{RT}{zF} \log (a_{C^+} / a^{o}_{C^+}) \qquad (10)$$

The response of the electrode is linearly dependent on surfactant activity up to the cmc. After this limit, the ratio between the free amphiphile activity and the surfactant analytical concentration tends to decrease, with the behaviour remaining linear (see Fig. 5.6). Comparison of activity data of free surfactant molecules and of their counterions allows the characterization of micellar ionization, although the influence of micellization upon the measured activities is not yet completely clear.

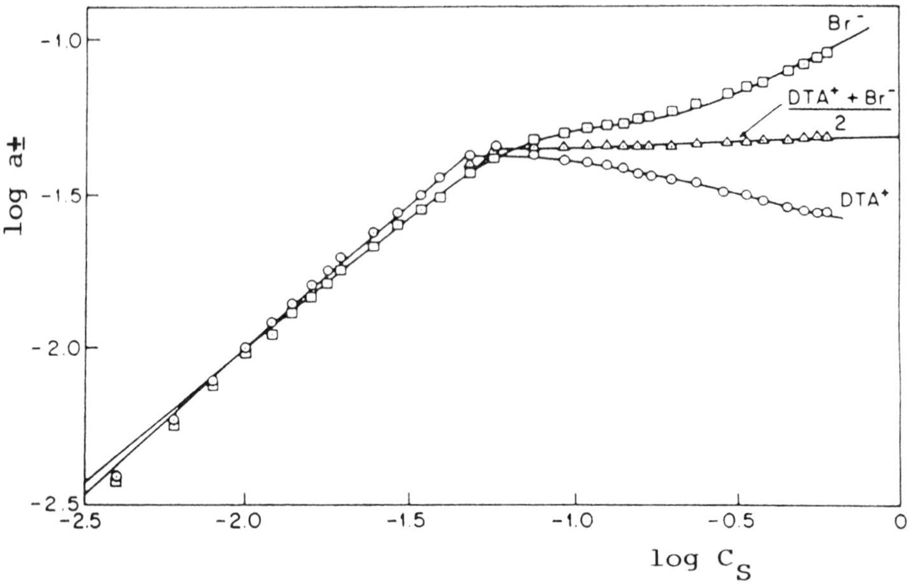

Fig. 5.6. Activity of bromide and dodecyltrimethylammonium ion as a function of DTAB concentration, at 25°C. Taken from Ref. [20], reproduced with permission.

Other important aspects of micellar physical chemistry have been investigated using surfactant-specific electrodes [21]. They include, for example, studies concerning the effect of ionic and non-ionic additives on micellar properties and investigations about premicellar association. Moreover, the investigation of surfactant adsorption onto solids (with the evaluation of the corresponding binding isotherms) was also accomplished following this approach.

5.2.1.2. Use of ion-selective electrodes to determine the effective degree of micelle dissociation

Surfactant ion-selective and counterion-selective electrodes are largely used in equilibria studies concerning amphiphiles. Typical emf measurements at the surfactant electrode are performed in respect to a standard electrode, using three different cells:

Cell 1:

Surfactant-selective electrode	Solution to be measured + salt	Electrode reversible to the salt coions

A constant amount of salt composed of the surfactant counterion and the corresponding coion (for example, for cationic surfactants having

Br^- as counterion, NaBr is used) is added, in order to work at constant coion concentration. From cell 1, the monomer surfactant concentration in the micellar solution, m_1, is measured as a function of the total surfactant concentration C_S.

The concentration of the free surfactant counterion, m_2, and the mean activity coefficients $\gamma_\pm$ are, in turn, measured using a second cell composed of:

Cell 2:		
Surfactant-selective electrode	Solution containing a constant amount of salt	Electrode reversible to surfactant counterions

For the above-mentioned cationic surfactants, the added salt is again NaBr, whereas the second electrode measures the bromide activity.

In another separate cell, the emf of a surfactant (or counterion) electrode against a reference electrode is measured:

Cell 3:		
Surfactant (or counterion) selective electrode	Solution containing constant amount of salt	Reference electrode

Palepu et al. [22] investigated in detail micellar solutions of tetradecylpyridinium bromide using this experimental approach. Three different treatments of potentiometric data arising from these investigations (giving comparable values of α) can be applied:

(i) When micelles are present and the aggregation numbers are large, the following equation [23] is valid:

$$\log m_1 \gamma_\pm = K - (1-\alpha) \log m_2 \gamma_\pm \quad (11)$$

Figure 5.7 shows that Eqn. (11) is satisfied for the reported cationic surfactant. The α value obtained from these plots was 0.21.

(ii) The second method examines the variation of cmc with added salt. In the presence of monovalent counterions, the following equation [23] can be applied:

$$\log \frac{\text{cmc}\,\gamma_\pm}{(\text{cmc})_0\,\gamma_\pm^0} = -(1-\alpha) \log \frac{\text{cmc} + m_3}{(\text{cmc})_0} \frac{\gamma_\pm^0}{\gamma_\pm^0} \quad (12)$$

here $(\text{cmc})_0$ is the cmc at zero added salt concentration, m_3 is the concentration of added NaBr. Moreover, according to Debye–Hückel theory:

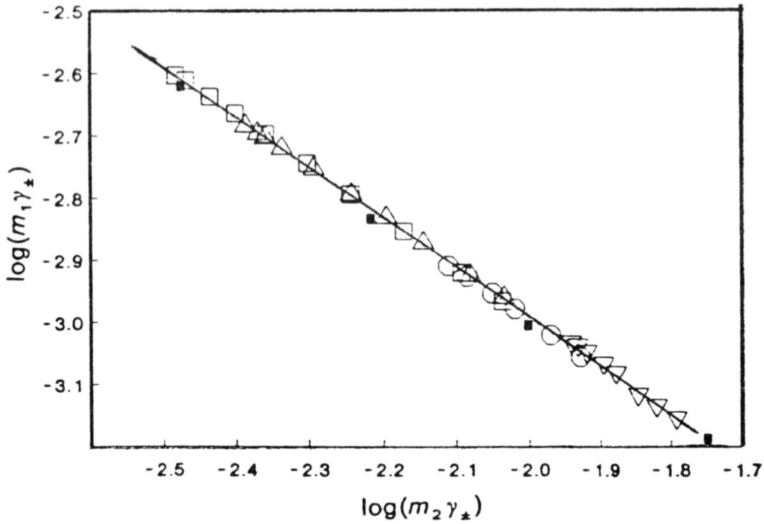

Fig. 5.7. Plots of log $m_1\gamma\pm$ vs. log $m_2\gamma\pm$ at different ionic strengths. Concentration of added NaBr: (□) 1×10^{-4} M; (Δ) 1×10^{-3} M; (O) 5×10^{-3} M, (∇) 1×10^{-2} M. Taken from Ref. [22], reproduced with permission.

$$\log \gamma_{\pm} = \frac{-A\sqrt{I}}{1 + \sqrt{I}} \tag{13}$$

with $I = (m_1 + C_{Br} + m_2)/2$ and $C_{Br} = C_S + m_3$ (C_{Br} is the total concentration of Br^- in the solution). From the plots according to Eqn. (12), shown in Fig. 5.8, $\alpha = 0.22$ was obtained.

(ii) Another method for the evaluation of α is based on measurements performed using cell 3, where the emfs of both the surfactant and bromide electrode relative to the reference electrode are measured in the same experiment. E_+ and E_- refer to the emfs relative to the double-junction reference electrode for S^+ and Br^-, respectively.

S_1 and S_2 are the slopes of the plots of E_+ and E_- against log C_S and log C_{Br}, respectively, measured below the cmc. In the micellar region, the following equation holds:

$$E_+/S_1 = \text{const.} - (1 - \alpha)\, E_-/S_2 \tag{14}$$

The corresponding plots, at two different electrolyte concentrations, are shown in Fig. 5.9. Two values of α (0.22 and 0.24) were obtained from these measurements.

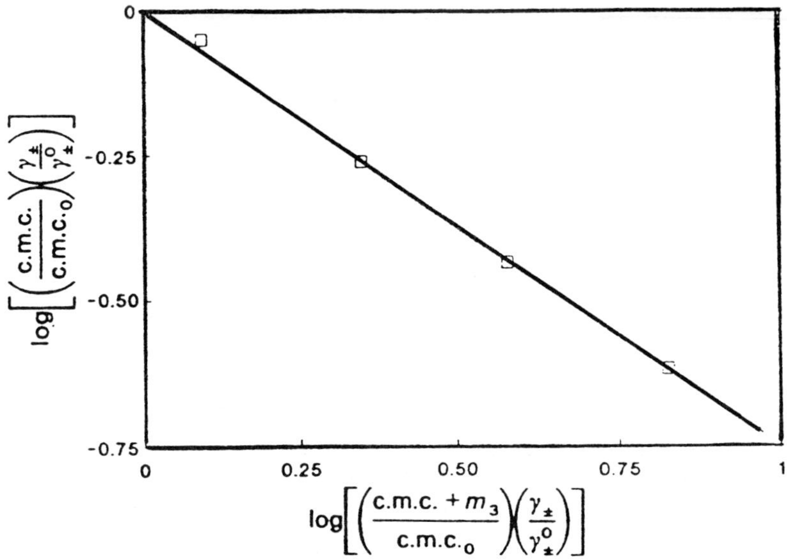

Fig. 5.8. Plots according to Eqn. (12) for DTAB. Taken from Ref. [22], reproduced with permission.

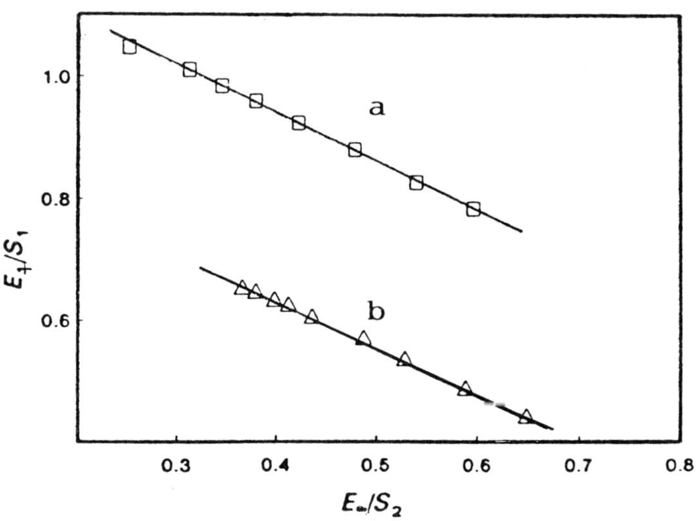

Fig. 5.9. Plots of E_+/S_1 vs. E_-/S_2 for: (a) NaBr 1×10^{-4} M; (b) NaBr 1×10^{-3} M. Taken from Ref. [22], reproduced with permission.

5.2.2. The measurement of pH in the presence of microaggregates

The effect of organized amphiphilic assemblies on acid–base equilibria in aqueous solution has been examined in detail in Chapter 3. We recall that the pK_a of a given acid in these systems can be related to that in bulk water by taking into account the surface potential of the aggregate and the transfer of the probe from the bulk solution to the aggregate–solution interface. Moreover, the pK_a values are estimated starting from the corrected pH, in turn evaluated taking into account the ion-exchange equilibria at the micellar surface.

5.2.2.1. pH measurements in micellar media

pH measurements in micellar media are possible, but different sources of error must be considered since: adsorption of the amphiphiles occurs on the glass electrode (thus changing the junction potentials); preferential accumulation of hydronium ions and hydroxyl ions take place at the surface of anionic and cationic micelles, respectively; the acid–base forms of a given acid (or basic) solute usually bind to a different extent to the aggregates.

A crucial point is the surfactant adsorption on the electrode because it can drastically alter the instrument response, giving rise in some cases to a continuous drift in the pH measurement. Detailed investigations performed on buffered solutions [24] showed that the evolution of the glass electrode response is strongly dependent on the nature of the surfactant present, being very relevant for SDS (see Fig. 5.10).

The formation of a precipitate of mixed dodecylsulfate of sodium and potassium has been clearly observed on the reference electrode (Hg_2Cl_2/KCl_{sat}), due to the diffusion of K^+ from this electrode. This inconvenience has been reduced by using Ag/AgCl as a reference electrode and an agar bridge containing NaCl 0.1 M.

On the other hand, the modification of pH due to the addition of SDS to standard acid solutions clearly shows a reduction of the hydrogen ion activity sensed by the glass electrode, thus indicating that accumulation of such species must occur in the micellar pseudophase (see Fig. 5.11). The pH of the bulk aqueous phase (pH_w) and the local pH at the aggregate–water interface (pH_m) can be estimated from the measured pH, after introducing a simple ion-exchange model [25,26] which accounts for the hydrogen ion exchange with Na^+ in the Stern layer of the micelles:

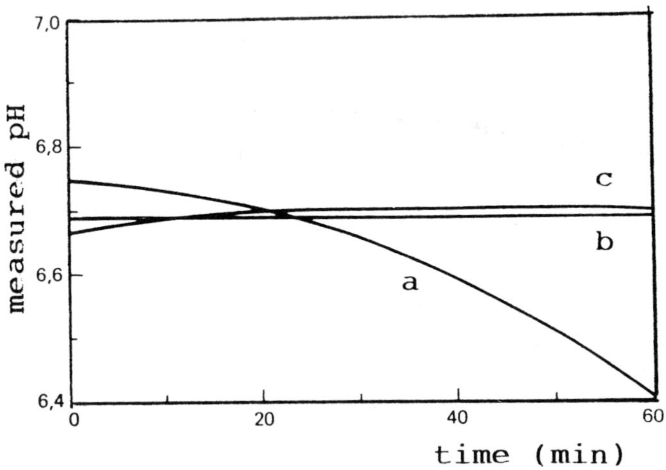

Fig. 5.10. Variation of pH response with time of a combined electrode immersed in a buffered solution of different surfactants. (a) SDS 3.47×10^{-2} M; (b) Brij 35 8.33×10^{-3} M; (c) HTAB 2.74×10^{-4} M. Buffer system: Na_2HPO_4 2.5×10^{-2} M/NaH_2PO_4 2.5×10^{-2} M. Reproduced with permission from Ref. [24].

$$H^+_{(w)} + Na^+_{(m)} \rightleftarrows H^+_{(m)} + Na^+_{(w)} \tag{15}$$

$$K_{H/Na} = \frac{C_{H^+(m)}\, C_{Na^+(w)}}{C_{H^+(w)}\, C_{Na^+(m)}} \tag{16}$$

The selectivity coefficient reported for this ion-exchange equilibrium is ca.0.8 [25]. The counterion concentrations in both pseudophases can be estimated from the following equations:

$$C_{Na^+(w)} = \alpha C_D + C_{H^+(m)} + \mathrm{cmc} \tag{17}$$

$$C_{Na^+(m)} = (1 - \alpha)C_D - C_{H^+(m)} \tag{18}$$

where α is the dissociation degree of the micelle ($\alpha \cong 0.2$ for SDS under the reported experimental conditions) [24].

Combination of Eqns. (16–18) allows the calculatation of $C_{H^+(m)}$, from which pH_w can be estimated through

$$pH_w = -\log\,(C_A - C_{H^+(m)}) \tag{19}$$

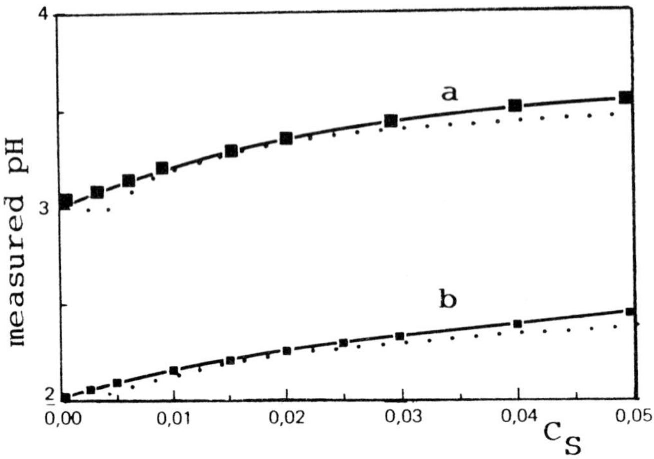

Fig. 5.11 Modification of pH of acid solutions as a fuction of SDS concentration. (a) HCl 1×10^{-3} M; (b) HCl 1×10^{-2} M. The points under the experimental curves were calculated using Eqn. (19). Taken from Ref. [24], reproduced with permission.

where C_A is the analytical concentration of the acid added to the SDS micellar solution. The dotted lines in Fig. 5.11, which represent the calculated pH_w at different surfactant concentrations, confirms that the measured pH essentially corresponds to that of the bulk aqueous phase.

The following practical considerations are valid when pH is measured in micellar solutions:

– When the formation of low solubility products from surfactant ions and reference electrode salts is possible, these electrodes have to be shielded with a conductive jelly.

– Glass electrodes can be used, although they essentially measure the aqueous phase pH.

– The pH of the micellar pseudophase may be very different (up to 2–3 pH units).

– Classical buffers stabilize only the aqueous phase pH.

– There are relevant effects of micelles on acid–base equilibria of usual indicators.

– The use of indicator papers should be avoided since mistakes up to about 3 pH units have been observed.

5.2.2.2. Acid–base titrations in micellar media

A great number of hydrophobic acids and bases, which are sparingly soluble in water, can be dissolved in aqueous micellar solutions and successively titrated, avoiding the use of more expensive and usually more toxic organic solvents. The effect of surfactant aggregates on acid–base equilibria was previously discussed in detail in Chapter 3.

From a practical point of view, cationic surfactants stabilize the anionic form of acids, increasing their apparent dissociation constant and facilitating the potentiometric (or visual) titration. Conversely, anionic surfactants are the best candidates for solubilization of hydrophobic bases since protonation is favoured in the micellar phase. The

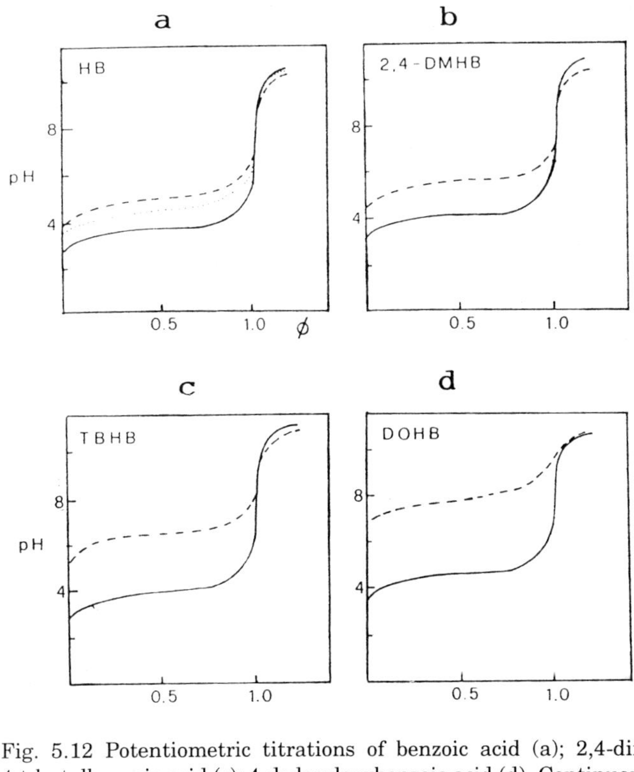

Fig. 5.12 Potentiometric titrations of benzoic acid (a); 2,4-dimethylbenzoic acid (b); 4-t-butylbenzoic acid (c); 4-dodecyloxybenzoic acid (d). Continuous lines: 0.1 M HPC (0.2 M HPC for compound d). Dashed lines: water (a), 0.1 M SDS (b and c), 0.2 M SDS (d). Acid concentration: 0.01 M. Reproduced from Ref. [27], with permission.

typical effect of both types of surfactants on the titration curves of different substituted benzoic acids [27] is shown in Fig. 5.12.

For the reported acids, pK_a shifts in the range 0.4–0.7 compared to the aqueous homogeneous solutions [28] were observed in HPC, reflecting the different electrostatic contribution to the binding constant, K_B, of each anion. It is evident that SDS micelles give rise to less defined titration curves because the corresponding $pK_{a(app)}$ values are enhanced. Very similar results have been obtained for other acidic compounds, such as phenols, in HTAB solutions [29].

The ΔpK_a data measured in different micellar media for more than 150 acids and bases, including carboxylic acids, phenols, sulphophthaleine indicators, aromatic and aliphatic amines, coumarines, heterocyclic derivatives and other miscellaneous compounds, are reported in Ref. 30. Table 5.2 summarizes the results for only some selected examples of acids and bases.

TABLE 5.2

Variation of pK measured in different surfactant systems for some acid and basic compounds

Analytes	Surfactant system	ΔpK	Ref.
Acids:			
Alkyl derivatives of benzoic acids	HPC, HTAB	–0.4 to –0.7	27
2-nitrobenzoic acid	DPC	0.07	35
3-nitrobenzoic acid	DPC	–0.4	
Acetic acid	HTAB	–0.1	24
	SDS	0	
Chloroacetic acid	SDS	0.10	
Dichloroacetic acid	HTAB	0	
Hexadecanoic acid	SDS	2.73	36
Octadecanoic acid	SDS	2.68	
Sulphonamides	HPC	<0	37
Phenol	HTAB	–0.3 to –0.5	38
Alkylphenols	SDS	>0	39
4-Chlorophenol	HTAB	–0.58	38
4-Methoxyphenol	HTAB	–0.28	
4-Nitrophenol	HTAB	–0.95	
	DTAB	–0.96 to –1.05	40
	Brij 35	0.71 to 0.93	
4-t-Butylphenol	DTAB	–1.42 to –1.53	
	$C_{12}E_8$	1.65 to 1.73	

Analytes	Surfactant system	ΔpK	Ref.
1-Naphthol	HTAB	–0.70	41
	SDS	0.81	
4-Chloro-1-naphthol	Triton X100	0.86 to 1.41	42
8-Hydroxyquinoline[a]	HTAB	–1.04	36
(pK_{a1})	SDS	0.7	
	Triton X100	0	
8-Hydroxyquinoline	HTAB	–0.23	
(pK_{a2})	SDS	0.62	
	Triton X100	0.09	
Bases[a]:			
4-Methylaniline	$C_{12}E_8$	0.43 to 0.66	40
	SDS	1.5 to 1.8	
4-Nitroaniline	SDS	1.15 to 1.58	43
Benzylamine	SDS	1.35	44
n-Octylamine	SDS	1.77	
Dioctylamine	SDS	1.03	45
4-Phenylbutylamine	SDS	1.61	44
4-Phenylazodiphenylamine	SDS	0.01 to 0.35	46
Diethazine	Septonex[b]	–2	47
	Triton X100	–1.6	
	SDS	0.7	
Chlorpromazine	Septonex	–2	
	Triton X100	–1.4	
	SDS	0.8	

a: pK_a variation of the protonated forms; *b*: 1-carbethoxypentadecyltrimethylammonium bromide.

For the majority of the investigated solutes, the differences between the pK_a values in pure water and the apparent pK_a values in micellar solutions can be mainly ascribed to the differences between the mean intrinsic solvent properties of the interfacial and bulk phases, with a contribution arising from the electrostatic effects when ionic aggregates are present (see Chapter 3).

The behaviour of several acid–base indicators has also been investigated in detail, since for most of them the spectrophotometric determination of pK_a is easy to perform. Specific interactions (including ion-pair formation) between the negatively charged sulfonate groups and cationic surfactant head groups are very important for sulphonephthalein indicators [31]. Similarly, specific molecular interactions operating between the cationic (protonated) azine molecules and the anionic surfactant head groups are crucial when acridine, neutral red and other azine indicators are considered [32].

Data concerning the acid–base properties of azo dyes, which belong to another important indicator group, have also been reported [33,34]. For some of these compounds, the changes in acid-base properties due to the presence of micelles have been adequately described by using a simple titration model [34]. The theoretical aspects related to the observed micellar effects on acids and bases have been already discussed in Chapter 3.

5.2.2.3. pH measurement in emulsions and microemulsions

The main problems related to the use of electrodes in emulsified media are the following: (i) a very slow stabilization of the glass electrode is observed in emulsions and a reference electrode shielded by a conductive jelly is often required; (ii) the hydrogen ion electrode seems to work better in some microemulsions; (iii) the results obtained working with a given emulsion (or microemulsion) cannot, in principle, be extended to other similar systems; (iv) the definition of pH in emulsions and microemulsions is (as for micelles) related to the measurement of hydrogen ion concentration in the aqueous phase.

In these systems the measured hydrogen ion concentration is defined as follows:

$$[H^+] = [H^+]_w (1 - \phi) \tag{20}$$

where $[H^+]$ and $[H^+]_w$ are the hydrogen ion concentrations per litre of emulsion and per liter of aqueous phase, respectively; ϕ is the volume fraction occupied by the organic phase.

Berthod and Saliba [48] have examined the evolution of the hydrogen electrode potential immersed in a five-components microemulsion (SDS/1–pentanol/n-dodecane/water + NaCl) as a function of the microemulsion composition. Figure 5.13A shows the compositions examined, whereas the corresponding curves E vs. pH are shown in Fig. 5.13B. Since the slopes of these plots are practically constant (57.1–57.9 mV/pH in the pH range 1–4), this implies that also in these systems the hydrogen electrode measures a potential mainly related to the hydrogen ion concentration in the aqueous phase.

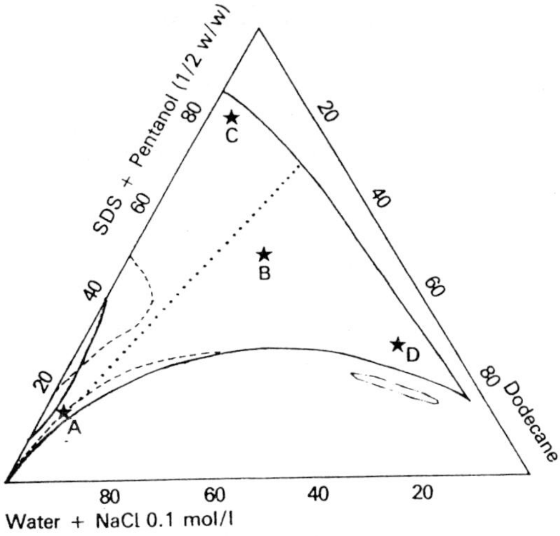

A

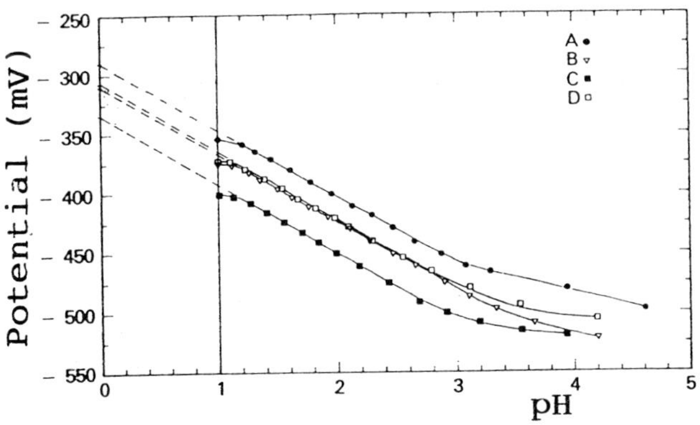

B

Fig. 5.13. (A) Phase diagram of the microemulsion. (B) Response of the hydrogen ion electrode at the indicated microemulsion compositions. Reproduced with permission from Ref. [48].

All the diffusion-controlled electrochemical techniques can, in principle, be used to investigate the behaviour of electroactive probes dissolved in micellar and emulsion systems. The information obtained through these measurements are strongly dependent on the nature of the probe, on probe partitioning between the phases, and on the electron-transfer processes occurring at the electrode (often covered by a layer of adsorbed amphiphile).

Polarography is the most important and widely applied voltammetric technique. Ilkovic proposed the following equation, which relates the measured diffusion current with the concentration of the electroactive analyte:

$$i_d = 708.1 \, n \, D^{1/2} \, C_p \, (m^4 t)^{1/6} \tag{21}$$

where i_d is the limiting diffusion current (usually measured in μA or mA), n is the number of electrons transferred at the dropping mercury electrode, D is the translational diffusion coefficient of the electroactive species ($cm^2 \, s^{-1}$), C_p is the stoichiometric concentration of electroactive species (mM), m is the flow of mercury at the dropping electrode (mg s^{-1}), and t is the lifetime of the mercury drops (s).

There are usually some preliminary assumptions when a probe is used in micellar environments: (i) it does not change the micellar size and shape; (ii) if the probe is partitioned between the two pseudophases, the effective C_p value should be calculated; (iii) a statistical distribution is assumed for $R < 0.8$ ($R = C_p/C_m$ and C_m indicates the concentration of micelles). The above assumptions are reasonable for dilute probe and surfactant solutions, and for amphiphiles having a long chain. In these cases, the linearity of i_d vs. C_p has been confirmed.

The detailed mechanism of electron-transfer between a micelle-solubilized electroactive analyte and the electrode is still not completely clear, since various concurrent processes are often simultaneously present in the system. However, the selection of suitable electroactive probes allows the investigation of the nature of these electrochemical reactions at least in certain well defined domains of the microheterogeneous phases.

Table 5.3 reports some examples of the variety of probes used in different voltammetric experiments, performed in micellar and emulsion media. The features of the most commonly used surfactant-modified voltammetric techniques and the information accessible from the experimental data will be examined in the following sections.

TABLE 5.3

Examples of solubilizates used as probes in voltammetry work

Compound	Features of the probe	Type of experiment and surfactant system used	Ref.
Ferrocene	oil soluble	Oxidation in dodecane-in-water emulsion and in Tween 20, SDS, benzthonium chloride micelles	49,50
Coumarines	partitioned	Oxidation in Triton X 405 micelles	51
Organic compounds (miscellaneous)	oil soluble and partitioned	Oxidation in hyamine micelles and MeCN/water/emulsions	52
Amphiphilic aromatic ketones	forming mixed micelles	Reduction and dimerization in HTAB-amphiphilic ketone mixed micelles	53
Nitrobenzene	oil soluble	Reduction in SDS, HTAC and Brij 35 micelles	54
Viologens	charged, tunable hydrophobicity	Reduction: in SDeS micelles; in Triton X100, HTAB, SDS; in HTAB/1-butanol/hexadecane microemulsions	55 56 57
Pyrene derivatives (PyCHO, $PyNO_2$)	oil soluble	Reduction in TTAB micelles	58
Tetrazolium blue	partitioned	Reduction and oxidation in Triton X100 micelles	59,60
Tetrathiofulvalene		Oxidation in HTAC micelles	61
Phenothiazines	oil soluble, tunable hydrophobicity	Oxidation in sodium octyl-benzene sulfonate/1-pentanol/water microemulsions	62
Benzenediols	hydrophilic-hydrophobic behaviour	Oxidation in SDS micelles	63
p-Phenylene-diamines	as for benzenediols, possible charge	Partioning studies in oil–water microemulsions	64,65
Ionic species:	most are hydrophilic and water-soluble		
Tris(1,10-phen-anthroline)Fe(II)		Oxidation in SDS micelles	66
Cd^{2+}, Tl^{+}, $Fe(CN)_6^{3-/4-}$		Redox behaviour in mineral oil/water/1-pentanol/Tween 40 microemulsions	67,68
1-dodecyl-4-cyano-pyridinium ion	oil soluble	Redox behaviour in water/octane/1-butanol/HTAB microemulsions	69

5.3.1. Voltammetric measurements at the DME electrode

The effect of surfactant monomers and aggregates on polarographic parameters has been carefully examined in several studies. In particular, the amphiphile adsorption at the electrode surface may play an important role in electrocatalytic processes and can be conveniently exploited since the accumulation of electroactive species is possible in the hemimicellar layers.

However, in some polarographic DME experiments, adsorption is clearly a negative factor. For example, non-ionic surfactants (such as Triton X100) can inhibit the reduction of several metal ions when the mercury drop is fully covered [70]. Several examples of polarographic systems sensitive to adsorption phenomena can be found in Ref. [5].

In other cases, as for water-insoluble anthraquinone dyes in non-ionic surfactant solutions [71], the dc and ac polarograms as well as the half-wave potentials of the probes were independent of the surfactant concentration, indicating that the processes were mainly diffusion controlled.

5.3.1.1. Dependence of i_d from surfactant and probe concentration

Generally, a linear dependence of i_d from C_p indicates a diffusion-controlled process. This dependence tends to break down for partitioned probes, when R > 0.7–0.8. Figure 5.14 shows the variation of i_d with C_p for different hydrophobic probes, in the presence and in the absence of added electrolyte.

A decrease of i_d with increasing C_S (total surfactant concentration) is usually also observed and this behaviour can be either attributed to probe partitioning, with the reduction of the probe diffusion coefficient, and/or to surfactant adsorption on the mercury electrode (with a corresponding more difficult electron-transfer). For highly hydrophobic probes (such as those reported in Fig. 5.14) which are completely bound to the micelles, the second effect seems to prevail and it is more relevant for cationic than for anionic amphiphiles [58]. In fact, a temperature increase reduces the adsorption layer, thus facilitating again the electron-transfer.

5.3.1.2. Dependence of the half-wave potential ($E_{1/2}$) on surfactant concentration

The position of the polarographic wave along the applied potential scale (measured by $E_{1/2}$) is largely dependent on surfactant effects at the electrode. Figure 5.15 shows the variation of $E_{1/2}$ with surfactant concentration for the hydrophobic electroactive probe N-hexadecyl-4-cyano-

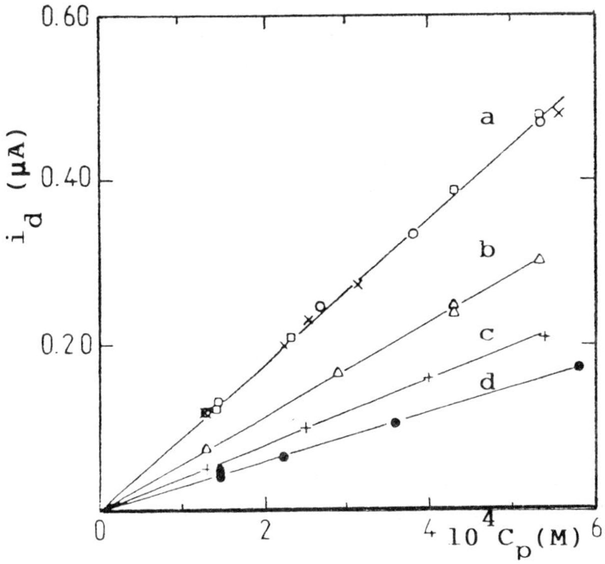

Fig. 5.14. Effect of the probe concentration on i_d in 0.07 M tetradecyltrimethylammonium bromide (TTAB) solution, at 25°C. (a) 1-pyrenecarboxaldehyde in the absence (O) and in the presence (×,□) of 0.1 M KBr; (b) 9-anthracenecarbonitrile; (c) N-hexadecyl-4-cyano-pyridinium bromide; (d) probe (c) + 0.1 M KBr. Reproduced with permission from Ref. [58].

pyridinium ion ($C_{16}Py+$). The measured variation of i_d is also shown, confirming its inverse dependence on C_S.

The nearly constant values of $E_{1/2}$ indicate the practically complete binding of the probe to the aggregates and a scarce influence of amphiphile adsorption onto the electrode. The observed decrease in i_d has been assigned to intermicellar interactions, since for completely bound compounds a constant value of the micellar diffusion coefficient is expected.

5.3.1.3. Determination of diffusion coefficients in micellar media

The diffusion of strongly bound electroactive species takes place with the host micelles, with probes and aggregates moving as a single kinetic entity. Since polarography measures a diffusion-controlled current, it should be possible to evaluate the apparent micellar translational diffusion coefficient (D_m) using suitable reducible (or oxidizable) probes which are completely or nearly completely solubilized by the aggregates. the Ilcovic equation can be applied for both reversible or irreversible electrochemical processes.

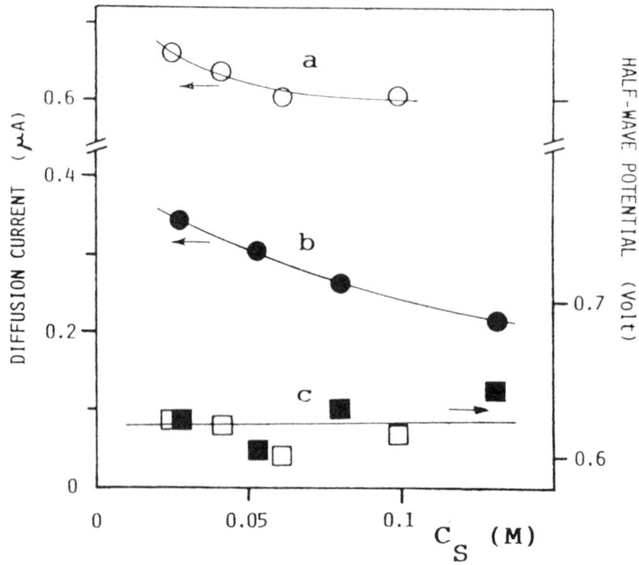

Fig. 5.15. Effect of surfactant concentration on $E_{1/2}$ and i_d for $C_{16}Py^+$. Curves (a) and (b): variation of i_d with DDAPS (dodecyldimethylammonium propane sulfonate) and TTAB concentration, respectively. Plot (c): variation of $E_{1/2}$ in DDAPS (□) and TTAB (■), respectively. Reproduced with permission from Refs. [72,73].

It must be recalled that D_m is an apparent diffusion coefficient, since it is affected by intermicellar interactions and other effects, such as surfactant adsorption on the mercury electrode surface. For example, 1,4,5,8-tetraaminoanthraquinone can be reversibly reduced on the dropping mercury electrode, the current is diffusion-controlled and the half-wave potential is not dependent on amphiphile concentration. Under the previous considerations, D_m can be calculated from the Ilkovic equation, giving a value of 3×10^{-7} cm^2 s^{-1} [74], which can be correlated with the aggregate size by the Stokes–Einstein equation:

$$D_m = kT/6\pi\eta r_m \qquad (22)$$

where k is the Boltzmann constant, T is the temperature (K), η is the viscosity of the bulk solvent and r_m is the radius of the diffusing micelle. Typical calculated micelle sizes correspond well to sizes determined using other techniques, such as SANS. For SDS micelles, a r_m of ca. 80 Å was estimated.

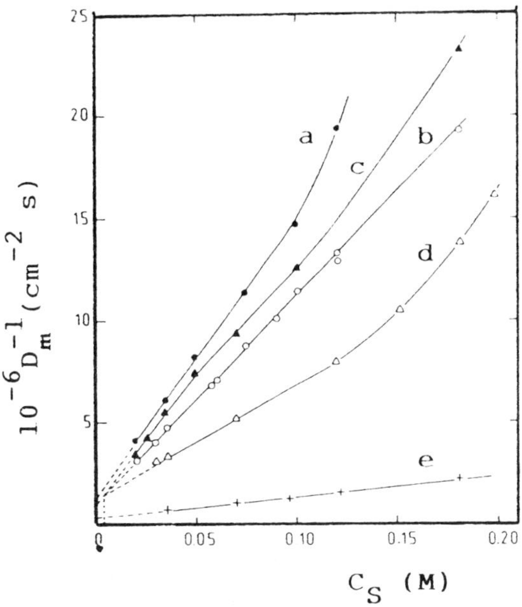

Fig. 5.16. Estimation of D_m in TTAB from polarographic data, at 25°C. (a) $PyNO_2$; (b) $PyNO_2$ + 0.1 M KBr; (c) $C_{16}Py^+$; (d) $C_{16}Py^+$ + 0.1 M KBr; (e) PyCHO (with and without 0.1 M KBr). R = 0.3. Reproduced with permission from Ref. [58].

The extrapolation of $1/D_m$ vs. C_S at the cmc gives D_m^o, the self-diffusion coefficient in the absence of intermicellar interactions (see Fig. 5.16). Although the estimation of this parameter is not very accurate, values of D_m^o comparable to those obtained using other techniques have been obtained. For example, data for 1-nitropyrene and $C_{16}Py^+$ extrapolate to the same value at the cmc ($D_m^o = 7.2 \times 10^{-7}$ cm^2 s^{-1}), which is comparable with the results obtained from laser light-scattering experiments [75].

On the basis of reported literature data it appears that the measurements of diffusion coefficients by polarographic MDE give the most reliable data for anionic and cationic surfactants at low amphiphile concentration and/or high temperature.

5.3.1.4. Investigation of microemulsion structure using oxygen polarograms

The oxygen reduction in microemulsions can give useful information about the structure of the microheterogeneous domain [76]. This probe

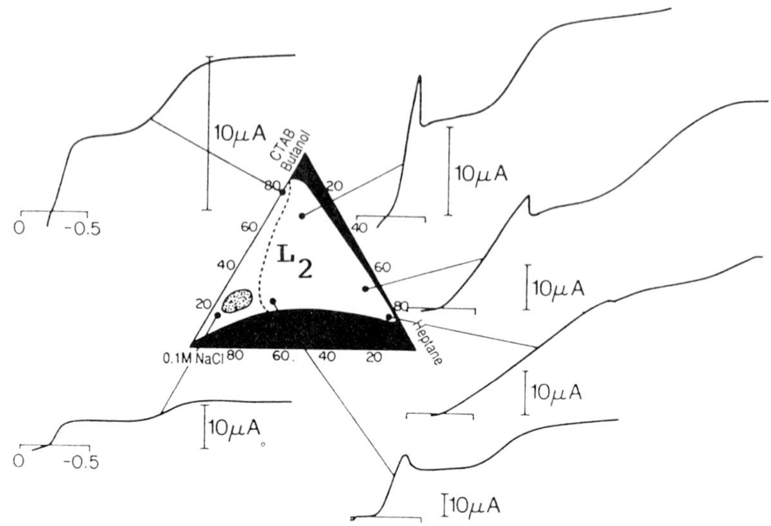

Fig. 5.17. Oxygen polarographic waves obtained at different compositions of a HTAB/1-butanol/heptane/water microemulsion. Drop time 1.5 s; scan rate 10 mV s^{-1}. Reproduced with permission from Ref. [78].

dissolves in either the aqueous and oily phases without altering the microemulsion structure and a typical polarographic maximum is detected on the first reduction wave, at well defined microemulsion compositions, when oxygen is present in a non-aqueous continuum environment (i.e., an L_2 structure). The surfactant adsorption onto the mercury drop has been excluded.

For example, the area of the ternary phase diagram of an SDS/1-pentanol/dodecane/water microemulsion, where the polarographic maxima were observed, corresponded well with the L_2 region of the microemulsion as determined by conductometry [77]. Figure 5.17 shows the phase diagram of a cationic microemulsion (HTAB/1-butanol/heptane/water) and the corresponding polarographic profiles of the oxygen reduction. The black area represents biphasic turbid systems.

It can be seen that the polarographic maxima are present in the polarograms corresponding to the L_2 area, whereas they do not appear outside this region, thus confirming the diagnostic capability of these measurements.

5.3.2. Measurements at the rotating disk electrode

The most practical form of an electrode for which a rigorous hydrodynamic treatment can be given is the rotating disk electrode (RDE). The limiting current for a reaction controlled only by mass transfer is given as follows (Levich equation):

$$i_L = 0.62\, n\, F\, A\, D^{2/3}\, \eta^{-1/6}\, \omega^{1/2}\, C_p \qquad (23)$$

where i_L is the limiting current (A), A is the electrode area (cm^2), η is the kinematic viscosity ($cm^2\, s^{-1}$), ω is the angular velocity (rad s^{-1}) of the electrode and C_p is the probe concentration. The angular velocity is equal to $2\pi N/60$, N being the number of rotations per minute.

For experiments with a fixed electrode in a particular solution, some of the terms of Eqn. (23) remain unchanged and can be included in a common constant factor, k:

$$i_L = k\, n\, D^{2/3}\, \omega^{1/2}\, C_p \qquad (24)$$

The diffusion coefficient can be simply obtained from the slope of the limiting current vs. $\omega^{1/2}$. The linearity of the plot eventually confirms a diffusion-controlled process.

For experiments performed in solutions containing micellar aggregates and in the presence of partitioned electroactive solutes, the following equation holds:

$$i_{L,m} = k\, n\, D_m^{2/3}\, \omega^{1/2}\, C_p \qquad (25)$$

where $i_{L,m}$ is the limiting current observed in the presence of aggregates and D_m is the corresponding apparent diffusion coefficient.

The observed current is contributed to by the solute from the aqueous and aggregate phase:

$$i_{L,m} = k\, n\, D_1^{2/3}\, \omega^{1/2}\, C_w + k\, n\, D_2^{2/3}\, \omega^{1/2}\, C_m \qquad (26)$$

D_1 and D_2 are the diffusion coefficients in the aqueous and aggregate phase, respectively. For the applicability of this equation, some conditions need to be satisfied:

(1) $D_1 \cong D_w$ (generally true, provided the viscosity is not very different).

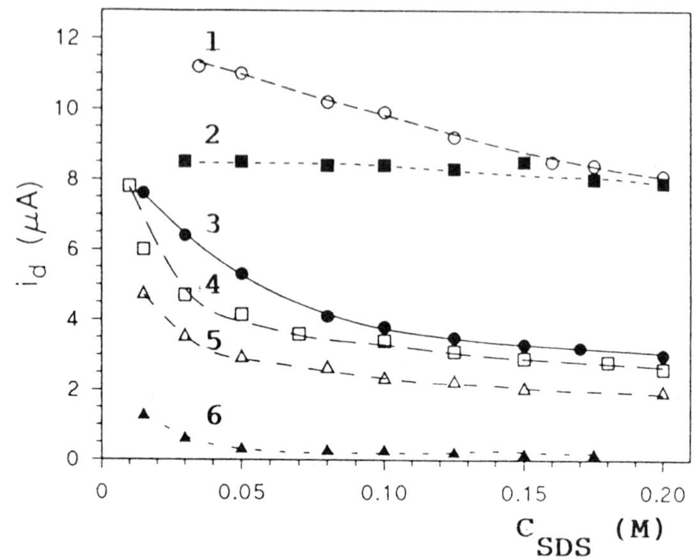

Fig. 5.18. Variation of i_d as a fuction of SDS concentration. (1) benzene-1,4-diol; (2) 2,5-dihydroxybenzene sulfonate; (3) 2-t-butylbenzene-1,4-diol; (4): 4-t-butylbenzene-1,2-diol; (5): 2,5-di-t-butylbenzene-1,4-diol; (6): 2,5-di(1,1,3,3-tetramethyl)butylbenzene-1,4-diol. Reproduced with permission from Ref. [63].

(2) The ratio C_m/C_w in the diffusion and hydrodynamic layers is not significantly perturbed by the steady-state distribution. Under certain conditions [65], this condition is also verified.

(3) The kinetics of solute diffusion and the mechanism of electron transfer at the electrode surface are not significantly affected by any surfactant film on the electrode surface.

Equations (23–26) (or their derived forms) have been succesfully applied to describe the RDE voltammetric behaviour of different probes, and the influence both of partitioning and medium effects on the electron-transfer processes occurring at the electrode have clearly been in evidence from these experiments. Among the examples reported in Table 5.3, those concerning 1,4-dihydroxybenzenes and 1,4-phenylenediamines have been chosen to illustrate micellar RDE voltammetry.

(i) *Oxidation of benzenediols*: The reversible two-electron oxidation of a group of 1,4-benzenediol derivatives having different binding constants to SDS aggregates constitutes an interesting example of RDE experiments in micellar media [63]. The measured diffusion current for the partitioned electroactive probes (at constant probe concentration) is reduced in the presence of SDS micelles, whereas practically no effect is observed in the case of hydroquinone sulfonate, which is electrostatically repelled from the anionic aggregates (see Fig. 5.18).

The observed linear dependence of $i_{L,m}$ on the square root of angular velocity indicates that the Levich equation is satisfied, under the reported experimental conditions, for the examined compounds. Linear plots of $i_{L,m}$ vs. C_p were also observed.

According to Eqns. (24) and (25), two limit cases are expected, one corresponding to a probe completely associated to the micelle (such as compound 5) for which the diffusion coefficient would correspond to D_m, and the other observed for unbound probes (such as compound 1). In the latter case, the diffusion coefficient obtained from the plots of $i_{L,m}$ vs. C_p would give essentially D_w. For partitioned compounds both contributions are present, Eqn. (26) holds and the slopes of the corresponding straight lines decrease with the increasing hydrophobicity of the examined compound.

The voltammetric measurements also allow the determination of the half-wave potential ($E_{1/2}$) of the electroactive probes. For hydroquinone derivatives, the presence of anionic micelles shifts toward higher values (in respect to water) the half-wave potential for their reversible two electron oxidation to the corresponding quinones:

$$Q + 2H+ + 2e- \leftrightarrow QH_2 \qquad (27)$$

The shift, which is proportional to the extent of binding of the probe to the aggregates, reflects different availability of the probe to the electron transfer at the electrode. For unbound compounds (such as hydroquinone sulfonate), practically no shifts of their $E_{1/2}$ values are observed, whereas for partitioned compounds the measured half-wave potential in the micellar solution, $E_{1/2(app)}$, is given by

$$E_{1/2(app)} = E_w^{o\prime} + \frac{RT}{nF} \ln \frac{(D_{red}^{2/3} + D_m^{2/3} P_{red} C_D)}{(D_{ox}^{2/3} + D_m^{2/3} P_{ox} C_D)} \qquad (28)$$

where $E_w^{o\prime}$ is the formal standard potential in the medium conditions, P_{red} and P_{ox} are the partition coefficients of the reduced and oxidized probe, respectively, C_D is the concentration of micellized surfactant, D_{red} and D_{ox} are the diffusion coefficients of the two forms in water.

In the absence of micelles, Eqn. (28) predicts that $E_{1/2(app)}$ tends to $E_{1/2w}$, whereas for compounds having very high partition coefficients,

$E_{1/2(app)}$ tends to $E_{1/2m}$. The half-wave potentials in micellar phase and in water are related through the equation:

$$E_{1/2m} = E_{1/2w} + \frac{RT}{nF} \ln \frac{(D_{red}^{-2/3} P_{red})}{D_{ox}^{-2/3} P_{ox})} \tag{29}$$

The measured variation of $E_{1/2}$ in the presence of aggregates reflects the preferential stabilization effect of the microaggregate on one of the forms (reduced or oxidized) of the redox couple.

(ii) *Oxidation of p-phenylenediamines*: Another approach to the analysis of RDE voltammetric data, reported for 1,4-diamino benzene derivatives [79], is based on the analysis of the slope of the Levich equation. Below the cmc, for dilute aqueous micellar solutions, the following equation holds:

$$S_0 = \frac{\partial i_L}{\partial \omega^{1/2}} = \frac{i_L}{\omega^{1/2}} \tag{30}$$

The ratio of the slopes in the presence and absence of micellar aggregates is in turn given by the equation:

$$\frac{S}{S_0} = \frac{C_p - C_m}{C_p} + \left(\frac{D_2}{D_1}\right)^{2/3} \frac{C_m}{C^p} \left(\frac{\eta}{\eta_o}\right)^{-1/6} \tag{31}$$

where S_0 is the slope measured in the absence of aggregates. Assuming that the density of the amphiphilic microaggregate equals that of the continuous phase and taking into account the volume fraction of the aggregates (ϕ_m), the distribution coefficient P can be written as:

$$P = \frac{C_m}{C_p - C_m} \frac{1 - \phi_m}{\phi_m} \tag{32}$$

Usually, $\phi_m << 1$. Then, by combining Eqns. (31) and (32):

$$\frac{S}{S_0} = \frac{I + P\phi_m (D_2 / D_1)^{2/3}}{1 + P\phi_m} \left(\frac{\eta}{\eta_o}\right)^{-1/6} \tag{33}$$

Generally $\eta_o \approx 1$. Provided ϕ_m, D_1 and D_2 are known, the partition coefficient can be derived from these measurements, according to Eqn. (33). Moreover, if $ø_m$, P and D_1 are known, $D_2 \approx D_m$ can be evaluated.

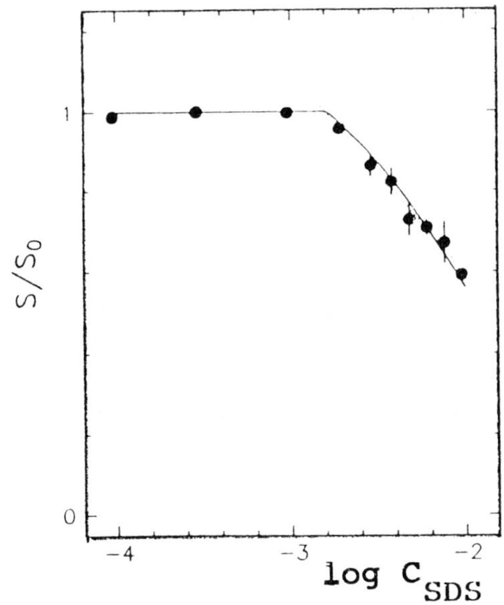

Fig. 5.19 Variation of the slopes ratio from oxidation of PPD ($C_P \cong 1\times10^{-4}$ M) at the RDE in aqueous SDS solution. Taken from Ref. [79], reproduced with permission.

The two electron oxidation of N,N-diethyl-3-methyl-p-phenylene-diamine (PPD) to the corresponding p-quinoneimine in SDS micellar solutions shows a decrease of the half-wave potential ($E_{1/2}$) with increasing surfactant concentration, indicating a more difficult electron-transfer. The linear dependence of $i_{L,m}$ vs. $\omega^{1/2}$ is typical of a diffusion-controlled process. The plot of the relative slopes as a function of SDS concentration fits Eqn. (33) well (see Fig. 5.19), a cmc value of ca. 1.6×10^{-3} M is obtained from the intersection of the two straight lines and P = 324 was calculated.

5.3.2.2. RDE in microemulsions

Electrochemical methods are useful to investigate the microemulsion structure, since the diffusion coefficients of the electroactive probes employed are related to the self-diffusion of the microenvironment (or phase) where the probe resides. Most of the reported studies concern oil–water microemulsions.

For hydrophilic electroactive probes, such as Tl^+, Cd^{2+}, $Fe(CN)_6^{3-}$), $Fe(CN)_6^{4-}$, a decrease in the measured polarographic D values with the

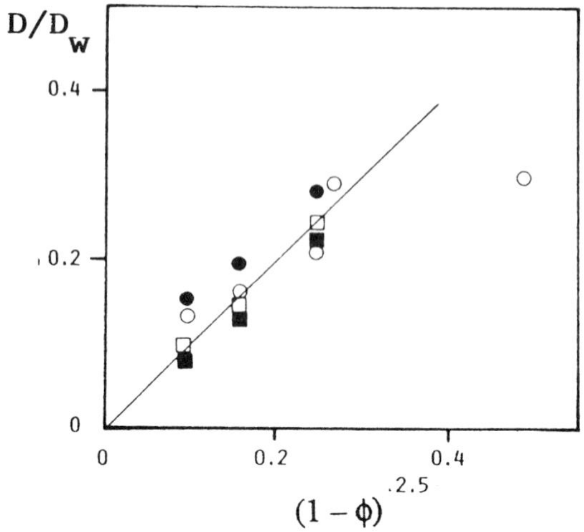

Fig. 5.20. Variation of the diffusion coefficient ratio (D/D_w) as a function of the disperse phase volume for Tl^+ (○), Cd^{2+} (●), $Fe(CN)_6^{3-}$ (□), $Fe(CN)_6^{4-}$ (■). Reproduced with permission from Ref. [72].

increasing oil phase volume (ϕ) of the microemulsion is usually observed, as can be expected for a species entirely located in the aqueous phase. The following equation holds:

$$D = D_w(1 - \phi)^n \quad (34)$$

where the exponent n is in the range 1.5–2.5 [6]. Figure 5.20 shows the agreement of experimental data concerning the above ionic probes in a water/sodium hexadecylsulfate (SHS)/1-pentanol/mineral oil microemulsion [80] with Eqn. (34), taking n = 2.5.

Conversely, the diffusion coefficients of the hydrophobic oil-soluble probe 1-dodecyl-4-cyanopyridinium ion ($C_{12}CPy^+$), measured in octane/1-butanol/HTAB cationic microemulsion and in mineral oil/1-pentanol/SHS anionic microemulsion, were used to estimate the diffusion coefficients of the oily droplets [69].

The analysis of rotating disk voltammograms (RDV) also gives useful information about the stabilization (or destabilization) of the electrode processes. For example, the variation in $i_{L,m}$ of some p-phenylenediamines when passing from an aqueous homogenous buffered solution to an oil–water cationic microemulsion (1.60% C_{18}DMB/1.88% 1-butanol/

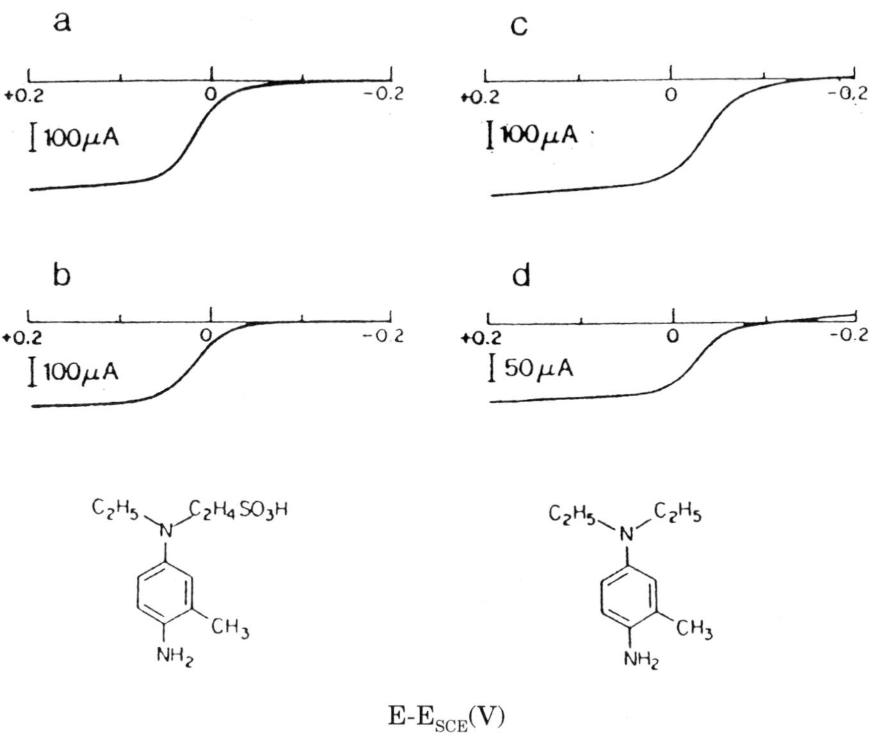

Fig. 5.21. RDV at 1000 rpm. (a) PPD1 in aqueous solution; (b) PPD1 in C_{18}DMB microemulsion; (c) PPD2 in aqueous solution; (d) PPD2 in C_{18}DMB microemulsion. Probe concentration $\cong 1\times10^{-3}$ M. C_{18}DMB: octadecyldimethylbetaine. Reproduced with permission from Ref. [81].

3.00% n-dodecane/98.7% water + carbonate buffer) is consistent with probe partitioning between the polar and non-polar phases of the microheterogeneous system [81]. The measured limiting current is, in fact, lower in the investigated microemulsions (see Fig. 5.21) and the observed $i_{L,m}$ decrease depends on the extent of probe partitioning. The structures of the examined probes are shown in the same figure.

Water-in-oil microemulsions have been, on the contrary, less investigated because it is difficult to perform electrochemical measurements with conventional electrodes in such low conductivity systems. The use of recently developed carbon-fibre ultramicroelectrodes should, in principle, be useful for these investigations.

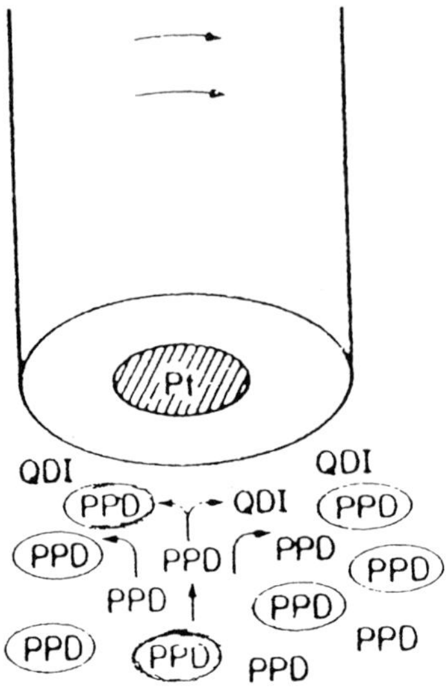

Fig. 5.22. Simplified representation of the RDE oxidation of p-phenylenediamines (PPD) to the corresponding quinonediimines (QDI) in an oil–water emulsion. PPD molecules incorporated in the oily drops are shielded against oxidation. Taken from Ref. [82], reproduced with permission.

5.3.2.3. RDE in emulsions

As for microemulsions, a large decrease of $i_{L,m}$ is generally observed for electroactive probes solubilized in oil–water emulsions. This effect has been mainly attributed to the shielding of the probe located within the dispersed oil droplets against the electrode processes, although a contribution from the higher viscosity of these systems cannot be excluded. A schematic representation of PPD oxidation to QDI at RDE surface in an oil-in-water emulsion is given in Fig. 5.22.

RDV curves are very similar to those obtained in microemulsions, the diffusion current is linearly proportional to $\omega^{1/2}$ (confirming the diffusion-controlled nature of the process) and, also for unstable and turbid oil–water emulsions, the estimation of probe partition coefficients has been reported [82].

5.3.3. Cyclic voltammetry in micellar and emulsion media

In this type of voltammetry, the voltage applied to the working electrode is scanned linearly to a predetermined limit where the direc-

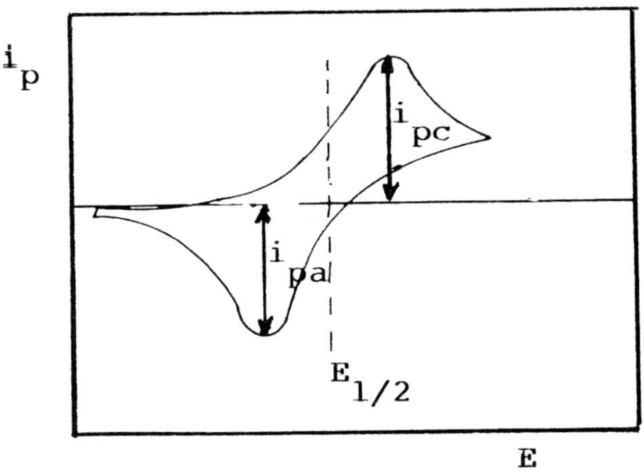

Fig. 5.23. Measured current vs. applied potential for one-electron reversible transfer at the electrode in a cyclic voltammetry experiment.

tion of the scan is reversed. The peak current for a reversible system is described by the Randles–Sevcik equation for the forward step of the first cycle:

$$i_p = 0.47\ n\ FAC_p \frac{(nF)^{1/2}}{(RT)^{1/2}} D^{1/2}\ \nu^{1/2} \tag{35}$$

In Eqn. (35) ν is the scan rate. For experiments with a fixed electrode, at a given temperature, Eqn. (35) simplifies to:

$$i_p = k\ n^{3/2}\ \nu^{1/2} D^{1/2}\ C_p \tag{36}$$

The diffusion coefficient D can be determined from the slope of the linear plots of i_p vs. $\nu^{1/2}$.

If the reaction is reversible, the difference between anodic and cathodic peak potential (ΔE_p) is equal to 59/n mV (at 25°C), and the ratio of anodic to cathodic peak current i_{pa}/i_{pc} is equal to 1. Figure 5.23 shows the typical CV profile for a reversible one-electron transfer.

Moreover, for a reversible reaction, $E_{1/2}$ is given by:

$$E_{1/2} = \frac{E_{pa} + E_{pc}}{2} \tag{37}$$

where E_{pa} and E_{pc} are the anodic and cathodic peak potential, respectively.

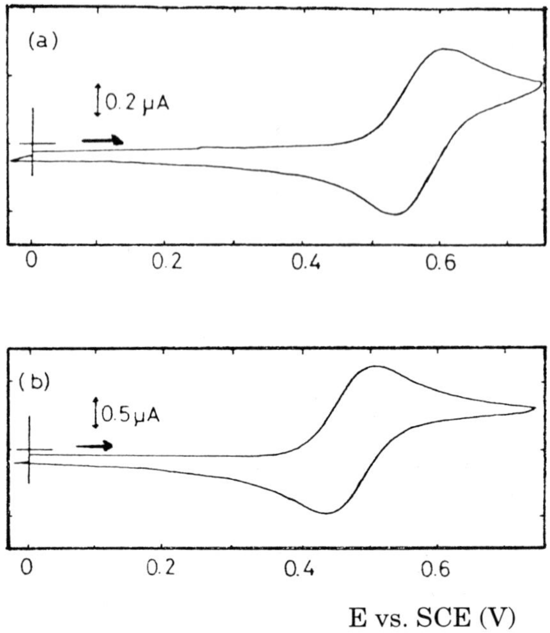

Fig. 5.24. Cyclic voltammograms obtained from 5×10^{-4} M $Os(5\text{-dmbpy})_3^{2+}$ in 0.5 M H_2SO_4. (a) Without SDS; (b) with 0.1 M SDS. Reproduced with permission from Ref. [83].

The addition of surfactants causes in some cases a change of the reversible half-wave potential, whereas in others, only small variations are observed. For example, the CV curves corresponding to the oxidation of the hydrophilic probe $Fe(H_2O)_6^{2+}$ in aqueous acid solution with and without SDS micelles are nearly identical, indicating that the cation must be essentially in the aqueous bulk [83].

On the contrary, for the couple $Os(5\text{-dmbpy})_3^{3+/2+}$ at a scan rate of 5 mV s^{-1}, $E_{1/2}$ is shifted towards more positive values (ca. 100 mV) in 0.1 M SDS, whereas the increase in surfactant concentration (from 0.1 to 0.2 M) practically does not change $E_{1/2}$ (see Fig. 5.25) [83]. The behaviour of the osmium complexes was qualitatively attributed to their solubilization in the micellar aggregates.

Changes in the microstructure of some oil–water microemulsions have been detected using ferrocene and ferrocene derivatives as electroactive probes. Working in HTAB/ 1-butanol/ n-octane/ water microemulsions Mackay et al. [57] have found that the diffusion coefficient

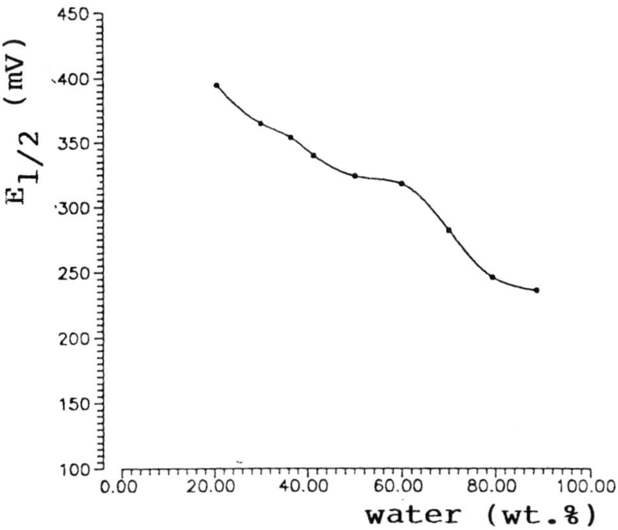

Fig. 5.25. Variation of $E_{1/2}$ for ferrocene (5×10^{-3} M) as a function of water content in a HTAB/1-butanol/n-octane/water microemulsion. Reproduced with permission from Ref. [57].

decreases when water content of the microemulsion increases. This behaviour, which is opposite to that expected on the basis of the increase of the oily droplet size with decreasing water percent, was assigned to the formation of a bicontinuous hydrocarbon/water structure.

From the measured ΔE_p (ca. 60 mV) and from the i_{pa}/i_{pc} ratio (ca. 1), a reversible electron-transfer with a fast kinetics was confirmed for ferrocene in the investigated microemulsions. The reaction potential (measured by the $E_{1/2}$ changes) is sensitive to the microemulsion composition, decreasing with increasing water percent (see Fig.5. 26).

This behaviour is consistent with a decrease in the diffusion coefficient of the reduced (Fc) form and the nearly constant diffusion coefficient of the water-soluble oxidized form (ferricenium ion, Fc^+), according to the equation:

$$E_{1/2} = E_w^{o\prime} + (RT/nF) \ln (D_{red}/D_{ox})^{1/2} \qquad (38)$$

Another interesting example of CV experiments performed in microemulsion media is the electrochemical reduction of N,N′-dimethyl-4,4′-bipyridine (methyl viologen) dication (MV^{2+}), investigated in

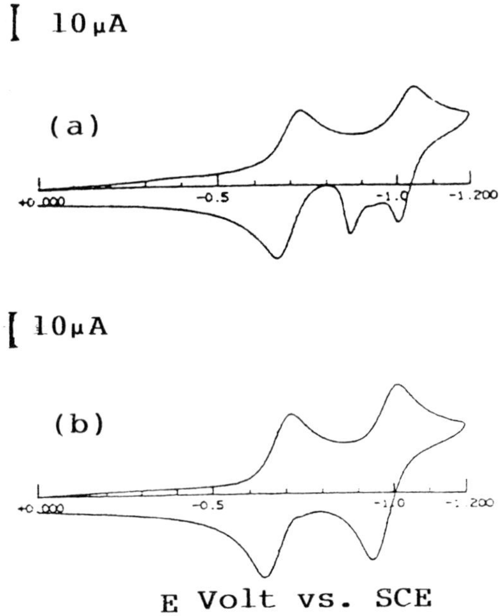

Fig. 5.26. CV curves for MV^{2+} at constant sweep rate (100 mV s^{-1}). (a) Homogeneous solution; (b) 2.3% HTAB/4.6% 1-butanol/2.3% n-octane/1.3% NaBr–89.5% water; *opposite page*: (c) 10% Triton X100/1.4% n-octane/1.0% NaBr–87.6% water; (d) 5.5% SDS/10.3% 1-pentanol/5.1% n-dodecane/1.0% NaBr–78.1% water. Adapted from Ref. [84], reproduced with permission.

cationic, anionic and non-ionic microemulsions [84]. The reduction proceeds via two reversible single electron-transfer steps (see Eqns. (39) and (40)) and the corresponding $E_{1/2}$ values, as well as the diffusion currents, are affected by the microemulsion composition.

$$MV^{2+} + e^- \leftrightarrow MV^{\cdot +} \quad (39)$$

$$MV^{\cdot +} + e^- \leftrightarrow MV \quad (40)$$

Figure 5.26 shows the cyclic voltammograms of 1×10^{-3} M MV^{2+} in aqueous 0.1 M NaBr and in different microemulsions. The CV profile at 100 mV s^{-1} in aqueous solution shows two peaks (one at a potential of –1.0 V and the other at –0.875 V) for the anodic reaction corresponding to the second step of MV^{2+} reduction. This implies the presence of two

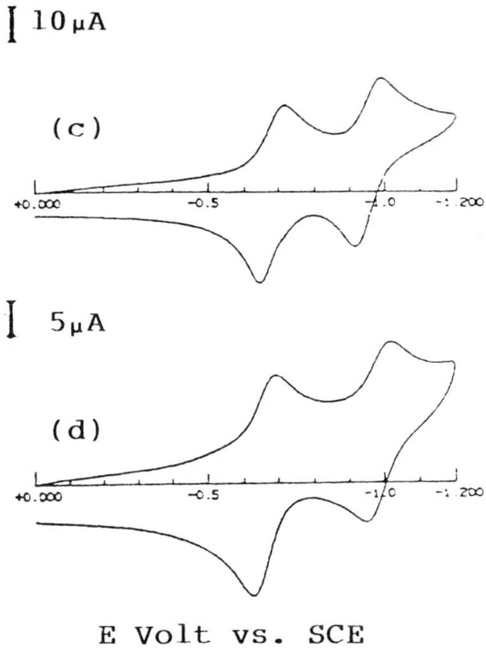

Fig. 5.26 continued. Caption on opposite page.

forms arising from this step. The position of the first peak is sensitive to changes in sweep rate, whereas the peak at –0.875 V is independent (in the range 10–200 mV s^{-1}).

The height of these peaks is also sensitive to the sweep rate: at 20 mV s^{-1} practically only the peak at –0.875 V remains, whereas at higher sweep rates (500 mV s^{-1}) the height of the peak at –1 V predominates. The reduction process becomes less reversible at higher sweep rates and the reduction step appears not to be diffusion-controlled. This behaviour was attributed to adsorption of the hydrophobic product MV onto the electrode.

A comparison of the CV curves in microemulsions indicates that in these systems the reduction of MV^{2+} occurs in two reversible one-electron steps, with a value of cathodic to anodic peak current ratio of ca. 1 and $\Delta E_p \cong 60$ mV. The plots of cathodic peak current vs. $v^{1/2}$ are linear for the first step in all the microemulsions.

By using the Randles–Sevcic equation, the diffusion coefficients of MV^{2+} were evaluated and found to be nearly equal to that in aqueous solutions for HTAB and Triton X100 microemulsions, being significantly lower for SDS microemulsion. This indicates a strong association of MV^{2+} to the anionic microemulsion droplets.

For the first reduction step, the peak potentials, the $E_{1/2}$ values and the peak currents are nearly the same in water and in cationic and non-ionic microemulsions, indicating that the reduction of MV^{2+} to $MV^{\cdot+}$ is not affected. In SDS the corresponding peak potential is shifted to less negative values and the peak current is decreased, suggesting the association of MV^{2+} with the anionic surfactant at the droplet surface and a stabilization of MV^{2+} against reduction in the negatively charged environment.

The second reduction step (from $MV^{\cdot+}$ to MV) shows a less negative $E_{1/2}$ value in cationic and non-ionic microemulsions. The highly hydrophobic product MV is more soluble in the non-polar domains of the microemulsion and, thus, its formation is facilitated.

5.3.4. Voltammetry in water–oil microemulsions

The development of an ultramicroelectrode disk ($r < 15\ \mu m$) extended the application range of voltammetry to highly resistive media such as water–oil microemulsions. The steady-state diffusion-controlled limiting current (i_L) of a probe bound to the aggregates is given by the following expression:

$$i_L = 4\ n\ F\ D\ C_p\ r \qquad (41)$$

The application of such peculiar electrodes to the investigation of water–oil AOT microemulsions has been recently reported [85–87].

5.4. ELECTROCATALYTIC REACTIONS IN ORGANIZED AMPHIPHILIC MEDIA

The heterogeneous electron-transfer from an electrode to an electroactive probe present in solution can be a kinetically limiting process. In current electroanalytical techniques the diffusion of the probe to the electrode is the rate-determining step and the electron-transfer rate does not usually influence the measured variable. The presence of amphiphilic adsorbed layers onto the electrode can drastically alter the above situation, allowing the introduction of a third component (the

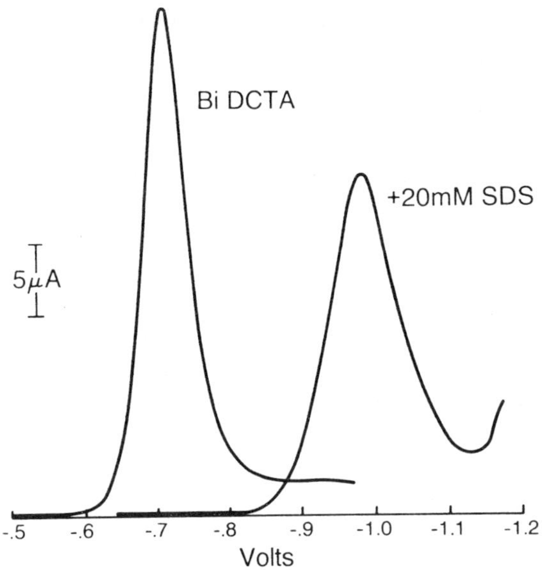

Fig. 5.27. DPV reduction of the Bi(III)-DCTA complex at the DME in the presence and in the absence of SDS. Scan rate 1 mV s^{-1}; pulse amplitude: 25 mV, drop time: 1 s. Reproduced with permission from Ref. [5].

catalyst) able to act as electron-transfer mediator between the electrode and the probe.

The presence of such adsorbed layers was hypothesized several years ago to explain the effects observed in differential pulse voltammetric (DPV) studies concerning the electroreduction of a bismuth(III)-DTCA complex (DTCA: 1,2-diaminocyclohexanetetraacetic acid) to Bi° [88]. In the presence of a cationic amphiphile (dodecylammoniun perchlorate) the corresponding reduction peak is shifted toward less negative potentials and a greater peak current is measured, thus indicating faster electron-transfer. On the contrary, the addition of SDS shifts the peak potentials towards more negative values and the peak becomes broader, suggesting inhibition of the electron-transfer (see Fig. 5.27). These two opposite effects were attributed to the formation of a surfactant layer onto the electrode.

In micellar electrocatalysis, either the aggregates and the surfactant adsorbed layers can solubilize the probe and/or the catalyst. As a consequence, molecules having low water solubility can approach the electrode when dissolved in micelles or microemulsions and, if their electron-transfer rates at the electrode are very slow, a suitable electron shuttle (catalyst) can intervene as mediator.

The presence of adsorbed surfactant films has been confirmed from different experimental data, including differential capacitance [74] and ellipsometry [89], among others.

Different examples of electrocatalytic systems can be mentioned. An example of such processes occurring in micellar solutions is the electron-transfer from the electrode to biomolecules (cytochrome-c) using ferrocene as catalyst [90]. More recently, important analytical and environmental applications have been reported since electrocatalysis can be exploited to degrade various organic pollutants. For example, the electrochemical reduction of aromatic halides and alkyl vicinal dibromides, or the dehalogenation of highly toxic PCBs, can be conveniently catalyzed in surfactant media [91,92]. Moreover, the use of water–oil microemulsions for the electrocatalytic treatment of oil-soluble contaminants is also possible, provided that electrodes capable of operating in highly resistive media are used.

5.4.1. Electrocatalysis in micellar media

An important example of an electrocatalytic process in micellar media is the reduction of aryl halides, which follows the steps indicated in scheme 1 [91]:

$$C + e^- \rightarrow R \qquad \text{(a)}$$

$$R + ArX \underset{k_d}{\overset{k_f}{\rightleftarrows}} ArX.- + C \qquad \text{(b)}$$

$$ArX.- \overset{k}{\rightleftarrows} Ar. + X- \qquad \text{(c)}$$

$$Ar. + R \rightarrow C + Ar- \qquad \text{(d)}$$

$$Ar- + H+ \rightarrow ArH \qquad \text{(e)}$$

Scheme 1

Compound C (catalyst) is an electron acceptor reduced at the cathode to the active donor R, which in turn can react with the compound to be reduced (ArX). The rate-determining step of the process is reaction (b), although in some cases reaction (c) can also be kinetically relevant. On the contrary, reactions (d) and (e) are usually fast. The main advantage of this approach is that the potential required to drive the reaction is that of reaction (b), which is less negative than that required for direct reduction of ArX.

The use of proper micellar aggregates to enhance the second order electron-transfer reactions (already discussed in detail in Chapter 3) is at the basis of most recent electrocatalytic applications. The favourable kinetic effect is in fact due to the increase in the local concentrations of catalyst and substrate either in the hemimicellar films formed onto the electrodes or in the micelles present in the solution (see Fig. 5.28). This effect becomes very relevant if both components are strongly bound to the aggregates.

Different hydrophobic mediators, giving rise to long-lived anion radicals, have been proposed for dehalogenation of aryl halides. Arenes such as 9-phenylanthracene (9-PA) or 1,2-dicyanobenzene (1,2-DCB), appear to be very efficient in aqueous HTAB solutions.

The cyclic voltammograms of 9-PA in 0.1 M HTAB shows that reversible diffusion-controlled processes are present at the scan-rates in the range 10 mV s^{-1}–50 V s^{-1}. Upon addition of 4-bromobiphenyl (4-BB) to these solutions, a further increased current is observed, related to the reduction of the substrate to biphenyl (see Fig. 5.29). The mediated reduction of the C–Br bonds has been attributed to the enhancement in the rate of electron-transfer (Reaction (b) in Scheme 1) due to the accumulation of reactants in the surfactant layer on the electrode, with a corresponding shift of the rate-determining step of the process to Reaction (c) (decomposition of the intermediate ion-radical) [93]. The ratio of cathodic peak current for the catalyst in the presence of substrate (i_c) to that in the absence of substrate (i_a) is usually indicated as the catalytic efficiency, at constant scan rate.

It must be noted that the positive surface potential of the aggregates and the hydrophobic interactions stabilize the 9-phenylanthracene anion radical (R) against competitive reactions, whereas in non-ionic micelles (where the favourable electrostatic effects are not present) the anion-radical coming from the catalyst is involved in other reactions, with lower yields of the reaction end-product (biphenyl).

The potential required to reduce 9-PA (in HTAB) to its anion-radical is –2.2 V vs.SCE: At this potential, the reduction of 4-BB is possible. Other catalysts, having less negative formal potentials (such as 1,2-dicyanobenzene) are necessary to dehalogenate PCBs.

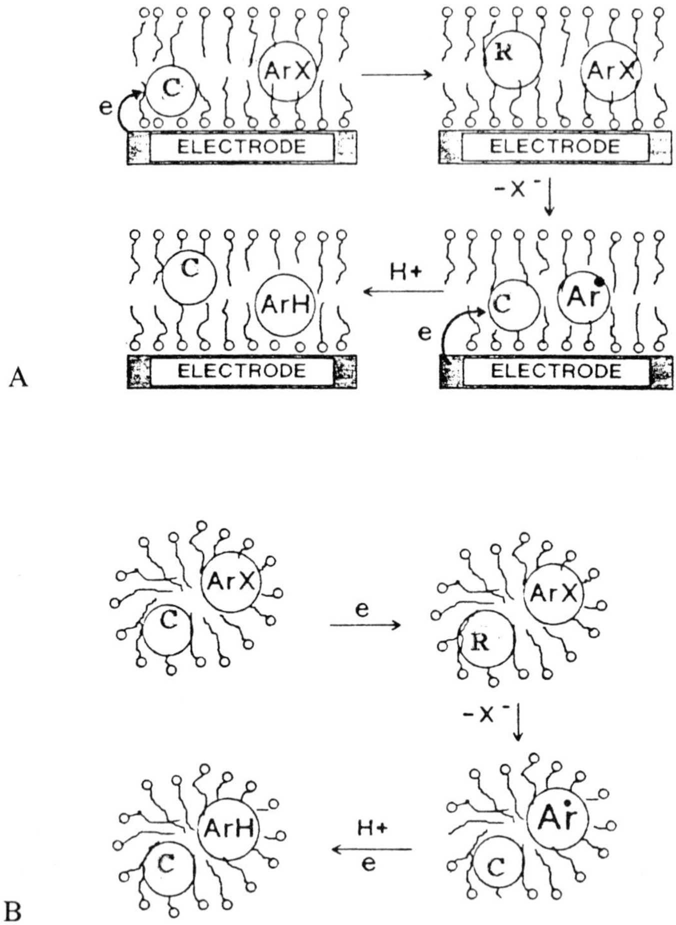

Fig. 5.28. Schematic representation of the electrochemical catalytic reduction of aryl halides in the presence of surfactant aggregates. (A) On the electrode surface; (B) in the micellar solution. Taken from Ref. [91], reproduced with permission.

Metal complexes of the type CoL_3^{2+} (L represents 2,2′-bipyridine or a derivative of this ligand), dissolved in micellar solutions, have also been proposed as efficient electrocatalysts. The decomposition of allyl chloride (CH_2=CH–CH_2Cl) to yield mainly 1,5-hexadiene has been achieved

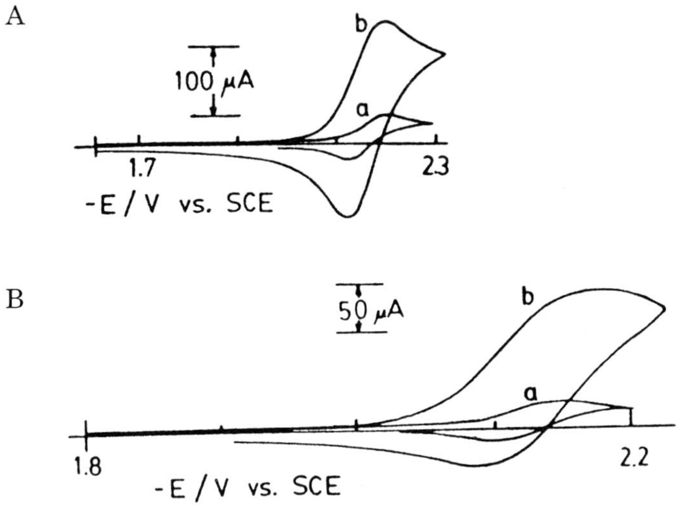

Fig. 5.29. Electrocatalytic reduction of 4-BB in different micellar systems. Scan rate: 0.5 V s^{-1}. (A) 0.1 M HTAB + 0.1 M TEAB. (a) 3.4×10^{-4} M 9-PA; (b) 3.4×10^{-4} M 9-PA + 7.9×10^{-4} M 4-BB. (B) 0.09 M Igepal CO-720 + 0.1 M TEAB. (a) 5×10^{-4} M 9-PA; (b) 5×10^{-4} M 9-PA + 8×10^{-3} M 4-BB. Taken from Ref. [90], reproduced with permission. TEAB: tetraethylammonium bromide.

in SDS and HTAB micellar systems using these complexes [94]. Figure 5.30 shows the typical CV curves obtained in cationic micellar solutions.

In some cases the observed kinetic effects may be relevant. Defining k_r as the ratio between the forward electron transfer rates (in M^{-1} s^{-1}) measured in the micellar system and in homogeneous solvents, respectively ($k_r = k_{mic}/k_{homog}$), $k_r = 3.3\times10^4$ was measured for dehalogenation of 4-BB in HTAB (the homogeneous reaction was performed in DMF) [93], whereas lower k_r values were measured for other reactions.

5.4.2. Electrocatalysis in microemulsions

Microemulsions with a high content of oil phase can solubilize higher amounts of non-polar electroactive substrates than micelles. However, since these microemulsions generally show a very high electric resistivity, voltammetric measurements cannot usually be performed. In some cases, when the water content is relevant and in the presence of suitable

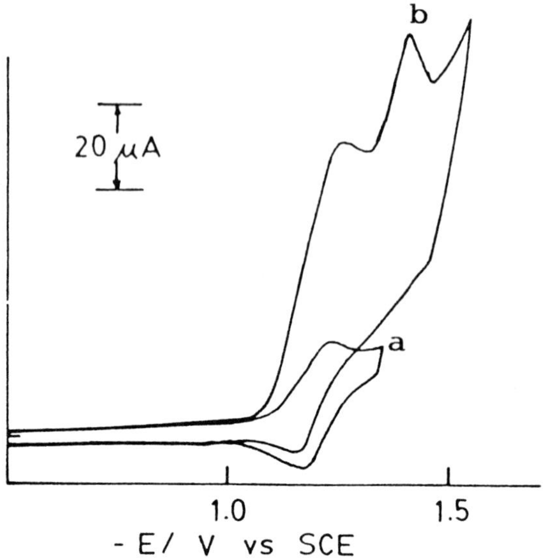

Fig. 5.30. CV profiles at constant scan rate (50 mV s^{-1}) for the systems. (a) 2×10^{-3} M tris(2,2′-bipyridyl)Co(II) in 0.1 m HTAB + 0.1 M KBr; (b) the same components of curve(a) + 2.5×10^{-2} M allyl chloride. Reproduced with permission from Ref. [94].

electrolytes (such as tetraethylammonium salts) more conductive microemulsions, with resistivity values in the range 500–1000 Ωcm, can be obtained. The development of suitable electrodes (see Section 5.3.4), including glassy carbon electrodes, has allowed workers to perform voltammetric measurements in these systems and to investigate catalytic reactions.

Reduction of vicinal dibromides (such as ethylene dibromide) to alkenes in microemulsion media is interesting both from the analytical and environmental point of view, since these compounds are still used as fumigants in agriculture. Model systems using water–oil microemulsions composed of AOT/water/isooctane (3.4/41.5/53.5 by weight) and vitamin B_{12} as catalyst have been investigated [95].

The CV data indicate the inner-sphere electron-transfer between the B_{12}Co(I) complex and the substrate as the rate-determining step. Diffusion data (estimated using the Randles–Sevcik equation) suggests that the catalyst resides in the water pool of the microemulsion, whereas the substrates (ethylene dibromide, 1,2-dibromobutane, trans-1,2-dibromocyclohexane) are dissolved in the isooctane continuous phase. A reported analytical application based on these studies is the determination of ethylene dibromide in gasoline by square-wave voltammetry in gasoline/AOT/water emulsions containing the same catalyst [95].

Bicontinuous microemulsions composed of didodecyldimethylammonium bromide (DDAB), dodecane and water have also been reported as suitable media for voltammetric oxidations and/or reductions of different substrates using either water-soluble ($Fe(CN)_6^{3-}$, $Ru(NH_3)_6^{3+}$, Vitamin B_{12}) or oil-soluble (ferrocene, pyrene, 9-phenylanthracene) redox mediators [96].

5.4.3. Surfactant-modified electrodes for electrocatalysis work

Research in electrocatalysis was also devoted to the preparation and study of surfactant-modified electrodes. The synthesis of amphiphilic electrocatalysts which can give rise to the formation of ordered films onto the electrodes is an important aspect of this work, which includes the investigation of both adsorption and micellization processes and their influence on the electrochemistry of such compounds.

Other relevant aspects of these studies concern the investigation of electrocatalytic processes occurring at the modified electrodes in homogeneous solutions and in organized surfactant media, as well as the analysis of electrocatalytic reactions involving different substrates and adsorbed catalysts. Table 5.4 briefly summarizes some of the reported experiments performed using surfactant-modified metallic electrodes.

More recent developments in the field include the use of clay-modified electrodes, which present a large surface area, a high cation-exchange capacity and favourable intercalating properties. The main advantage of such electrodes is the rapid and easy surfactant coverage, whereas the formation of Langmuir–Blodgett monolayers onto metals is more difficult and tedious. A detailed description of the preparation and performances of clay-based electrodes can be found in Refs. [91, 103–105].

Coating of electrodes with octadecylsilyl-groups (ODS) is another promising way of facilitating the incorporation of redox amphiphiles and/or hydrophobic substrates within the electrode-adsorbed layer microenvironment [106].

TABLE 5.4

Examples of electrode coating with unreactive and electroactive amphiphiles

Electrode	Adsorbed surfactant	Type of investigation	Ref.
Gold	Hexadecylmercaptan monolayers + N-octadecyl-4-pyridinium-4′-pyridyl bromide and N-octadecyl-4,4′-bipyridiumim dibromide	Incorporation of the redox probe into the monolayers, CV experiments, electrocatalytic reduction of O_2	92
Gold and platinum	Amphiphile derivative of MV^{2+}	Formation of monolayers	98
Gold	Co(II)hexadecyltetrapyridyl porphirin amphiphile	Study of self-assembly and electrocatalytic O_2 reduction to H_2O_2	99
Gold	Cationic redox amphiphile (Ferrocenylmethyl)heptyl-dimethyl-ammonium bromide	Adsorption and desorption of reduced and oxidized forms and corresponding $E_{1/2}$ shifts	100
Gold	N-methyl-N′-octadecyl-viologen	Study of self-assembling (formation of monolayers and bilayers at the electrode)	101
Glassy carbon	Phosphatidylcholine	Enrichment of redox amphiphiles (ferrocene and MV^{2+} derivatives) from aqueous solution to the lipid layer	102

REFERENCES

1 L. Meites, *Polarographic Techniques*. Interscience, New York, 1955.
2 G.E.O. Proske, *Anal. Chem.*, 24 (1952) 1834.
3 W.L. Hinze, in: K.L. Mittal (Ed.), *Solution Chemistry of Surfactants*. Plenum Press, New York, 1979. Vol. 1, pp. 108–110.
4 E. Pelizzetti and E. Pramauro. *Anal. Chim. Acta*, 169 (1985) 1.
5 G.L. McIntire, *Crit. Rev. Anal. Chem.*, 21 (1990) 257.
6 R.A. Mackay, *Coll. Surfaces A*, 82 (1994) 1.
7 J.F. Rusling, in: A.J. Bard (Ed.), *Electroanalytical Chemistry*. Dekker. New York, 1994. Vol. 18, pp. 1–88.

8 R.A. Mackay and J. Texter (Eds), *Electrochemistry in Colloids and Dispersions*. VCH, New York, 1992, pp. 546.
9 B.W. Barry and R. Wilson, *Coll. Polym. Sci.*, 256 (1978) 251.
10 M.J. McDowell and C.A. Kraus, *J. Am. Chem. Soc.*, 73 (1951) 2173.
11 S. Diekmann and J. Frahm, *J. Chem. Soc. Faraday I*, 75 (1979) 2199.
12 P. Mukerjee and K.J. Mysels, *Critical Micelle Concentrations of Aqueous Surfactants Systems.* NSRDS-NBS 36. U.S. Gov. Printing Office, Washington, DC, 1971.
13 H.C. Evans, *J. Chem. Soc.*, (1956) 579.
14 C. Botre, V.L. Crescenzi and A. Mele, *J. Phys. Chem.*, 63 (1959) 650.
15 M. Saito, Y. Moroi and R. Matuura, in: K.L. Mittal and B. Lindman (Eds.), *Surfactants in Solution*. Plenum Press, New York, 1984, Vol. 2, pp. 771–788.
16 P.C. Shanks and E.I. Franses, *J. Phys. Chem.*, 96 (1992) 1794.
17 P.A. Winsor, *Trans. Faraday Soc.*, 44 (1948) 376.
18 M. Bourrel and R.S. Schechter, *Microemulsions and Related Systems.* Dekker, New York, 1988, Ch. 1.
19 M. Bourrel and R.S. Schechter, *Microemulsions and Related Systems.* Dekker, New York, 1988, Ch. 4.
20 K. Kale, E.L. Cussler, D.F. Evans et al., *J. Phys. Chem.*, 84 (1980) 593.
21 R. Zana (Ed.), *Surfactant Solutions. New Methods of Investigation.* Dekker, New York, 1987, Ch. 9.
22 R. Palepu, D.G. Hall and E. Wyn-Jones, *J. Chem. Soc. Faraday Trans.*, 86 (1990) 1535.
23 D.G. Hall, *J. Chem. Soc. Faraday Trans. I*, 177 (1981) 1121.
24 A. Berthod and C. Saliba, *Analusis*, 13 (1985) 437.
25 J.B.S. Bonilha, H. Chaimovich, V.G. Toscano and F.H. Quina, *J. Phys. Chem.*, 83 (1979) 2463.
26 F.H. Quina, M.J. Politi, I.M. Cuccovia, S.M.M. Franchetti and H. Chaimovich, *J. Phys. Chem.*, 84 (1980) 361.
27 E. Pelizzetti and E. Pramauro, *Anal. Chim. Acta*, 117 (1980) 403.
28 G. Kortum, W. Vogel and K. Andrussov, *Dissociation Constants of Organic Acids in Aqueous Solutions*. Butterworth, London, 1961.
29 C.A. Bunton and L. Sepulveda, *J. Phys. Chem.*, 83 (1979) 680.
30 O.A. El Seoud, *Adv. Coll. Interface Sci.*, 30 (1989) 1.
31 C.J. Drummond, F. Grieser and T.W. Healy, *J. Chem. Soc. Faraday Trans. I*, 85 (1989) 537.
32 C.J. Drummond, F. Grieser and T.W. Healy, *J. Chem. Soc. Faraday Trans. I*, 85 (1989) 551.
33 C.J. Drummond, F. Grieser and T.W. Healy, *J. Chem. Soc. Faraday Trans. I*, 85 (1989) 561.
34 J.S. Esteve Romero, G. Ramis Ramos and M.C. Garcia Alvarez-Coque, *J. Coll. Interface Sci.*, 141 (1991) 44.

35 S. Harada, H. Yano, T. Yamashita, S. Nishioka and T. Yasunaga, *J. Coll. Interface Sci.*, 110 (1986) 272.
36 A.L. Underwood, *Anal. Chim. Acta*, 140 (1982) 89.
37 E. Pelizzetti and E. Pramauro, *Anal. Chim. Acta*, 128 (1981) 273.
38 G. Meyer, P. Viout and B. Tchoubar, *Tetrahedron Lett.* (1974) 2487.
39 E. Pramauro and E. Pelizzetti, *Anal. Chim. Acta*, 126 (1981) 253.
40 C.J. Drummond, F. Grieser and T.W. Healy, *J. Chem. Soc. Faraday Trans. I*, 85 (1989) 521.
41 C.M. Harris and B.K. Selinger, *Z. Phys. Chem. N. F.*, 134 (1983) 65.
42 L.K. Tong and M.M. Glesmann, *J. Am. Chem. Soc.*, 79 (1957) 4305.
43 F. Dorion and R. Gaboriaud, *J. Chim. Phys.*, 78 (1981) 555.
44 T. Yamashita, H. Yano, S. Harada and T. Yasunaga, *J. Phys. Chem.*, 88 (1984) 2671.
45 T. Yamashita, M. Sumino, H. Yano, S. Harada and T. Yasunaga, *Bull. Chem. Soc. Jpn.*, 57 (1984) 2352.
46 G. Charbit, F. Dorion and R. Gaboriaud, *J. Chim. Phys.*, 81 (1984) 187.
47 P. Rychlovsky and I. Nemcova, *Talanta*, 35 (1988) 211.
48 A. Berthod and C. Saliba, *Analusis*, 14 (1986) 414.
49 J. Georges and A. Berthod, *J. Electroanal. Chem.*, 175 (1984) 143.
50 M. Gemes and M. Thomalia, *Electrochim. Acta*, 26 (1981) 829.
51 J.M. Pingarron Carrazon, A. Gordon Vergara, A.J. Reviejo Garcia and L.M. Polo Diez, *Anal. Chim. Acta*, 216 (1989) 231.
52 T.C. Franklin and M. Iwunze, *Anal. Chem.*, 52 (1980) 973.
53 C. Mousty, B. Cheminat and G. Mousset, *Am. Chem. Soc.*, 54 (1989) 5377.
54 G.L. McIntire, D.M. Chiappardi, R.L. Casselberry and H.N. Blount, *J. Phys. Chem.*, 86 (1982) 2632.
55 P.A. Quintela and A.E. Kaifer, *Langmuir*, 3 (1987) 769.
56 A.E. Kaifer and A.J. Bard, *J. Phys. Chem.*, 89 (1985) 4876.
58 R.A. Mackay, S.A. Myers, L. Bodalbhai and A. Brajter Toth, *Anal. Chem.*, 62 (1990) 1084.
58 R.E. Verrall, S. Milioto and A. Giraudeau, R. Zana, *Langmuir*, 5 (1989) 1242.
59 M.I. Viseu, M.L.S. Simoes Goncalves, S.M.B. Costa and M.I.C. Ferreira, *J. Electroanal. Chem.*, 282 (1990) 201.
60 M.I. Viseu, M.L.S. Simoes Goncalves, S.M.B. Costa and M.I.C. Ferreira, *J. Electroanal. Chem.*, 282 (1990) 215.
61 M.J. Eddowes and M. Gratzel, *J. Electroanal. Chem.*, 163 (1984) 31.
62 E. Dayalan, S. Qutubuddin and A. Hussan, *Langmuir*, 6 (1990) 715.
63 E. Fisicaro, C. Minero and E. Pelizzetti, in: R.A. Mackay and J. Texter (Eds.), *Electrochemistry in Colloids and Dispersions*. VCH, New York, 1992, Ch. 2.
64 T. Matsubara and J. Texter, *J. Coll. Interface Sci.*, 112 (1986) 421.
65 E. Dayalan, S. Qutubuddin and J. Texter, *J. Coll. Interface Sci.*, 143 (1991) 423.

66 D. Lemordant, *J. Electroanal. Chem.*, 219 (1987) 125.
67 R.A. Mackay, C. Hermansky and R. Agarwal, in: M. Kerker (Ed.)., *Colloids and Interface Science*. Academic Press, New York, 1976. Vol. II, pp. 289–303.
68 R.A. Mackay, *Adv. Coll. Interface Sci.*, 15 (1981) 131.
69 R.A. Mackay, N.S. Dixit, R. Agarwal and R.P. Seiders, *J. Dispersion Sci. Technol.*, 4 (1983) 397.
70 R. Guidelli and M.L. Foresti, *J. Electroanal. Chem.*, 77 (1977) 73.
71 S. Hayano and N. Shinozuka, *Bull. Chem. Soc. Jpn.*, 42 (1969) 1469.
72 E. Pelizzetti, E. Fisicaro, C. Minero, V. Maurino and E. Pramauro, *J. Surface Sci. Technol.*, 6 (1990) 107.
73 R. Zana and R.A. Mackay, *Langmuir*, 2 (1986) 109
74 N. Shinozuka and S. Hayano, in: K.L. Mittal (Ed.), *Solution Chemistry of Surfactants*. Plenum Press, New York, 1979, Vol. 2, pp. 599–623.
75 S.J. Candau, E. Hirsch and R. Zana, *J. Phys.*, 45 (1984) 1263.
76 A. Berthod and J. Georges, *J. Coll. Interface Sci.*, 106 (1985) 194.
77 A. Berthod in: H.L. Rosano and M. Clausse (Eds.), *Microemulsion Systems*. Dekker, New York, 1987, Ch. 20.
78 A. Berthod, in: R.A. Mackay and J. Texter (Eds.), *Electrochemistry in Colloids and Dispersions*. VCH, New York, 1994, Ch. 11.
79 J. Texter, F.R. Horch, S. Qutubuddin and E. Dayalan, *J. Coll. Interface Sci.*, 135 (1990) 263.
80 R.A. Mackay in: I.D. Robb (Ed.), *Microemulsions*. Plenum Press, New York, 1982.
81 E. Dayalan, S. Qutubuddin and J. Texter, *J. Coll. Interface Sci.*, 143 (1991) 423.
82 J. Texter, *J. Coll. Interface Sci.*, 120 (1987) 389.
83 Y. Ohsawa, Y. Shimazaki and S. Aoyagui, *J. Electroanal. Chem.*, 114 (1980) 235.
84 E. Dayalan, S. Qutubuddin and A. Hussam, *Langmuir*, 6 (1990) 715.
85 F. Rusling, *Coll. Surfaces*, 48 (1990) 173.
86 Z. Wang, A. Owlia and J.F. Rusling, *J. Electroanal. Chem.*, 270 (1989) 407.
87 A. Owlia, Z. Wang and J.F. Rusling, *J. Am. Chem. Soc.*, 111 (1989) 5091.
88 N. Gundersen and E.J. Jacobsen, *J. Electroanal. Chem. Interfacial Electrochem.*, 20 (1969) 13.
89 J.B. Hayter, M.W. Humphreys, R.J. Hunter and R. Parsons, *J. Electroanal. Chem.*, 56 (1974) 160.
90 Y. Fujihira, T. Kuwana and C.R. Hartzell, *Biochem. Biophys. Res. Commun.*, 61 (1974) 488.
91 J.F. Rusling, *Acc. Chem. Res.*, 24 (1991) 75.
92 Q. Huang and J.F. Rusling, *Environ. Sci. Technol.*, 29 (1995) 98.
93 J.F. Rusling, C.N. Shi, D.K. Gosser and S.S. Shukla, *J. Electroanal. Chem.*, 240 (1988) 201.

94 G.N. Kamau and J.F. Rusling, *J. Electroanal. Chem.*, 240 (1988) 217
95 J.F. Rusling, T.F. Connors and A. Owlia, *Anal. Chem.*, 59 (1987) 2123.
96 M.O. Iwunze, A. Sucheta and J.F. Rusling, *Anal. Chem.*, 62 (1990) 644.
97 K.A. Bunding Lee, R. Mowry, G. McLennan and H.O. Finklea, *J. Electroanal. Chem.*, 246 (1988) 217.
98 A. Diaz and A.E. Kaifer, *J. Electroanal. Chem.*, 249 (1988) 333.
99 D.A. Van Galen and M. Majda, *Anal. Chem.*, 60 (1988) 1549.
100 J.J. Donohue and D.A. Buttry, *Langmuir*, 5 (1989) 671.
101 C.A. Widrig and M. Majda, *Langmuir*, 5 (1989) 689.
102 O.J. Garcia, P.A. Quintela and A.E. Kaifer, *Anal. Chem.*, 61 (1989) 979.
103 B. Brahimi, P. Labbe and G. Reverdy, *J. Electroanal. Chem.*, 267 (1989) 343.
104 C. Shi, J.F. Rusling, Z. Wang, W.S. Willis, A.M. Winiecki and S.L. Suib, *Langmuir*, 5 (1989) 650.
105 N. Hu and J.F. Rusling, *Anal. Chem.*, 63 (1991) 2163.
106 J.F. Rusling and H. Zhang, *Anal. Chim. Acta*, 235 (1990) 307.

Chapter 6

Micellar chromatography

In the past decade, aqueous micellar solutions have been used as mobile phases in a variety of liquid chromatographic techniques, gaining the increasing attention of many analytical chemists. Most of these applications concern reverse-phase HPLC, but thin layer chromatography and gel permeation are also of interest. Micelles can also act as liquid "pseudostationary phases" in micellar electrokinetic capillary chromatography, providing a new powerful separation technique for many neutral and charged compounds.

Moreover, organic solutions containing reversed micelles have been applied as structured eluents in normal-phase liquid chromatography and, recently, in supercritical fluid chromatography.

In all the above techniques there is a common feature: the microheterogeneous micellar environment allows the organization of partitioned solutes in different regions of the aggregates through hydrophobic and electrostatic interactions, offering an alternative to the use of organic and hydro-organic eluents. When normal micelles are used, the hydrocarbon nature of their inner structure permits the solubilization of hydrophobic species in aqueous eluting phases and the simultaneous separation of these compounds from hydrophilic solutes can be achieved.

In traditional liquid chromatography the retention of each analyte is mainly governed by solute–solvent interactions (with secondary contributions arising from solute–stationary phase interactions), whereas in micellar chromatography there is a more complex distribution pattern of solutes between micellar aggregates, the bulk solution, and the stationary phase.

The extension of partition theory of Martin and Synge to micellar liquid chromatography has been developed by Armstrong and Nome [1] assuming a three phase model to describe the chromatographic system.

This model, which accounts for the observed behaviour of a great variety of compounds eluted with aqueous micelles, allows the correlation of partition equilibria and retention with surfactant concentration in the mobile phase through quite simple basic formulas.

Several advantages of employing aqueous micellar eluents have been demonstrated: enhanced selectivities, lower costs, more safety, specific detection modes, possibility of direct injection of biological samples, and fast gradient elution capabilities. The peculiar features of these new reverse-phase systems have been analyzed in some comprehensive reviews [2–8].

This chapter will summarize the basic concepts underlying micellar chromatography, with particular attention to the role of surfactant molecules and their aggregates upon the retention mechanisms. The performances and limits of each technique will be examined in the light of existing theories.

6.1. MICELLAR HPLC

The first report on the use of micellar mobile phases concerned the separation of some ribonucleic acids using gel permeation chromatography [9]. Starting from this initial work, aqueous micellar phases were successively employed in TLC and, later, in HPLC.

The theoretical background of micellar liquid chromatography (MLC), also termed pseudophase liquid chromatography (PLC), has been mainly developed working on HPLC. Thus, the following sections will mainly concern this technique.

6.1.1. Partition equilibria in MLC: The three-phase partition model

Two types of analytes will be separately considered: (i) neutral molecules, and (ii) dissociated compounds.

6.1.1.1. MLC elution of neutral solutes

In conventional RP-HPLC the retention of uncharged solutes is governed by its partition coefficients between the mobile and the stationary phase, whereas in MLC, also termed pseudophase liquid chromatography (PLC), three competitive partition equilibria regulate the solute distribution, as indicated in Fig. 6.1.

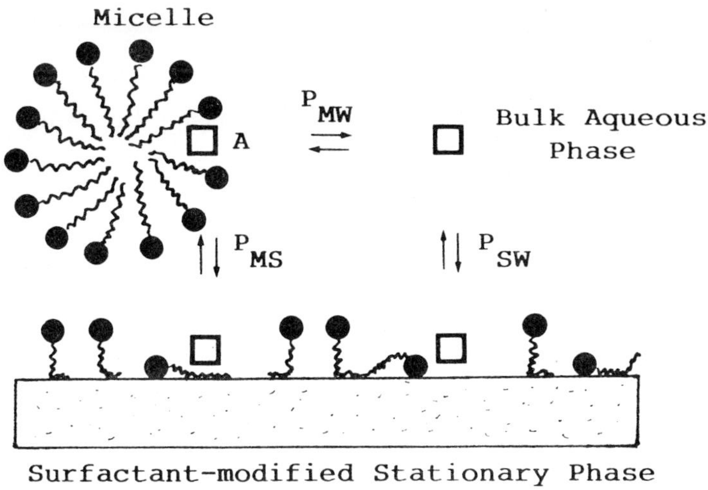

Fig. 6.1. Three phase distribution model of MLC. P_{MW}, P_{SW}, and P_{MS} are the partition coefficients of the analyte, A, between the indicated phases.

According to the Armstrong–Nome assumptions, the elution behaviour of a great variety of solutes can be explained in terms of column and micellar parameters and of the partition coefficients between the involved phases.

As in traditional HPLC, the column is divided in many theoretical plates. The following parameters have the usual meaning: V_s = volume of the stationary phase, V_m = volume of the mobile phase (void volume of the column), V_t = total volume of the column, V_e = elution volume of a given analyte. The mass balance of solutes in each theoretical plate accounts for their distribution between the three phases depicted in Fig. 6.1.

The final form of the derived chromatographic equation is the following

$$\frac{V_s}{(V_e - V_m)} = \frac{\overline{V}(P_{MW} - 1)\, C_m}{P_{SW}} + \frac{1}{P_{SW}} \qquad (1)$$

where $\overline{V}$ is the partial molar volume of the surfactant and C_m is the molar concentration of the micellized amphiphile.

A modified form of Eqn. (1) describes the variation of the capacity factor, k′, as a function of the surfactant concentration

$$\frac{1}{k'} = \frac{\overline{V}(P_{MW} - 1)\, C_m}{\phi P_{SW}} + \frac{1}{\phi P_{SW}} \tag{2}$$

where $k' = (V_e - V_m)/V_m$; ϕ is the chromatographic phase ratio, defined as V_s/V_m.

A similar equation can be derived taking into account the binding constants of solutes to the stationary phase and the "concentration" of binding sites in this phase:

$$\frac{1}{k'} = \frac{K_B C_m}{K_S \phi C_s} + \frac{1}{K_S \phi C_s} \tag{3}$$

where K_B and K_S are the solute binding constants to micelles and stationary phase, respectively. C_s represents the concentration of the stationary phase sites.

It can be demonstrated that the partition coefficient P_{SM} is not independent, but it arises from the ratio of the other two partition coefficients ($P_{SM} = P_{SW}/P_{MW}$). The complete derivation of Eqn. (1) is reported in Ref. [1].

In the absence of micelles, the partition equation valid for homogeneous solvents holds

$$V_e = V_m + P_{SW}\, V_s \tag{4}$$

A number of assumptions have been made in deriving Eqns. (1–3), which are indicated below:

– The compounds which interact positively (bind) with the micelles exhibit a decrease in retention increasing the surfactant concentration in the mobile phase.

– There is only one form of the solute present over the surfactant concentration range investigated.

– The solute forms a sort of complex with the micelle having 1:1 stoichiometry.

– The solutes which do not bind to the micelles show a constant retention (their elution volume is ca. equal to the void volume).

– The micellar parameters (cmc, size and shape of aggregates, aggregation number, etc.) do not change in the presence of solutes within the surfactant concentration range investigated.

– Above the cmc, the adsorption of surfactant onto the stationary phase is assumed to be constant because the concentration of free

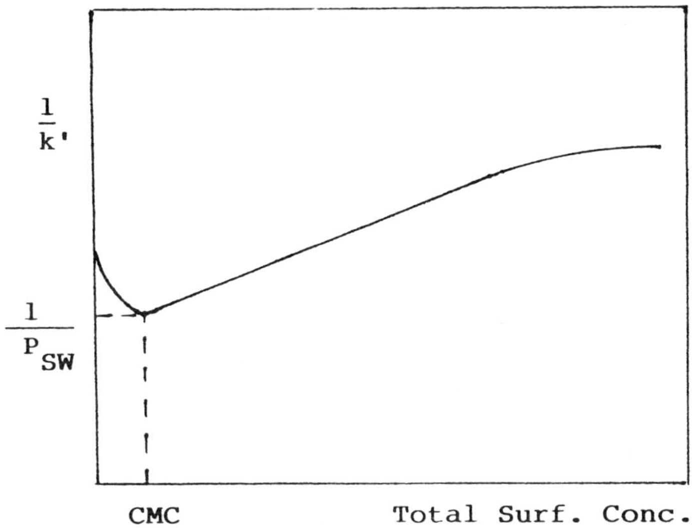

Fig. 6.2. Plots of 1/k′ vs. total surfactant concentration for partitioned solutes.

amphiphile molecules remains unchanged. This implies that the term P_{SW} does not vary with surfactant concentration.

Many uncharged solutes eluted with micellar solutions exhibit the retention dependence shown in Fig. 6.2, as predicted by the Armstrong–Nome model. At higher surfactant concentrations, their retention volumes approach the column void volume.

According to the model, retention of solutes decreases with increasing micelle concentration in the mobile phase, but the rate of decrease varies markedly depending on the solute nature. Moreover, inversions in the retention order of analytes are observed.

Figure 6.3 shows this particular effect, which can be attributed to the presence of two competing equilibria, namely solute solubilization in micelles (regulated by the binding constant), and solute–stationary phase interaction (dependent on P_{SW}). For example, nitrobenzene is eluted before 2-naphthol using 0.04 M SDS, while the elution order is reversed when working with surfactant concentrations higher than 0.1 M.

When the micelle concentration is increased, the fraction of solute in the mobile phase increases largely for strongly bound compounds, which exhibit steeper slopes in figure. Little or no effect is expected on the P_{SW} term if a surfactant saturated stationary phase is assumed.

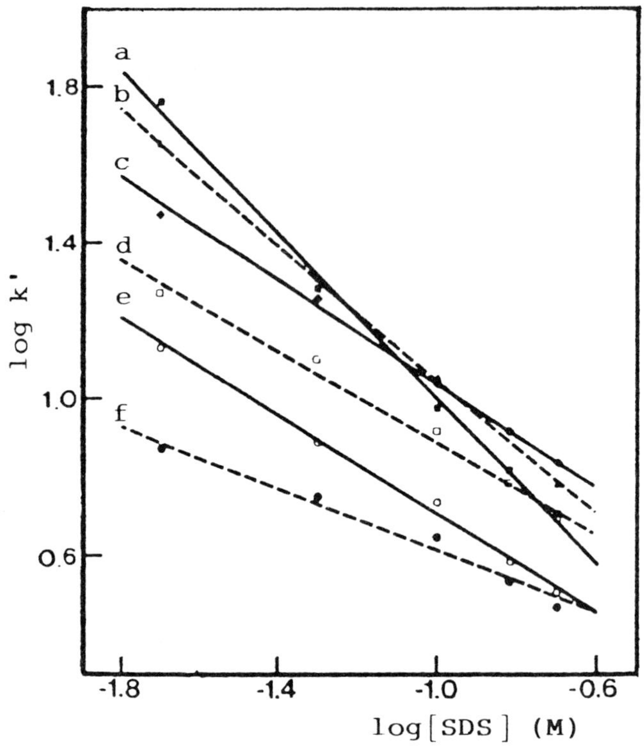

Fig. 6.3. Effect of SDS concentration on retention of some aromatic compounds: (a) 2-naphthol; (b) toluene; (c) nitrobenzene; (d) benzene; (e) 4-nitroaniline; (f) phenol. Figure taken from Ref. [10]. Reprinted with permission.

The analysis of chromatographic data for a variety of organic compounds eluted with micellar mobile phases can be performed using the Armstrong–Nome approach. Table 6.1 lists a series of undissociated solutes which obey to the partition model and that can be separated using MLC.

6.1.1.2. MLC elution of ionizable species

A more complex approach is necessary to describe the chromatographic behaviour of charged solutes in the presence of ionic micelles.

The influence of dissociation equilibria of weak acids and bases on binding constants with micelles has been investigated by Arunyanart and Cline Love [16]. For a monoprotic acid HA and its conjugate base

TABLE 6.1

Selected examples of HPLC separations of neutral organic solutes using aqueous micellar mobile phases

Compounds	Mobile phase	Column	Ref.
Phenols, PNA hydrocarbons	SDS	RP-C_{18}	11
Dihydroxybenzenes, nitroaromatics, naphthalene	SDS	RP-C_{18},-CN	1]
Phenol, naphthol, benzene, nitro-aromatics, arylamines	SDS, DTAB	RP-C_1	10
Alkylphenols, catechols, quinols	SDS	RP-C_{18}	12
Aromatic hydrocarbons, 1-bromo-naphthalene	SDS	RP-C_{18}	13
Aniline derivatives	HTAB	RP-C_{18}	14
Toluene, caffeine, benzoic acid, HPC[a], SOBS[b], BTAB[c]	SDS, HTAB	Silica, RP-C_1, -C_8, -C_{18}, -CN	15

a: Hexadecylpyridinium chloride; *b*: sodium octylbenzene sulfonate; *c*: benzyltrimethylammonium bromide.

A^-, both present in the column, the equilibria scheme shown in Fig. 6.4 accounts for the solute transfer between the phases and for acid dissociation.

Taking into account the equilibrium constants of each process and assuming that the direct transfer of solutes from micelles to stationary phases is negligible, the model predicts well the dependence of retention upon pH and surfactant concentration when both the acid and its conjugate base (or a base and its conjugated acid) are present. For a weak acid HA, the following equation holds:

$$k' = k'_{(HA)}B + k'_{(A^-)}C \qquad (5)$$

where $k'_{(HA)}$ and $k'_{(A^-)}$ are the limiting capacity factors of the neutral and dissociated form, respectively. The terms B and C are, in turn, given by the equations: $B = 1 + K_{B(HA)}C_m$, and $C = (1 + K_{B(A^-)})K_{a(app)}[H^+]^{-1}$. When only one form is present, Eqn. (5) reduces to Eqn. (2).

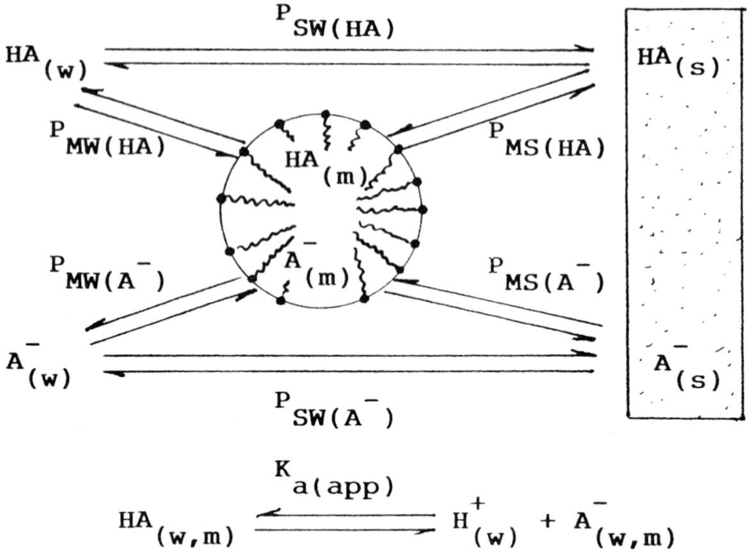

Fig. 6.4. Distribution scheme of a dissociated monoprotic acid in MLC.

According to Eqn. (5), a sigmoidal-type dependence of k′ on pH (at constant surfactant concentration) must be observed in the presence of ionized solutes (see Fig. 6.5).

6.1.2. Deviations from the Armstrong–Nome model

On the basis of their behaviour in the presence of micelles, the analytes can be classified into three groups [17]:

(1) Solutes which bind to the micelles (binding compounds), having $P_{MW} > 1$.
(2) Solutes which are not repelled from the micelles, but exhibit $P_{MW} \approx 1$ (non-binding solutes).
(3) Solutes which are excluded from the micellar phase, characterized by $P_{MW} < 1$ (antibinding compounds).

Analytes which deviate from the Armstrong–Nome model generally belong to the groups 2 or 3. Moreover, among the binding solutes, a parabolic retention behaviour has been observed in a few cases.

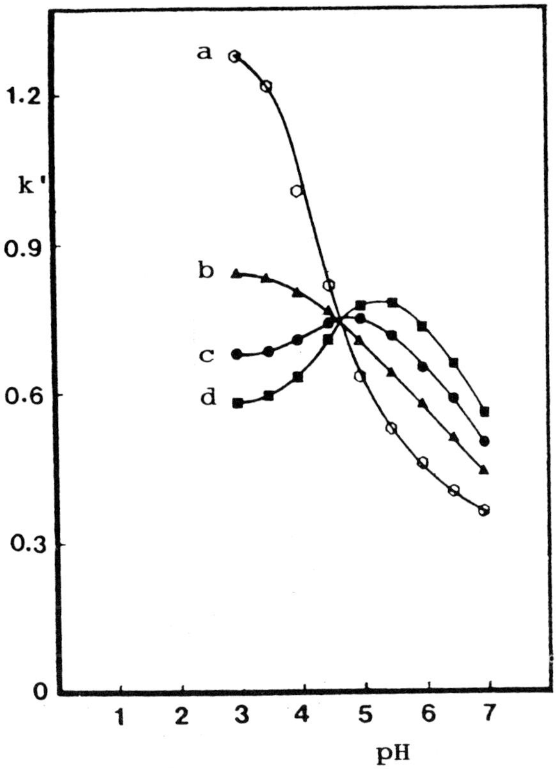

Fig. 6.5. Retention variation for benzoic acid as a function of pH. Column: RP-CN. Flow rate: 2.0 ml/min. SDS concentrations: (a) 0.02 M; (b) 0.05 M; (c) 0.08 M; (d) 0.10 M. Phosphate buffer. Figure taken from Ref. [16]. Reprinted with permission.

Figure 6.6 shows the dependence of retention parameters on micelle concentration exhibited by the different classes of compounds cited above. When plots with negative slopes are obtained, negative binding constants, which have no physical meaning, but indicate repulsive interactions between solutes and micelles, must be assumed.

Moderately hydrophobic compounds having the same electrical charge as the aggregates can show antibinding behaviour, which can be explained on the basis of a simple exclusion mechanism. For example, 2-naphthalenesulfonate and 2-naphthol-6-sulfonic acid are repelled from the SDS micellar surface and excluded from the aggregates in the mobile phase. When the number of micelles is increased, the solute fraction in the

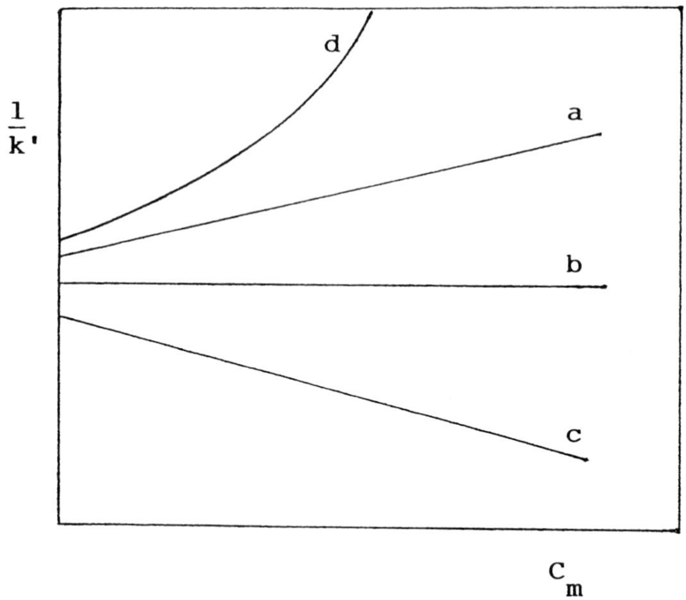

Fig. 6.6. Retention behaviour of solutes in MLC. (a): Binding; (b): non-binding; (c): antibinding; (d): anomalous binding.

stationary phase becomes larger and the retention increases. These repulsive interactions tend to disappear increasing the ionic strength.

Lines of zero slope indicate that there are no interactions between the solute and the micelle (non-binding compounds). If attractive and repulsive forces are present, they compensate one another.

The deviations indicated as anomalous binding (Fig. 6.6, curve _d_), have been attributed to a simultaneous interaction between one solute molecule and more than one micelle. Some bulky macromolecules, such as proteins, exhibit this behaviour [18]. A proper retention equation has been derived [2] starting from Eqn. (3)

$$\frac{1}{k'} = \frac{1}{K_S \phi C_s} + \frac{K_B C_m}{K_S \phi C_s} + \frac{K_B K'_B C_m^2}{K_S \phi C_s} \qquad (6)$$

where K'_B is the binding constant between the analyte and the second aggregate.

6.1.3. Selectivity and retention mechanisms in MLC

6.1.3.1. Selectivity

Selectivity is defined in LC as the difference between the retention of two given solutes. It is measured by the selectivity factor, defined as the ratio between the capacity factors of the examined solutes.

Due to the unique nature of the mobile phase, the solute–micelle interactions play a very important role in selectivity. It is possible to regulate not only the hydrophobic effects (as in traditional HPLC), but also electrostatic attraction–repulsion, hydrogen-bonding, dipole–dipole forces and other more specific interactions by changing the micelle concentration and/or the aggregate charge.

Changes in pH and ionic strength of micellar eluents can be used to adjust the retention of analytes having acid or basic character. The presence of additives can also alter the solute binding constants and influence selectivity.

Selectivity is strongly dependent on the eluent strength, which is defined in HPLC by the following expression [19]:

$$\ln(k'_1/k'_2) = a(n^o\phi_2 - n^o\phi_1) \qquad (7)$$

where k'_1 and k'_2 are the capacity factors of a solute measured in a given column using the eluents 1 and 2, respectively. The term $\underline{a}$ represents the solute molecular area, whereas n^o is defined as the eluent strength parameter.

Taking methanol as a reference solvent ($k'_1 = 1$), it has been found that micellar mobile phases are weaker eluents (see Table 6.2). The

TABLE 6.2

Eluent strength of some micellar solutions

Mobile phase	n^o
Methanol	1.000
Methanol water (90–10 v/v)	0.888
SDS 0.01 M	0.208
SDS 0.05 M	0.321
SDS 0.10 M	0.390
HTAB 0.01 M	0.199
HTAB 0.05 M	0.336

Data taken from Ref. [20].

eluent strength can thus be adjusted by changing the micelle concentration and/or the nature of the surfactant.

Recently, a more detailed model was developed which explains the dependence of the solvation capacity of organic solvents in MLC and predicts the extent of interaction between solutes and micelles [21]. The model accounts for the selectivity variation observed when the volume fraction of the organic solvent is changed (at constant micelle concentration) and for the effect of surfactant concentration at constant solvent content.

In some cases, as for substituted benzenes and aminoacids, it was found that these effects are opposite. Good separations can thus be obtained by varying the surfactant/organic solvent ratio.

6.1.3.2. Retention mechanisms in MLC

In conventional RP-HPLC, studies on the retention mechanism are mainly based on the analysis of chromatographic behaviour of homologous series of compounds bearing the same functional moiety and a tuned hydrophobic substituent, which is usually an alkyl chain.

Within these series, a linear relationship exists between log k′ and the number of carbon atoms of the substituent. The regular increase of retention associated with the addition of a methylene group in the hydrocarbon chain allow us to split the retention dependence into two contributions, one arising from the common polar group (termed specific retention, β) and the other coming from the increasing number of methylene groups (non-specific or hydrophobic contribution, α).

The following expression is valid in RP-HPLC

$$\log k' = \log \beta + n_C \log \alpha \tag{8}$$

where n_C is the number of methylene groups in the normal alkyl chain.

In contrast with Eqn. (8), a linear relationship between k′ and n_C has been observed in work with ionic [22] and non-ionic micellar eluents [23]. A parabolic equation relates log k′ with n_C in these cases, as shown in Fig. 6.7.

An equation has been derived from the basic definition of k′ by Hinze and Weber [24] to explain the nonlinear relationship between log k′ and n_C. The solvophobic basis [25] for the distribution of solutes is, of course, the same in micellar chromatography as in other similar physical systems such as reversed phase liquid chromatography. What makes the micellar system different is the two-phase nature of the mobile

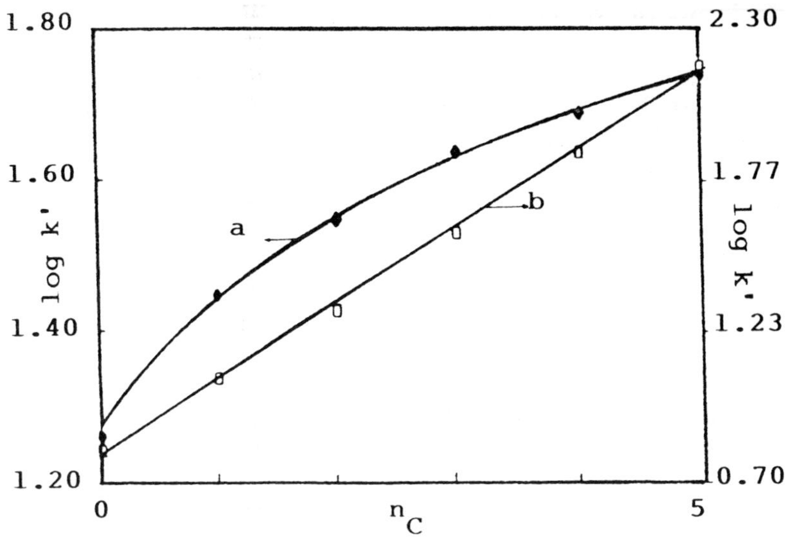

Fig. 6.7. Comparison of retention dependence on n_C in MLC and in conventional RP-HPLC systems. Analytes: alkylbenzenes. (a): 0.072 M HTAB; (b): 60% methanol in water. Figure taken from Ref. [22]. Reprinted with permission.

phase. The influence of this on k′ is more easily appreciated by using a simple equation for k′:

$$1/k' = V_W/V_S P_{WS} + V_X/V_S P_{XS} \qquad (9)$$

where the volumes V are for W, solvent part of mobile phase; S, stationary phase; and X, micellar phase, and the partition coefficients are for the processes (WS) solvent part of the mobile phase to stationary phase, (XS) micelles to stationary phase. In this expression, the two processes contributing to retention, WS and XS, each have a term.

As phases X and S are chemically similar, P_{XS} does not vary too much with solute hydrophobicity. On the other hand, P_{WS} varies greatly. Highly hydrophobic solutes have a large P_{SW}, so the first term in Eqn. (1) becomes negligible. Then, k′ becomes proportional to P_{XS}. The dependence of P_{XS} on the number of methylene groups is small [24]. For more water soluble solutes, P_{WS} is small, the first term is large and dominant, and k′ becomes proportional to P_{WS}. Thus, as the solute's hydrophobicity increases, the dependence of k′ on n_C decreases, and a linear dependence of log k′ on n is not expected.

6.1.4. Efficiency in MLC

One of the major limitations of MLC is its low efficiency, which has been initially attributed to the poor wettability of common non-polar stationary phases by the aqueous surfactant solutions. In several cases, the efficiency can be improved by adding suitable organic modifiers to the mobile phase and raising the working temperature.

In the absence of added modifiers, the elution peaks obtained when working with aqueous micellar eluents are broader than those observed using hydroorganic solvents. Moreover, they often show a skewed profile. A typical MLC elution pattern, obtained using an aqueous surfactant mobile phase, is shown in Fig. 6.8.

The evaluation of MLC efficiency, the influence of solute mass transfer processes within the column and the effect of physical and chemical factors will be separately examined.

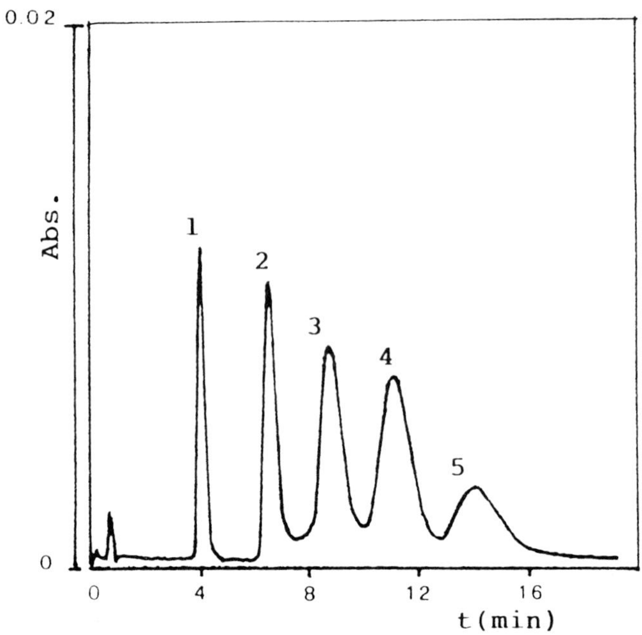

Fig. 6.8. MLC elution of some polymethylphenols. Stationary phase: RP-C_{18}: Eluent: 0.09 M SDS. Flow rate: 2.5 ml/min. Peaks: (a) phenol; (b) 4-methylphenol; (c) 3,5-dimethylphenol; (d) 2,4,5-trimethylphenol; (e) 2,3,5,6-tetramethylphenol. Figure taken from Ref. [12]. Reprinted with permission.

6.1.4.1. Efficiency evaluation

For Gaussian elution peaks, the efficiency is simply related to the number of theoretical plates, N, through the equation:

$$N = (t_R/\sigma)^2 \tag{10}$$

where t_R is the retention time and σ is the standard deviation. For a perfect Gaussian peak profile, the peak width at 60% of peak height ($W_{0.6}$) equals 2σ, then Eqn. (10) can be written as follows:

$$N = 4(t_R/W_{0.6})^2 \tag{11}$$

In micellar HPLC, most solutes exhibit a tailored peak shape. In these cases, the simple method for calculation of plate number as described by Foley and Dorsey [26] can be applied:

$$N = \frac{41.7(t_R/W_{0.1})^2}{\{(B/A) + 1.25\}} \tag{12}$$

The peak width is measured at 10% of the peak height ($W_{0.1}$). The vertical line traced through t_R divides the peak width into two segments: A (to the left) and B (to the right). Then, $W_{0.1} = A + B$, and B/A is a measure of the peak asymmetry.

6.1.4.2. Influence of molecular diffusion in the mobile phase

The diffusion coefficients of analytes are strongly affected by the presence of surfactants. According to Armstrong and coworkers [27], the observed diffusion coefficient of a solute in the presence of micelles (D_{exp}) can split into two contributions, one due to the diffusion coefficient in bulk water (D_w), measured below the cmc, and the other arising from the diffusion coefficient of the aggregate (D_m):

$$D_{exp} = \alpha\, D_m + (1 - \alpha)D_w \tag{13}$$

where α is the percent of solute bound to the micelles, $(1 - \alpha)$ is the fraction of solute present in the bulk phase. The observed diffusion coefficient can be rewritten as follows:

$$D_{exp} = \frac{D_m}{1 + K} + \frac{D_w}{1 + 1/K} \tag{14}$$

TABLE 6.3

Variation of the diffusion coefficient of some solutes with surfactant concentration

Compound	D ($\times 10^6$ cm^2 s^{-1}) at SDS concentrations			
	0.0 M	0.1 M	0.3 M	0.5 M
2-naphthol	6.10	0.90	0.64	0,57
4-nitroaniline	7.46	2.28	1.22	0.90
4-nitrophenol	6.92	2.70	1.69	1.11
2,6-naphthalene-disulfonate	6.71	4.00	3.52	3.06
4,5-dihydroxy-naphthalene-2,7-disulfonate	6.14	3.49	2.96	2.52

Data taken from Ref. [27].

where $K = N_{AG}(1 - C_m\overline{V})/P_{MW}C_m\overline{V}$, and N_{AG} is the micelle aggregation number. Equations (13,14) are valid if cmc, aggregation number and geometry of micelles do not vary in the presence of the diffusing analytes.

Binding compounds exhibit a decrease of their diffusion coefficients with increasing surfactant concentration, which is faster than those of solutes repelled or excluded from the micellar pseudophase (see Table 6.3). For completely bound solutes, D approaches the diffusion coefficient of micelles (ca. 6×10^{-7} cm^2 s^{-1}).

Although unbound and strongly bound solutes exhibit very different diffusion coefficients, their corresponding MLC efficiencies are not significantly different, thus indicating a minimal influence of transport processes within the mobile phase.

Other investigators [28] explained the diminished efficiency of MLC in terms of a slow solute transfer from the micelles, combined with a slow solute desorption from the surfactant-modified adsorbent. The contribution of the first process becomes relevant only for very hydrophobic compounds, exhibiting very slow exit rates from micelles, whereas the mass transfer from/to the stationary phases always appear to play a crucial role.

6.1.4.3. Effect of mass transfer at the stationary phase

The adsorption of surfactant molecules can alter the thickness and microviscosity of the bonded-phase, influencing the solute retention and lowering the mass transfer rate. Moreover, the direct transfer of solutes from the micelles to the surfactant-coated stationary phase can be rate-determining in some particular cases. The influence of these processes upon efficiency will be separately considered.

Surfactant adsorption modes

A systematic study of the adsorption isotherms of SDS and HTAB on stationary phases of various polarities was performed by Berthod and coworkers, both in the absence [29] and in the presence [30] of organic additives and salts. If the adsorbed amount of these surfactants on hydrophobic stationary phases (i.e., RP-C_{18}) is plotted against the total amphiphile concentration, a typical H-isotherm profile is observed, with the adsorption plateau reached just above the cmc. The maximum quantity of adsorbed surfactant is approximately the same for SDS and HTAB (see Fig. 6.9).

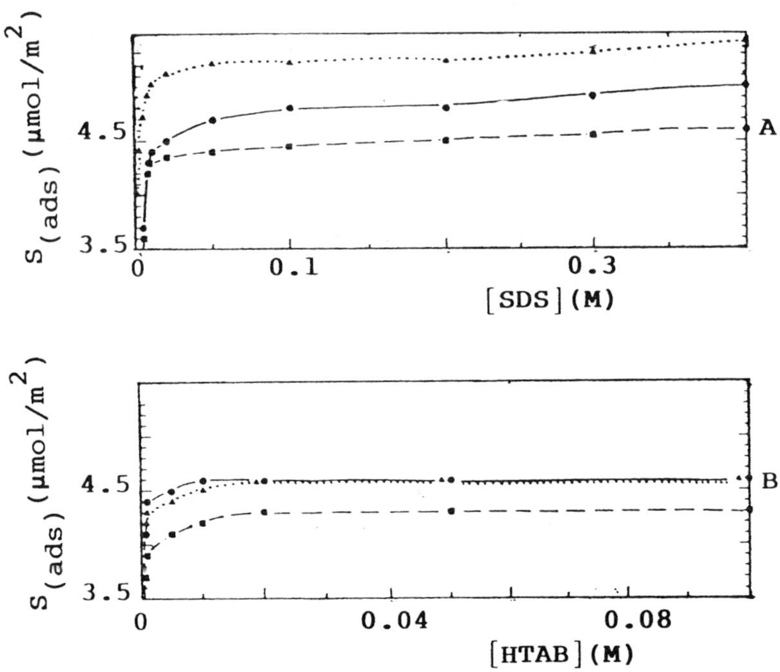

Fig. 6.9 Adsorption isotherms of SDS (A), and HTAB (B), on ODS Hypersil stationary phases. (———): aqueous solution, (···): NaCl 0.1 M solution; (- - -): methanol–water (5–95% v/v). Figure taken from Ref. [30]. Reprinted with permission.

Working with more polar bonded phases or with silica, further surfactant adsorption occurs above the cmc. A similar behaviour is also shown by non-ionic surfactants on hydrophobic packings [31], where the amount of loaded amphiphile can be as great as that of the bonded phase itself. Clearly, the partition coefficient P_{SW} is not constant in these cases.

Different adsorption mechanisms can be invoked to explain the above observations:

– Hydrophobic adsorption, when the interaction occurs through the non-polar hydrocarbon chains of silanized silica and surfactant, respectively.

– Mixed adsorption, due to silanophilic interactions with the charged surfactant head groups and simultaneous hydrophobic interactions through the tails.

The adsorption of SDS on C_{18}-bonded stationary phases is an example of purely hydrophobic mechanism, whereas HTAB can be adsorbed onto the same stationary phases according to a mixed adsorption scheme. The features of these different interactions are depicted in Fig. 6.10.

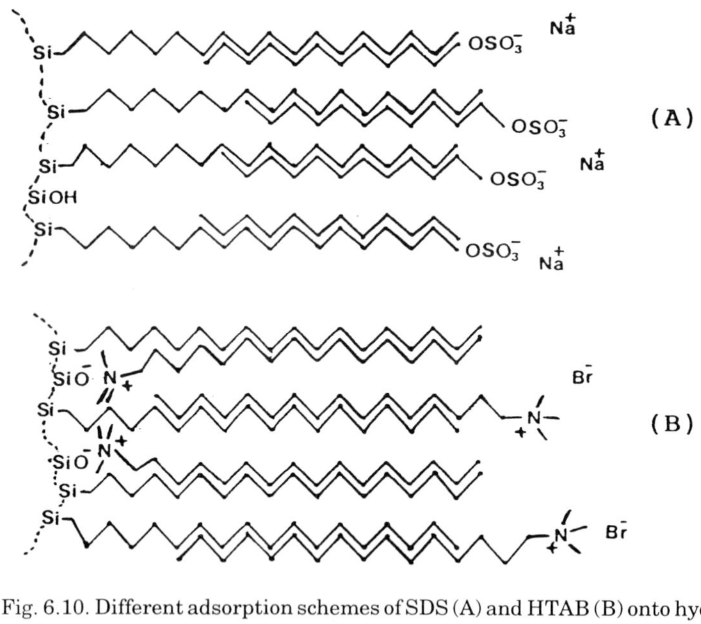

Fig. 6.10. Different adsorption schemes of SDS (A) and HTAB (B) onto hydrophobic RP-C_{18} stationary phases. Taken from Ref. [5]. Reprinted with permission.

According to this model, the experimental data can be qualitatively explained. For example, ionic surfactants are more adsorbed on C_1- than on C_{18}-bonded phases due to a mixed interaction mechanism, whereby the silica OH groups are more accessible to surfactant ionic heads.

Since the surfactant-modified stationary phase becomes more or less hydrophobic depending on the adsorption mode, analyte retention is also influenced.

The effect of surfactant adsorption on column efficiency can be evaluated considering the last term of the Snyder–Kirkland equation [32]:

$$H = C_e d_p + \frac{d_p^2 u C_m}{D_m} + \frac{D_m C_d}{u} + \frac{C_{sm} d_p^2 u}{D_m} + \frac{C_s d_f^2 u}{D_s} \qquad (15)$$

The C values are constant plate height coefficients related to different mass transfer processes: eddy diffusion (e), mobile phase mass transfer (m), longitudinal diffusion (d), stagnant mobile phase mass transfer (sm), and stationary phase mass transfer (s). d_p is the diameter of the packing particles, d_f is the thickness of the stationary phase layer, D_m is the solute diffusion coefficient in the mobile phase, D_s is the solute diffusion coefficient in the stationary phase, u is the mobile phase velocity.

The thickness of the stationary phase layer, d_f, strongly depends on surfactant adsorption. In particular, if adsorption is relevant, the viscosity of the adsorbed layer increases, with a corresponding decrease of the diffusion coefficient D_s. When the amount of surfactant coated onto the packing material increases above the cmc (e.g., Brij 35 on RP-C_{18} columns [31]), both d_f and D_s change with surfactant concentration, leading to a further decrease of efficiency.

Direct transfer from micelles to the stationary phase

One of the basic assumptions of the MLC theory is that the term P_{MS} is small enough to be considered negligible and this implies that only the solute transfer from the bulk pseudophase to the surfactant-modified stationary phase (and vice versa) plays a significant role in retention. A lot of published data concerning MLC is consistent with this approach, although for very hydrophobic compounds it is conceivable that the exit from the micelle would be a difficult process.

In order to explain the chromatographic behaviour of water-insoluble compounds, a direct transfer model based on sorption of solute-contain-

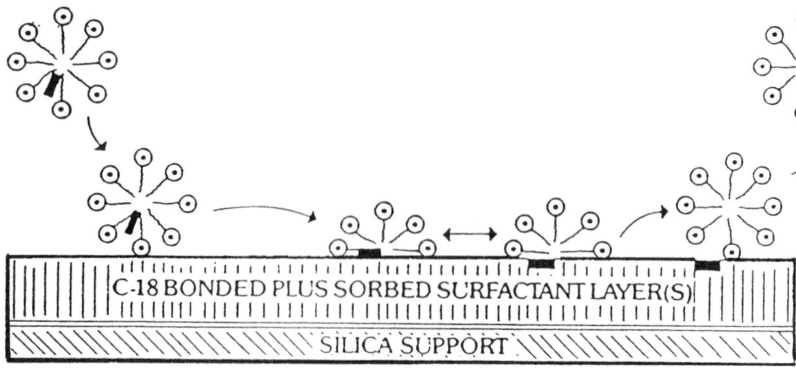

Fig. 6.11. Direct transfer model for water-insoluble compounds. Reproduced from Ref. [23] with permission.

ing aggregates onto the hemimicellar structured layer has been proposed by Hinze and coworkers [23] (see Fig. 6.11).

A modified multipartition equation has been derived assuming that these solutes are strongly bound to the aggregates. Consequently, the term $P_{MW} - 1$ in Eqn. (2) tends to P_{MW}. Moreover, since P_{MS} is given by P_{SW}/P_{MW}, the following simplified retention equation holds:

$$\frac{1}{k'} = \frac{\bar{V}C_m}{\phi P_{SM}} + \frac{1}{\phi P_{SW}} \qquad (16)$$

If the examined solutes are water insoluble, P_{SW} tends to a very high value, then, the capacity factor can be expressed as follows:

$$k' = \frac{\phi P_{SM}}{VC_m} \qquad (17)$$

Equation (17) predicts a linear variation of k′ with $1/C_m$ (with a zero intercept) for water insoluble compounds, which are assumed to follow the direct transfer model. The validity of the above assumptions has been confirmed experimentally for a series of homologous lipophilic compounds [23].

Recent contributions to the understanding of mass transfer in mobile and stationary phases in MLC have been published [33,34]. The effects of organic modifiers on surfactant loading and on mobile phase properties have been recognized as major factors, although their relative importance is controversial.

6.1.4.4. Influence of experimental parameters on MLC efficiency

Effect of additives

The amount of sorbed surfactant on the stationary phase decreases in the presence of organic additives, such as aliphatic alcohols, alkanes, alkylnitriles, alkanediols, tetrahydrofuran, etc. This fact has been largely exploited to improve the column efficiency [20,35,36].As shown in Fig. 6.12, efficiencies comparable to hydro-organic eluents can be obtained.

Aliphatic alcohols are often used as additives, the observed efficiency improvement being inversely proportional to their polarity. For example, the number of theoretical plates for benzene eluted on a RP-C_{18} column using 5×10^{-2} M SDS, increases from ca. 1000 (when 10% methanol is

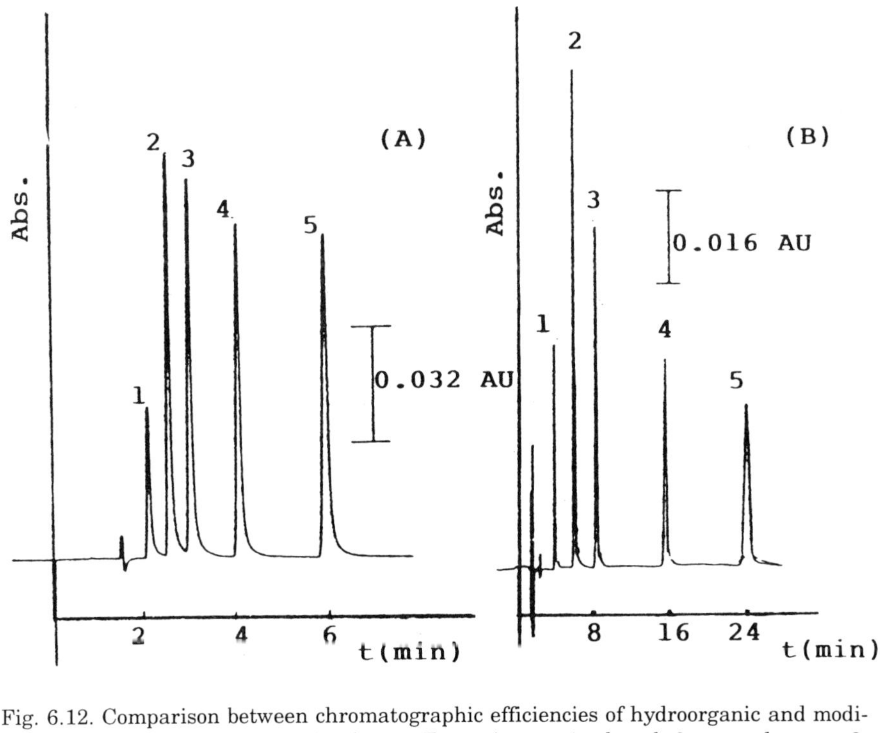

Fig. 6.12. Comparison between chromatographic efficiencies of hydroorganic and modifier-containing surfactant mobile phases. Test mixture: 1: phenol; 2: acetophenone; 3: nitrobenzene; 4: benzene; 5: toluene. Column: ultrasphere ODS. Flow rate: 1.5 ml/min. Eluent: (A): 70/30 methanol/water; (B). 10/90 1-propanol/0.1 M SDS solution. Figure taken from Ref. [35]. Reprinted with permission.

added) to more than 2200 (if 10% of 1-propanol is present) [35]. The beneficial effects can be attributed to the displacement of part of the adsorbed surfactant from the stationary phase due to the modifier adsorption.

A smaller film thickness around the solid particles can be obtained in these conditions, with a corresponding higher fluidity of the modified stationary phase. The solute transfer process is then facilitated.

Moreover, the presence of organic modifiers can favour the exit of hydrophobic solutes from the micelles because the bulk phase becomes less polar and the solubility of the analytes is enhanced. If the added modifier is very hydrophilic (e.g. methanol) it does not associate with the micelles, but increases the surfactant self-diffusion and, thus, the solute exit rates.

It has been demonstrated by Tomasella et al. [37] that the organic modifier shifts the equilibrium of the solutes away from the micelles and

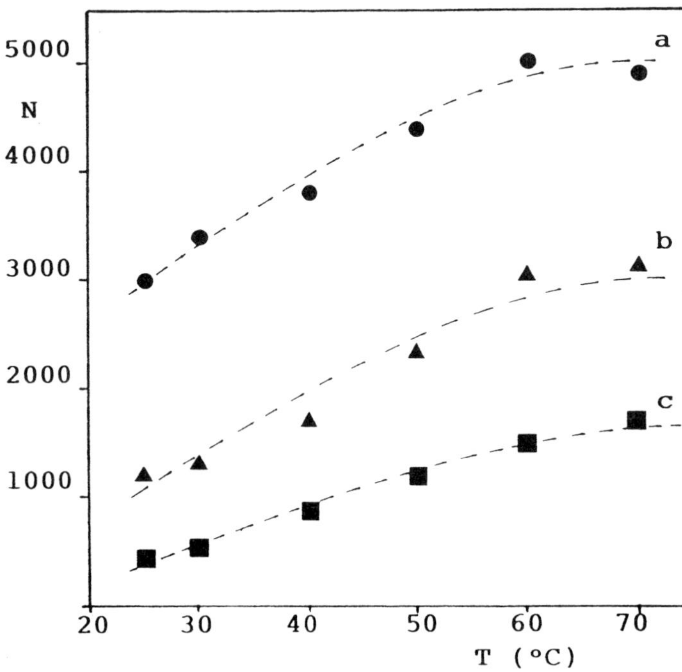

Fig. 6.13. Thermal effects on MLC efficiency. (a): benzene; (b): naphthalene; (c): anthracene. Column: RP-C_1. Mobile phase: 0.1 M DTAB. Flow rate: 2.0 ml/min. Data taken from Ref. [28].

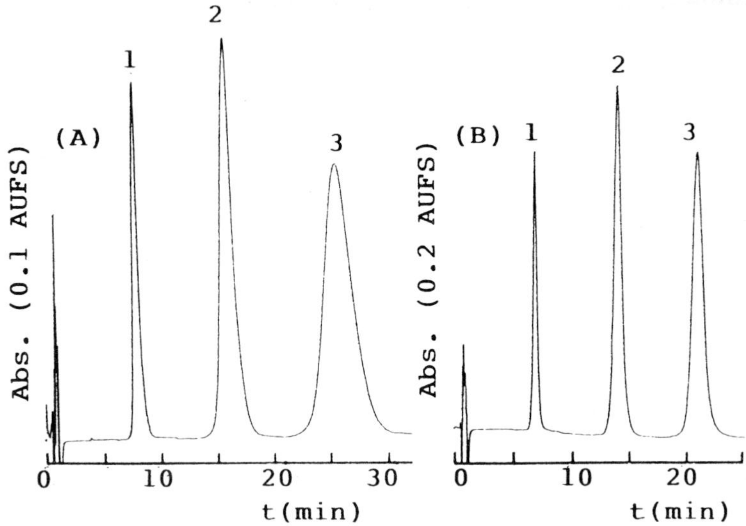

Fig. 6.14. Elution curves of benzene (1), naphthalene (2) and anthracene (3) in 0.1 M SDS. Column: RP-C_8. Flow rate: 2.0 ml/min. (A): 25°C; (B): 60°C. Figure taken from Ref. [28]. Reprinted with permission.

the stationary phase and toward the bulk aqueous phase. Consequently, the additive alters both the efficiency and the chromatographic selectivity.

Temperature effects

If the column temperature increases, all the mass transfer processes are improved because their overall rate constants are increased and the medium viscosity is lowered. The positive effects of temperature on MLC efficiency are shown in Fig. 6.13, where the number of theoretical plates (N) is reported as a function of T for some aromatic hydrocarbons eluted with DTAB.

In Fig. 6.14 the elution profiles of the same analytes (obtained by working with SDS at different temperatures) are compared.

A shortening of column life is observed in MLC when working at elevated temperatures (higher than 50°C). This deleterious effect can be attributed to the increase of silica solubility in the mobile phase.

The use of thermostated precolumns packed with silica and placed between the pump and the injector, can help to diminish these dissolution effects.

6.1.5 Micellar HPLC of organic and inorganic anions

In cases where the compounds to be separated are ionizable hydrophilic species, ion-pair and ion-interaction chromatography are often employed.

When ionic micellar solutions are used as mobile phases, the adsorption of amphiphile monomers onto the stationary phase confers an ion-exchange capability to this phase, and electrostatic attractions between the charged surface and the oppositely charged analytes can be exploited to regulate the retention (ion–interaction mechanism).

The efficient separation of several hydrophilic dithiocarbamate salts (largely applied as pesticides) and their degradation products has been achieved using cyano-bonded columns and micellar HTAB solutions as eluents [38–40]. Some common inorganic anions, usually present in natural and polluted waters, can be similarly separated using aqueous HTAC micellar solutions and octadecyl bonded silica (ODS) as stationary phase [41].

The elution order is the same as that observed in ion-exchange experiments using strongly basic anionic exchangers, namely: $I^- > NO_3^- > Br^- > NO_2^- > IO_3^-$ (see Fig. 6.15).

If the surfactant concentration is well above the cmc and the stationary phase loading is constant, the ion-exchange capacity of the column does not vary by changing the number of micelles in the eluent. The addition of organic modifiers, such as alcohols, reduces the cmc values and speed up the analysis.

This MLC technique, coupled with a simple UV detector, can be applied to separate several anions present in aqueous samples, and represents an alternative to the use of ion chromatography.

The characteristic features and performances of MLC and ion-pair chromatography (IPC) have also been compared [42]. From these studies MLC appears as a good alternative to IPC in order to achieve efficient and rapid separation of several ionic analytes.

6.1.6. Capabilities and performances of micellar HPLC

The unique features of micellar phases allow the practitioners to design and conduct chromatographic experiments using detection systems and operation modes which are not usually compatible with traditional hydro-organic mixtures. Moreover, information about the solute–aggregate interactions can be obtained from MLC retention data.

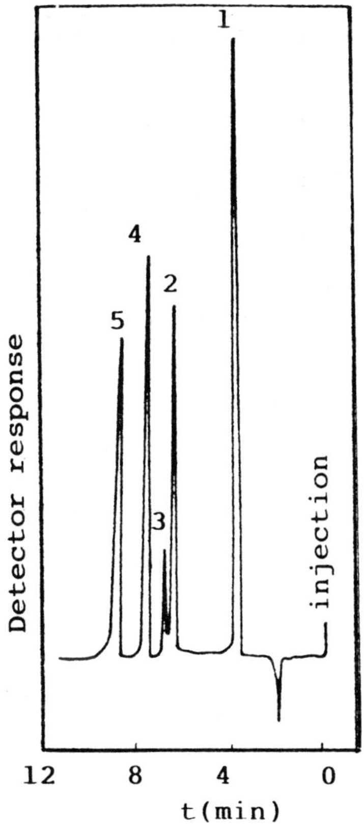

Fig. 6.15 Separation of inorganic anions using MLC. Column: C_{18} bonded silica. Eluent: 0.136 M HTAC. Flow rate: 1.5 ml/min. Detection: UV at 205 nm. Analytes: (1) iodate; (2) nitrite; (3) bromide; (4) nitrate; (5) iodide. Figure taken from Ref. [41]. Reprinted with permission.

The following aspects of MLC are largely exploited and will be examined in more detail: (i) gradient elution capabilities; (ii) possibility of direct injection of complex samples into the column; (iii) compatible detection systems; (iv) analysis of solute–micelle partition equilibria from MLC data.

6.1.6.1. Gradient elution in MLC

The limited use of gradient elution in traditional LC is due to the solvation structure changes of the stationary phase with solvent

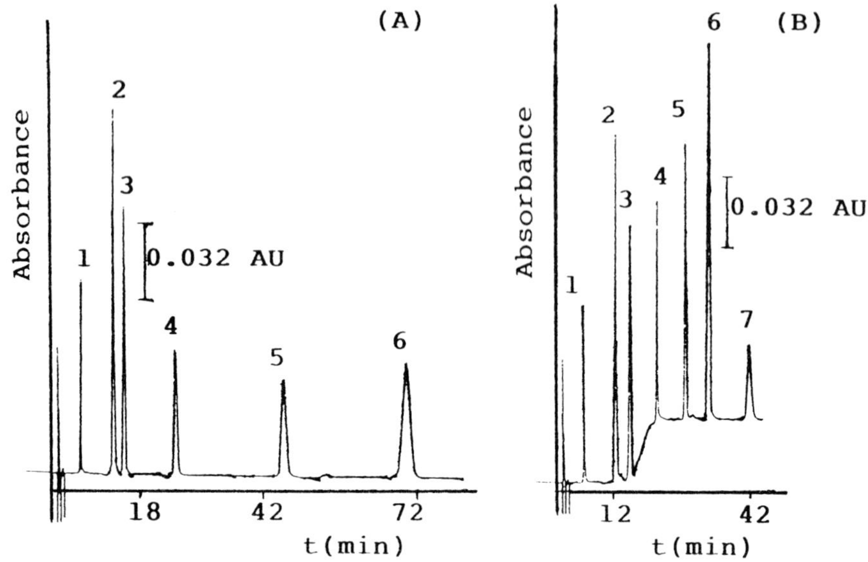

Fig. 6.16. Separation of seven-component test mixture. Ultrasphere ODS column; mobile phase: 0.05 M SDS plus 3% 1-propanol; flow rate 1.5 ml/min. (A): isocratic elution. (B): From 12 min, 2 min-linear gradient to 0.20 M SDS. Peaks: (1) phenol; (2) acetophenone; (3) nitrobenzene; (4) benzene; (5) toluene; (6) ethylbenzene; (7) anthracene. Figure taken from Ref. [36]. Reprinted with permission.

composition during the experiments. After a gradient separation, many column volumes of the initial mobile phase (usually 10–30) are required to re-equilibrate the stationary phase prior to the start at a new analytical run. This step is very time-consuming.

Using proper micellar eluents coupled with suitable adsorbents, rapid gradient elutions can be performed and re-equilibration becomes unnecessary.

Dorsey and coworkers [35,36] first recognized this unique MLC capability and attributed the favourable results to the constancy of the amount of surfactant adsorbed onto the stationary bonded phases above the cmc. Although this assumption is not always valid, it is certainly true in some cases (see Section 6.1.5). In these cases the re-equilibration steps can be avoided.

If gradient elution is performed with micellar eluents containing organic modifiers (e.g. alcohols), very efficient separations can be achieved in shorter times. Figure 6.16 illustrates the MLC separation of the components of a test mixture using both isocratic and gradient elution modes.

Another fundamental advantage offered by gradient elution MLC is the possible application of electrochemical detection methods.

It is known that the electrochemical detectors are usually incompatible with gradient elution in conventional HPLC techniques due to the changes in conductance, pH and viscosity of the mobile phase during the elution steps. In these hydroorganic systems, the Faradaic current arises from the electroactive analyte, while the non-Faradaic contributions (including the noise) depend on the nature and concentration of the mobile phase.

During the elution, the residual current changes continuously, thereby causing relevant shifts of the detection baseline. These effects are crucial when the applied potential is high ($\geq$ 1.2 V); then, very careful buffering and conductance balancing are required over the organic solvent/water gradient range.

These strict experimental requirements cannot always be satisfied when working with hydroorganic eluents because the solubility of salts and buffers is generally low in these solvents. Moreover, the conductance vs. composition curves are not linear, but show a minimum corresponding to the maximum viscosity of the mixture.

When ionic micellar solutions are used, the changes in the eluent properties can be more easily compensated. In fact, the conductance of the mobile phase is proportional to the amphiphile concentration, thus it can be balanced by dissolving suitable supporting electrolytes. If non-ionic micellar eluents are employed, the conductivity is exclusively due to the amount of dissolved electrolytes.

On the basis of the above considerations, it appears that gradient elution MLC is compatible with electrochemical detection and can be applied in current analytical work.

Figure 6.17 shows that effective separations are obtained using the technique. Note the very low residual current shift reported in this example (less than ca. 10 nÅ) during the gradient elution.

6.1.6.2 Application of MLC to the analysis of complex samples

Micellar eluents allow the direct injection of samples which are usually altered in the presence of organic solvents.

For example, the development of sensitive methods for HPLC analysis of metabolites and drugs in biological fluids requires a very careful elimination of proteins in order to avoid column clogging and matrix interferences. A prior extraction or precipitation step is necessary to

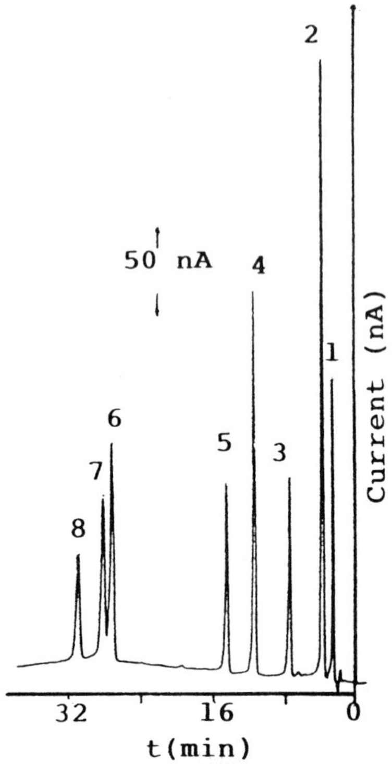

Fig. 6.17. Separation of phenolic compounds using MLC with gradient elution-electrochemical detection. Eluent (I): 0.05 M SDS, 3% 1-propanol, phosphate buffer (pH 2.5), $NaClO_4$; eluent (II): 0.112 M SDS, 3% 1-propanol, phosphate buffer (pH 2.5). Gradient: from (I) to (II) in 12 min. Peaks: 1 = hydroquinone; 2 = resorcinol; 3 = catechol; 4 = phenol; 5 = 4-nitrophenol; 6 = 2-nitrophenol; 7 = 4-chlorophenol; 8 = 4-bromophenol. Figure taken from Ref. [20]. Reprinted with permission.

separate the proteins from the sample, thus increasing the analysis time and the possibility of error.

Micellar HPLC provides a unique solution to these problems by solubilizing the proteins via a surfactant coating, making possible the direct injection of the samples onto the columns.

Most of the reported MLC separations, based on the direct injection approach, concern serum and urine samples. Table 6.4 lists some examples of drug monitoring in these fluids. (See also Refs. [126,127].)

TABLE 6.4

MLC analysis of biological samples using direct injection

Analyzed compounds	Experimental conditions	Samples/Ref.
Propranolol, quinidine, morphine	0.03 M SDS/0.03 M SDS + 10% 1-propanol; 10 μm RP-C_{18} (μ-Bondapak); Fluorimetric detection.	Serum, urine [43]
Acetaminophen, phenobarbital, quinidine, chloramphenicol, acetylsalicylic acid, phenytoin, carbamazepine, procainamide, theophylline	0.02 M SDS; 5 μm LC-CN (Supelcosil); pH 7; Detector: UV 254 nm.	Serum [44]
Cephalosporins	0.08 M SDS; 5 μm RP-C_{18} (Nucleosil); pH 3.3; Detector: UV 260 nm.	Serum [45]
Proline and other amino acids	0.03 M SDS + 10^{-3} M Cu^{2+} + 8% 1-propanol; 5 μm RP-C_{18}; pH 5.3; Detector: photodiode array, UV 235 nm.	Urine [46]
Methotrexate	0.1 M SDS; 5 μm RP-C_{18} (Lichrospher); pH 5.2–5.7; Detector: photodiode array, UV 254 nm.	Serum, urine [47]
6-Mercaptopurine, 6-thioguanine and their metabolites	0.04–0.2 M SDS; 10 μm RP-C_{18} (Lichrosorb); pH 3–7; Detector: UV 320/340 nm.	Serum [48]
Illegal drugs in sport: testosterone, oxandrolone, spironolactone, etc.	0.05 M SDS + 3% 1-butanol; 5 μm RP-C_{18} (Nucleosil); Detector: UV 254 nm.	Urine[49]
Dopamine	0.01 M SDS + 3% 1-propanol + 1 mM EDTA; 5 μm RP-C_{18} (Micropak); pH 4.15; Detector: amperometric.	Urine[50]

6.1.6.3. Detection modes in MLC

The UV-VIS detector, which is very popular in LC, can be applied to monitor the absorbance changes of the eluate when aliphatic-type surfactants are used as mobile phases. Working with aromatic amphiphiles, the use of UV detectors is certainly more difficult.

The presence of micelles usually modifies the spectra of analytes displacing their absorption maxima toward longer wavelengths and, in several cases, increasing the molar absorptivity coefficients. Sensitization effects in both UV and VIS regions have been previously examined (see Chapter 4).

Relevant advantages are related to the use of luminescence detectors in MLC, since their sensitivity is strongly enhanced due to the stabilization of the emitting excited states within the aggregates. For example, the detection limits of PNA compounds using fluorescence detection can be lowered by an order of magnitude with respect to homogeneous solvents and, moreover, their LDR are extended.

Phosphorescence detection is also possible in MLC by adding a proper metal ion (usually Tl^+) to increase the lifetime of the excited triplet states. Efficient degassing of micellar mobile phases or other suitable chemical treatments are required to eliminate the dissolved oxygen (which is a strong phosphorescence quencher).

The compatibility of electrochemical detectors with aqueous micellar eluents has been previously examined.

6.1.6.4. Determination of partition coefficients using MLC

As previously indicated in Chapter 2, the use of MLC represents a viable and valuable method for determination of partition data in micellar systems. For example, from the slope and intercept of the plots according to Eqns. (1) or (2), both the partition coefficients (P_{MW} and P_{SW}) and the binding constant, K_B, can be readily calculated.

The accuracy of K_B and P determinations depends on the retention equation used [2]. For moderately hydrophobic solutes, Eqn. (1) is applied (see Fig. 6.18).

The evaluation of K_B for more hydrophobic compounds is severely limited by the uncertainty of the intercept value, given by the term $1/P_{SW}$ (which tends to zero). When more polar stationary phases are used, P_{SW} decreases, allowing a better estimation of K_B.

However, for analytes having P_{MW} values in the range 10^3–10^4, the application of retention equation (17) would give better results. Accord-

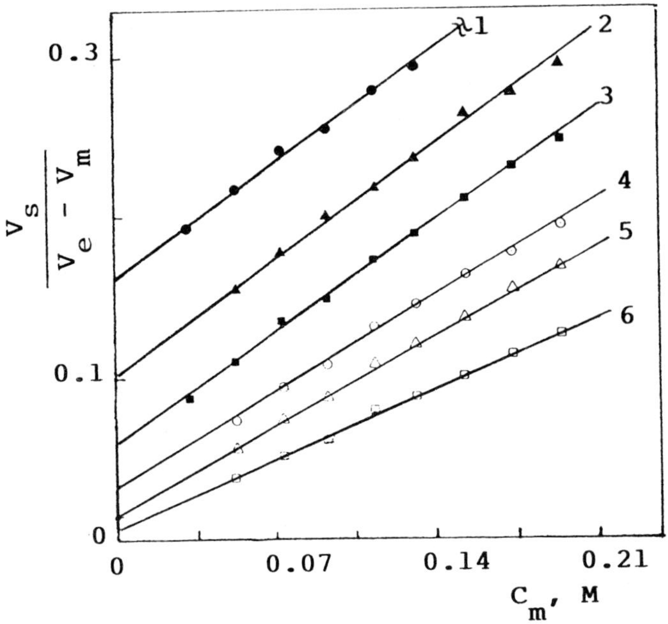

Fig. 6.18. Plots of chromatographic parameter $V_s/(V_e - V_m)$ of some hydroxybenzenes as a function of micellized SDS concentration, according to Eqn. (1). Solutes: 1 = catechol; 2 = phenol; 3 = 4-methylcatechol; 4 = 4-methylphenol; 5 = 4-ethylphenol; 6 = 4-n-propylphenol. Figure taken from Ref. [12]. Reprinted with permission.

ing to Borgerding et al. [23], a very accurate determination of the chromatographic phase ratio (ϕ) is necessary using this approach.

Two main advantages of K_B evaluation using MLC must be emphasized: moderately pure analyte samples are required in these experiments, and the measurements can be accomplished quite rapidly, also in cases where other techniques fail.

It was demonstrated [51] that the binding constants of several hydroxyaromatic compounds, measured using MLC, compare well with those obtained using other direct and indirect experimental methods, including monitoring of spectral changes, variation of acid-base dissociation constants, ultrafiltration, solubility measurements and kinetic analysis of fast reactions. A survey of partition data obtained using MLC is given in Table 6.5. (See also Refs. [128,129].)

TABLE 6.5

Binding constants of some aromatic solutes to SDS and HTAB micelles measured using the MLC technique

Compound	K_B (M^{-1})	
	SDS	HTAB
Phenol	9[a]	71[a]
	10[b,c]	
4-methyl-	24[c]	
4-ethyl-	63.5[c]	
4-n-propyl-	117[c]	
4-t-butyl-	258[c]	
3,5-dimethyl	59.5[c]	
	49.5[a]	248.5[a]
2,4,5-trimethyl-	133[c]	
2-hydroxy-(catechol)	7[c]	
4-hydroxy-(quinol)	3.5[d]	
4-nitro-	17[e]	
4-cyano-	16[e]	
4-methoxy-	15[e]	
1-naphthol	105[a]	
Benzene	19[a]	40[a]
	20[b]	
	17[f]	24[f]
	15.5[g]	27[g]
Toluene	57[f]	124.5[f]
	54[b]	
	36–40[g]	70–75[g]
Nitrobenzene	23[a]	37[a]
Chlorobenzene	68[a]	157[a]
Benzonitrile	7[g]	10.5[g]
Naphthalene	290[a]	
Aniline		28[h]
4-methyl-		52[h]
4-fluoro-		42[h]
4-chloro-		130[h]
Acetophenone[f]	20	22[f]
Propiophenone[f]	34	53[f]
Butyrophenone[f]	57	130.5[f]
Valerophenone[f]	99	457[f]

a: From Ref. [52]; *b*: from Ref. [10]; *c*: from Ref. [12]; *d*: from Ref. [1]; *e*: from Ref. [53]; *f*: from Ref. [54]; *g*: from Ref. [55]; *h:* from Ref. [14].

Aqueous micellar mobile phases have been applied in TLC, but a serious limitation arises due to the instability of most common TLC stationary phases. For example, silica gel plates are not compatible with aqueous solutions of ionic and non-ionic surfactants because excess water causes the detachment of the layer from the plate support.

Acceptable chromatographic performances, without streaking effects, can be obtained working with TLC plates of polyamide and, under certain experimental conditions, with alumina.

The use of bonded reverse-phase layers, which can withstand higher water contents in the eluent, extends the potential applications of the technique. In some RP-TLC experiments, performed with ionic micellar solutions and alkyl bonded stationary phases, the plate stability is further improved by adding proper amounts of inorganic salts.

6.2.1. Main features of MTLC

A typical feature of this chromatographic technique is the presence of a double solvent front, the upper one mainly consisting of a very dilute aqueous surfactant solution, while the lower one contains the micellar aggregates and is usually termed the micellar front.

This phenomenon, often observed, can be explained taking into account the possibility of adsorption of surfactant molecules onto the stationary phases, with a corresponding decrease of the amphiphile concentration below the cmc range. Thus, practically no micelles are present in the initial solution moving up the plate. The micelle-containing mobile phase starts to elute the plate, initiating the second solvent front, only after saturation of the sorbent.

Consequently, more hydrophobic solutes tend to migrate in the micellar-rich eluent, while hydrophilic analytes will be present in the middle region placed between the two solvent fronts (see Fig. 6.19). In the presence of weak adsorption interactions, the micellar eluent composition does not change significantly during the development run, thus, a single front solvent is observed.

A common and interesting feature of MTLC is its ability to separate compounds having very different solubilities in water while performing a single chromatographic run. When normal micelles are used in TLC, the increase of R_f with increasing surfactant concentration in the eluent can be related to the extent of binding of the analytes to the micelles

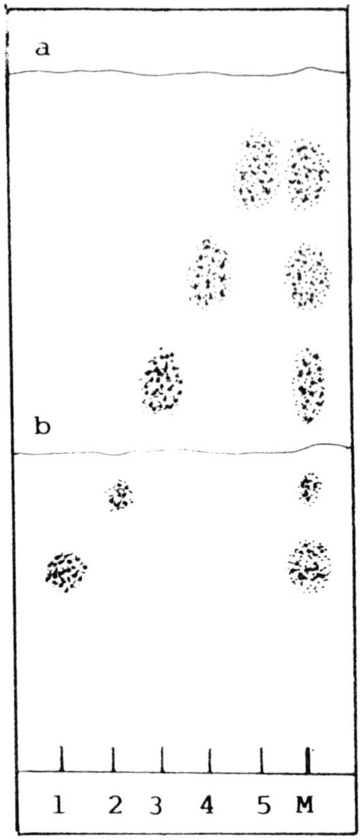

Fig. 6.19. Typical solvent front splitting in MTLC. Eluent: 0.2 M HTAC + 0.2 M NaCl. Plate: RP-ODS. a: solvent front; b: micellar eluent front. Analytes: 1 = 2-nitrophenol, 2 = phenol, 3 = catechol, 4 = resorcinol, 5 = hydroquinone, M = mixture. Figure taken from Ref. [56]. Reprinted with permission.

through a partition equation [17] which is similar to those previously shown for micellar HPLC:

$$\frac{R_f}{1 - R_f} = \frac{\bar{V}(P_{MW} - 1)C_m}{P_{SW}\phi} + \frac{1}{P_{SW}\phi} \qquad (18)$$

With this equation, the binding constants of solutes to micelles can be estimated from simple and cheap TLC experiments, provided the concentration of micelles in the eluent remains constant during the development. However, when the stationary phase exhibits extended surfactant adsorption, this condition is not satisfied, and it becomes necessary to determine the effective C_m value along the plate.

Relevant differences between the surfactant concentration in the reservoir and in the plate have, in fact, been observed in some experiments. For example, during the elution of polyamide plates with 0.3 M SDS, the average SDS concentration in the plate (determined using scanning densitometry) drops to ca. 0.2 M [17]. On the contrary, if the same plates are eluted with cationic micellar solutions, the surfactant concentration remains practically constant, thus indicating lower adsorption effects.

From the above considerations, the use of MTLC to evaluate K_B appears to be more limited than micellar HPLC.

6.2.2. Examples of MTLC separations

MTLC separations largely concern aqueous micellar mobile phases (see Table 6.6). In a few cases, also reversed micelles have been used.

A comparison of the chromatographic performances of MTLC with those of traditional TLC reveal, however, some disadvantages which must be carefully examined when working with micellar eluents.

TABLE 6.6

Chromatographic separations using aqueous MTLC

Investigated analytes	TLC plate	Surfactant	Ref.
Aromatic hydrocarbons	Polyamide	SDS	57
Chlorinated pesticides	Polyamide, alumina	SDS, HTAB	58
Phenols	Polyamide	SDS	59
Aromatic amines	RP-C18	HTAC	
Naphthalene derivatives	Polyamide	SDS	60
4-nitrophenol	Polyamide	SDS	61
4-chlorophenol		HTAB	
DDE[a]		HTAC	
Food dyes	Polyamide	SDS	61
Indicators	Alumina polyamide	SDS	61
Caffeine from coffee	Alumina	HTAC	61
	polyamide	SDS	
Benzamide, 2-naphthol, biphenyl	Polyamide	SDS	61
Fatty acids	Polyamide	SDS	62

a: 2,2-bis(4-chlorophenyl)-1,1-dichloro-ethylene.

Disadvantages related to the use of micellar eluents in TLC

Due to the higher viscosity of micellar solutions, long development times are required. Usually, many hours are necessary to complete the elution of a 20×20 cm chromatographic plate in most MTLC experiments. Above certain concentration limits (e.g. ca. 0.6 M for SDS, ca. 0.1 M for HTAB) the surfactant solutions are too viscous to be applied.

The dissolving ability of micellar eluents is lower than that of hydroorganic solvents, thus only small quantities of sample must be applied on the plate.

The detection of spots using the common reagents can in some cases be more difficult because of the presence of surfactants.

The separation efficiency (measured by the spot size) obtained with aqueous micellar eluents is generally lower than that of organic solvents. If reversed micelles are used (see Section 6.4), the efficiency becomes comparable.

Advantages of MTLC

As for micellar HPLC, the main advantages of MTLC are related to the lower costs, higher safety and peculiar selectivities offered by the surfactant-based eluents.

Moreover, the applications of MTLC in teaching laboratories are of particular interest since a great number of practical demonstrations can be performed with practically no risks for the students.

6.3 MICELLAR GEL PERMEATION CHROMATOGRAPHY (MGPC)

The first report concerning the use of micellar mobile phases was presented by Herries and coworkers [63], who applied size exclusion chromatography as a suitable technique for evaluating the partition coefficients of solutes in micellar media. The development of MGPC as a viable and useful separation method was initiated some years later [9].

6.3.1. Micelle exclusion separation of organic solutes

The basic principles of the technique are the following: a suitable gel is chosen so that its pore sizes are smaller than the micelle diameter, and the aggregates are thus completely excluded from the stationary phases.

Sephadex G, a bead-formed gel prepared by cross-linking dextran with epichlorohydrin, is largely used in aqueous media. This gel is

TABLE 6.7

Properties of some Sephadex G polymers

Sephadex type	Mean particle diameter (μm)	Fractionation range (dalton)
G-10	40–120	0–700
G-15	40–120	0–1500
G-25 (C)[a]	100–300	1000–5000
G-50 (C)[a]	100–300	1500–30000
G-100[a]	40–120	5000–150000
G-200[a]	40–120	5000–600000

(C): Coarse particles; *a*: gels from which most micelles are completely excluded.

available in different particle size grades (see Table 6.7), some of which are able to exclude the usual micellar aggregates.

The analytes to be separated interact with the micelles and their retention volumes are dependent upon the extent of binding. Compounds which do not bind to the aggregates can move in the aqueous bulk mobile phase and can easily penetrate into the gel (free permeation), thus they have longer elution times. The analytes completely bound to the micelles are, on the contrary, excluded from the gel phase and their retention volumes approximately correspond to the column void volume. For partitioned compounds, retention volumes between the above limits are expected.

Figure 6.20 qualitatively illustrates the elution behaviour of solutes in MGPC as a function of their binding constants. If micelles are excluded from the gel matrix, the compounds completely bound to the aggregates ($K_B > K_{B1}$) are eluted at the void volume, also indicated as exclusion volume (V_E). Unbound solutes ($K_B < K_{B2}$) exhibit total permeation and are eluted at the corresponding volume (V_{PT}). For partitioned solutes, the elution volume inversely depends on K_B.

The first application of MGP was the separation of some transfer RNAs (ribonucleic acids), exploiting the selective interactions of these analytes with HTAC micelles. The tRNA–micelle aggregates are excluded from Sephadex G-100 gel particles, whereas the hydrated tRNAs can permeate [9].

Amino acids have been separated using micellar SDS and Sephadex G-25 columns on the basis of electrostatic as well as hydrophobic interactions with the aggregates [64].

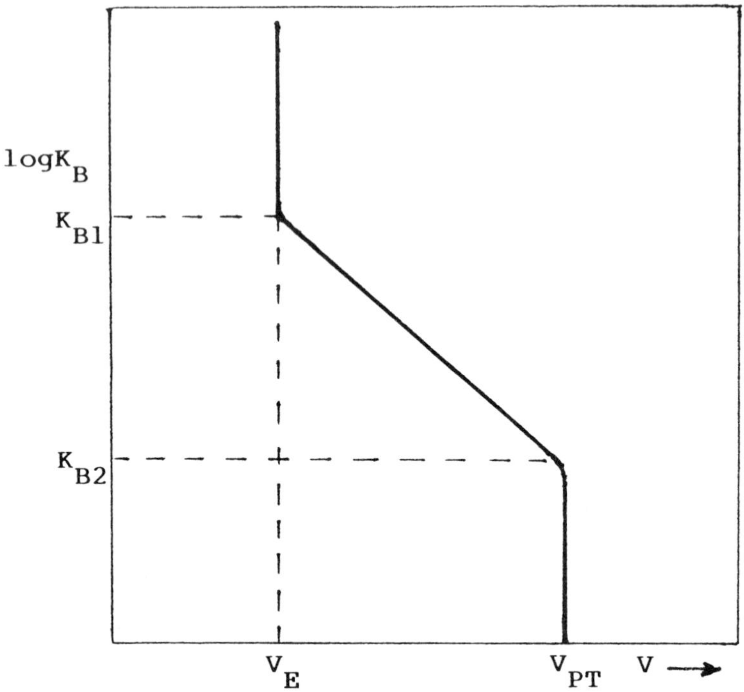

Fig. 6.20. Variation of the elution volume of solutes in MGPC as a function of their binding constants to micelles. Completely excluded aggregates are assumed.

The partition coefficients of several nucleosides and nucleotides in micellar solutions of sodium dodecanoate have been determined using MGP [65].

The separation of the nucleosides adenosine and uridine (having a very similar structure) using the same micellar solution has also been reported [2]. Due to its stronger interaction with the aggregates, uridine is eluted more quickly than adenosine, whereas no appreciable separation is observed using aqueous buffers as mobile phases (see Fig. 6.21).

Basic equations describing retention in MGP have been derived by Herries [63]:

$$\frac{V_i}{V_e - V_0} = \frac{\overline{V}(P_{MW} - 1)C_m}{k'K_d} + \frac{1}{k'K_d} \tag{19}$$

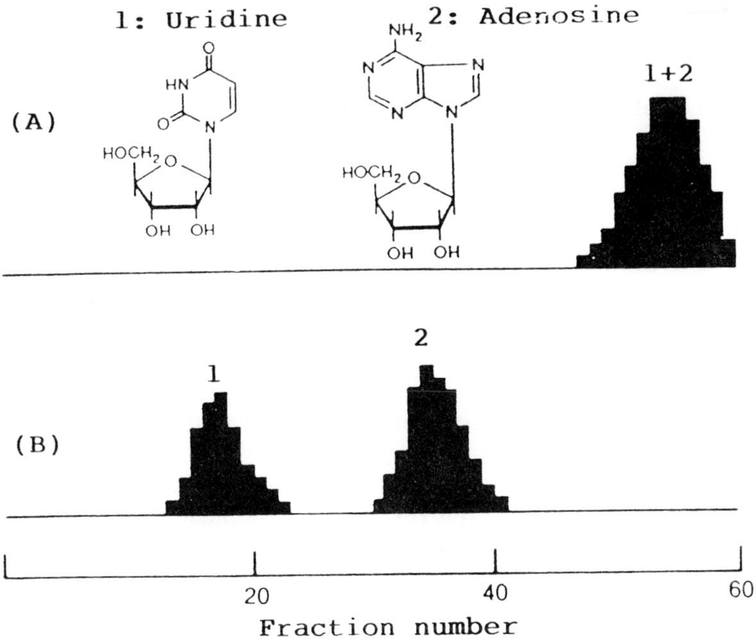

Fig. 6.21. Elution pattern of uridine and adenosine on a Sephadex G-25 column. (A): Aqueous buffer solution; (B): same buffer containing sodium dodecanoate 0.09 M. Figure taken from Ref. [2.] Reprinted with permission.

where K_d is the molecular sieving constant, V_i is the volume of the internal stationary phase and k′ is a proportionality factor.

Taking into account the number of micelles present in the external volume V_o, Eqn. (19) can be rewritten as follows:

$$\frac{V_i}{V_e - V_0} = \frac{Q\overline{V}(P_{MW} - 1)}{k'K_d} + \frac{1}{k'K_d} \tag{20}$$

where Q is the number of micelles in the external volume divided by the total number of micelles in the column.

In the absence of micelles, the classic gel filtration equation is obtained:

$$V_e = V_o + PV_i \tag{21}$$

Gel filtration with micellar solutions can also be applied as a suitable analytical tool for the study of amphiphile aggregation equilibria [66-68].

6.3.2. Micelle exclusion chromatography of inorganic anions and cations

In this technique, introduced by Okada [69], the retention of ionic analytes is regulated by two equilibria: the partition between the ionic micelles and the bulk external solution, and the electrostatic interaction between the adsorbed surfactant monomer and the inner solution within the stationary phase.

The micelles are not completely excluded from this phase, but they can partially permeate. This is because the gel used has a pore-size cutoff higher than the mean diameter of the hydrated ionic micelles present in the eluent.

It must be emphasized that, in the absence of surfactants, the analytes (anions or cations) are not retained in the packing gel. In the presence of micellar eluents, only surfactant monomers and solute ions can diffuse in the inner solution. If a proper stationary phase is chosen, the amount of adsorbed surfactant is approximately constant above the cmc, and the adsorbed molecules play the role of an ion-exchanger.

Two elution steps can be recognized in these experiments: the first one is due to the exit of the external solution (containing the micelles) from the column, and the second is due to the simultaneous exit of this solution and the internal solution, containing only the monomeric surfactant. At low surfactant concentrations, a variation of the effluent conductance reflects the change of eluate composition.

The following retention equation holds:

$$\frac{1}{(V_R - V_e)} = \frac{(P_{MW} - 1)\bar{V}C_m}{(V_i + V_s P_{SW})} \tag{22}$$

where V_R is the measured retention volume; V_e, V_i and V_s are the volumes of the external solution, inner solution and stationary phase, respectively.

6.3.2.1. Separation of inorganic anions

The combination of both ionic interaction and size-exclusion chromatography allows the retention order of solutes to be controlled by simply changing the surfactant concentration. For example, using HTAC micelles and a poly(vinyl alcohol) gel having a molecular weight cut-off of ca. 40,000 dalton, the following retention orders are observed [69]:

$I^- < Br^- < IO_3^- < NO_3^- < NO_2^-$, for 0.05 M HTAC
$I^- < IO_3^- < Br^- < NO_3^- < NO_2^-$, for 0.02 M HTAC
$IO_3^- < I^- < Br^- < NO_2^- < NO_3^-$, for 0.01 M HTAC

These orders are different from those observed in conventional anion-exchange chromatography or in ion-interaction chromatography with cationic micelles (Section 6.1.5), indicating that retention is here dependent on a different mechanism which accounts for size exclusion effects.

The nature of micelle counterions is another important factor because the ion-exchange equilibria with like charged analytes at the micelle–water interface strongly depends on the counterion binding. Increasing the hydrated anion radius decreases the counterion binding.

The addition of salts to the mobile phase causes increased adsorption of surfactant onto the stationary phase.

The technique can be also applied to the separation and analysis of some organic anions present in mixtures [70].

6.3.2.2. Separation of transition-metal ions

By using anionic SDS micelles and the same type of stationary phases previously mentioned, the separation of several cations can be accomplished [71].

Also in this case, the retention behaviour expected on the basis of a simple ion-exchange mechanism is drastically altered. For example, Pb(II) is strongly retained on cation-exchangers, while is rapidly eluted in the presence of SDS micelles because the strong binding to the aggregates is more important than the interaction with the adsorbed monomers.

Using auxiliary complexing agents, such as organic acids, the separation of metal ions can be improved. Depending on the stability constants of these complexes (usually low for carboxylic and hydroxycarboxylic acids), the elution orders can be controlled.

Cations which form labile complexes, such as Mn(II), are strongly retained because a relevant fraction of free metal ion exists in the solution, whereas the elution of cations that form more stable complexes with these acids (e.g. Cu(II)) is faster.

When a given metal ion is exclusively present as an anionic complex in the mobile phase, it is rapidly eluted, before the neutral forms. For example, the tartrate complexes of Fe(III) undergo electrostatic repulsion

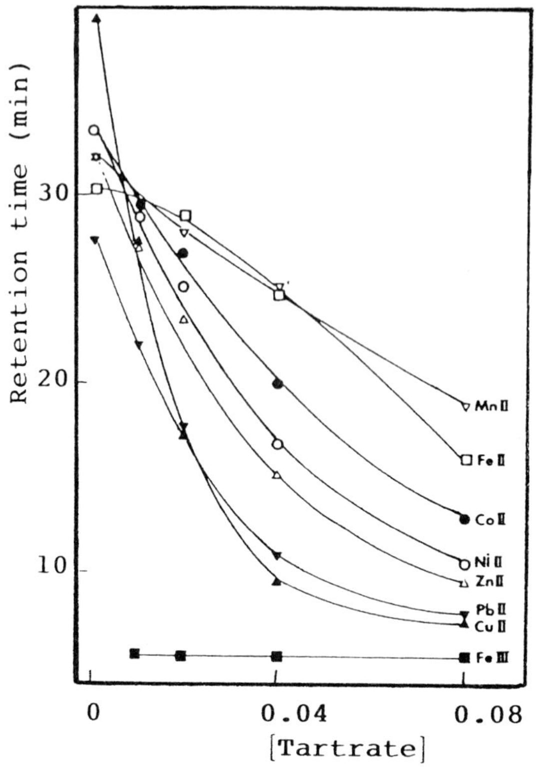

Fig. 6.22. Variation of retention of transition metal ions as a function of the auxiliary ligand concentration. Mobile phase: 0.05 M SDS, pH: 4.4. Figure taken from Ref. [71]. Reprinted with permission.

both from the adsorbed SDS (stationary phase) and from the micelles present in the external phase (see Fig. 6.22).

The retention equation valid in these systems is the following:

$$\frac{1}{(V_R - V_e)} = \frac{(P'_{MW} - 1)\overline{V}C_m}{(V_i + V_s P_{SW}\alpha)} + \frac{1}{(V_i + V_s P_{SW}\alpha)} \tag{23}$$

where P'_{MW} is the partition coefficient of the metal ion (present in all the species) between the micelle and the bulk external solution, and α is the fraction of free cation, which is the only species attracted by the SDS molecules adsorbed onto the stationary phase.

6.4. APPLICATION OF REVERSED MICELLES AND OIL–WATER MICROEMULSIONS IN MLC AND SFC

To date, the applications of reversed micellar eluents in LC have been scarcely investigated, although some considerations based on the unique properties of the reversed micelle structure indicate that these systems could be exploited in both normal- and reversed-phase HPLC and TLC separations of polar compounds. Moreover, recently published reports have extended the use of these structured organic-rich phases to the SFC domain, opening new promising analytical perspectives.

6.4.1. Use of reversed micelles in TLC

A few examples of TLC separations of polar compounds (e.g., aminoacids and nucleosides) using reversed micelles of AOT in n-hexane as mobile phases, have been reported in the literature [57,58].

The partitioning of the hydrophilic analytes to the aggregates mainly depends on their solubility in the inner water pool, while partitioning to the pure hydrocarbon solvent is usually negligible.

Within a series of related compounds (e.g. aminoacids), more polar solutes exhibit higher R_f values because, at a given surfactant/water ratio, their fraction in the mobile phase is larger. The R_f values usually increase with the water content up to a limit given by the aggregate stability.

It should be noted that the reported TLC experiments were performed on polyamide and silanized silica gel, that is, with non-polar or slightly polar stationary phases.

6.4.2. Normal-phase HPLC with reversed micelles

Normal-phase chromatography based on the use of polar stationary phases could be an adequate separation technique for polar compounds. However, the main limitation of polar eluents, when used in the presence of strongly polar adsorbents, is the difficulty to control solute retention by changing the water/organic solvent ratio. This is due to the fact that not only does the eluent strength change, but also the activity of the stationary phase because water preferentially adsorbs on the active sites of the sorbent.

Moreover, any variation in water content of an organic pure solvent due to humidity absorption from the local environment can strongly

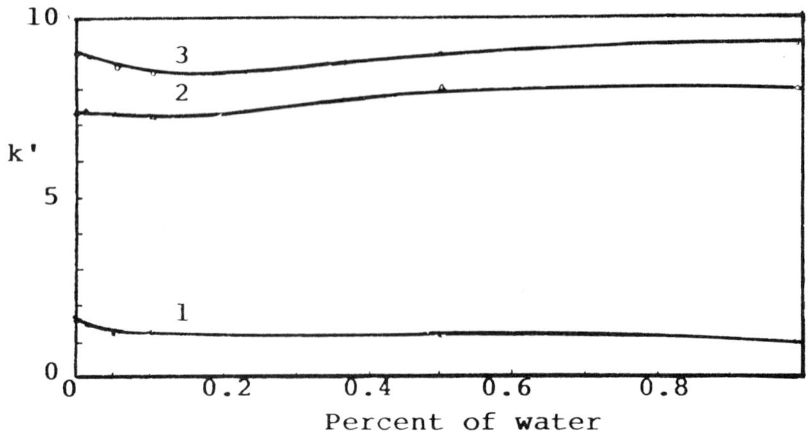

Fig. 6.23. Capacity factors of polar aromatic solutes as a function of water content. Mobile phase: reversed micelles of AOT (5×10^{-2} M) in n-hexane. Stationary phase: Ultrasil-NH_2. Compounds: 1 = 2,4-dinitrotoluene; 2 = phenol; 3 = naphthol. Figure taken from Ref. [72]. Reprinted with permission.

alter the analyte retention. Consequently, the reproducibility of retention data in these experiments is usually low.

Conversely, elutions performed using reversed micelles in apolar solvents indicate that, within certain water concentration ranges, the dependence of solute retention on water content is practically constant. This behaviour can be attributed to the hydration of the surfactant ionic groups, located in the inner part of the aggregates.

For example, if AOT reversed micelles in apolar solvents are considered, the first water molecules are bound tightly to the sodium counterion and to the AOT polar head. Therefore, practically no free water is available to adsorb on the polar stationary phase and retention is almost constant (see Fig. 6.23).

The more or less pronounced k′ decrease observed at very low water contents can be attributed to a strong adsorption of water molecules on the active sites of the stationary phase (usually silica, alumina, or amino bonded phases). Above a certain limit, which depends on the nature of the surfactant and sorbent, water is mainly captured by the aggregates to complete the hydration of their ionic groups. Thus, the sorbent activity remains practically constant and k′ does not vary significantly.

A different behaviour is observed when the water content becomes large, that is, when free water molecules are able to promote the

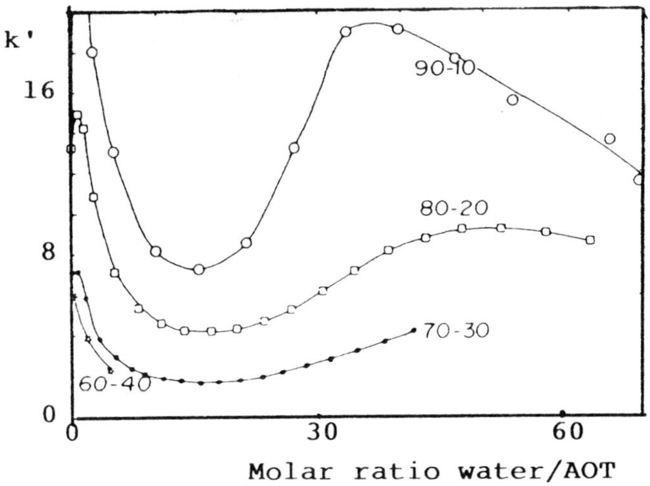

Fig. 6.24. Retention of 4-nitrobenzoic acid as a function of $M_{W/AOT}$. Stationary phase: 10 μm silica. Mobile phase: water-AOT-n-heptane (the numbers on the curves indicate the heptane to AOT ratios). Figure taken from Ref. [73]. Reprinted with permission.

water-pool growth, approaching the solvating properties of pure water. In this case, the solubility of polar analytes increases with the water droplet size and, thus, k′ is expected to vary.

Normal-phase HPLC experiments, performed using silica columns and reversed micelles of AOT in n-heptane as mobile phases, clearly show that the capacity factor is strongly dependent on the water to AOT molar ratio ($M_{W/AOT}$) [73]. The eluent compositions, chosen within the L2 region of the ternary phase diagram, correspond to aggregates having different structures.

Figure 6.24 shows the variation of k′ as a function of $M_{W/AOT}$ for 4-nitrobenzoic acid. For a given AOT concentration, if the water percent is increased, the water droplet size also increases and the number of reversed micelles remains constant. The maximum water to surfactant ratio (M_{max}) defines the maximum size of the water core of the aggregates, larger amounts of added water result in formation of a two-phase system.

If the total amount of AOT and water is varied, keeping $M_{W/AOT}$ constant, the number of reversed micelles also vary but not the droplet size.

The analysis of the complex dependence of k′ upon $M_{W/AOT}$ needs further investigations. A number of experimental parameters, such as the dead volume and the pressure drop in the column, also vary with

$M_{W/AOT}$ because they are very sensitive to the viscosity of the mobile phase, which is in turn dependent on the aggregate structure.

The efficiencies of HPLC separations using reversed micelles are lower than those obtained with hydro-organic solvents. Since the surfactant adsorption onto the stationary phase is usually low and practically constant above certain water to surfactant ratios (this limit is ca. 10 for AOT), the kinetic effects concerning the solute transfer from/to the reversed micelles are probably dominant. Due to the preferential location of polar solutes in the inner water core of the aggregates, the exit and enter rate constants are lower than those observed with normal micelles, where the solutes penetrate less deeply into the organized structure.

6.4.3. Use of oil–water microemulsions in MLC

In addition to AOT reversed micelles, L2 microemulsions (without any alcohol cosurfactant) can also be used as normal LC mobile phases [73], although poor efficiencies have been observed using these eluents.

More recently, low viscosity L1 microemulsions containing 1-pentanol, SDS, heptane and water have been used to elute alkylbenzenes on a C_{18} column [74]. These microemulsions exhibit a very high solubilization capacity toward the hydrophobic terms of the series, even when in presence of large amounts of water (up to ca. 90% w/w) in the eluent.

However, a drastic decrease of efficiency has been observed upon increasing the percent of hydrocarbon in the microemulsion (probably due to structural changes of the eluent). Consequently, the practical use of these mobile phases is still a subject of debate.

6.4.4. SFC with reversed micelles

The use of supercritical fluids in chromatography allow separation efficiencies to be obtained which are intermediate between those of LC and GC. The properties of these fluids, including their viscosity and diffusion rates, vary between those of gases and liquids and can be readily controlled by changing the pressure and temperature of the mobile phase.

Apparatus and methods suitable for SFC routine work have been developed in recent years, although some limitations of the technique have been recognized. In particular, the analysis of highly polar and thermally labile compounds cannot be easily performed.

TABLE 6.8

Critical temperatures and pressures of some SFC eluents

Solvent	T_c (°C)	P_c(atm)
Water	374.1	217.6
Ammonia	132.4	111.3
Methanol	239.4	79.9
Acetonitrile	274.8	47.7
Tetrahydrofuran	267.0	51.2
Carbon dioxide	31.0	72.8
Nitrous oxide	36.4	71.5
Ethane	32.2	48.2
Propane	97.0	43.3
n-butane	152.0	37.5
n-pentane	196.5	33.3
n-hexane	234.2	29.3

Data taken from Refs. [75,76].

As indicated in Table 6.8, the use of polar supercritical fluids (e.g. water or ammonia) is difficult because their critical temperatures are too high (>200°C) and, under these conditions, the alteration of solutes through solvolytic reactions and/or by thermal decomposition can occur. The use of less polar mobile phases having lower critical temperatures (e.g. CO_2 and mixtures containing CO_2 and polar modifiers) can be more suitable, but their ability to dissolve polar analytes is lower.

Generally, the technique cannot be applied to hydrophilic compounds that are sparingly soluble in organic solvents, thus many important organic and biological molecules cannot be handled by SFC.

The formation of reversed micelles in supercritical fluids, demonstrated by Gale and coworkers some years ago [77], can widely expand the SFC applications. Some of these aggregates are unsuitable for the separation of polar analytes due to their small dimensions (e.g. AOT in supercritical CO_2 or N_2O). In other cases, structures with proper aggregation numbers and water contents can be obtained (e.g. AOT in supercritical alkanes), which can readily dissolve a variety of hydrophilic analytes.

Reversed micelles can be used in SFC over a wide range of experimental conditions. In order to have a single-phase fluid system, a well defined pressure is required, which depends on the water and surfactant concentrations.

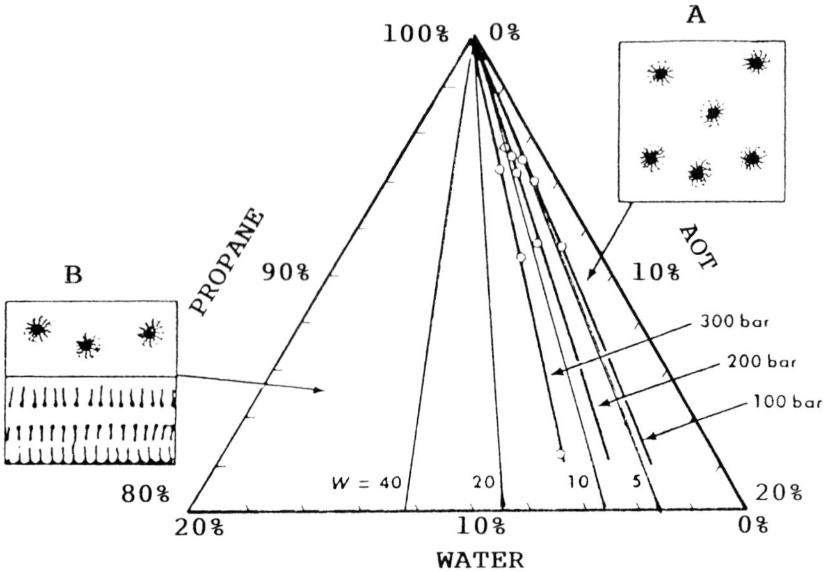

Fig. 6.25. Phase diagram for AOT–water–supercritical propane. A: reversed-micelles domain (isotropic liquid phase). B: Two-phase system (reversed micelles plus gaseous propane). Figure taken from Ref. [78]. Reprinted with permission.

The maximum water-to-surfactant ratio (M_{max}) increases with the applied pressure, thus indicating that pressure (or fluid density) variations provide a means for modifying the eluent strength.

Figure 6.25 shows a part of the ternary phase diagram for propane–AOT–water. Each boundary line defines the existence domain of reversed micelles and that of a two-phase liquid–gas system, respectively. By simply changing the applied pressure, at a fixed temperature, the reversed-micellar region can be expanded or reduced.

The preliminary results obtained working with AOT reversed micelles in supercritical alkanes [79] demonstrated that rapid and effective separation of quite polar organic compounds is possible. Figure 6.26 illustrates an example of such separations.

On the basis of the above initial studies, the use of SFC with reversed micellar phases could offer some potential advantages:

– The chromatographic efficiencies of these systems are higher than those obtained using the same reversed-micellar eluents in HPLC. The higher diffusion rates and lower viscosities of supercritical fluids favour

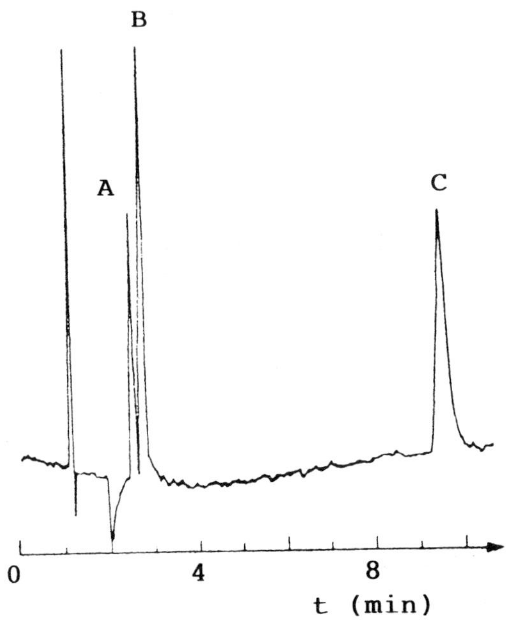

Fig. 6.26. Separation of hydroxyaromatic solutes using SFC. Mobile phase: reversed micelles of AOT (5×10^{-2} M) in supercritical propane; $M_{W/AOT} = 5$. Column: microbore silica. Temperature: 110°C; pressure 250 bar. Peaks: (A) phenol, (B) 2-naphthol, (C) resorcinol. Figure taken from Ref. [79]. Reprinted with permission.

all the mass-transfer processes.

– The separation of hydrophilic compounds, which are difficult to analyze using GC or traditional SFC, can be achieved using reversed micelles in supercritical fluids.

– The selectivity could be adjusted by manipulating pH, ionic strength, and other more specific parameters such as the fluid density (by simply changing the applied pressure).

On the other hand, a number of limitations have to be considered. For example, the presence of relevant amounts of surfactants in the fluid phase does not allow the use of FID detectors (very popular in SFC). For the same reason, the direct fluid injection in SFC-MS coupled systems becomes difficult. Moreover, the development of particular pumping systems which allow the continuous variation of the water-to-surfactant ratio under supercritical conditions is required.

Although the unique solvating characteristics of both the supercritical fluids and reversed micelles have been clearly recognized, further developments require a better understanding of the effects of physical and chemical variables on the phase behaviour of these ternary systems.

6.5. MICELLAR ELECTROKINETIC CAPILLARY CHROMATOGRAPHY (MECC)

This new powerful surfactant-based separation technique, introduced by Terabe and coworkers [80,81], can be considered as a combination of MLC and capillary zone electrophoresis (CZE).

In CZE, charged solutes can be efficiently and rapidly separated using narrow-bore capillaries filled with aqueous buffered solutions, at the ends of which a high electric potential (usually tens of kV) is applied. Since a strong electroosmotic flow of the electrolyte solution occurs under these conditions, neutral solutes are also transported, although there is a poor separation between them.

Thus, CZE is unable to separate uncharged analytes and, moreover, is not suitable for separation of ionizable species possessing similar electrophoretic mobilities, such as small ions.

When ionic micelles (i.e., SDS) are added to the aqueous conducting solution, a relative motion is produced between the two pseudophases, the electroosmotic flow of the electrolyte being usually faster than the opposite electrophoretic migration of the aggregates. As a result, the direction of flow of all the components is the same, but micelles migrate slowly.

In these conditions, a chromatographic system is operative in which two distinct phases, the aqueous (mobile) and micellar (pseudostationary), are present within the capillary column and migrate at different velocities toward the electrode with the same charge as the micelles (see Fig. 6.27).

Neutral solutes partitioned between these two phases will move more or less slowly depending on their binding constants to the micelles, whereas, if charged analytes are present, they will in addition undergo electrophoretic migration in the opposite direction of the electroosmotic flow (micelle-like charged analytes) or in the flow direction if they act as micelle counterions.

In Fig. 6.28, a typical MECC experimental set-up is shown.

The retention times of partitioned neutral solutes are limited within the range: $t_0 \leq t_R \leq t_m$, where t_0 and t_m are the retention times of free and completely bound solutes, respectively.

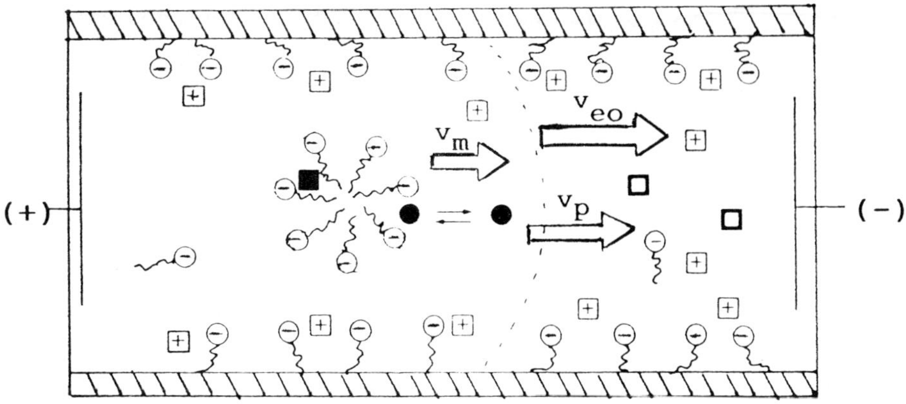

Fig. 6.27. Transport of neutral analytes and anionic micelles within the capillary. Solutes: (■) completely bound (v_m), (□) unbound (v_{eo}); (●) partitioned ($v_m < v_p < v_{eo}$). Micelle counterions: (⊞).

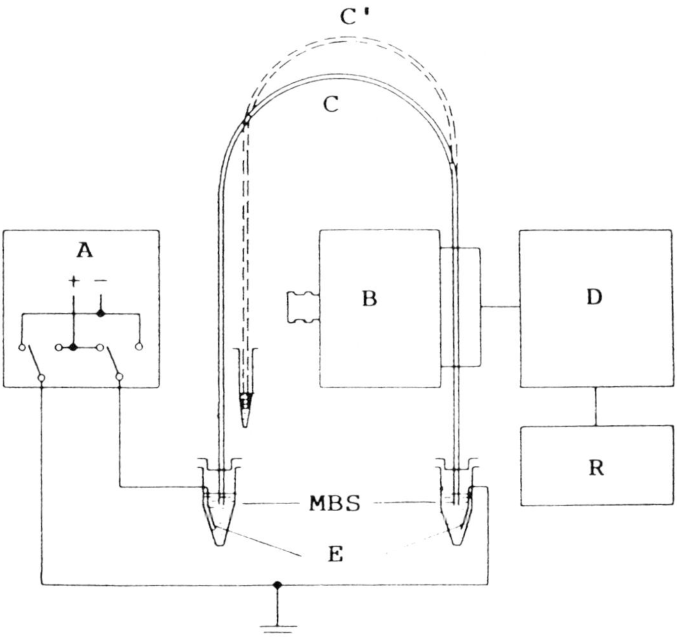

Fig. 6.28. MECC experimental device. A: Regulated high-voltage dc power supply; B: UV detector; C: fused-silica capillary (separation position); C′: capillary in sample application position; D: detector amplifier; E: platinum-wire electrodes; (MBS): micellar buffered solution; R: recorder. Figure taken from Ref. [82]. Reproduced with permission.

The capabilities of the MECC technique have to be analyzed by studying the effect of the experimental parameters upon the capacity factor, the effective resolution of the sample components, and the separation efficiency.

6.5.1. Capacity factor, resolution, and elution ranges in MECC

6.5.1.1. Capacity factor

The following definition of k′, according to Terabe [80], is based on the assumption that micelles migrate in the same direction of the electroosmotic flow. This corresponds to the situation commonly observed in MECC.

$$k' = \frac{t_R - t_0}{t_0(1 - t_R/t_m)} \tag{24}$$

When t_m is very large, Eqn. (24) becomes equivalent to the classic definition of k′.

In typical experiments performed with SDS micelles, methanol has been chosen as a reference neutral solute present only in the aqueous phase (t_0), whereas the completely bound lipophilic dye 1-[4-(phenylazo)phenylazo]-2-naphthol (Sudan III) has been used to estimate the elution time of the aggregates (t_m).

The samples are introduced at the end of the capillary in contact with the anode and the migration times of both markers are determined under the same conditions of polarity. The elution range is given by $t_0 - t_m$ (see Fig. 6.29). For cationic micelles migrating in the direction of the electroosmotic flow, the samples and the markers are introduced at the cathode end.

This mode of operation allows the separation of analytes showing very different k′ values, within a short analysis time.

In order to increase the MECC selectivity, the retention times of the analytes can be properly manipulated by changing the pH or by introducing selected additives, as indicated in the examples cited below.

pH changes: The capacity factors of weak acids (such as phenols) decrease with an increase of pH in the presence of anionic micelles [83]. The corresponding retention times are lower because the anions become more soluble in the aqueous bulk solution.

Complex formation: The addition of selected metal ions to samples containing anionic oligonucleotides improves the separation of closely

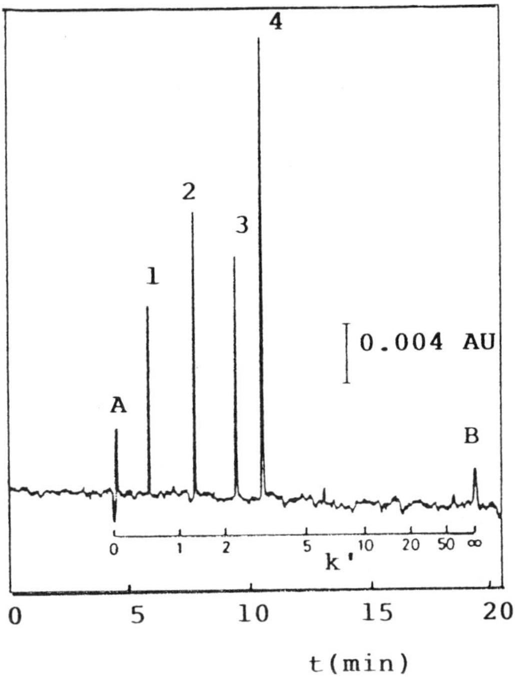

Fig. 6.29. MECC elution range of phenolic solutes. Markers: A (methanol) and B (Sudan III). Analytes: (1) phenol, (2) 4-methylphenol, (3) 2,6-dimethylphenol, (4) 4-ethylphenol. SDS 0.05 M; phosphate–borate buffer (pH 7); capillary: 50 μm i.d. × 500 mm; 20 kV–33 μA; UV detection at 220 nm. Figure taken from Ref. [80]. Reprinted with permission.

related molecules (in the presence of SDS micelles) through the formation of complexes which are differentially bound to the aggregates [84]. Similarly, the partitioning of cationic and neutral catechols in SDS solutions can be altered through complexation with boric acids, due to the formation of neutral and anionic derivatives, respectively [85].

Ion-pair formation: The use of oppositely charged large ions can be a useful way to alter the retention times of ionic analytes. Improved resolution of mixtures containing ionic antibiotics and aromatic carboxylates has been achieved in MECC by adding tetraalkylammonium ions to the SDS carrier solution [86].

6.5.1.2. Resolution

The resolution, R_s, between two compounds 1 and 2 is defined in chromatography by:

$$R_s = \frac{2(t_{R2} - t_{R1})}{(w_1 + w_2)} \tag{25}$$

where t_{R1} and t_{R2} are the retention times of the eluted compounds (the subscript 2 indicates the last eluted solute), w denotes the baseline peak width.

Assuming Gaussian peak shapes and $w_1 = w_2$, the following equation defines resolution in MECC [81]:

$$R_s = \frac{\sqrt{N}}{4} \frac{(\alpha - 1)}{4} \frac{k'_2}{\alpha(1 + k'_2)} \frac{(1 - t_o/t_m)}{\{1 + (t_o/t_m)k'_1\}} \tag{26}$$

where N is the plate number and α is the separation factor, defined as the ratio k'_2/k'_1. For closely eluting peaks, the approximation: $k'_1 = k'_2 = k'$ is usually made. Equation (26) is valid only if the net migration of micelles is in the direction of the electroosmotic flow.

Resolution optimization in MECC has been the subject of recently published papers [87,88]. In particular, the experimental variables can be properly adjusted and/or varied by using a suitable statistical design [88], thus reducing the number of experiments necessary to find the best conditions.

6.5.1.3. Peak capacity and elution range in MECC

The peak capacity, n, is defined as the maximum number of peaks that can be resolved within a given retention time range. If peaks separated by 4σ resolution are considered (σ is the standard deviation of the peaks), n is given by the Giddings equation [89]:

$$n = 1 + \frac{\sqrt{N}}{4} \ln (t_m/t_o) \tag{27}$$

If the ratio t_m/t_o is increased, the peak capacity also increases and the number of compounds that can be resolved for a given capillary length becomes larger.

In order to increase t_m, the flow velocity of micelles v_m has to be minimized:

$$\vec{v}_m = \vec{v}_{mep} + \vec{v}_{eo} \tag{28}$$

According to Eqn. (28), v_m can be reduced by increasing the electrophoretic migration (v_{mep}), or by decreasing the electroosmotic velocity (v_{eo}). These two components can be separately calculated as follows:

$$v_{eo} = -\frac{\varepsilon\zeta_1 E}{\eta} \tag{29}$$

$$v_{mep} = \frac{\varepsilon\zeta_2 E}{\eta} \tag{30}$$

where ε is the liquid permittivity, η is the viscosity of the solution, E is the applied electrical field, ζ_1 is the zeta potential for the inner surface of the tube and ζ_2 is the zeta potential of the micelle. The minus sign in Eqn. (29) means that when ζ_1 is negative (as for SDS solutions) the liquid flows toward the cathode.

The use of smaller micelles (e.g., by shortening the surfactant alkyl chain) allows v_{mep} to be increased. However, since the cmc also increases, more concentrated surfactant solutions are necessary to fill the capillaries and the corresponding higher column currents can cause heat dissipation problems.

The alternative way to reduce the micelle migration is to decrease the electroosmotic velocity. The inner surfaces of the fused silica capillaries

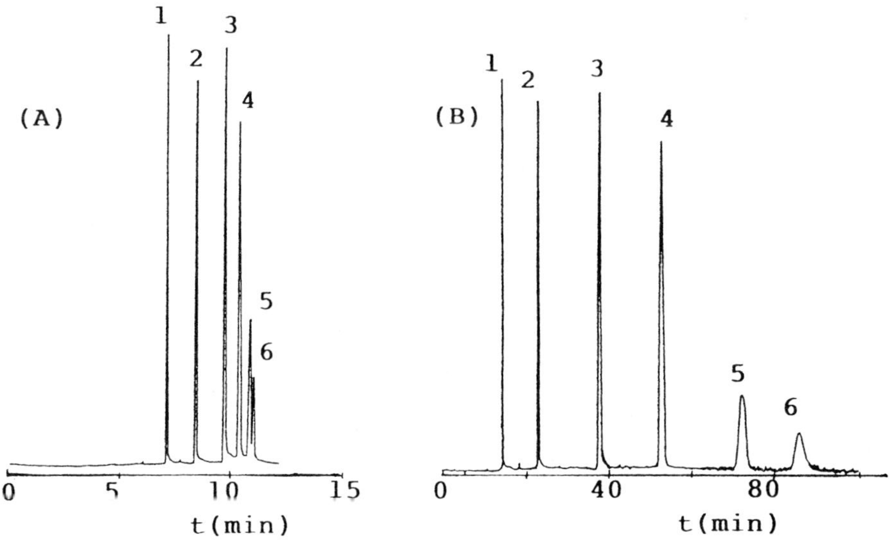

Fig. 6.30. Separation profiles of NBD derivatives of some test amines. (A) Fused-silica capillaries; (B) silanated silica capillaries. Peaks: (1) ethylamine, (2) n-propylamine, (3) n-butylamine, (4) cyclohexylamine, (5) n-hexylamine, (6) coumarin 540 A (lipophilic dye). Capillary: 50 μm i.d. × 650 mm long. Solution: 0.025 M SDS + 0.005 M Na_2HPO_4. Applied voltage: 20 kV. Figure taken from Ref. [90]. Reprinted with permission.

can be properly modified in order to reduce ζ_1. For example, pH changes can alter the zeta potential at silica-solution interfaces, but these changes can also modify the capacity factors of ionizable solutes.

A more interesting approach is based on the alteration of the inner surface of the capillary by silanation [90]. As shown in Fig. 6.30, the MECC elution range can be significantly increased by working with these modified capillaries.

Although the above treatment method improves the elution range, the efficiency decreases because the modified capillary wall can interact with the solutes, introducing a slow mass-transfer step in the separation process.

6.5.2 Column efficiency in MECC

A very important feature of most MECC separations is that, under the proper conditions, column efficiencies higher than 100,000 plates/m can be obtained. These efficiencies are comparable to those yielded by GC with open-tubular glass capillaries.

However, in MECC there are also several kinetic processes which can contribute to the solute band broadening. Some of them are related to extracolumn factors, such as injection and detection modes, whereas others are dependent on diffusion and mass transfer of solute molecules within the capillaries (intracolumn factors).

6.5.2.1. Extracolumn contributions to MECC band broadening

Because of the small volumes of the capillary columns (typically few μl), a very careful injection of the sample is necessary. The use of automated electroinjectors, which allows the reproducible introduction of very low sample volumes, improves the MECC efficiency [91]. Also detection volumes (determined by the length of the flow cell used) have to be minimized in order to increase the MECC efficiency.

The relative contributions to the plate height arising from the above factors and from the time constants of detector and recorder have been analyzed in detail [92].

6.5.2.2. Intracolumn effects on the plate height

The following contributions to the plate height, H, are important in MECC [92]:

$$H = H_l + H_{sd} + H_{aq} + H_T + H_{ep} \quad (31)$$

The influence of each term of Eqn. (31) on MECC efficiency varies as indicated below:

H_l (longitudinal diffusion): the magnitude of this contribution to the plate height is proportional to $1/v_{eo}$.

H_{sd} (sorption–desorption kinetics): related to the enter/exit rates of solutes to/from the micelles, contribution of this term to band broadening is significant only for strongly bound solutes, which have very slow desorption rates.

H_{aq} (intermicelle mass transfer in the aqueous phase): in MECC, the open capillaries behave as packed columns because the micelles can be considered as uniformly sized packing particles. Smaller interparticle diffusion distances (obtained by increasing the micelle concentration), larger diffusion coefficients of solutes and narrow-bore capillaries are expected to improve the mass transfer kinetics. Although some experimental results are consistent with this simple mechanism [93], other interpretations are also possible. The influence of this term increases directly with v_{eo}.

H_T (radial temperature gradient effect on v_{eo}): Joule heating within the capillary produces a radial temperature gradient, the temperature being higher along the tube axis. The contribution of these thermal effects to band broadening are negligible under the usual experimental conditions.

H_{ep} (electrophoretic dispersion of the micelles): this term is related to the existence of aggregates having a different number of surfactant molecules and to the effect of analyte incorporation upon the micelle size and mobility. H_{ep} increases linearly with v_{eo} and is higher for solutes exhibiting a slow exit rate from micelles.

The relative significance of each contribution to the total plate height largely depends on the experimental conditions. The longitudinal diffusion appears to be the dominant factor when the electroosmotic flow is slow, whereas contributions from sorption-desorption kinetics and from micellar heterogeneity become important at higher flow velocities (i.e., at high applied voltages).

Table 6.8 summarizes the qualitative effects of usual experimental parameters on the plate height.

6.5.3. Advantages and limitations of MECC separations

The use of the technique is generally limited to compounds which are at least partially soluble in the aqueous mobile phase. Highly hydrophobic

TABLE 6.8

Influence of some experimental parameters on MECC efficiency[a]

Experimental parameter varied	Effect on H
Higher applied potential	H increases
Increased C_m	H decreases
Increased capillary i.d.	H increases
Lower ionic strength	H decreases
Electroinjection:	
– Increased injection time	H increases
– Increased injection voltage	H increases

a: Experimental data reported in Refs. [91,93].

compounds tend to be nearly completely associated to the aggregates and are, thus, scarcely resolved.

On the basis of an increasing amount of experimental data, the following advantages and disadvantages of MECC (compared with CZE, HPLC and MLC techniques) can be mentioned:

Advantages:

– Separation of neutral and ionic compounds is allowed, whereas CZE is primarily restricted to ionic compounds.

– MECC is a much more efficient technique than MLC.

– MECC gives higher resolution than MLC and HPLC within a shorter time.

– Very low sample volumes are needed (if compared with HPLC).

– In comparison to open capillary LC, it is not necessary to bond stationary phases to the capillary walls. Thus, the risks of column occlusion are minimized.

– Amperometric detection is possible, with a corresponding high analysis sensitivity.

– Biological molecules are particularly amenable to MECC separation.

Disadvantages:

– There is generally a limited elution range (if compared with HPLC).

– In the presence of high applied voltages (> 30 kV) the heat dissipation becomes less efficient and a transverse temperature gradient acts within the column.

– Poor reproducibility of peak area and peak height if manual sample injection is performed.

– Automated injection techniques (i.e. electroinjection) are necessary in order to obtain maximum efficiencies.

6.5.4. Particular cases of MECC operation

Figure 6.27 represents the situation commonly encountered in most MECC systems, with micellar migration in the direction of the electroosmotic flow. Under certain conditions, micelles can migrate in the direction opposite to the electroosmotic flow, thus establishing new separation modes regulated by different retention equations. The basic condition satisfied in these cases is the following: $|v_{mep}| > |v_{eo}|$.

According to Gareil [94], two different situations have to be considered, depending on the absolute value of the solute capacity factor:

6.5.4.1. $k' > t_m / t_o$

In this case capacity factor and resolution are defined as follows:

$$k' = \frac{t_R + t_o}{t_0\{(t_R/t_m) - 1\}} \tag{32}$$

$$R_s = \frac{\sqrt{N_1}}{4} \frac{(\alpha - 1)}{\alpha} \frac{k_1'}{(1 + k_1')} \frac{1(+ t_o/t_m)}{\{k_2'(t_o/t_m) - 1\}} \tag{33}$$

When solutes behave as indicated in this case, migration in the direction of the aggregates occurs. The injection end must be the cathode with anionic micelles and the anode with cationic micelles; t_R and t_m are determined under the same migration conditions (same polarity), whereas t_o can be measured after reversing the polarity of the injection end. This separation mode allows to reverse the elution order of solutes and may be exploited when analytes are present in very different proportions.

6.5.4.2. $k' < t_m / t_o$

The corresponding equations for k′ and R_s are:

$$k' = \frac{t_R - t_o}{(1 + t_R/t_m)} \tag{34}$$

$$R_s = \frac{\sqrt{N_2}}{4} \frac{(\alpha - 1)}{\alpha} \frac{k_2'}{(1 + k_2')} \frac{(1 + t_o/t_m)}{\{1 - (t_o/t_m)k_1'\}} \tag{35}$$

The situation indicated in this case corresponds to the solute migration in the direction of the electroosmotic flow. The sample is introduced at the cathode end with positive micelles and at the opposite end with negative aggregates; t_0 and t_R can be measured simultaneously, whereas t_m can be determined after reversing the capillary polarity. In this case, the solution migration range is not limited by t_m, thus the operation mode allows a better resolution between similar solutes (while increasing the analysis time).

6.5.5. Examples of MECC separations

A great number of effective separations and determinations of a variety of analytes can be performed using MECC. Several examples of such applications are listed in Table 6.10, where the experimental conditions of the runs are also reported.

Recent progress in the field include the use of peculiar surfactants. For example, a series of N-D-gluco-N-methylalkanamides (MEGA) can react in basic media with boric acid through their polyhydroxy groups, forming anionic borate complexes. The corresponding anionic micelles exhibit adjustable surface charge density and allow a wide variation of retention of neutral and charged analytes by changing the pH and/or the concentration of borate [95]. The investigated analytes include herbicides having very different structures, aromatic amines and hydrocarbons, barbiturates and dansyl amino acids.

Other recent developments concern the use of MECC in chiral separations. In particular, chiral micelle polymers have been designed and synthesized for this purpose since it was found that the elimination of dynamic equilibrium between monomers and micelle enhances the chiral recognition [96].

In addition to the examples listed in Table 6.10, other recent applications of MECC have been reported. They include, among others, the following determinations (the micellar system used is indicated between parenthesis):

– Creatinine, uric acid and antipyrine in human plasma and urine (SDS) [106].

– Metal ions as β-diketonato complexes (SDS) [107].

– Chiral separation of amino acids as naphthalene-2,3-dicarboxaldehyde derivatives (cyclodextrin-modified SDS) [108].

– Chiral separation of amino acids as dansyl derivatives (cyclodextrin-modified taurodeoxycholic acid micelles) [109].

TABLE 6.10

Examples of MECC separations

Analytes	N	Exper. conditions	Ref.
Acetylacetone, cresols, xylenols, chlorophenols, phenol, ethylphenol	15	SDS 0.05 M; B: phosphate–borate (pH 7); 25 kV, 28 μA; Det.: UV 270 nm; Cap.: 50 μm i.d., 650–900 mm long; t: 20 min; T: 25°C.	80
Resorcinol, phenol, 4-nitroaniline, nitrobenzene; 2-naphthol	6	SDS 0.05 M; B: phosphate–borate (pH 7), 15 kV, 26 μA; Det.: UV 210 nm; Cap.: 50 μm i.d., 500 mm long, t: 15 min; T: 35°C.	81
Chlorinated phenols	20	SDS 0.07 M; B: phosphate–borate (pH 7); 10 kV, 28 μA; Det.: UV 220 nm; Cap.: 50 μm i.d., 500 mm long; t: 30 min; T: 35°C.	83
Substituted purines	6	SDS 0.05 M; B: Na_2HPO_4 0.01 M + $Na_2B_4O_7$ 0.006 M; 17.5 kV, 30 μA; Det.: UV 280 nm; Cap.: 75 μm i.d., 1000 mm long; t: 45 min; electroinjection.	91
NBD[a] derivatives of cyclohexylamine and ethylamine	2	SDS 0.01 M; Na_2HPO_4 0.003 M; 15 kV, 10 μA; Det.: Fluorescence, 540 nm; Cap.: 75 μm i.d., 750 mm long; t: 27 min; electroinjection.	93
NBD derivatives of cyclohexylamine and alkylamines	5	SDS 0.025 M; Na_2HPO_4 0.005 M; 20 kV; Det: Fluorescence, 540 nm; Cap.: silanated-silica, 50 μm i.d. 650 mm long; t: 80 min.	90
Pyrimidine bases and nucleosides	9	SDS 0.3 M; B: phosphate–borate (pH 7); 5.9 kV, 30 μA; Det.: UV 210 nm; Cap.: 50 μm i.d., 500 mm long; t: 35 min; T: 35°C.	84
Oligonucleotides (polythymidine mixture)	12	SDS 0.05 M; B: phosphate–Tris(pH 7) + urea 7 M; 20 kV, 10 μA; Det.: UV 260 nm; Cap.: 50 μm i.d., 450 mm long; t: 40 min; T: 25°C; 5 mM Mg(II) added to B.	84
Components of antipyretic analgesics	6	SDS 0.05 M; B: phosphate–borate (pH 11); 100 μA; Det.: UV 214 nm; Cap.: 100 μm i.d., 800 mm long; t: 20 min.	97
Catechols	4	SDS 0.01 M; B: phosphate–borate (pH 7); 13 kV, 8 μA; Det.: electrochemical; Cap: 52 μm i.d., 650 mm long; t: 20 min.	98
Catechols (as borate complexes)	7	SDS 0.01 M; B: phosphate–borate (pH 7); 20 kV, 7 μA; Det.: electrochemical; Cap.: 26 μm i.d., 643 mm long; t: 12 min; electroinjection.	85

continued

TABLE 6.10 (*continuation*)

Analytes	N	Exper. conditions	Ref.
Water-soluble vitamins	11	SDS 0.05 M; B: phosphate–borate (pH 9); 25 kV; Det.: UV 210 nm; Cap.: 50 μm i.d., 500 mm long; t: 15 min.	99
Aromatic carboxylic acids	4	SDS 0.05 M; B: phosphate–borate (pH 9) + TMAB[b] 0.06 M; 20 kV; Det.: UV 210 nm; Cap.: 50 μm i.d., 650 mm long; t: 20 min.	86
Zwitterionic antibiotics	9	SDS 0.05 M; B: phosphate–borate (pH 9) + TMAB 0.04 M; other conditions as indicated above.	78
Phenylthiohydantoin-amino acids	22	SDS 0.05 M; B: phosphate–borate (pH 9); 10 kV, 14 μA; Det.: UV 260 nm; Cap.: 50 μm i.d., 500 mm long; t: 30 min; T: 35°C.	100
PAR[c] complexes of transition metals	4	SDS 0.02 M; B: phosphate–borate (pH 7); 16.5 kV, 25 μA; Det.: VIS 500 nm; Cap.: 50 μm i.d., 500 mm long; t: 30 min.	101
Dansyl[d] derivatives of methylamine and methyl-d_3-amine	2	SDS 0.025 M + 20% methyl alcohol; B: phosphate–borate–bicarbonate; 30 kV 25 μA; Det.: Fluorescence, 470 nm; Cap.: 50 μm i.d., 1050 mm long; t: 80 min.	102
Penicillin antibiotics Cephalosporins	7 9	SDS 0.15–0.30 M; B: phosphate–borate (pH 9); 20 kV; Det.: UV 210 nm; Cap.: 50 μm i.d., 500 mm long; T: ambient.	103
Derivatives[e] of amino acid enantiomers	8	SDS 0.0125 M + SDV[f] 0.0125 M + 10% methanol; B: phosphate–borate (pH 7); 10.5 kV, 26 μA; Det.: UV 254 nm; Cap.: 50 μm i.d., 400 mm long; t: 50 min.	104
Racemic dansylated amino acids	6	TDC[g] 0.005 M; B: phosphate (pH 3); 50 μA; Det.: UV 210–220 nm; Cap.: 50 μm i.d., 500 mm long; T: 40°C; t: 70 min.	105

N: number of resolved compounds; B: buffer; Det: detector; Cap: fused-silica capillary; *a*: NBD = 4-chloro-7-nitrobenzofuran (derivatizing fluorescent label); *b*: TMAB = tetramethylammonium bromide; *c*: PAR = 4-(2-pyridylazo)-resorcinol; *d*: Dansyl = 5-dimethylamino-1-naphthalenesulfonyl-; *f*: SDV = sodium N-dodecanoyl-L-valinate; *g*: TDC = taurodeoxycholate.

– Illicit drugs in human biological fluids: heroin and impurities, opium alkaloids, amphetamines, benzodiazepines, etc. (SDS) [110]; opioids, amphetamines, benzoylecgonine and methaqualone (SDS) [111]; barbiturates, hypnotics, amphetamines, metabolites of cocaine, etc.(SDS) [112].

– Components of the antibiotic Augmentin (amoxycillin and derivatives),(SDS)[113].

Non-ionic micellar systems have also been used in some MECC separations. For example, peptidic molecules have been separated using buffered Tween 20 solutions [114]. The binding interactions between the micelles and the hydrophobic (and charged) peptides have been exploited, although in these cases the aggregates act mainly as a pseudostationary phase. Other MECC data are given in Refs. [130–133].

6.6. USE OF AMPHIPHILIC STRUCTURES AS STATIONARY PHASES IN CHROMATOGRAPHY

The ability of surfactants to adsorb onto solid surfaces can be, in principle, exploited to prepare modified stationary phases for LC and GC having different polarities, electrical charges, and chemical structures.

The formation of adsorbed structured monolayers (hemimicelles) and bilayers (admicelles) on certain inorganic solids was recognized by Fuerstenau [115] and, more recently, confirmed by Nunn and coworkers [116]. Since solute molecules from the mobile phase partition into the immobilized surfactant layer, a separation analogous to conventional LC occurs, which is termed admicellar-enhanced chromatography (AEC).

The binding process in which solutes are retained in a structured amphiphilic region (hemimicelle or admicelle) formed reversibly by adsorption of surfactant molecules onto the packing particles is called adsolubilization (see Fig. 6.31). The amount of analyte adsolubilized is proportional to the extent of surfactant adsorption until surface saturation is reached.

Practical applications of surfactant-modified stationary phases in current GC and LC have been reported in the literature. Some examples are indicated below.

Borosilicate glass capillary columns have been coated with cationic surfactants (HTAB) to provide a charged thin film of lipophilic stationary phase. In work with HTAB-containing mobile phases, the separation of drugs and metabolites present in biological samples has been achieved [117,118].

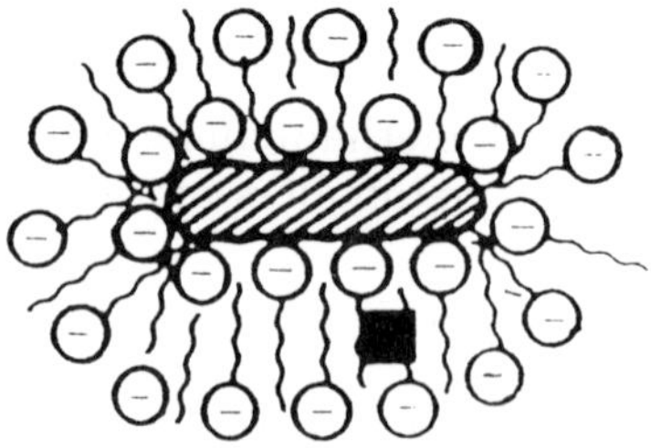

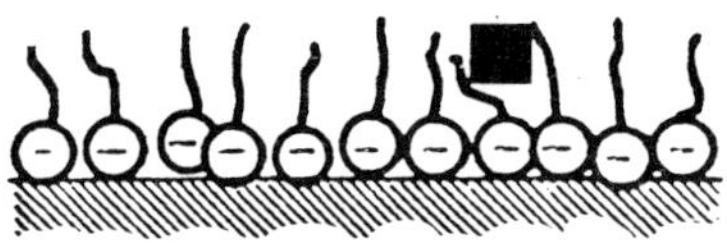

Fig.6.31. Adsolubilization by hemimicelles and admicelles.

Packing materials containing bonded polyoxyethylene phases, which are suitable for GC and LC work, have been obtained starting from silicic supports and non-ionic surfactants [119]. Alumina can also be covered with monolayers, bilayers and bulk liquid layers of Triton X-100, showing a solute retention dependent on the surfactant loading [120].

Macroreticular hydrophobic resins loaded with alkylamine amphiphiles improve the separation of gold, platinum and palladium, which are present as chlorocomplexes [121]. The retention of copper as an anionic C-nitroso-R-salt complex using the same stationary phases has also been reported [122].

Chromatographic papers have been also modified by adsorbing amphiphiles. For example, reversed-phase separation of some rare-earth elements has been conducted on Whatman 1 paper treated with hydrophobic alkylamines [123].

The main problem with using the above mentioned surfactant-modified stationary phases in LC is that the amphiphile is slowly eluted from the support, thus giving different retention performances with time.

In order to overcome this difficulty, stable polymerized thin-films can be obtained by reaction between the adsorbed amphiphiles and suitable adsolubilized compounds. For example, a polyion complex composed of dialkyammonium bilayers and poly(styrene sulfonate) has been deposited onto inert solid particles and performs well as a stationary phase for GC [124]. Immobilized bilayers of this type could be potentially useful in chromatography.

Due to these limitations, admicellar-enhanced chromatography has been scarcely applied until now as an analytical technique, whereas its use in stream treatment processes appear to be more interesting and promising [125]. The capabilities and performances of AEC as a new surfactant-based extraction technique will be examined in Chapter 7.

REFERENCES

1 D.W. Armstrong and F. Nome, *Anal. Chem.*, 53 (1981) 1662.
2 D.W. Armstrong, *Sep. Purif. Methods*, 14 (1985) 213.
3 J.G. Dorsey, *Adv. Chromatogr.*, 27 (1987) 167.
4 M.G. Khaledi, *Biochromatography*, 3 (1988), 20.
5 W.L. Hinze and D.W. Armstrong (Eds.), *Ordered Media in Chemical Separations*. ACS Symp. Series 342, American Chemical Society, Washington DC, 1987, Chapters 1,3–6.
6 A. Berthod and J.G. Dorsey, *Analusis*, 16 (1988) 75.
7 M.G. Khaledi, *Trends Anal. Chem.*, 7 (1988) 293.
8 M.F. Borgerding, R.L. Williams Jr., W.L. Hinze and F.H. Quina, *J. Liq. Chromatogr.*, 12 (1989) 1367.
9 D.W. Armstrong and J.H. Fendler, *Biochim. Biophys. Acta*, 418 (1977) 75.
10 P. Yarmchuk, R. Weinberger, R.F. Hirsch and L.J. Cline Love, *Anal. Chem.*, 54 (1982) 2233.
11 D.W. Armstrong and S.J. Henry, *J. Liq. Chromatogr.*, 3 (1980) 657.
12 E. Pramauro and E. Pelizzetti, *Anal. Chim. Acta*, 154 (1983) 153.
13 M. Arunyanart and L.J. Cline Love, *Anal. Chem.* 56 (1984) 1557.
14 R. Graglia, E. Pramauro and E. Pelizzetti, *Ann. Chim. (Rome)*, 74 (1984) 41.
15 A. Berthod and A. Roussel, *J. Chromatogr.*, 449 (1988) 349.
16 M. Arunyanart and L.J. Cline Love, *Anal. Chem.*, 57 (1985) 2837.
17 D.W. Armstrong and G.Y. Stine, *J. Am. Chem. Soc.*, 105 (1983) 6220.
18 R.A. Barford and B.J. Sliwinski, *Anal. Chem.*, 56 (1984) 1554.
19 H. Colin, J.C. Diez Masa, G. Guiochon, T. Czajkowska and I. Miedziak, *J. Chromatogr.*, 41 (1978) 167.
20 J.G. Dorsey, M.G. Khaledi, J.S. Landy and J.L. Lin, *J. Chromatogr.*, 316 (1984) 183.
21 A.S. Kord and M.G. Khaledi, *Anal.Chem.*, 64(1992)1894.

22 M.G. Khaledi, *Anal. Chem.*, 60 (1988) 876.
23 M.F. Borgerding, F.H. Quina, W.L. Hinze, J. Bowermaster and H.M. McNair, *Anal. Chem.*, 60 (1988) 2520.
24 R. Kaliszan, *Quantitative Structure Chromatographic Relationships*. Wiley, New York, 1987, Ch.7.
25 W.L. Hinze and S.G. Weber. *Anal.Chem.*, 63 (1991) 1808.
26 J.P. Foley and J.G. Dorsey, *Anal. Chem.*, 55 (1983) 730.
27 D.W. Armstrong, T.J. Ward and A. Berthod, *Anal. Chem.*, 58 (1986) 579.
28 P. Yarmchuk, R. Weinberger, R.F. Hirsch and L.J. Cline Love, *J. Chromatogr.*, 283 (1984) 47.
29 A. Berthod, I. Girard and C. Gonnet, *Anal. Chem.*, 58 (1986) 1356.
30 A. Berthod, I. Girard and C. Gonnet, *Anal. Chem.*, 58 (1986) 1362.
31 M.F. Borgerding, W.L. Hinze, L.D. Stafford, G.W. Fulp Jr. and W.C. Hamlin Jr., *Anal. Chem.*, 61 (1989) 1353.
32 L.R. Snyder and J.J. Kirkland, *Introduction to Modern Liquid Chromatography*, 2nd ed. Wiley, New York, 1979.
33 R. Bailey and R.M. Cassidy. *Anal.Chem.*, 64 (1992) 2277.
34 M.J. Medina Hernandez and M.C. Garcia Alvarez-Coque. *Analyst*, 117 (1992) 831.
35 J.G. Dorsey, M.T. De Echegaray and J.S. Landy, *Anal. Chem.*, 55 (1983) 924.
36 J.S. Landy and J.G. Dorsey, *J. Chromatogr. Sci.*, 22 (1984) 68.
37 F.P. Tomasella, J. Fett and L.J. Cline Love, *Anal. Chem.*, 63 (1991) 474.
38 G.F. Kirkbright and F.G.P. Mullins, *Analyst*, 109 (1984) 493.
39 F.G.P. Mullins and G.F. Kirkbright, *Analyst*, 111 (1986) 1273.
40 F.G.P. Mullins and G.F. Kirkbright, *Analyst*, 112 (1987) 701.
41 F.G.P. Mullins and G.F. Kirkbright, *Analyst*, 109 (1984) 1217.
42 A.S. Kord and M.G. Khaledi. *Anal.Chem.*, 64 (1992) 1901.
43 M. Arunyanart and L.J. Cline Love, *J. Chromatogr.*, 342 (1985) 293.
44 F.J. De Luccia, M. Arunyanart and L.J. Cline Love, *Anal. Chem.*, 57 (1985) 1564.
45 J. Haginaka, J. Wakai, H. Yasuda and T. Nakagawa, *Anal. Chem.*, 59 (1987) 2732.
46 I.Z. Atamna, G.M. Muschik and H.J. Issaq, *J. Liq. Chromatogr.*, 12 (1989) 1085.
47 F. Palmisano, A. Guerrieri, P.G. Zambonin and T.R.I. Cataldi, *Anal. Chem.*, 61 (1989) 946.
48 P. Menendez Fraga, E. Blanco Gonzalez and A. Sanz Medel, *Anal. Chim. Acta*, 212 (1988) 181.
49 A. Berthod, J.M. Asensio and J.J. Laserna, *J. Liq. Chromatogr.*, 12 (1989) 2621.
50 Y. Qu, P. Hu and P.L. Zhu. *J.Liq.Chromatogr.*, 14 (1991) 2755.
51 E. Pelizzetti and E. Pramauro, *J. Phys. Chem.*, 88 (1984) 990.
52 M.L. Marina, S. Vera and A.R. Rodriguez, *Chromatogr.*, 28 (1989) 379.

53 E. Pramauro, C. Minero, G. Saini, R. Graglia and E. Pelizzetti, *Anal. Chim. Acta*, 212 (1988) 171.
54 M.G. Khaledi, E. Peuler and J. Ngeh-Ngwainbi, *Anal. Chem.*, 59 (1987) 2738.
55 M.A. Garcia, S. Vera and M.L. Marina. *Chromatographia*, 32 (1991) 148.
56 D.W. Armstrong and K.H. Bui, *J. Liq. Chromatogr.*, 5 (1982) 1043.
57 D.W. Armstrong and M. McNeely, *Anal. Lett.*, 12 (1979) 1285.
58 D.W. Armstrong and R.Q. Terrill, *Anal. Chem.*, 61 (1979) 2160.
59 D.W. Armstrong, *Am. Laboratory*, 13 (1981) 14.
60 D.W. Armstrong, in: K.L. Mittal and E.J. Fendler (Eds.), *Solution Behavior of Surfactants: Theoretical and Applied Aspects*. Plenum Press, New York, 1982, p.1273.
61 D.W. Armstrong, K.H. Bui and R.M. Barry, *J. Chem. Ed.*, 61 (1984) 457.
62 F.M. Menger and D.W. Doll, *J. Am. Chem. Soc.*, 106 (1984) 1109.
63 D.G. Herries, W. Bishop and R.M. Richards, *J. Phys. Chem.*, 64 (1964) 1842.
64 F. Maley and D.U. Guarino, *Biochem. Biophys. Res. Com.*, 77 (1977) 1425.
65 J. Nagyvary, J.A. Harvey, F. Nome, D. Armstrong and J.H. Fendler, *Precambrian Res.*, 3 (1976) 509.
66 N. Funasaki, S. Hada and S. Neya, *J. Phys. Chem.*, 92 (1988) 7112.
67 N. Funasaki, S. Hada and S. Neya, *Bull. Chem. Soc. Jpn.*, 62 (1989) 1725.
68 N. Funasaki, S. Hada and S. Neya, *Bull. Chem. Soc. Jpn.*, 62 (1989) 2485.
69 T. Okada, *Anal. Chem.*, 60 (1988) 1511.
70 T. Okada, *Anal. Chim. Acta*, 230 (1990) 9.
71 T. Okada, *Anal. Chem.*, 60 (1988) 2116.
72 M.A. Hernandez Torres, J.S. Landy and J.G. Dorsey, *Anal. Chem.*, 58 (1986) 744.
73 A. Berthod, O. Nicolas and M. Porthault, *Anal. Chem.*, 62 (1990) 1402.
74 A. Berthod and M. De Carvalho. *Anal.Chem.*, 64 (1992) 2267.
75 W. Asche, *Chromatogr.*, 11 (1971) 411.
76 R.C. Reid, J.M. Prausnitz and T.K. Sherwood, *The Properties of Gases and Liquids,* 3rd Ed. McGraw-Hill, New York, 1977.
77 R.W. Gale, J.L. Fulton and R.D. Smith, *J. Am. Chem. Soc.*, 109 (1987) 920.
78 R.D. Smith, R.W. Gale and J.L. Fulton, *LC-GC*, 6 (1988) 134.
79 R.W. Gale, J.L. Fulton and R.D. Smith, *Anal. Chem.*, 59 (1987) 1977.
80 S. Terabe, K. Otsuka, K. Ichikawa, A. Tsuchiya and T. Ando, *Anal. Chem.*, 56 (1984) 113.
81 S. Terabe, K. Otsuka and T. Ando, *Anal. Chem.*, 57 (1985) 834.
82 Z. Prusik, V. Kasicka, S. Stanek, G. Kuncova, M. Hayer and J. Vrkoc, *J. Chromatogr.*, 390 (1987) 87.
83 K. Otsuka, S. Terabe and T. Ando, *J. Chromatogr.*, 348 (1985) 39.
84 A.S. Cohen, S. Terabe, J.A. Smith and B.L. Karger, *Anal. Chem.*, 59 (1987) 1021.

85 R.A. Wallingford and A.G. Ewing, *J. Chromatogr.*, 44 (1988) 299.
86 H. Nishi, N. Tsumagari and S. Terabe, *Anal. Chem.*, 61 (1989) 2434.
87 H.T. Rasmussen, L.K. Goebel and H.M. McNair. *J.High Res.Chromatogr.*, 14 (1991) 25.
88 J. Vindevogel and P. Sandra. *Anal.Chem.*, 63 (1991) 1530.
89 J.C. Giddings, *Anal. Chem.*, 39 (1967) 1027.
90 A.T. Balchunas and M.J. Sepaniak, *Anal. Chem.*, 59 (1987) 1466.
91 D.E. Burton, M.J. Sepaniak and P. Maskarinec, *Chromatogr.*, 21 (1986) 583.
92 S. Terabe, K. Otsuka and T. Ando, *Anal. Chem.*, 61 (1989) 251.
93 M.J. Sepaniak and R.O. Cole, *Anal. Chem.*, 59 (1987) 472.
94 P. Gareil, *Chromatogr.*, 30 (1990) 195.
95 J.T. Smith, W. Nashabeh and Z. El Rassi, *Anal.Chem.*, 66 (1994) 1119.
96 J. Wang and I.M. Warner, *Anal.Chem.*, 66 (1994) 3773.
97 S. Fujiwara and S. Honda, *Anal. Chem.*, 59 (1987) 2773.
98 R.A. Wallingford and A.G. Ewing, *Anal. Chem.*, 60 (1988) 258.
99 H. Nishi, T. Tsumagari, T. Kakimoto and S. Terabe, *J. Chromatogr.*, 465 (1989) 331.
100 K. Otsuka, S. Terabe and T. Ando, *J. Chromatogr.*, 332 (1985) 219.
101 T. Saitoh, H. Hoshino and T. Yotsuyanagi, *J.Chromatogr.*, 469 (1989) 175.
102 M.M. Bushey and J.W. Jorgenson, *Anal. Chem.*, 61 (1989) 491.
103 H. Nishi, N. Tsumagari, T. Kakimoto and S. Terabe, *J. Chromatogr.*, 477 (1989) 259.
104 A. Dobashi, T. Ono, S. Hara and J. Yamaguchi, *Anal. Chem.*, 61 (1989) 1984.
105 S. Terabe, M. Shibata and Y. Miyashita, *J. Chromatogr.*, 480 (1989) 403.
106 M. Miyake, A. Shibukawa and T. Nakagawa, *J. High Res. Chromatogr.*, 14 (1991) 181.
107 K. Saitoh, C. Kiyohara and N. Suzuki, *J. High Res. Chromatogr.*, 14 (1991) 245.
108 T. Ueda, F. Kitamura, R. Mitchell and T. Metcalf, *Anal. Chem.*, 63 (1991) 2979.
109 G. Okafo, C. Bintz, S.E. Clarke and P. Camilleri, *J. Chem. Soc. Chem. Commun.,* (1992) 1189.
110 R. Weinberger and I.S.Lurie, *Anal. Chem.*, 63 (1991) 823.
111 P. Wernly and W. Thormann, *Anal. Chem.*, 63 (1991) 2878.
112 P. Wernly and W. Thormann, *Anal. Chem.*, 64 (1992) 2155.
113 G.N. Okafo and P. Camilleri, *Analyst*, 117 (1992) 1421.
114 N. Matsubara and S. Terabe, *Chromatographia*, 34 (1992) 493.
115 D.W. Fuerstenau, *J. Phys. Chem.*, 60 (1956) 981.
116 C.C. Nunn, R.S. Schechter and W.H. Wade, *J. Phys. Chem.*, 86 (1982) 3271.
117 A. Balchunas, M. Capacci, M. Maskarinec and M. Sepaniak, *J.Chromatogr. Sci.*, 23 (1985) 381.

118 M.J. Capacci and M.J. Sepaniak, *J. Liq. Chromatogr.*, 9 (1986) 3365.
119 P.P. Wickramanayake and W.A. Aue, *Can. J. Chem.*, 64 (1986) 470.
120 K. Naito, H. Ogawa, S. Moriguchi and S. Takei, *J. Chromatogr.*, 299 (1984) 73.
121 H. Matsunaga and T.M. Suzuki, *Nippon Kagaku Kaishi*, 7 (1986) 859.
122 A. Yoshinaga and Y. Gohshi, *Bunseki Kagaku*, 35 (1986) 789.
123 C.G. Yeole and V.M. Shinde, *Analyst*, 108 (1983) 1102.
124 K. Taguchi, K. Hiratani and Y. Okahata, *J. Chem. Soc., Chem. Commun.*, (1986) 364.
125 J.H. Harwell and E.A. O'Rear, in: J.F. Scamehorn and J.H. Harwell (Eds.), *Surfactant-Based Separation Processes*. Dekker, New York, 1989, Ch. 7, p. 155.
126 R.M. Villaneuva Camañas, J.M. Sanchis Mallols, J.R. Torres Lapasiò, and G. Ramis Ramos, *Analyst*, 120 (1995) 1767.
127 L. Carretero, J.M. Vadillo and J.J. Laserna, *Analyst*, 120 (1995) 1729.
128 C. Garcia Pinto, J.L. Pèrez Pavòn and B. Moreno Cordero, *Analyst*, 120 (1995) 53.
129 H. Zov, Y. Zhang and P. Lu, *Anal. Chim. Acta*, 310 (1995) 461.
130 S. Yang and M.G. Khaledi, *Anal. Chem.*, 67 (1995) 499.
131 K. Jinno and Y. Sawada, *J. Liq. Chromatogr.*, 18 (1995) 3719.
132 E.S. Ahuja, E.L. Little, K.R. Nielsen and J.P. Foley, *Anal. Chem.*, 67 (1995) 26.
133 P.B. Wright and J.G. Dorsey, *Anal. Chem.*, 68 (1996) 415.

Chapter 7

Surfactant-based extraction and concentration methods

The role of surfactant-based separation techniques is growing in analytical chemistry due to the unique properties of amphiphile molecules and their organized assemblies, which have found a great number of useful applications. The main features of some new analytical methods, such as micellar chromatography and micellar electrophoresis, have been examined in previous chapters and the advantages offered by the corresponding determinations in terms of increased sensitivity and selectivity, have been discussed.

The ability of surfactant molecules and aggregates to promote the solute phase transfer in traditional liquid–liquid extractions and to act as specific solvents for the analytes can also be exploited in several extraction and preconcentration procedures. Particularly important is the use of aqueous micellar solutions, which can replace the more expensive and (in most cases) dangerous organic solvents, allowing efficient separations to be performed in aqueous media and under mild conditions. The development of suitable processes, capable of separating the enriched surfactant phase from the bulk solution is a further critical step for these micellar extractions.

In addition to the analytical applications, the use of surfactant-based methods for the treatment of aqueous streams and solid matrices to remove organic and inorganic contaminants, and for the recovery of valuable products, are promising new areas of great environmental and technological importance.

Fundamental and applied research in these fields have been extensively developed in the last ten years, as demonstrated by the increasing number of books and reviews including the basic features and practical applications of these new separation techniques [1–10].

Surfactant-assisted transport of solutes across liquid membranes, solute enrichment in coacervates, foam-based separations, cloud-point extractions, micellar ultrafiltration and solubilization in surfactant adsorbed layers are the most important and promising concentration–extraction schemes. Their features and performances will be examined in the following sections.

7.1. SURFACTANT-MEDIATED LIQUID–LIQUID EXTRACTIONS

Many surfactant-mediated extraction systems have been described in the literature, especially in the area of metal ion separation and for the preconcentration and removal of organic pollutants. The purification of biomolecules [11] is another very important field of application.

There is a variety of processes depending upon the nature of the analyte mixture and the extracting surfactant system employed. A broad division into two groups is generally accepted: processes involving non-polar solvent–surfactant systems (including reverse micellar extractions), and methods based on aqueous surfactants (normal micelles). Processes involving phase separations (i.e., cloud-point phenomena and coacervation) are included in the second group.

7.1.1. Surfactant-assisted extractions with non-polar solvents

Along with the extraction of metals and organic compounds in different liquid phases, the concentration of biomolecules into organic solvents with retention of their biological activity is a fundamental task in recently developed biotechnological procedures. In many cases, the extraction yield of the analytes can be greatly improved by introducing suitable amphiphiles into the extraction system.

Some examples of surfactant applications in liquid–liquid extractions will be described in this section, while the main aspects of bioseparations will be examined in detail in Chapter 8.

7.1.1.1. Separation of metal ions

The liquid–liquid extraction of metal ions in the presence of surfactants is a widely applied separation technique [12–14]. Depending on the conditions and the type of species present, the surfactant can act as either an ion-pair or phase-transfer agent. Moreover, it can participate in the formation of aggregates in the non-polar phase (e.g., reverse micelles or water–oil microemulsions), which usually play a crucial role in the extraction process [14]. The nature and properties of the aggregates as well as the binding and the catalytic effect on the ions to be extracted, can be greatly affected by changes in the medium conditions, such as pH, ionic strength, or temperature [15,16].

The presence of surfactants (either as phase-transfer agents, reverse micelles or microemulsions) can favourably alter the interfacial properties, facilitating the interphasic transfer of the species to be extracted. In fact, the amphiphiles influence the rate of extraction, which in turn depends on the kinetics of transfer of the species from the bulk water phase across the water-oil interface to the bulk organic phase.

When water-in-oil microemulsions form they have a water content which depends on the surfactant concentration and on the co-surfactant chain length. In some cases, the observed increase in the extraction rate in these systems can be explained in terms of the increased interfacial area of the microemulsion, faster mass transfer, and the concentration of the reagents in the dispersed droplets.

The above effects have been observed during the extraction of gallium from aqueous alkaline solutions into kerosene [17], where the lipophilic ligand 7-(1-ethenyl-3,3,5,5-tetramethylhexyl)-8-hydroxyquinoline accumulates at the droplet surface of the water–oil microemulsions obtained after addition of long-chain carboxylates and linear alcohols to the extraction systems. In this extraction, the dispersed aqueous phase (containing the cations to be removed) provides the hydrophilic core of the microemulsion (see Fig.7.1). Neutral chelate species and/or ion-pairs formed from cationic chelates and carboxylate counterions are efficiently transferred to the non-polar bulk solvent.

There are many examples of surfactant-assisted extractions of metals from aqueous and solid matrices where the use is reported of hydrophobic complexing agents having an amphiphilic behaviour. In some cases, the amphiphilic ligands can form neutral chelates (or ion-pairs) with the metal ions, which are extracted into the organic phase (cyclohexane, heptane, xylene, dodecanol, kerosene, chloroform, carbon tetrachloride, etc.) [18–22]. In other cases, anionic metal complexes can be extracted through the formation of suitable ion-pairs with amphiphilic cations (e.g., long-chain protonated amines or quaternary ammonium salts) [23–25]. Furthermore, polyoxyethylene-type surfactants are effective extractants of several metals, including transition metals [26], alkali and alkaline earth metals [27,28], and lanthanide ions [29]. These surfactants form cationic complexes with the metals

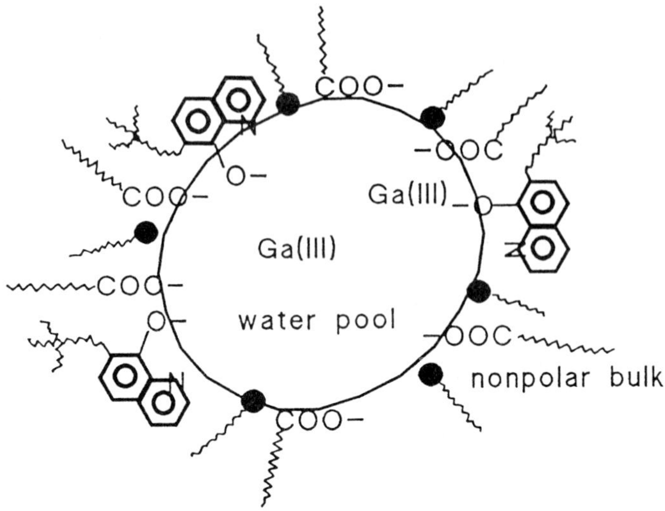

Fig. 7.1. Schematic representation of the distribution of reagents in the extracting water–oil microemulsion. Adapted with permission from Ref. [17].

through their polyoxyethylene groups and, in the presence of suitable anions, a neutral ion-pair is formed and transferred from water into organic solvents. An extended list of surfactant-mediated metal extractions can be found in Ref. [14].

Most surfactant-based extraction schemes reported in the literature [14,30,31] involve the use of either cationic or anionic surfactants, whereas the application of zwitterionic or non-ionic surfactants was reported in only a few cases.

After the extraction, the analytes accumulated in the organic phase can be back-extracted into the aqueous phase. This stripping step is usually performed by stirring the loaded organic surfactant medium with an aqueous acid solution having a relatively high organic/aqueous volume ratio (10:1, or even higher). Efficient back-extractions can be achieved, allowing the metal recovery with high concentration factors [32].

7.1.1.2. Extraction of organic solutes

Ionizable organic species can also be extracted from aqueous media or solid matrices through the formation of ion-pairs with suitable amphiphilic counterions. For example, various benzoic acid derivatives of pharmaceutical interest have been extracted in chloroform as ion-pairs with tri-n-octylamine [33]. The formation of zero-charge ion-pairs between the cationic amphiphile and the anionic solutes in the aqueous phase is followed by their transfer across the interface to the organic solution.

Other similar extraction schemes based on alkylammonium surfactants have been applied to transfer food dyes [34], pharmaceutical dyes [35], steroid sulfates [36] and ligninsulfonates [37] from aqueous solutions or biological fluids to organic phases.

7.2. EXTRACTIONS WITH AQUEOUS NORMAL MICELLES

Aqueous micellar media have been employed to extract a variety of biological products from cell membranes and microbial cultures [14,38]. During the solubilization of membrane components, the selectivity can be controlled by adjusting experimental parameters such as pH, ionic strength and amphiphile concentration.

The extraction of various organic components from hydrocarbons and solid matrices (soils, sediments, etc.) is also possible — thus exploiting the ability of aqueous micelles to exhibit differential solubilizing capabilities and rates with towards different compounds. For example, selectivity factors of ca. 10 were reported for the extraction of aromatic compounds, such as benzene, from hexane using cationic (HPC) or anionic (SDS) micelles [39,40].

Other examples of such micellar extractions include: recovery of phenols from carbolic oils [41], separation of polycyclic aromatic isomers from solid mixtures [42], washing of polychlorinated biphenyl from sandy materials using biodegradable alkyl poly(oxyethylene) glycols [43], removal of polycyclic aromatic hydrocarbons from soil and sediment samples using SDS [44], and the application of dodecanol ethoxylates for extraction of liquid hydrocarbons from environmental samples [45].

This area of separation science appears to hold great potential for soil remediation treatments and for large-scale extraction schemes since, as will be discussed in the following section, the coupling of micellar extraction/solubilization steps with concentration/enrichment procedures (cloud-point phase separation, or coacervation) is feasible.

7.2.1. Extractions based on thermal phase separation (cloud-point extractions)

Turbidity is often observed when aqueous solutions of several nonionic or zwitterionic surfactants are heated or cooled. This phenomenon,

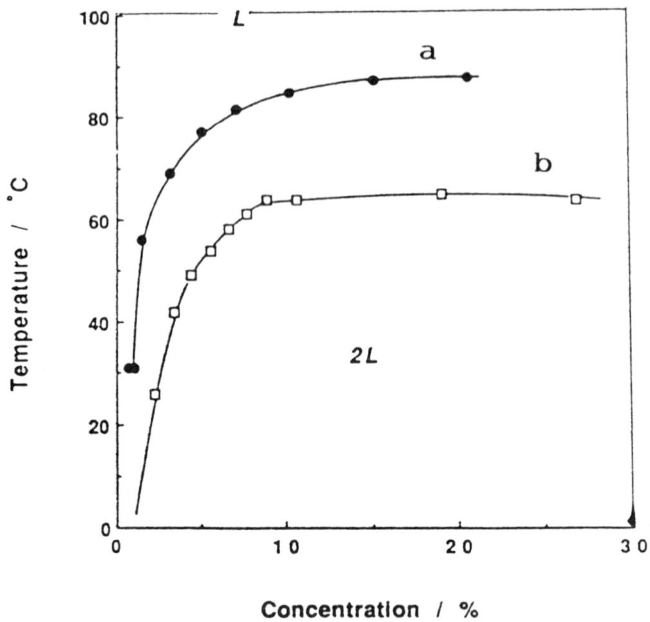

Fig. 7.2. Phase diagrams of aqueous solutions of zwitterionic amphiphiles. a: $C_{10}APSO_4$; b: C_9APSO_4. The corresponding formulae are shown in Table 7.1. Figure taken from Ref. [48], reproduced with permission.

which occurs in a narrow temperature range, is either due to the diminished solubility of the amphiphile in water or to a sharp increase in the micelle aggregation number [46,47]. The corresponding critical temperature is called the "cloud point" and depends on the amphiphile's nature and concentration.

A lower (or higher) consolute boundary is observed in the corresponding phase diagram, depending on the nature of the surfactant used. For example, when non-ionic surfactants are employed the solution separates into two isotropic liquid phases above the cloud point, whereas turbidity appears below the cloud temperature in aqueous solutions of several zwitterionic amphiphiles [48] (see Fig.7.2).

A typical phase diagram for a polyoxyethylene alkyl ether dissolved in water is shown in Chapter 1 (Section 1.13.2). At a given surfactant concentration, below the cloud point, the system is isotropic and transparent. Upon heating above the cloud point the system becomes turbid, and phase separation occurs, with the formation of a surfactant-rich

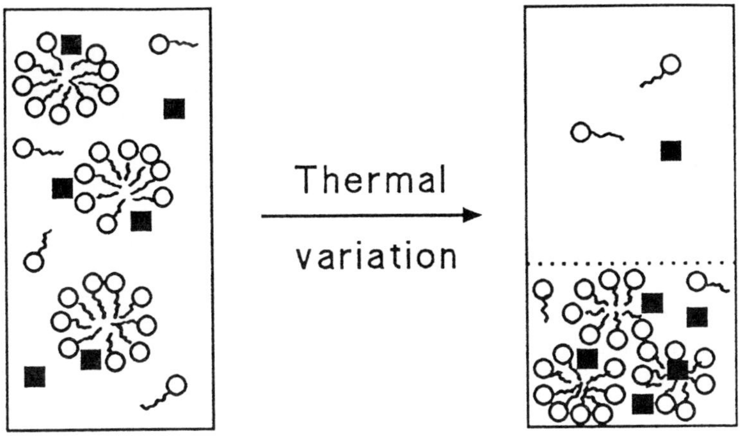

Fig. 7.3. Simplified scheme representing a cloud-point extraction.

liquid phase and an aqueous phase in which the surfactant concentration is in the cmc range. The accumulation of micelle-bound solutes takes place in the low-volume, surfactant-rich liquid.

In any cloud-point system the phase separation occurs spontaneously, at the working temperature, with a rate depending on the densities of the liquid phases. When the densities of the liquids are not very different, centrifugation is usually used to speed up the process.

The extraction procedure usually involves the addition of a proper surfactant (at concentration higher than the cmc) to an aqueous solution containing the analyte to be concentrated, and then changes of temperature, eventually followed by centrifugation. The accumulation of the analyte generally occurs in the lower phase since the surfactant-rich liquid is usually more dense, but the reverse situation can readily be obtained by increasing the density of the aqueous phase — for example by dissolving inert salts. A schematic representation of a cloud-point extraction is shown in Fig. 7.3.

The concentration factor in these extractions is fixed by the volume fraction of each phase, which is in turn dependent on the amphiphile concentration. The volumes of the extracting phase of typical alkyl- and alkylaryl-polyoxyethylene surfactants (present in the concentration range 0.5–2% w/v) represent about 1–10% of the initial solution volume, with corresponding concentration factors in the range 10–100. These factors are very sensitive to the surfactant type and structure [9].

7.2.1.1. Selection of surfactants for cloud-point extractions

Most work on cloud-point extractions has been performed using alkylphenyl poly(oxyethylene) ethers and poly(oxyethylene glycol) monoethers. The second class of amphiphiles is preferred in some cases since the absence of aromatic groups minimizes interferences during the analyte determinations by UV or fluorescence spectroscopy. More recently, zwitterionic surfactants (e.g., ammonium alkylsulfates) have been successfully applied in several separations of biological compounds [9,48].

By using suitable surfactants (or surfactant mixtures) and adjusting the experimental conditions (added solutes, buffers, ionic strength) phase separation can be obtained in the room temperature range. Table 7.1 lists the cloud-point temperatures of several aqueous surfactant solutions. The effect of additives is also indicated in some of the examples reported.

7.2.1.2. Cloud-point extraction of organic compounds

The cloud-point extraction has been applied for preconcentration and removal of a variety of organic compounds present in aqueous samples. These include organic pollutants, such as pesticides, aromatic hydrocarbons, PCBs, phenols and aliphatic alcohols, and biochemical and clinical species such as proteins, enzymes, porphyrins, hormones or vitamins.

The extraction yields are, in most cases, comparable to those obtained using classical extraction systems and sensitized determinations (i.e., fluorimetric, phosphorimetric, potentiometric, etc.) of the analytes can be performed directly on the enriched surfactant solution.

The cloud-point approach can also be applied to treat contaminated environmental phases, especially soils and water streams. In fact, non-ionic surfactant solutions have been proposed for washing contaminated soils [43,61–64], exploiting their solubilization capability and their interfacial tension lowering power. The treatment can be followed by disposal or reuse of the dilute wash solution obtained after the cloud-point separation step.

Another interesting application of micellar solutions is in the storage of aqueous samples prior to analysis, in order to avoid the loss of the organic hydrophobic analytes by sorption on the container walls [65]. Also, in this case the use of proper non-ionic surfactants could be further combined with a direct cloud-point extraction step. A more extended examination of the theoretical basis and applications of the cloud-point technique can be found in Ref. [9].

TABLE 7.1

Cloud-point temperatures of some non-ionic and zwitterionic surfactants[a]

Surfactant	Additive	Cloud point (°C)	Ref.
Non-ionic			
C_6E_4		59–66	49
C_8E_4		40	46
C_8E_6		71–76	
C_8E_8		96	
$C_{10}E_4$		21	
$C_{12}E_4$		2	49
		4	46
	3×10^{-3} M HTAC	>85	9
	1×10^{-3} M SDS	78–81	
$C_{12}E_8$		71–78	49–51
	40% Glycerol	62	50
	6 M Urea	>100	52
$C_{12}E_{23}$ (Brij 35)		100	53
Triton X-114		23–25	54
Triton X-100		65	55
	0.5 M NaCl	56	56
	0.5 M Urea	68	57
PONPE 7.5		5	58
	0.1 M KSCN	35	59
	0.5 M KSCN	41.5	
Zwitterionic			
$C_{10}AESO_4$		77	60
$C_{12}AESO_4$		>120	
$C_{10}APSO_4$		90	
$C_{12}APSO_4$		120	
$C_{14}APS$		12	

a: Surfactant concentration range 0.1–5%w/v. $C_nAESO_4 = C_nH_{2n+1}–(CH_3)_2N^+(CH_2)_2OSO_3^-$; $C_nAPSO_4 = C_nH_{2n+1}–(CH_3)_2N^+(CH_2)_3OSO_3^-$; $C_nAPS = C_nH_{2n+1}–(CH_3)_2N^+(CH_2)_3SO_3^-$.

Table 7.2 lists several examples of cloud-point extractions of organic analytes. Additional data can be found in Ref. [194].

As a general rule the cloud-point extractions are more effective when the analytes are uncharged, although hydrophobic ionized compounds can also be extracted. For a given surfactant system there is a threshold

TABLE 7.2

Cloud-point extractions of organic compounds

Analytes extracted	Surfactant system used	Ref.
Environmental pollutants		
Pesticides		
Endrin, DDT, lindane, aldrin, chloropyrifos, methoxy-chlor, chlordane, BHC	PONPE-7.5	66
2,4-D, 2,4,5-T	$C_{12}E_8/C_{12}E_{4.2}$	67
Parathion	Triton X-114	68
Phenols		
Phenol	$C_{12}E_8/C_{12}E_{4.2}$	67
4-t-Butylphenol	IGEPAL CA-520	69
Aromatic hydrocarbons		
Fluorene, fluoranthene, benzo[a]pyrene, pyrene, benzo[e]pyrene	IGEPAL CO-630	66
Chloroaromatic compounds		
4-chlorophenol, 3,5-dichlorophenol, 2,4,5-trichlorophenol, 3-chlorobiphenyl, 3,3′-dichlorobiphenyl, 2,3,4,5-tetrachlorophenol, pentachlorophenol	$C_{12}E_8/C_{12}E_{4.2}$	67
Aromatic amines		
4-chloroaniline, 2,4-dinitroaniline, 3,5-dinitro-aniline, 2,4,5-trinitroaniline, 2-chloro-4-nitro-aniline, 4-chloro-2-nitroaniline	C_8E_3	70
Biochemical and clinical analytes		
Polypeptides from rat organs	Triton X-114	71
Lipophilic vitamins (A and E)	Triton X-114	68
Vitamin E	$C_{10}APSO_4$	48
Haemoglobin (from bacteriorhodospin)	Triton X-114	72
Human membrane glycoproteins and phosphoproteins	Triton X-114	73
Porphyrins and metalloporphyrins	Triton X-114	74,75
Estriol, β-estradiol, progesterone	$C_{10}APSO_4$	48
Polyphenol oxidase	Triton X-114	76
Pyruvate oxidase	Triton X-114	77
ATP-citrate lyase	Triton X-100	78

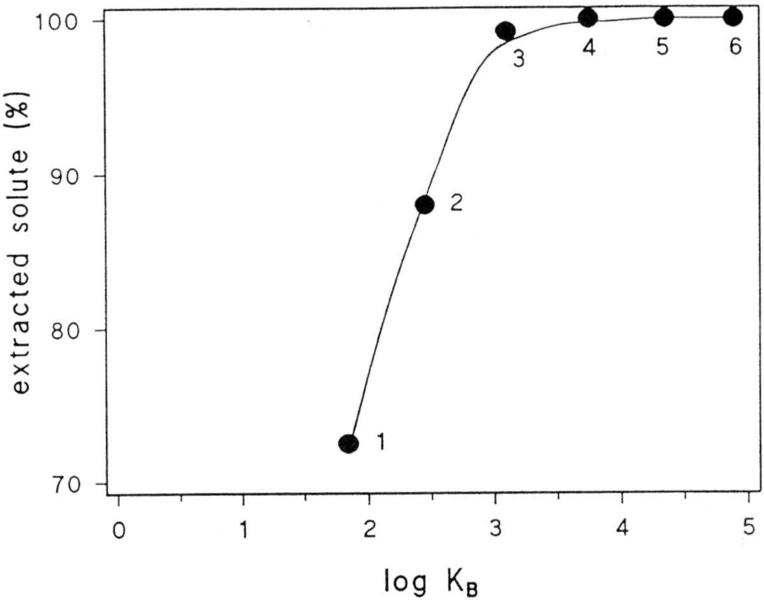

Fig. 7.4. Dependence of %E on the analyte–micelle binding constant for chlorophenols. (1) phenol; (2) 4-chlorophenol; (3) 3,5-dichlorophenol; (4) 2,4,5-trichlorophenol; (5) 2,3,4,5-tetrachlorophenol; (6) pentachlorophenol. Data taken from Ref. [67], with permission.

value of the binding constant which ensures the quantitative recovery of the analyte in the low-volume surfactant-rich phase. For example, a systematic study performed on chlorophenols [67] indicated that complete recovery of the analytes can be achieved when $K_B \geq 1000\ M^{-1}$, under the reported experimental conditions (see Fig. 7.4).

Other environmental pollutants, such as various polycyclic aromatic hydrocarbons, also exhibit very high binding constants to common non-ionic micelles [79], and can thus be quantitatively removed from aqueous wastes using the cloud-point approach [80]. Concentration factors comparable to those obtained using classical organic solvents have been achieved in these experiments.

Several experimental factors can be varied in order to maximize the extraction efficiency. For example, enhanced extraction of chlorophenols has been obtained by increasing the surfactant hydrophobicity in experiments performed with aqueous solutions of alkylpoly(oxyethylene) ethers having the same number of oxyethylene units and different alkyl chain lengths [70]. This effect results mainly from to the increase of the surfactant volume ratio and thus has a negative effect upon the concentration factor.

The cloud-point temperatures are clearly dependent on the surfactant structure, as shown in Table 7.1. Moreover, the presence of additives alters the cloud-point, and the analyte itself can reduce this temperature.

As expected, the pH is relevant when the analytes exhibit acidic or basic properties (e.g., aromatic amines, phenols or carboxylic acids) but its effect is negligible for neutral compounds such as hydrocarbons. The phase separation times (roughly 30 min at the working temperature) and the centrifugation times also need to be optimized. Detailed evaluations of all the factors affecting cloud-point extractions can be found in Refs. [9] and [70].

7.2.1.3. Cloud-point extraction of metal ions

Most investigations in this field concern the extraction of sparingly water-soluble chelate complexes. The separation of Ni(II)–TAN (TAN = 1-(2-thiazolylazo)-2-naphthol) complexes in Triton X-100 micellar solutions at temperatures above ca. 70°C was the first reported experiment [81]. Cloud-point extractions of transition metals using similar ligands (PAR, PAN) and other polyoxyethylene-type non-ionic surfactants have been reported since [82–88]. The application of this extraction step prior to the flow injection analysis of trace metals has also been proposed recently [54].

One of the major drawbacks of cloud-point extractions is, however, the relatively low partition coefficients exhibited by many neutral chelates in the two-phase system, with values of consequent partial recoveries of these analytes. This is because the extracting liquid is still highly hydrated and polar.

In order to increase the extraction yield, efforts were devoted to the development of reactive mixed micelles formed from common (unreactive) ionic or non-ionic surfactants and suitable amphiphilic ligands [89]. These ligands, which contain a selective chelating group and a tuned hydrophobic moiety (usually an alkyl chain of proper length), give rise to the formation of tightly bound complexes and allow one to perform more selective and efficient extractions of the target metals.

A systematic study has been performed on a series of acyl derivatives of 4-aminosalicylic acids (PAS-C_n) (see Fig. 7.5) dissolved in mixed micelles of Triton X-100 and $C_{12}E_{4.2}$ [polyoxyethylene (4.2) dodecyl

PAS-C_n Y-PAS-C_n

Fig. 7.5. Structures of amphiphilic ligands obtained from 4-aminosalicylic acid. C_n represents the group, $C_{n-1}H_{2n-1}$–CO.

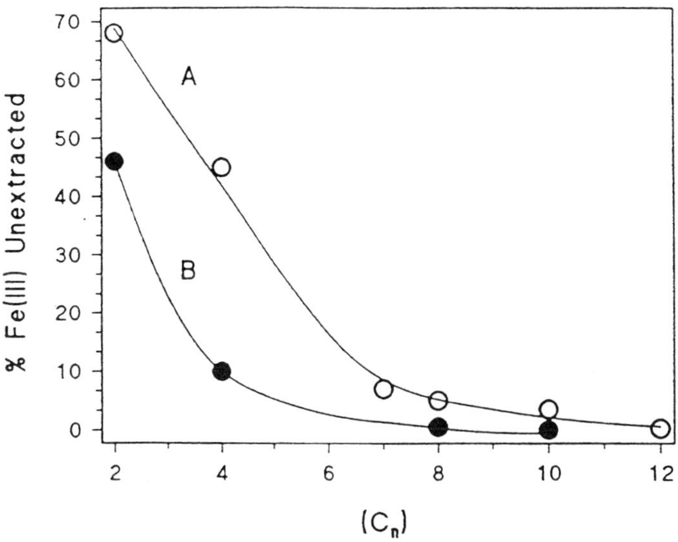

Fig. 7.6. Variation of the extraction efficiency as a function of the ligand hydrophobicity. (A) PAS-C_n. (B) Y-PAS-C_n. Experimental conditions: $C_{Fe(III)} = 1\times10^{-4}$ M; $C_{ligand} = 2\times10^{-3}$ M; Triton X-100 = 1% (w/v); $C_{12}E_{4.2}$ = 1% (w/v); $NaNO_3$ = 5% (w/v); pH = 3.5; T = 35°C. Reproduced with permission from Ref. [9].

ether], used as model systems for the extraction of iron(III) [90]. These ligands spontaneously form chelating micelles only in aqueous alkaline solution, since dissociation of both the carboxylic and phenolic groups is necessary to balance the hydrophobic moiety [91]. In neutral and acidic media, they readily form mixed aggregates with the host non-ionic surfactants used. A high binding constant of the complexes appears to be the fundamental factor in these extractions, although the kinetic aspects of chelate formation cannot be disregarded. The ligand binding constant and the complex stoichiometry, which is often different from that observed in homogeneous media, are also important factors.

For example, the ligand PAS-C_4 ($K_B \cong 500\ M^{-1}$ in Brij 35) is largely bound to non-ionic micelles, whereas the positively charged 1:1 iron(III)-PAS-C_4 complex, formed in the described extraction system, exhibits a much lower K_B value (ca. 70 M^{-1}) [92]. Thus, the metal cannot be quantitatively extracted in the surfactant-rich phase after clouding.

The use of related anionic ligands (Y-PAS-C_n) having the same chelating group but a different structure allows improvement of the extraction yield, since the corresponding 1:1 complexes exhibit a formal zero charge and are more tightly bound to the non-ionic micelles. It must be noted that ligands possessing the same alkyl chain length in both series have nearly the same binding constant to the micelles. The corresponding extraction performances are shown in Fig. 7.6.

Table 7.3 reports some examples of effective cloud-point extractions of metal chelates using non-ionic polyoxyethylene-type surfactants. The experimental details can be found in the corresponding references.

7.2.2. Separations based on coacervation

Aqueous solutions of ionic surfactants, proteins, synthetic polymers, and some microemulsions can affect separations into two liquid layers at constant temperature and under proper conditions. This phenomenon, termed coacervation, occurs when species of opposite charge are added to aqueous solutions of ionic amphiphiles. For example, if ammonium thiocyanate is added in the correct amounts to micellar solutions of HTAB, a coacervate is obtained [94] where analytes bound to the aggregates can be accumulated. In some cases, the term coacervation has also been used to describe cloud-point separations [95].

Coacervation usually occurs through different steps. Firstly, micelles aggregate to form "clusters" which in turn coalesce to give rise to microscopic droplets. Further coalescence brings about the formation of a continuous surfactant-rich phase and, finally, this phase can precipitate or flocculate upon addition of a critical concentration of electrolytes (cec). Consequently, separations analogous to those described in the previous section should be possible, although only a few examples have been reported so far [96,97].

TABLE 7.3

Cloud-point extractions of metal chelates

Analyte	Ligand	Surfactant system	%E	Ref.
Ni(II), Fe(IIII)	TAC	Triton X-100		82,84
Ni(II), Fe(II), Zn(II), Cd(II)	TAC	PONPE 7.5	≈100	87
Zn(II)	PAN	PONPE 7.5	≈100	81
U(VI)	PAN	Triton X-114	98	54
Zn(II), Cd(II), Cu(II), Ni(II)	PAMP	PONPE 7.5	≈100	83,85,86
Cu(II), Fe(III), Zn(II)	Thiocyanate	PONPE 7.5	72–97	88
Fe(III)	PAS-C_2	Triton X-100 +	30	90
	PAS-C_4	$C_{12}E_{4.2}$	55	
	PAS-C_8		95	
	PAS-C_{10}		≈100	
	Y-PAS-C_4		90	9
	Y-PAS-C_8		≈100	
Zr(IV), U(VI)	Arsenazo I	Tween 40	>96	93

Arsenazo: 2-(1,8-dihydroxy-3,6-disulfo-2-naphthylazo)-benzenearsonic acid trisodium salt; TAC: 2-(2-thiazoylazo)-4-methylphenol; PAN: 1-(2-pyridylazo)-2-naphthol; PAMP: 2-(2-pyridylazo)-5-methylphenol.

The use of coacervates has some disadvantages compared with the cloud-point technique, mainly because the solid phase must often be redissolved. However, this area is practically unexplored and still open to further applications.

7.3 .SOLUTE EXTRACTIONS USING EMULSIONS AND MICROEMULSIONS

The application of emulsions and microemulsions in extraction processes has a major practical interest, especially in the field of tertiary oil recovery [98–101]. Microemulsions, which are thermodynamically more stable, can extract organic material (oil) from emulsion droplets under rapid mixing conditions, with the diffusion-controlled collisions between the two types of aggregates being the rate-determining step [102].

Microemulsions may lead, in some cases, to the formation of new phases (including liquid crystals) during the extraction. The examination of these rather complex systems needs careful attention since the relative volumes and compositions of the phases vary with time [103]. Without entering into the details of processes having a predominantly industrial interest, two cases of emulsion-based extractions will be briefly mentioned as interesting examples of applications of microaggregates in separation science and for environmental purposes.

The first one concerns the separation of oleic and stearic acids from their mixtures. Upon addition of proper surfactants, the liquid fatty acid (oleic acid) is dissolved, giving rise to a surnatant water–oil emulsion, whereas the stearic acid (solid) remains in the aqueous phase. Sedimentation (usually with centrifugation) allows a sharp separation between the oil and the water phase, while further separation of (molten) stearic acid from water can be obtained by heating the dispersion [104].

The second example concerns the use of emulsions for preconcentration or removal of volatile organic compounds from exhaust air samples. Figure 7.7 shows a scheme of a laboratory gas washing bottle used to analyze the air coming from an automotive painting plant [105].

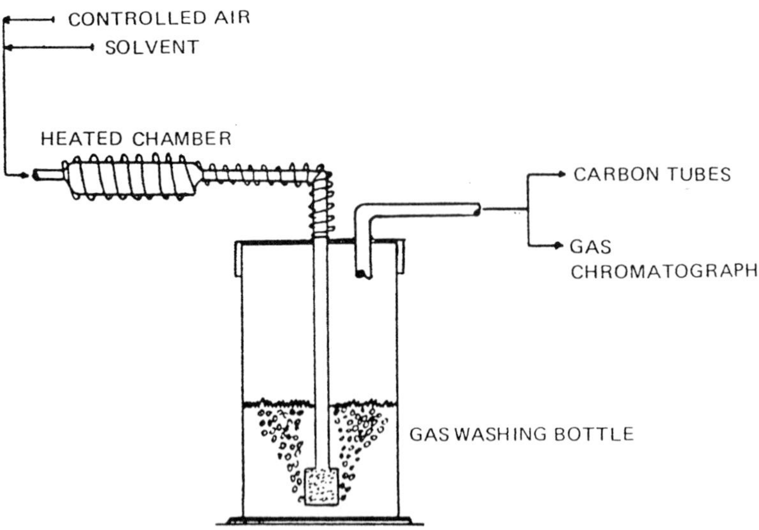

Fig. 7.7 Simplified scheme of an emulsion-based gas washing (or VOC absorption) system. Reproduced with permission from Ref. [105].

Various types of emulsifiers have been used to form pH-sensitive emulsions working in the pH range 8–10; they include dodecylbenzenesulfonic acid, oleic acid, polyoxyethylated fatty alcohols, stearic acid, palmitic acid, and lauric acid. The oil-soluble solvents (such as toluene) from the paints are dissolved in the oil droplets while other components (such as methyl ethyl ketone) are also partially dissolved in the continuous aqueous phase. A small pH variation breaks the emulsion, forming an oil phase, an aqueous phase and a sludge containing about 50% of water and 50% of oil with solids. The dangerous aromatic solvents removed from the air are concentrated in the sludge and partially in the oil phase. After the removal of solvents as sludge or from the liquid phases, the emulsion can be reformed by readjusting the pH. This emulsion-based stripping method can be exploited for the analytical control of polluted air samples and as a viable method for removal of volatile pollutants from gaseous streams. Applications of water–oil microemulsions and reverse micelles for extraction and purification of biomolecules are also of analytical concern and, moreover, have relevant practical interest. These extractions will be examined in detail in Chapter 8.

7.4. USE OF AMPHIPHILIC MOLECULES AND THEIR ASSEMBLIES IN LIQUID MEMBRANES

The use of surfactant-containing liquid membranes for the separation and purification of analytes was first proposed by Li [106]. In addition to their applications in separation techniques, these membranes have also been used as models of biological systems. The transport of the target species in such experiments is usually facilitated by the incorporation of a mobile carrier in the liquid membrane [107–109], which often is an amphiphilic molecule.

When the source phase is an aqueous solution containing the species to be separated, the surfactant is dissolved in an immiscible organic solvent and the receiving phase is again aqueous. The surfactant acts as a transport vehicle, either in monomeric form (with the formation of neutral ion-pairs with the charged analyte) or aggregated (by the formation of reversed micelles or water–oil microemulsions). The term emulsion liquid membrane (ELM) is generally used to indicate this second type.

When the analytes are dissolved in non-polar organic solvents, the liquid membrane is an aqueous solution containing amphiphilic monomers or micelles and the receiving phase is another immiscible organic solution. In some cases, supported liquid membranes (SLM) have been obtained by introducing a porous solid where the liquid is retained by adsorption or through capillary forces.

The tendency of a given solute to permeate the liquid membrane is given by its solubility and diffusion coefficient in that liquid, the product of these two terms being a measure of the analyte permeability. Proper selection of the liquid membrane can thus allow effective separations of solutes having very different permeabilities, usually with high flux rates.

7.4.1. Use of amphiphilic carriers in homogeneous liquid membranes

The transport of a given analyte across a liquid membrane (LM) can occur only if the solute is, at least partially, soluble in the LM and if there is a driving force towards the mass transfer. When the solute is insoluble or sparingly soluble in the LM, a carrier can be used which reacts reversibly with the analyte at the membrane/source-solution interface to form a lipophilic complex or ion-pair. Figure 7.8(A) shows a schematic example of transport of amino acids from basic to acidic solutions through a toluene LM, using a cationic amphiphile (N^+Cl^-). The reverse situation (transport from acidic to basic solutions) is illustrated in Fig. 7.8(B), where an anionic amphiphilic carrier (K^+DNNS^-) was used [110].

Similarly, the transport of fatty acids from a heptane solution (source phase) to another hydrocarbon (receiving) phase has been reported [111]. An aqueous LM containing HTAB micelles was used in this case.

The separation of metal ions using LM is also possible. Hydrophobic and amphiphilic macrocyclic ligands have been employed as carriers in these experiments [89]. The transport mechanism usually involves the cation uptake at the liquid membrane-source interface, with a corresponding proton release from the carrier, or the simultaneous transfer of both the metal ion and its co-anion when un-ionizable carriers are present (see Fig. 7.9). The metal stripping takes place in the acid receiving phase.

Crown ether carboxylic acids and phosphonic acid monoalkyl ethers are two examples of ionizable amphiphilic ligands used in transport of alkali metals across organic liquid membranes [112]. The hydrogen-ion gradient between the basic source phase and the acidic receiving phase provides the driving force in these experiments.

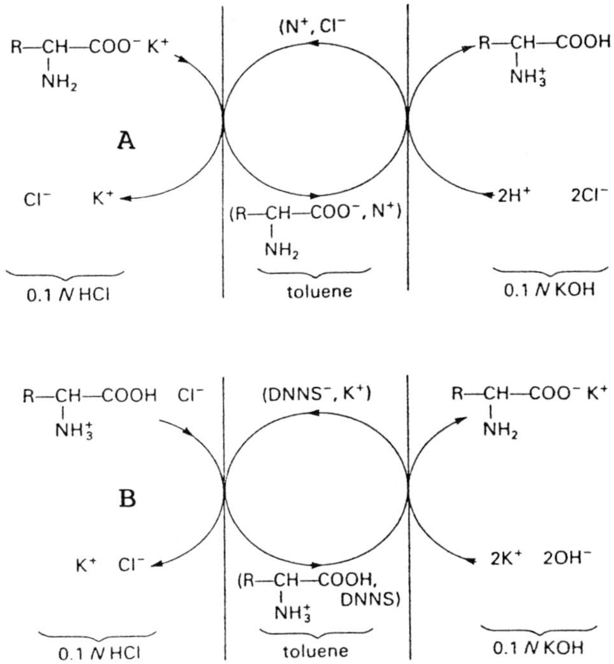

Fig. 7.8. Transport of amino acids through a toluene LM. (A) N^+Cl^- carrier: Aliquat 336 = tricaprylmethylammonium chloride; (B) K^+DNNS^- carrier: potassium dinonylnaphthalene sulfonate. Taken from Ref. [110], reproduced with permission.

Dicyclohexano-18-crown-6 (DC18C-6) is an example of a widely used unionized lipophilic ligand which allows the transport and separation of a variety of metals — Zn(II), Cd(II), Hg(II), Au(I), Ag(I), Pd(II). The presence of co-anions is essential since the diffusing species within the LM must be neutral ion-pairs [113].

The extraction of U(VI) using liquid membranes containing the ligands HDEHP and Kelex 100 [114,115], the transport of gold and silver cyanide complexes into a basic receiving solution through a decanol LM containing Kryptofix 22 DD [116], the extraction of alkaline earth nitrates using a chloroform membrane containing DC18C6 [117], the recovery of Co(II) using a paraffin LM containing di-2-ethylhexylphosphoric acid (D2EHPA) together with non-ionic polyamine surfactants [118], are other examples of ligand-mediated LM separations.

A survey of metal extractions using the LM approach, together with a

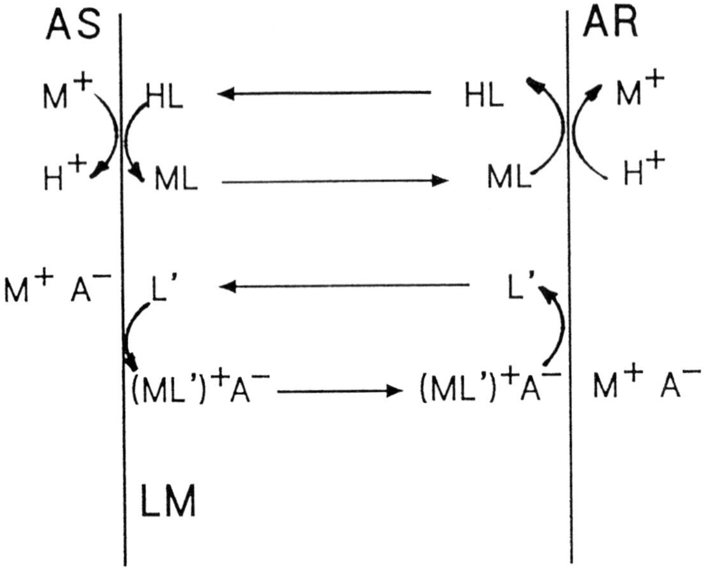

Fig. 7.9. Simplified scheme of a monovalent cation transport across a lipophilic liquid membrane. AR: aqueous receiving phase; AS: aqueous source phase; LM: hydrophobic liquid membrane. HL and L′ represent ionizable and un-ionizable ligands, respectively.

detailed description of some pilot and large scale processes of industrial interest, can be found in Ref. [119]. The structures of some widely used ionizable and neutral extractants are shown in Fig. 7.10.

The Reusch–Cussler equation (Eqn. 1) [120] describes the cation flux through the lipophilic barrier when crown-ether un-ionized carriers (L′) having oxygen, sulphur or nitrogen donor atoms, are involved in the metal transfer:

$$F_{M^+} = \frac{P_{(M^+,A^-)}\, K_f D_{(ML^+,A^-)}}{l}\, C_L C_{M^+}^{\ 2} \tag{1}$$

where F_{M^+} is the flux of the analyte (usually measured in mol h^{-1}), $P_{(M^+,A^-)}$ is the partition coefficient of the ion-pair between the aqueous

Opposite: Fig. 7.10. Structures of some hydrophobic and amphiphilic ligands used in LM metal ion extractions.

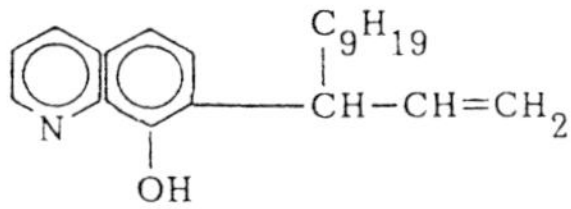

Kelex 100

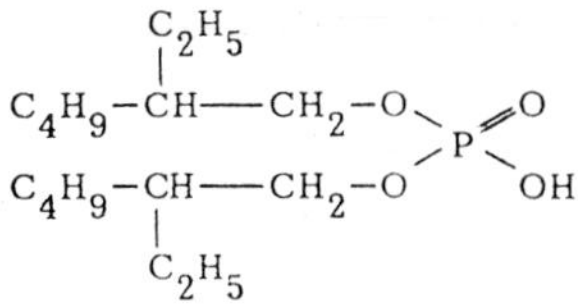

D2EHPA

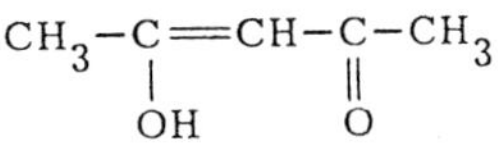

Acetylacetone

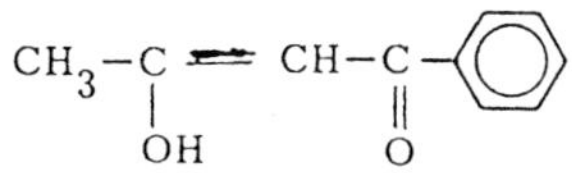

Benzoylacetone

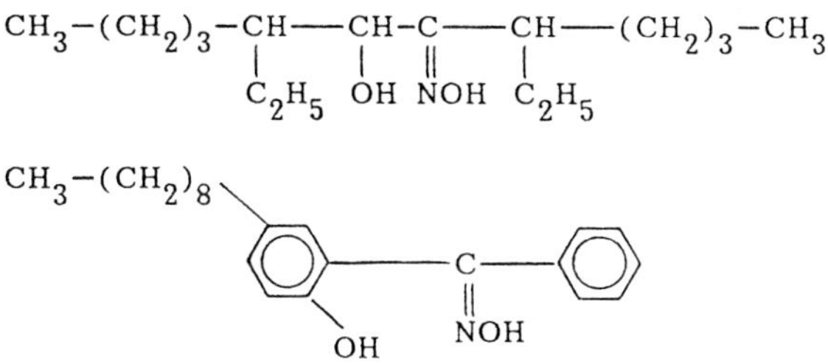

Hydroxyoximes

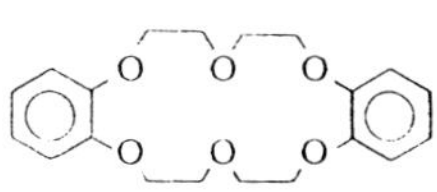

DB18 C6

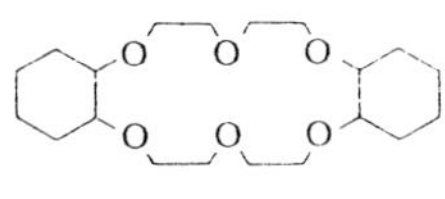

DC18 C6

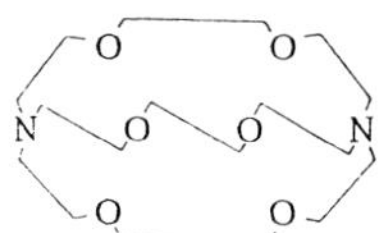

Kryptand (2,2,2)

source (or receiving) phase and the LM, K_f is the formation-constant of the complex in the organic phase, $D_{(ML^+,A^-)}$ is the diffusion coefficient of the neutral ion-pair formed between the cationic complex ML^+ and the counter-ion A^-, l is the length of the diffusion path, C_L is the concentration of the lipophilic carrier in the LM and C_{M^+} is the cation concentration in the source phase.

Detailed investigations concerning the experimental factors influencing the metal ion flux and the development of more complex flux models to describe the analyte transport can be found in Refs. [121,122].

7.4.2. Separation of solutes using emulsion liquid membranes

Selective separations of organic compounds and metal ions from aqueous sources can be performed using a number of types of ELM obtained in the following simple way. Emulsification of the inner receiving phase in an immiscible solvent (usually in the presence of amphiphilic emulsifiers) is performed in high speed stirrers. The resulting

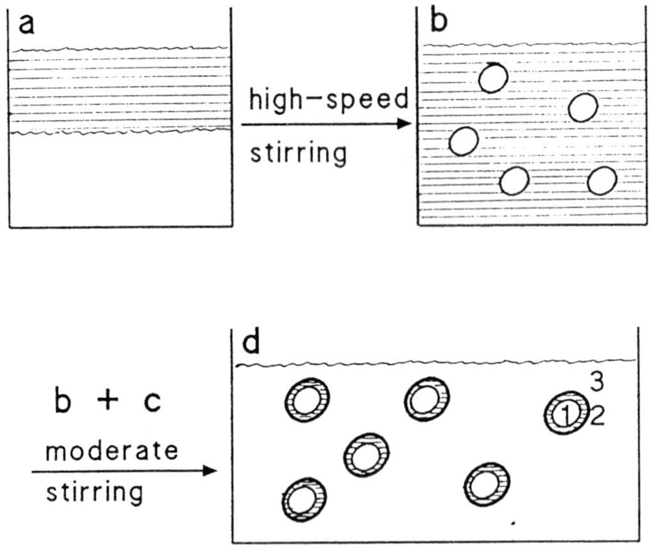

Fig. 7.11. Preparation of an emulsion liquid membrane. (a) Aqueous receiving phase (AR) + oil + surfactant; (b) water–oil emulsion; (c) aqueous source phase (AS); (d) ELM (1 = AR, 2 = LM, 3 = AS).

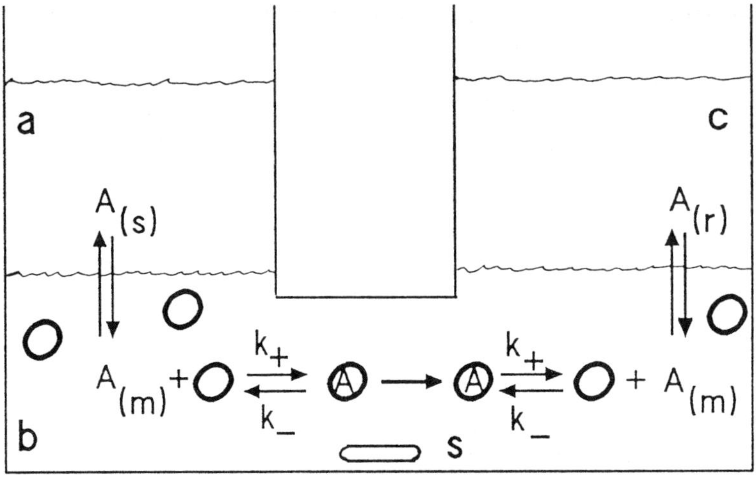

Fig. 7.12. Simplified scheme of a microemulsion-based separation cell. (a) Oily source phase; (b) oil–water microemulsion (liquid membrane); (c) organic receiving phase; (s) stirrer. The globules represent the oil–water droplets. Solubilization model proposed in Ref. [123].

emulsion is, in turn, dispersed in the continuous source phase under conditions of low shear (see Fig. 7.11). After the extraction, the emulsion must be broken (usually by thermal or electrical treatment) into its components to recover the receiving phase.

Problems related to the emulsion stability can be overcome by using more stable microemulsions. Figure 7.12 is a schematic representation of the analyte transport across a microemulsion liquid membrane and the experimental system used.

Basically, a lipophilic analyte, A, is transferred from the organic source phase to the interfacial region of the ELM, where it is dissolved in an oil–water microemulsion droplet and subsequently transported by diffusion to the membrane/receiving solution interface. After exit from the host droplet, the analyte diffuses to the bulk of the organic receiving phase.

In these systems, the flux F of the transported analyte is given by Eqn. (2) [123]:

$$F = \frac{D_m KPC_d}{l} \frac{C_s^0}{1 + PKC_s^0} \qquad (2)$$

where D_m is the diffusion coefficient of the microemulsion droplet in the aqueous liquid membrane, K is the ratio between the solute entry rate constant (k_+) and the exit rate constant (k_-); P is the partition coefficient of the analyte between the oil phase and the continuous aqueous phase of the microemulsion, C_d is the total droplet concentration, l is the diffusion layer thickness and C_s^0 is the analyte concentration in the source phase.

Plots of 1/F versus $1/C_s^0$ were found to be linear in transport experiments of pyrene, anthracene and perylene (from an oil source phase to an oil receiving phase) through a liquid membrane composed of an oil–water microemulsion (SDS/pentan-1-ol/n-dodecane/water). This

TABLE 7.4

Selective extractions of some metals using ELM

Analyte/source phase	Liquid membrane/ amphiphilic carrier	Receiving phase	Ref.
Pb(II)/aqueous nitrate	Chloroform/DC18C6	Aqueous $Na_4P_2O_7$	128
Cd(II), Hg(II), Zn(II)/ aqueous SCN^-, I^-, Br^- or Cl^-	Toluene/DC18C6	Aqueous $Mg(NO_3)_2$ or MgS_2O_3	129
Lanthanoids/aqueous solution, pH 3	Kerosene/2-ethylhexyl phosphonic acid/Span 80	Aqueous HCl/LiCl	130
Tl(I)/aqueous solution of different Tl(I) salts	Toluene/DC18C6/	Aqueous $Li_4P_2O_7$	131
Alkali metal picrates/ aqueous solution	DC18C6 and C_{12}18C6 in water–oil microemulsion: n-decane/$C_{12}E_4$/water/1-hexanol	Aqueous solution	125
Co(II)/aqueous solution, pH≈5	Kerosene/PC-88A/ Span 80	Aqueous HCl	132

Span 80 = sorbitan monooleate; PC-88A = 2-ethylhexylphosphonic acid mono-2-ethylhexyl ester; C_{12}18C6 = dodecano-18-crown-6.

confirmed the validity of the model in which the analyte diffuses within the droplets and a fast solubilization–desolubilization of solutes occurs [123]. A similar model can also describe the transport of hydrophilic (ionic) species through LM composed of water–oil microemulsions [124,125].

The crucial role of surfactant emulsifiers in ELM behaviour has been clearly recognized in several experiments. In addition to the widely used amphiphile Span 80, other surfactant compounds have been examined. In particular, glutamic acid dioleyl esters were found to be good candidates for the preparation of highly stable ELM [126].

Some metal separations based on emulsion and/or microemulsion LM are listed in Table 7.4. Examples of practical applications in large-scale plants, including efficient and cheap demetallation of industrial wastes, can be found in Ref. [127].

Applications of ELM for biomaterial separations in the areas of biomedicine and biotechnology have also been reported [110]. The main features of these methods and the importance of studies on membrane-mimetic processes for biological modelling will be examined in Chapter 8.

7.5. MICELLAR-ENHANCED ULTRAFILTRATION

One of the most difficult problems arising from environmental pollution is the removal of toxic organic and inorganic species present in small or trace quantities in aqueous streams. Membrane-separation processes are not, in principle, technically viable since membranes which allow the water permeation but reject small organic molecules (or inorganic ions) are not available. However, the incorporation of contaminants into long macromolecules or aggregates can allow this limitation to be overcome, as demonstrated some years ago by Dunn et al. [133]. In fact, micellar enhanced ultrafiltration (MEUF) allows the separation of dissolved solutes through their association to added micellar aggregates which are subsequently removed from the bulk solution by ultrafiltration.

An important point is that the membrane pore-size has to be small enough to block the aggregates (and their guest solutes) in the retained fraction (retentate) and large enough to allow acceptable flux rates. Figure 7.13 schematically represents the MEUF process.

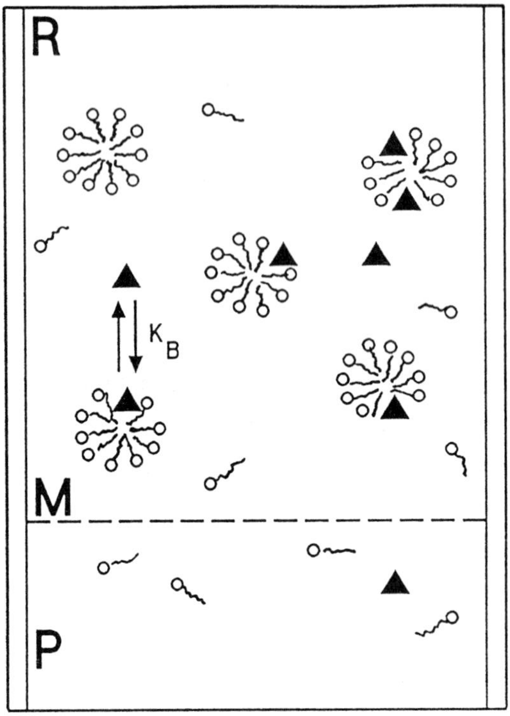

Fig. 7.13. Schematic representation of the MEUF procedure. R = retentate; M = membrane; P = permeate; ▲ = partitioned solute.

MEUF has been successfully applied to remove organic and inorganic solutes from aqueous streams and, more recently, the method has been proposed as a viable preconcentration step in some analytical determinations. Selective separations are possible because the extent of solubilization depends on both the structure of the substrate and the amphiphile. Another selectivity factor is introduced when electrostatic interactions are present between solutes and micelles.

If the micelles are completely rejected by the membrane, they accumulate in the retentate phase and the permeate contains only low amounts of surfactant, near the cmc. The efficiency of MEUF is usually measured by the rejection coefficient (or rejection factor), R, defined as follows:

$$R = 1 - C_p/C_0 \quad (3)$$

where C_p and C_0 are the solute concentrations in the permeate and in the initial (feed) solution, respectively.

Ultrafiltration, which usually involves the pressure facilitated passage of solutes through the membrane barrier, gives rise to a concentration gradient, referred to as concentration polarization (or gel-layer formation). This solute layer acts as a secondary membrane, reducing the flux rate. Moreover, the concentrations of the rejected solute and surfactant species are higher than in the bulk retentate.

The concentration polarization can be reduced by sample dilution during filtration, by efficient stirring next to the membrane surface or by sample flow parallel to the membrane. Figure 7.14 shows the configuration of a typical UF stirred cell, often used in laboratory experiments.The flux is enhanced by increasing the stirrer speed. For any given surfactant and solute composition a speed limit exists below which the flux sharply drops, indicating the formation of the gel layer.

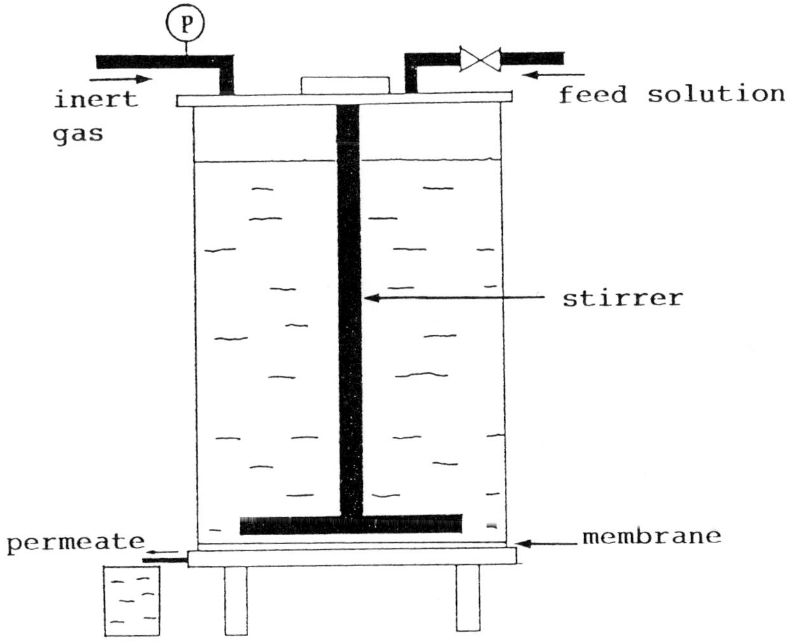

Fig. 7.14. Scheme of a MEUF stirred cell.

7.5.1. Separation of organic compounds using MEUF

Several organic compounds of environmental interest, present at the trace level in aqueous streams, have been separated using the MEUF approach. They include aromatic hydrocarbons [134], alkylphenols [133], aliphatic alcohols [135], chlorophenols and chlorinated pesticides [67], and aromatic amines [136] (see Table 7.5).

TABLE 7.5

Concentration of different organic solutes in the retentate phase using MEUF[a]

Analyte, concentration in the feed solution	Micellar system used	% R	Ref.
Aromatic hydrocarbons			
Benzene, 1×10^{-3} M	0.05 M HPC	89.9	134
	0.05 M SDS	80.4	
Phenols			
Phenol, 1×10^{-2} M	0.05 M HPC	73.0	137
1×10^{-4} M	0.02 M HTAB	55.0	67
1×10^{-4} M	0.02 M SDS	31.0	
t-Butylphenol, 3.2×10^{-3} M	0.032 M HPC[b]	97.6	133
3.2×10^{-2} M	0.32 M HPC[b]	99.7	
4-Chlorophenol, 1×10^{-4} M	0.02 M HTAB	95.0	67
	0.02 M SDS	64.0	
3,5-Dichlorophenol, 1×10^{-4} M	0.02 M HTAB	≥99.9	
	0.02 M SDS	97.0	
Pentachlorophenol, 5×10^{-5} M	0.02 M HTAB	≥99.9	
	0.02 M SDS	≥99.9	
Chlorinated biphenyls			
3-chlorobiphenyl, 2×10^{-5} M	0.02 M HTAB	≥99.9	
	0.02 M SDS	≥99.9	
3,3′-Dichlorobiphenyl, 2×10^{-5} M	0.02 M HTAB	≥99.9	
	0.02 M SDS	≥99.9	
Chlorinated pesticides			
2,4-D acid, 1×10^{-4} M	0.02 M HTAB	99.7	
	0.02 M SDS	97.5	
2,4,5-T acid, 1×10^{-4} M	0.02 M HTAB	≥99.9	
	0.02 M SDS	99.8	

Analyte, concentration in the feed solution	Micellar system used	% R	Ref.
DDT	0.02 M HTAB	≥99.9	
	0.02 M SDS	≥99.9	
Aromatic amines			
Aniline, 2×10^{-4} M	0.02 M SDS	34.0	136
	0.02 M HTAB	38.0	
4-Chloroaniline, 2×10^{-4} M	0.02 M SDS	61.0	
	0.02 M HTAB	85.0	
4-Bromoaniline, 2×10^{-4} M	0.02 M SDS	68.0	
	0.02 M HTAB	91.0	
4-Isopropylaniline, 2×10^{-4} M	0.02 M SDS	86.0	
	0.02 M HTAB	91.0	
4-t-Butylaniline, 2×10^{-4} M	0.02 M SDS	86.0	
	0.02 M HTAB	94.0	
Aliphatic alcohols			
1-Hexanol, 7×10^{-3}	0.07 M HPC	78.0	135
1-Heptanol, 7×10^{-3} M	0.07 M HPC	95.0	
1-Octanol, 7×10^{-3} M	0.07 M HPC	98.0	

a: Cellulose acetate membranes, MWCO: 10,000 dalton; b: MWCO: 1000–5000 dalton. T: 25–35°C. Solution pH: 3 (chlorophenols); 2 (chlorinated pesticides); 8–8.5 (aromatic amines); unmodified pH for the other solutes (≈5–6). Most reported pressures within the UF cell are in the range 2–5 bar.

Irrespective of the micellar system used, quantitative accumulation of the analytes in the retentate can be achieved when the solute–micelle binding constants are higher than ca. 1000 M^{-1}, a threshold limit which is comparable to that found in cloud-point extractions. The binding constants are, in turn, largely dependent on the nature of the surfactant and solute and on the presence of electrostatic effects. For example, the extent of partitioning of undissociated polar compounds, such as phenol, may be increased by choosing surfactants of proper molecular structure. Related effects on the ultrafiltration yield can also be obtained by modifying the solution pH and ionic strength (when charged solutes and ionic aggregates are present). Moreover, the characteristics of the membranes applied, and the effects induced by solutes, additives and micelles may influence the MEUF performance and must, therefore, be taken into account. Most of the above-mentioned experimental factors exert their influence also in MEUF separations of inorganic analytes.

7.5.1.1 Effect of pH

Changes in the pH of the solution can drastically alter the rejection of ionic solutes. For example, the removal of the carboxylic herbicide 2,4,5-T (2,4,5-trichlorophenoxyacetic acid) from contaminated waters can be performed in acid media (pH ≈ 2) using anionic SDS micelles. The ionization of this compound at higher pH values increases the hydrophilic character of the molecule and introduces electrostatic repulsion between micelles and substrate, thus reducing the rejection factor. In contrast, HTAB micelles are also effective at higher pH values because the negative effect of ionization on K_B is compensated by the electrostatic attraction and by the specific interactions operating between the surfactant head-groups and the aromatic ring of the pollutant (see Fig. 7.15).

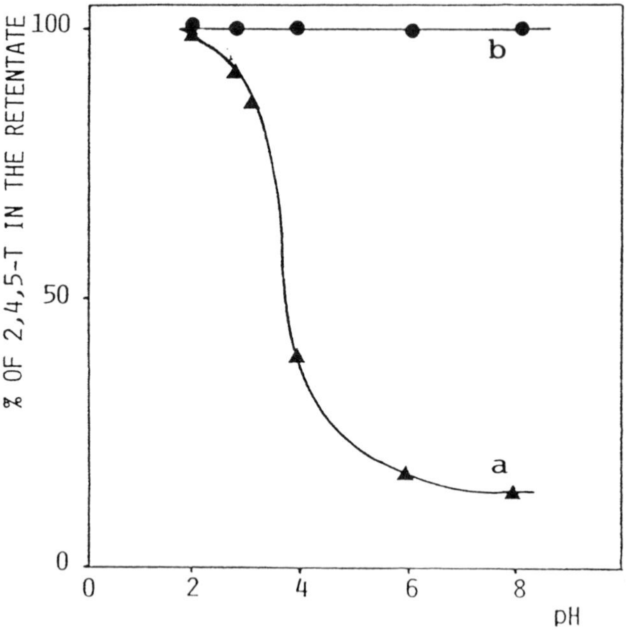

Fig. 7.15. Effect of pH on MEUF of 2,4,5-T acid. (a) 0.02 M SDS; (b) 0.02 M HTAB. Figure adapted, with permission, from Ref. [67].

As a consequence of their protonation in acid media, aromatic amines show higher rejection in acid SDS solutions, where electrostatic attraction is relevant: in basic solutions, HTAB micelles are more effective since uncharged aromatic solutes are more strongly bound to cationic aggregates.

As expected, the influence of pH is negligible for neutral solutes and for highly hydrophobic ionizable compounds (such as pentachlorophenol), where the hydrophobic contribution to the binding constant largely prevails.

7.5.1.2. Effect of surfactant concentration

The effect of surfactant concentration on the MEUF yield is relevant only for partitioned solutes, but does not alter significantly the bound fraction of highly hydrophobic substrates. For example, the rejection of aniline at pH 8 varies from ca. 17% to ca. 33% using SDS at 0.01 M and 0.04 M, respectively, whereas the % R of 2-chloro-4-bromoaniline changes from ca. 95 to 100 in the same SDS concentration range [136]. Of course, the added surfactant concentration must be minimized when MEUF is applied for environmental cleaning. However, when concentrated surfactant solutions are used, peculiar effects are observed. In particular, the surfactant concentration in the permeate increases (leakage of micelles) an effect that was formerly attributed to the formation of submicellar aggregates (dimers, trimers, etc.) [133]. This phenomenon occurs above a certain critical surfactant concentration, depending on the amphiphile structure and on the membrane pore-size (for HPC ca. 0.25 M leakage occurs when the solution passes through a 10,000 dalton cellulose acetate membrane).

Another explanation has been tentatively proposed in a more recent study, where the ultrafiltration of cationic (DTAB) and non-ionic (Igepal CO) micelles has been investigated in detail [138]. The rapid increase of viscosity at higher surfactant concentrations (see Fig. 7.16) indicates that the micelles become longer and more rod-like and this hypothesis is supported by the corresponding sharp decrease in the micellar diffusion coefficient observed in the same concentration range.

There is some contradictory evidence, since longer rod-like micelles exhibit a greater tendency to permeate through a cylindrical hole than do smaller ellipsoidal or spherical micelles. One possible explanation may be that the alignment of rod-like aggregates by fluid-shear facilitates their flow through the cylindrical holes. Experiments performed with polystyrene polymers longer than the pore diameter of the UF

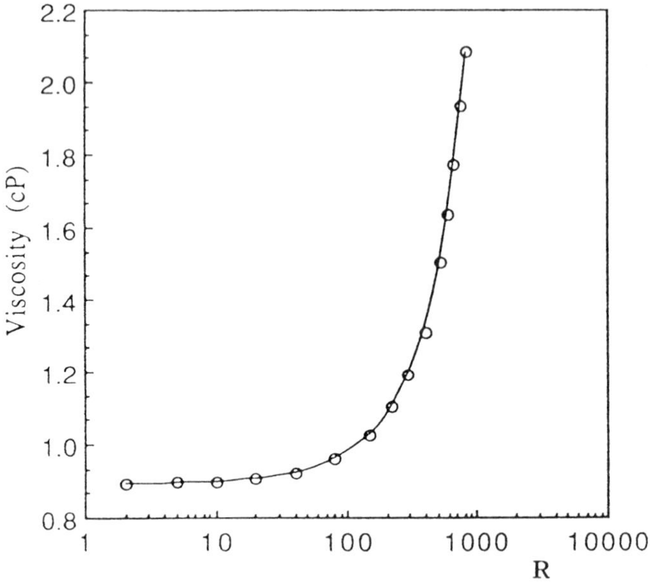

Fig. 7.16. Variation of solution viscosity with the surfactant concentration ratio, R. Surfactant: Igepal CO-850; R = C_S/cmc. Reproduced with permission from Ref. [138].

membrane showed polymer permeation at high flux rates and complete rejection at low solvent velocities [139]. The results are, however, still open to discussion.

7.5.1.3. Effect of ionic strength

The effect of ionic strength is another important factor when charged substrates and ionic aggregates are present. Because of shielding of the electrostatic solute-micelle attraction, the rejection coefficients of aniline derivatives in SDS (at pH 3) are reduced by the addition of inert salts. The reduction in ultrafiltration efficiency is evident for aniline, whereas more lipophilic compounds, such as alkylanilines, show only minor effects because the hydrophobic contribution to K_B largely predominates (see Fig. 7.17). Moreover, the addition of inorganic salts generally increases the rejection of ionic aggregates because the micelle grows and the cmc decreases. No significant ionic-strength effects have been observed in MEUF separations of uncharged solutes and/or in the presence of non-ionic aggregates.

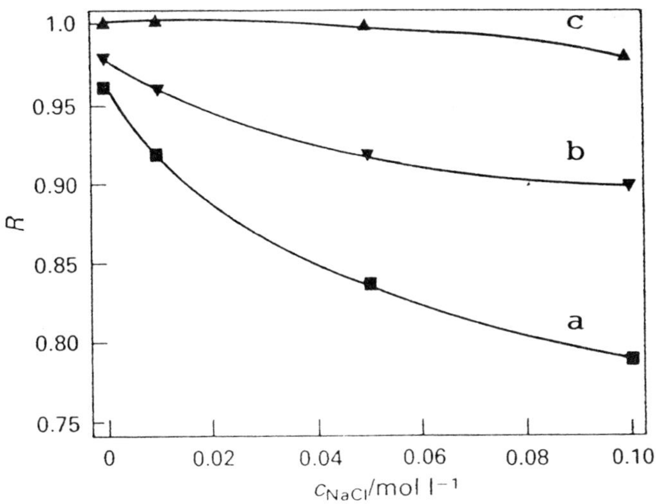

Fig. 7.17. Effect of ionic strength on MEUF separation of aromatic amines. Conditions: 0.02 M SDS; pH 3; amine concentration: 1×10^{-4} M. (a) aniline; (b) 4-methylaniline; (c) 4-isopropylaniline. Reproduced with permission from Ref. [136].

7.5.1.4. Effect of membrane pore-size

The effect of membrane pore-size upon MEUF performance has been clearly recognized since the first reported experiments. On the basis of the aggregation numbers of most common ionic and non-ionic micellar aggregates, and even disregarding the contribution of hydration to the micellar weight (and size), membranes with MWCO in the range 5,000–10,000 dalton usually allow complete micellar rejection. This is valid for modest surfactant concentrations, that is under conditions of negligible micellar leakage. However, one must also consider that (for a given membrane pore-size) increasing amounts of dissolved solute usually increase the host micelle rejection. This effect has been clearly observed in studies concerning phenol solubilization in cationic DTAB and non-ionic Igepal-CO micelles, and has been attributed to the increase of aggregate size (the shape becomes ellipsoidal) upon solute incorporation [138].

7.5.2. MEUF separation of inorganic ionic species based on electrostatic solute–micelle attraction

MEUF separations of inorganic anions and cations from aqueous solutions have been performed by exploiting the electrostatic attraction which occurs between these analytes and oppositely charged micelles. Although quite efficient for the removal of multicharged species, these systems are scarcely selective and very sensitive to the ionic strength.

Several equilibria need to be considered in order to describe the properties of these systems. The condition of electrical neutrality applied to the permeate stream requires that the sum of the products of ion concentration and electrical charge for the individual species must be zero. Moreover, the thermodynamic activity of each electrolyte in the permeate when the solution passes through the membrane must be equal to the activity of the same electrolyte in the retentate. solution.

A quantitative treatment of MEUF data for metal ions removed using SDS micelles has been proposed by Scamehorn and co-workers [140], starting from Oosawa's theory of polyelectrolytes [141]. The treatment assumes that the bound cations remain in the retentate whereas the free metal ions can permeate. In the absence of added salts, for a system composed of SDS micelles and Cu^{2+} as target cation, the following species must be taken into account: Cu^{2+} (analyte), Na^+ (surfactant counterion), DS^- (amphiphilic anion) and Cl^- (copper co-ion, assuming the addition of $CuCl_2$).

The binding of the metal ions to the anionic micelles can be described using the following equations:

$$\ln \frac{[Na^+]_b}{[Na^+]_f} = \ln \frac{\phi}{1-\phi} + P(1-\beta)(1-\phi^{1/2}) \tag{4}$$

$$\ln \frac{[Cu^{2+}]_b}{[Cu^{2+}]_f} = \ln \frac{\phi}{1-\phi} + 2\,P(1-\beta)(1-\phi^{1/2}) \tag{5}$$

The subscripts b and f indicate the bound and free ions, respectively. ϕ is the volume fraction of solution where the bound cations are located (approximately corresponding to the Stern layer region around the micelle). P is a dimensionless potential parameter. The term β, which represents the fraction of the micellar charge neutralized by the bound cations is, in turn, given by:

$$\beta = ([Na^+]_b + 2[Cu^{2+}]_b)/C_D \tag{6}$$

If added monovalent salt (e.g., NaCl) is added to the solution and assuming that the anion of the added salt is the same co-ion of the analyte, P can be estimated from P_0 (value of P in the absence of added electrolytes) through the equation:

$$P = P_0 \{1 + a\ [NaCl]^{1/2}\} \tag{7}$$

where a is a constant.

Application of the electroneutrality condition to both the retentate and permeate solutions, gives rise to the following equations:

$$2[Cu^{2+}]_r + [Na^+]_r = [DS^-]_r + [Cl^-]_r \tag{8}$$

$$2[Cu^{2+}]_p + [Na^+]_p = [DS^-]_p + [Cl^-]_p \tag{9}$$

The free ions are assumed to be in equilibrium across the UF membrane, so the activities of any cation–anion pair are equal on both sides. Moreover, in the absence of added salts and for dilute solutions, the activity coefficients can be considered close to unity. Under these assumptions:

$$[Cu^{2+}]_r[Cl^-]^2_r = [Cu^{2+}]_p[Cl^-]^2_p \tag{10}$$

$$[Cu^{2+}]_r[DS^-]^2_r = [Cu^{2+}]_p[DS^-]^2_p \tag{11}$$

$$[Na^+]_r[Cl^-]_r = [Na^+]_p[Cl^-]_p \tag{12}$$

The amphiphile is present in the retentate as free monomer or in a micelle. The concentration of surfactant present as micelles can, in turn, be calculated used the expression:

$$(\ln [Na^+]_f + 2[Cu^{2+}]_f)\beta + \ln [DS^-] = (1 + \beta) \ln cmc \tag{13}$$

The free metal-ion concentration in the permeate can be evaluated using Eqns. (3–13), provided P is known. The model fits the experimental data quite well over a wide range of conditions, particularly for low metal-ion concentrations. Figure 7.18 shows the predicted and measured Cu^{2+} concentration in the permeate as a function of the analyte concentration in the retentate. Comparable rejection coefficients have been measured, under the same experimental conditions, for cations having the same charge. The value of % R for Co^{2+}, Zn^{2+}, Cu^{2+} and Cd^{2+} using SDS micelles varies in a narrow range (98.9–99.5%) depending on the tendency of these ions to give complexes with their counterions.

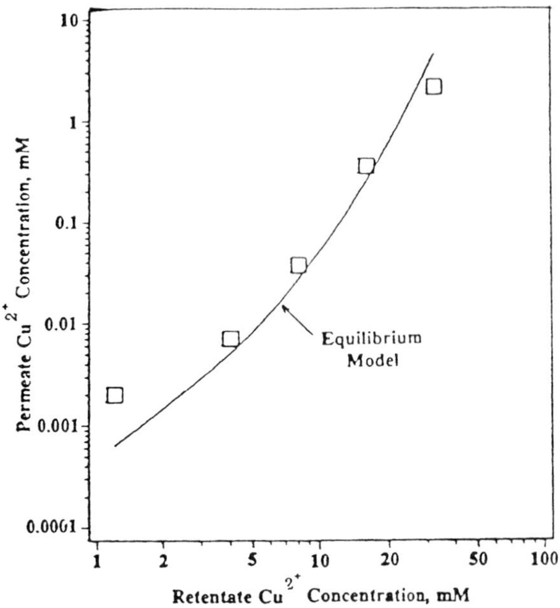

Fig. 7.18. Variation of $[Cu^{2+}]_p$ as a function of $[Cu^{2+}]_r$ with 0.08 M SDS. 1000 MWCO membrane. Reproduced with permission from Ref. [140].

Similar rejection values have also been determined for divalent anions in the presence of cationic aggregates; for example, % R = 99.4 has been reported for CrO_4^{2-} in MEUF separations using HPC micelles [142].

A very important feature of MEUF separations using ionic aggregates is their recognized capacity to improve the ion retention at very low analyte concentrations. This dependence is opposite to that observed using traditional separation techniques such as precipitation. The use of surfactants having a low cmc value is preferred, but it is known that ionic amphiphiles usually exhibit higher cmc's. In some cases, the addition of small amounts of non-ionic surfactants allows the formation of mixed aggregates having lower cmc values, with corresponding better MEUF performances [143].

The presence of added salts (usually monovalent electrolytes) tends to reduce the MEUF efficiency since the effective micellar charge is shielded and the electrostatic potential at the aggregate/water interface is reduced. Very strong competitive effects have been observed in the presence of salts of multivalent cations (or anions).

7.5.3. Use of polyelectrolytes in MEUF

In principle, polyelectrolytes can be used instead of ionic micelles to remove charged solutes. A potential advantage related to the use of polymeric macro-ions is that they are still effective at very low concentrations, whereas micellar systems are practically ineffective below the cmc. Moreover, a polyelectrolyte-free permeate is obtained whereas surfactant monomers are always present in MEUF.

Good results have been obtained, for example, in removing multivalent cations (such as Cu^{2+}) from aqueous streams using dilute solutions of poly(styrenesulfonate) [144]. At relatively low polyelectrolyte: metal concentration ratios (e.g., between 3:1 and 10:1) very efficient retention has been reported, with the permeate concentration of the target metal reduced to about one thousandth of the initial concentration in the feed solution. Although these separations show efficiencies comparable to those obtained using ionic micelles, polyelectrolytes have not been much used in MEUF, basically because they are more expensive.

7.5.4. Use of ligand micelles in MEUF

The use of chelating mixed micelles which are able to react with a limited number of metal ions, was proposed some years ago as a more selective approach to MEUF-based demetallation [145]. The feasibility and potential practical utility of this method has been demonstrated in recent reports [146–152], where the successful application of a variety of mixed-ligand aggregates has been described.

Hydrophobic ligands or their derivatives (see Fig. 7.19), obtained by introducing a tuneable alkyl chain, are usually dissolved in unreactive host-surfactant aggregates and successively added to the aqueous waste (or sample) containing the metals.

Cationic surfactants are, in principle, preferred since the uncomplexed cations are repelled from the aggregates. However, rejection of unbound multivalent cations has been observed in some cases [147] — this effect being attributed to the formation of a charged surfactant layer next to the membrane surface.

Several factors must be considered in order to predict the MEUF performances. The extent of metal complexation, which can vary significantly in the microheterogeneous media (see Chapter 3), is an important parameter since it defines the minimum ligand excess necessary, which in turn is limited by the ligand solubility in the host aggregates.

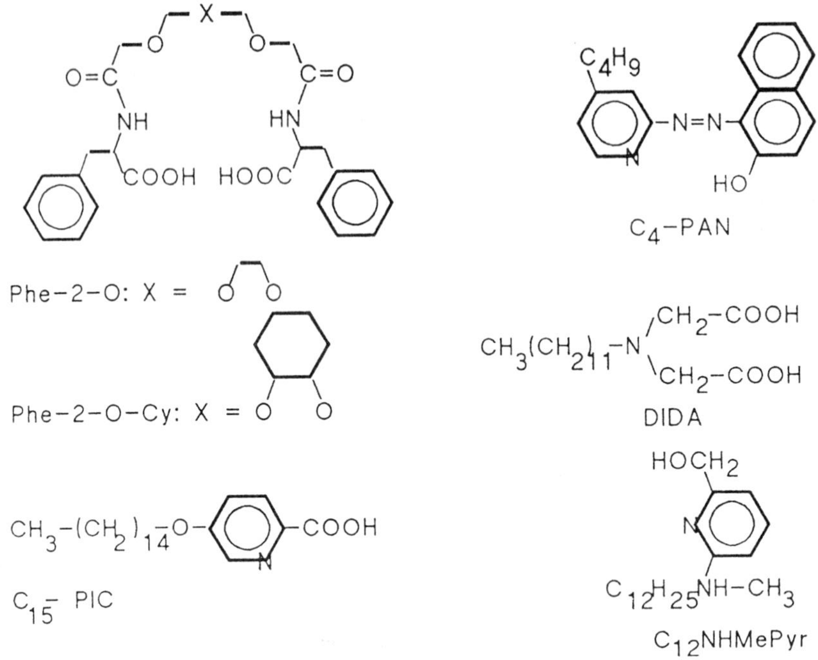

Fig. 7.19. Structures of some hydrophobic and amphiphilic ligands used in MEUF experiments.

Molecules giving rise to the formation of very stable chelates, such as 8-hydroxyquinoline, PAR or PAN, and their lipophilic derivatives, are particularly suitable.

Some examples of metal preconcentrations and separations using ligand micelles are listed in Table 7.6.

The stoichiometry of the complexes is often lower than observed in homogeneous media because of the steric constraints operating when the ligand is located within the aggregates. For example, the ligand PAN reacts with Ni(II) in Triton X-100 micelles, forming a 2:1 chelate (as in water–dioxane mixtures), whereas the hydrophobic homolog C_4-PAN forms a 1:1 complex with the same metal [10].

The kinetics of complex formation must also be considered since the reaction site and rate are strongly dependent on ligand binding (see Chapter 3). It is known that highly hydrophobic ligands react very slowly with metal ions when incorporated in neutral or cationic

TABLE 7.6

Concentration and separation of metal ions by MEUF using chelating micelles

Metal	Ligand/host micelle/solution pH	% R	Ref.
3×10^{-4} M Cu(II)[a]	7.85×10^{-3} M DIDA/0.157 M HPC/pH 7	97.3	146
1×10^{-4} M Fe(III)	1×10^{-3} M Y-PAS-C_4/0.01 M $C_{12}E_8$/pH 3.5	84.0	147
	1×10^{-3} M Y-PAS-C_8/0.01 M $C_{12}E_8$/pH 3.5	>99.9	
5×10^{-5} M UO_2^{2+}	1×10^{-3} M Phe-2-O/0.04 M HTAB/pH 7	66.0	148
	1×10^{-3} M Phe-2-O-Cy/0.04 M HTAB/pH 7	81.0	
	1×10^{-3} M Phe-2-O + 1×10^{-3} M TOPO/0.04 M HTAB/pH 7	98.0	
	1×10^{-3} M Phe-2-O-Cy + 1×10^{-3} M TOPO/0.04 M HTAB/pH 7	>99.0	
1×10^{-4} M Cu(II)	1×10^{-4} M C_4NHMePyr/0.025 M HTAB/pH 5	<10.0	149
	1×10^{-4} M C_{10}NHMePyr/0.025 M HTAB/pH 5	39.0	
	1×10^{-4} M C_{16}NHMePyr/0.025 M HTAB7 pH 5	73.0	
	5×10^{-4} M C_{16}NHMePyr/0.025 M HTAB/pH 5	>99.0	
2.2×10^{-4} M Al(III)	2×10^{-3} M LG/0.02 M HTAB/pH 5	98.5	150
1×10^{-4} M Ni(II)	5×10^{-4} M C_{15}-PIC/0.025 M HTAB/pH 7	95.0	151
	5×10^{-4} M C_{15}-PIC/8×10^{-3} M HTAB + 1.2×10^{-2} M $C_{12}E_6$/pH 7	>98.0	
5.1×10^{-5} M Ni(II)[b]	6×10^{-4} M C_4-PAN/0.02 M Triton X100/pH 3	66.3	152
5.1×10^{-5} M Co(II)[b]		>99.0	
4.7×10^{-5} M Cu(II)[b]		>99.0	
5.5×10^{-5} M Mn(II)[b]		30.0	
4.6×10^{-5} M Zn(II)[b]		11.8	
7.5×10^{-5} M Ca(II)		<5.0	

a: Separated from Ca(II); *b*: metals separated from their binary mixtures and from Ca(II). Hydrophilic cellulosic membranes (MWCO = 10,000 dalton). LG = lumogallion.

micelles, and that in some cases the reaction becomes strongly inhibited and does not go to completion even after some hours.

Certain useful information can be obtained from kinetic data. For example, the binding constants between chelates and micelles can be determined and the conditions under which bulk or interfacial complexation takes place can be evaluated [92]. Moreover, the separation of metal ions based on the time for complex formation becomes possible by using properly selected ligands [153,154]. The kinetic aspects of these reactions and their influence on MEUF performances have recently been examined and discussed by Tondre and his co-workers in some detailed and elegant contributions [149,151,155,156].

It is possible to predict the efficiency of MEUF for a given metal/ligand system reacting in micellar media if all the equilibria on both sides of the UF membrane are considered. A successful model has been developed for the system Cu^{2+}/Ca^{2+}/DIDA/HPC by introducing the mass balance, activity balance and electroneutrality conditions (previously described in Section 7.5.2), and taking into account the ligand ionization and metal complexation. An extended description of this model can be found in Ref. [157].

7.5.4.1. Influence of ligand hydrophobicity and structure on metal rejection

Within certain limits, an increase in ligand hydrophobicity is beneficial since the bound chelate fraction also increases (see Fig. 7.20). Moreover, complete ligand rejection is desired when MEUF is applied in wastewater treatment, whereas this aspect is less important in

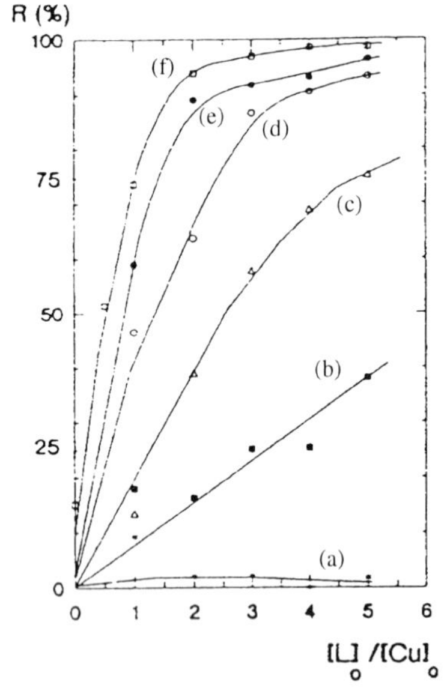

Fig. 7.20 Variation of Cu(II) recovery in the retentate as a function of the alkyl chain length of the ligand $C_nNHMePyr$, at different ligand–metal ratios. (a) n = 4; (b) n = 8; (c) n = 10; (d) n = 12; (e) n = 14; (f) n = 16. $[Cu^{2+}]_{in} = 1\times10^{-4}$ M; 0.025 M HTAB; pH 5. Reproduced with permission from Ref. [149].

analytical work. However, since the lipophilic ligands are less soluble in the micelles, the ligand excess may be insufficient for quantitative chelate formation. This is particularly true for complexes having moderate stability.

In some cases, as in the removal of U(VI) from wastewater, the structure of the bulky complexes does not allow their effective binding to the aggregates. In fact, there is experimental evidence indicating the ligand location in the outer interfacial region of the host micelle [148]. The formation of ternary complexes by introduction of the auxiliary ligand TOPO (trioctylphosphine oxide) increases the U(VI) rejection in these experiments, probably because the ternary complexes are more tightly bound to the host HTAB micelles through the TOPO alkyl chains.

Complexing molecules having the chelating moiety oriented at the micelle/bulk interface and a tuned alkyl chain placed far from the complexing groups seem to be the best candidates for selective MEUF work. Taking into account the fact that high stability constants are also necessary, compounds such as the alkyl derivatives of PAN [152] are among the most effective for metal ion separations.

7.5.4.2. Effect of pH on metal rejection

As expected on the basis of chelate formation from ionizable ligands, the MEUF performances are usually pH-dependent and different rejection vs. pH curves are obtained, depending on the tendency of each cation to react with the ligand (see Fig. 7.21).

Examination of Fig. 7.21 suggests that, by properly choosing the working pH, the selective enrichment of target components can be performed using simple or multistage MEUF. For example, by working at pH 3 with 1×10^{-4} M C_4-PAN in 6.6×10^{-3} M Triton X-100, the initial Co(II)/Zn(II) ratio in the feed aqueous solution (0.5 mg/l of each metal) varies until 0.5/0.01 mg/l (in the retentate) after three repeated ultrafiltrations. The corresponding R values at this pH are ca. 1 for Co(II) and ca. 0.3 for Zn(II). Separation of copper from manganese and from zinc can also be obtained using the same chelating aggregates. However, the preconcentration in the retentate phase of all the above mentioned transition metals can be achieved by operating in the pH range 6–7.5. Similar considerations are also valid for the other chelating systems reported in Table 7.6.

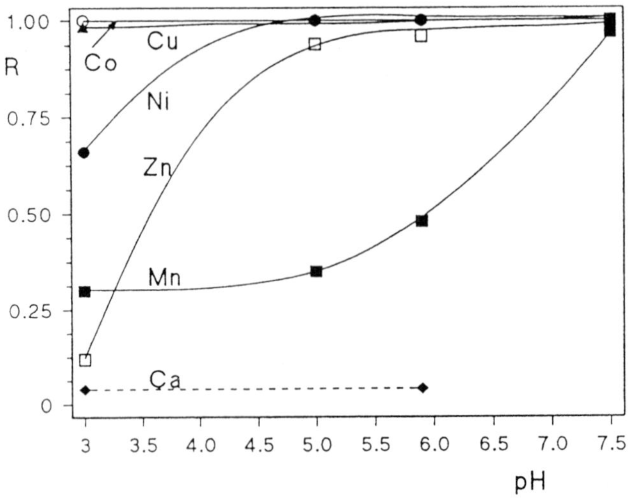

Fig. 7.21. Variation of % R with pH. Metal concentration (feed solution): 3 mg l^{-1}. C_4-PAN: 6×10^{-4} M; Triton X-100: 0.02 M. Cellulosic membranes (MWCO = 10,000 dalton). N_2 pressure: 300 kPa. Taken from Ref. [152], reproduced with permission.

7.5.5. Advantages and limits of MEUF-based separations of metal ions

When MEUF is used as a preconcentration technique, nearly the same concentration performance can be achieved as in liquid–liquid extractions using larger amounts (100–150 ml) of more toxic and expensive organic solvents, by the addition of moderate quantities of surfactants (5–10 g/l) to the aqueous samples. Moreover, the concentration factors can be increased by selecting amphiphiles with low cmc values and MEUF can be conveniently coupled with micelle-based determination methods without changing the matrix composition.

Simultaneous removal of heavy metals (or anionic species) and organic pollutants can be performed in a single MEUF step, provided micelles having a charge opposite to the target ions are used. Moreover, effective separation of metal ions from their mixtures is feasible using chelating micelles combined with a multistage ultrafiltration.

The main disadvantage of MEUF is related to the small solubilization capacity of the micellar phase, which represents a very low volume fraction of the solution. As a consequence, the amount of ligand excess present in the system is also limited. Moreover, the recovery and purification of the metals from the surfactant-rich retentate is generally more difficult than from organic solvents.

7.6. SEPARATIONS USING ADMICELLAR CHROMATOGRAPHY

The basis of the new separation method of admicellar chromatography is the tendency of certain surfactants, at concentrations below the cmc, to form bilayer aggregates at solid–liquid interfaces. These two-dimensional aggregates, called hemimicelles by Fuerstenau [158], and successively termed admicelles by Harwell and his co-workers [159,160], are similar to Langmuir–Blodgett films and can easily (and reversibly) be formed on a wide variety of solids. For example, a typical SDS admicelle is formed on the surface of solid alumina at low pH values, through a combined adsorption mechanism involving Coulombic attraction of the surfactant-sulfonate groups for charged sites on the oxide surface, followed by a favourable tail–tail hydrophobic interaction between adsorbed amphiphiles. The presence of such aggregate structures has been made clear by co-adsorbing suitable dyes (e.g., pinacyanol chloride) which show a colour indicating a typical non-polar environment [161] (see Fig.7. 22).

Adsolubilization may be described as a process in which the analytes are partitioned between the bulk solution and the admicellar structures. The amount of adsolubilized analyte is proportional to the extent of surfactant adsorption until surface saturation is reached. At surfactant concentrations well above the cmc, the competition between adsolubilization and micellar solubilization must be considered.

Since the admicelles act as two-dimensional solvents, capable of solubilizing the lipophilic species, this phenomenon can be exploited to

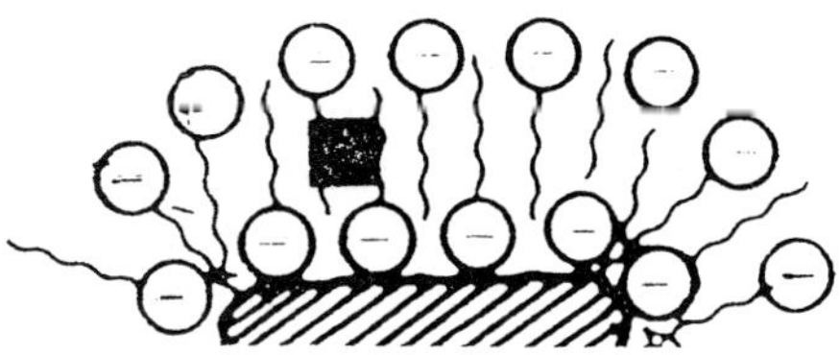

Fig. 7.22. Representation of an admicellar structure showing the location of an adsolubilized molecule (■) within the non-polar region.

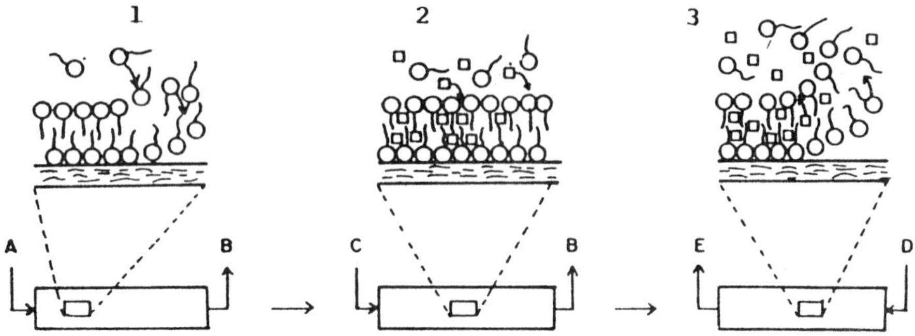

Fig. 7.23. Simplified scheme of an admicelle-based separation. Figure taken from Ref. [162], reproduced with permission.

develop a new separation technique. For example, admicellar-enhanced chromatography (AEC) is based on the use of surfactant-coated stationary phases in conventional liquid chromatography.

A scheme representing the AEC separation of the components of a binary mixture (operating in an equilibrium batch mode) is shown in Fig. 7.23. Suitable admicelles are obtained by running a surfactant solution through a packed column (step 1). Then the mixture is introduced (keeping the surfactant concentration constant) and the lipophilic analytes preferentially dissolve in the admicelles, where their enrichment takes place (step 2). After saturation of the admicellar phase with the removed target solutes, the bed is backflushed in order to desorb the surfactant and to recover the adsolubilizate (step 3).

An example of such separations is the selective recovery of heptanol isomers using SDS admicelles formed on an alumina bed [161]. In fact, the adsolubilization curves of 2-methyl-2-hexanol, 3-heptanol and 1-heptanol indicate a preferential uptake of 1-heptanol, which can be separated from the other isomers by working at low surfactant concentrations (see Fig. 7.24).

Other studies have confirmed the potential utility of admicelle-based separations. For example, the adsolubilization of azo-dyes in mixed anionic–non-ionic admicelles formed onto α-alumina demonstrated the higher solubilization capability of these adsorbed aggregates [163]. Admicelles of lithium dodecylsulfate and lithium perfluoroctanesulfonate (on alumina) have also been examined as suitable candidates for the separation of normal and fluorinated alcohols [164]. Preliminary investigations concerning the separation of the enantiomers of camphor

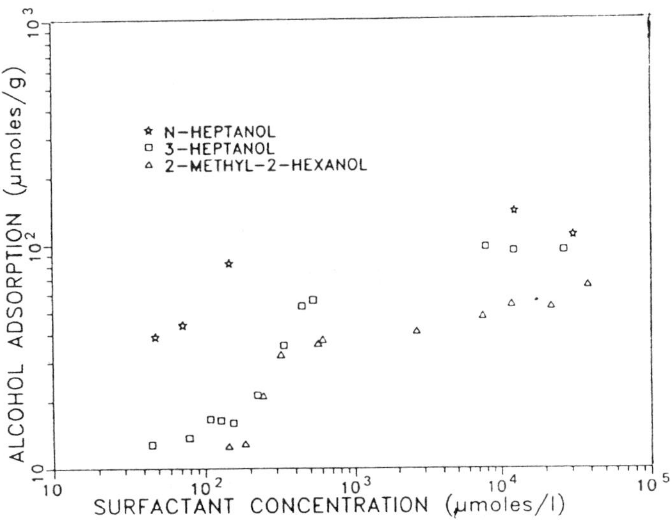

Fig. 7.24. Alcohol adsolubilization in the admicellar phase as a function of SDS concentration. Solution pH: 4; 0.15 M NaCl; room temperature. Reproduced with permission from Ref. [162].

using admicelles of sodium deoxycholate on aluminum oxide have also been mentioned [165]

Thin films obtained from the polymerization of adsolubilized styrene in SDS admicelles (on alumina) have been investigated as interesting new two-dimensional solvents [166,167]. The surface modification of the inorganic support significantly increases its affinity for organic solutes, making possible the removal of pollutants such as alkylphenols from aqueous streams.

Recent detailed studies on the adsolubilization of contaminants (naphthalene and naphthalene derivatives) in SDS admicelles formed onto alumina [168] confirm the increasing interest of researchers into the possible applications of the technique for wastewater treatment.

7.7. MISCELLANEOUS SURFACTANT-BASED SEPARATIONS

In addition to the techniques examined in the previous sections, some other surfactant-based procedures are of economic interest in industry. Although their applications in analytical chemistry are rare, these methods could undergo development and some basic knowledge of them is worthwhile.

7.7.1 Separations based on foams

Froth flotation is a separation process in which bubbles of air sparged into a column of water adhere to small particles of a mineral. The particles are then floated to the top of the column where they collect in the froth, which can be skimmed off to complete the removal of the particulate matter. Two types of reagents are usually required: reagents adsorbed on the particle to regulate the wettability and reagents to control foaming or frothing. Selective enrichment of the target solid particles can be obtained by preferential adsorption of surfactants at the solid–liquid interface.

Ore flotation is the most important surfactant-based separation, with major applications in the minerals industry. Some excellent literature covers this field [169–171].

In bubble separation, the same basic idea as applied in froth flotation is extended to the removal of dissolved solutes, through their adhesion to the surface of the ascending bubbles (foam fractionation). In some cases the method has been applied to remove dissolved surfactants, although ion separation is more interesting from the analytical point of view. In fact, different colligend ionic species of environmental concern (e.g., Zn^{2+}, Cd^{2+}, CrO_4^{2-}, PbO_2^{2-}) can form ion pairs with ionic surfactants adsorbed at the bubble–solution interface (see Fig. 7.25A). Complexes with reactive surfactants can also be formed and then extracted in the foam. For example, the surface-active compound 4-dodecyldiethylenetriamine forms complexes having a 2:1 ligand/metal ratio with Cd(II), Ni(II) and Cu(II), which can be effectively removed upon adsorption to the ascending bubbles [172]. Another example is the flotation of trace amounts of uranium ions by using 2-(alkylamino)propionohydroxamic acids as surface-active ligands [173].

Precipitate flotation can also be obtained using the above approach, provided that the colloidal solid particles are charged. For example, hydroxides such as $Al(OH)_3$ or $Fe(OH)_3$ obtained in the presence of an excess of precipitant (OH^-) are negatively charged and can thus be attracted by cationic surfactants present at the bubble/solution interface (Fig. 7.25B).

Other separation techniques based on bubble adsorption do not involve foaming. They include bubble fractionation (using a gas for

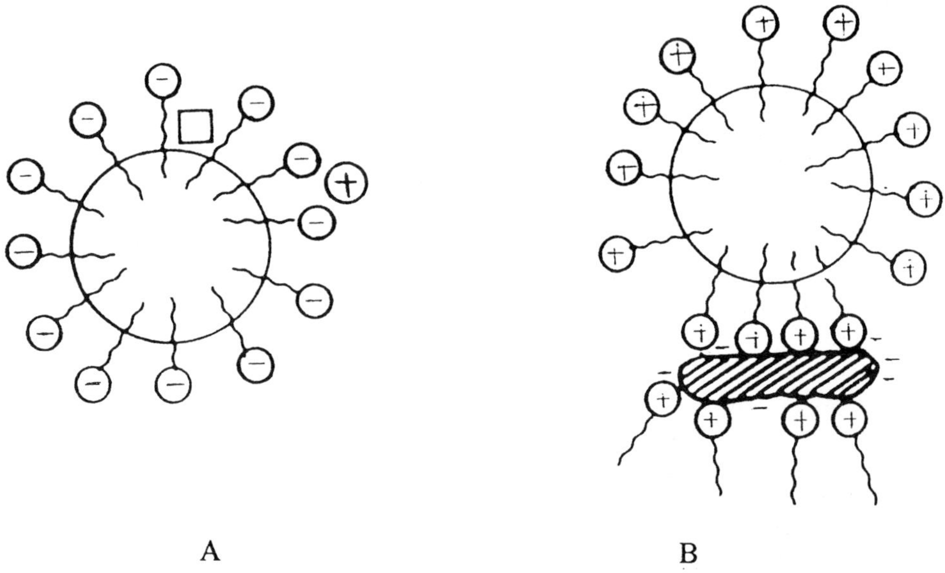

Fig. 7.25. Scheme of the foam flotation process. (A) Solutes retained by electrostatic attraction and hydrophobic interactions. (B) Flotation of charged solid particles.

adsorption and deposition of a concentrated solution at the top of the column) and solvent sublation (with use of another liquid phase to adsorb and concentrate the solute at the top of the column). In most of these processes the dispersed phase consists of air bubbles and the continuous phase, generally water, contains the species being adsorbed [174–178].

Several factors can influence the separation efficiency in foam flotation. They include the surface activity of the amphiphile, its complexation ability towards the species to be separated, the concentration of the species, the flow rates of the phases, and the pH [178]. For example, chlorophenols can be removed from basic aqueous solutions (pH 9–10) by exploiting their electrostatic and specific interactions with HTAB molecules. The pollutant recovery in the separated foam is strictly related to its hydrophobicity, being only partial for 2-chlorophenol (35%) and 2,4-dichlorophenol (73%), whereas pentachlorophenol is nearly completely removed (≈99%) [179,180]. Examples of separation of inorganic cations and anions of environmental interest can be found in Refs. [177,178,181–183].

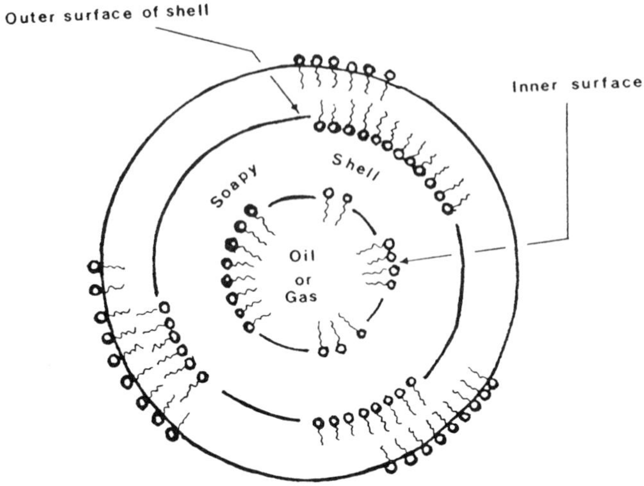

Fig. 7.26. Typical structure of a gas aphron. Reproduced with permission from Ref. [185].

The use of minute micro-gas dispersion (in the order of 10 μm) or the so-called gas aphrons also show promise since they have very large surface areas, exhibit long residence times in liquid pools, and can be easily transported by pumping [184,185]. These structures, which consist of gas bubbles encapsulated in a thin soapy film (see Fig.7. 26), allow the almost complete separation of different adsorbed solutes, including ionic species attracted by the oppositely charged surfactant (ion flotation). Good results have also been obtained with minute dispersed liquid droplets, termed colloidal liquid aphrons.

7.7.2. Surfactant-enhanced carbon regeneration (SECR)

Although this technique is not used for the direct separation or concentration of analytes from samples it has recently been proposed as an effective regenerating procedure for spent sorbents used in water treatments. The solubilization of the adsorbed solutes by micelles is exploited in this method. Concentrated surfactant solutions are required to regenerate the exhausted carbon beds but, nevertheless, the treatment is cheap compared with other reactivation procedures such as thermal or solvent regeneration, and faster than most biological treatments. An advantage of SECR is that no heating is required, and the successive washing cycles with water are sufficient to complete the removal of the adsorbed surfactant.

The recovery of the pollutant 4-tert-butylphenol (TBP) from loaded activated carbon after treatment with 0.4 M HPC under mild conditions has been reported by Scamehorn and his co-workers [186,187]. At the TBP concentrations usually present in wastewaters, approximately 50,000 pore volumes of the contaminated liquid can be treated before carbon saturation (one pore volume corresponds to the volume of liquid retained by the loaded adsorbent). The fraction of TBP desorbed from the exhaust carbon as a function of the volume (expressed as pore volume) of the regenerant HPC is shown in Fig. 7.27. Because of the adsorption of HPC, breakthrough occurs at ca. 1.5 pore volumes, and after ca. 10 pore volumes, about 50% of the contaminant has been removed. After passage of ca. 150 pore volumes of 0.4 M HPC the TBP removal is about 80%. The removal of surfactant from the sorbent is accomplished by flushing with water.

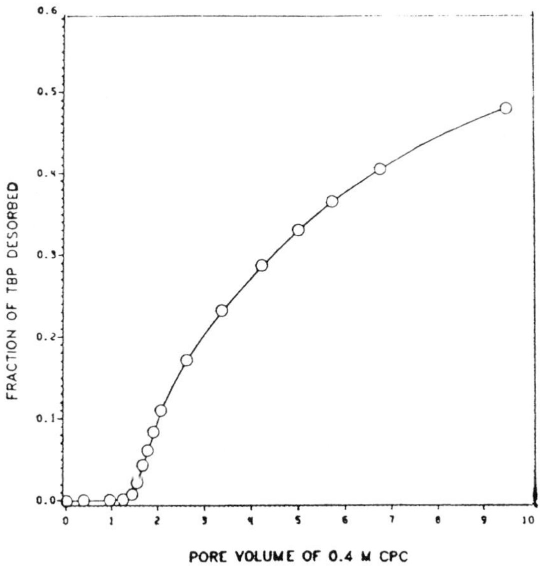

Fig. 7.27. Desorption of TBP from granular activated carbon (Calgon Filtrasorb 300). Column: 1 m long × 2.5 cm i.d. Pollutant loading: 0.15 g TBP/g of carbon. T: 30°C. Surfactant: 0.4 M HPC. Taken from Ref. [186], reproduced with permission.

A very important feature of SECR is that small volumes of the regenerant solutions are able to dissolve relevant amount of adsorbate, thus allowing one to obtain high concentration factors (in the range 10^3–10^4). The combination of adsorption steps with functionalized adsorbents, followed by the use of SECR, may also have useful applications in selective separation methods.

7.7. POSSIBLE FUTURE DEVELOPMENTS OF SURFACTANT-BASED SEPARATIONS

It is clear from the previous sections that the synthesis of novel surfactant systems, such as fluoroamphiphiles or with a "functional" polar head (complexing, redox or chiral) could aid the development of more selective procedures based on possible specific interactions between substrates and aggregates.

Of the methods we have examined, some already have commercial interest and others could have great potential for future practical use. Some emerging interesting possibilities can be envisaged for the use of surfactants in separation science. The use of surfactants for enantiomeric enrichment can be mentioned among these. There is already evidence that micellar catalytic effects can be different for reactants that are geometric, positional or optical isomers of each other [4]. However, these effects are generally too small to render the approach useful. The use of chiral micelles [15] or mixed micelles in which at least one component is a chiral molecule [188] may remarkably enhance the selectivity.

Another potential application of surfactant-based methods is in the field of isotope separations based upon differences in their magnetic moments. The enrichment factors obtained in bulk solvents are very low, but Turro and his co-workers have demonstrated the use of micelles to improve the separation [189–192]. The technique is based on the radical-pair model of chemically induced dynamic nuclear polarization, CIDNP [190,193], which predicts that the reactivity of chemical processes involving diradical species depends on the hyperfine interactions of the orbitally uncoupled electron of the radical pair with nuclear spin (magnetic isotope effect) or laboratory magnets (magnetic field effect). Magnetic isotopes can be separated from non-magnetic ones in appropriate photochemical reaction systems. In these separations the surfactant aggregate provides a suitable microenvironment in which there is restricted diffusion of the radical pair and a high local viscosity (hindered fluidity). Hydrophobic probes are conveniently solubilized within the aggregates.

An example of such isotopic enrichments is that based on the photolysis of dibenzyl ketone (DBK) in HTAB micelles. Initially the photolysis results in the formation of a triplet radical pair within the micelle. The triplet radical pair containing ^{13}C undergoes nuclear hyperfine coupling intersystem crossing to a singlet radical pair that regenerates the starting material, DBK, by recombination (very low amounts of 1-phenyl-4-methylacetophenone are also obtained). The ^{12}C nuclei, being non-magnetic, cannot undergo nuclear hyperfine coupling and react more slowly. The DBK molecule participating to the recombination cycle thus becomes enriched in ^{13}C. The presence of micelles enhances the encounter frequency of the radical pairs after the triplet–singlet conversion compared with that occurring in homogeneous solvents. Although potentially useful, the practical employment of this technique for separation and enrichment of isotopes still needs further investigation.

It must be emphasized that several practical applications of surfactants in chemical separations are actually at work, whereas mechanistic studies are often lacking owing to the complicated nature of the processes involved and the scarce knowledge of the aggregate's behaviour under every experimental condition. For the above reasons, the rational design of new efficient and selective surfactant-based separation methods will require more detailed investigation and understanding of all the physicochemical aspects of these processes.

REFERENCES

1 D.W. Armstrong, *Sep. Purif. Methods*, 14 (1985) 213.

2 W.L. Hinze and D.W. Armstrong (Eds.), *Ordered Media in Chemical Separations.* ACS Symp. Ser. 342, ACS, Washington DC, 1987.

3 M.J. Rosen (Ed.), *Surfactants in Emerging Technologies.* Dekker, New York, 1987.

4 W.L. Hinze, *Ann. Chim. (Rome)*, 77 (1987) 167.

5 D.T. Wasan, M.E. Ginn and D.O. Shah (Eds.), *Surfactants in Chemical/Process Engineering.* Dekker, New York, 1988.

6 J.F. Scamehorn and J.H. Harwell (Eds.), *Surfactant-Based Separation Processes.* Dekker, New York, 1989.

7 E. Pelizzetti, V. Maurino, C. Minero and E. Pramauro, in: D.M. Bloor and E. Wyn-Jones (Eds.), *The Structure, Dynamics and Equilibrium Properties of Colloidal Systems*. Kluwer, Dordrecht, 1990, p. 325.
8 E. Pramauro and E. Pelizzetti. *Coll. Surfaces*, 48 (1990) 193.
9 W.L. Hinze and E. Pramauro, *Crit. Rev. Anal. Chem.*, 24 (1993) 133.
10 E. Pramauro and A. Bianco Prevot, *Pure Appl. Chem.*, 67 (1995) 551.
11 C.H. Suelter, *A Practical Guide to Enzymology*. Wiley, New York, 1985.
12 W.L. Hinze, in: K.L. Mittal (Ed.), *Solution Chemistry of Surfactants*. Plenum Press, New York, 1979, Vol. I, p. 78.
13 E. Pelizzetti and E. Pramauro, *Anal. Chim. Acta*, 169 (1985) 1.
14 W.L. Hinze, in: W.L. Hinze and D.W. Armstrong (Eds.), *Ordered Media in Chemical Separations*. ACS Symp. Ser. 342, ACS, Washington DC, 1987, p. 2.
15 J.H. Fendler, *Membrane Mimetic Chemistry*. Wiley, New York, 1982.
16 P.L. Luisi and L.J. Magid, *CRC. Crit. Rev. Biochem.*, 20 (1986) 409.
17 P. Fourre, D. Bauer and J. Lemerie, *Anal. Chem.*, 55 (1983) 662.
18 S.N. Bhattacharya and K.M. Ganguly, *Radiochimica Acta*, 39 (1986) 199.
19 A. Rajab, D. Pareau, J.P. Moulin and A. Chesne, *Bull. Soc. Chim. France* (1987) 29.
20 H. Yamada, K. Adachi, Y. Fuji and M. Mizuta, *Solvent Extr. Ion Exch.*, 4 (1986) 1109.
21 D.D. Ensor, G.R. McDonall and C.G. Pippin, *Anal. Chem.*, 58 (1986) 1814.
22 K. Ohashi, S. Nakata, M. Katsume, K. Nakamura and K. Yamamoto, *Anal. Sci.*, 1 (1985) 467.
23 S. Umetani, M. Matsui, H. Kawaro and T. Nagai, *Anal. Sci.*, 1 (1985) 55.
24 J. Caello, J. M. Madariaga, M. Muhammed, M. Valiente and H. Iturriaga, *Polyhedron*, 11 (1986) 1845.
25 R. Paimin and R.W. Cattrall, *Aust. J. Chem.*, 36 (1983) 1017.
26 T. Suzuki, N. Murakami and T. Sotobayashi, *Bull. Chem. Soc. Jpn.*, 53 (1980) 1453.
27 Y. Sakai, H. Nakamura, M. Takagi and K. Ueno, *Bull. Chem. Soc. Jpn.*, 59 (1986) 381.
28 Y. Sakai, M. Shinmura, H. Otsuka and M. Takagi, *Bull. Chem. Soc. Jpn.*, 60 (1987) 545.
29 I. Yoshida and R. Takashita, *Anal. Sci.*, 2 (1986) 53.
30 N.N. Li (Ed.), *Recent Developments in Separation Science*. CRC Press, Florida, 1986, Vol. VIII.
31 J. Wisniak and A. Tamir (Eds.), *Liquid–Liquid Equilibrium and Extraction*. Elsevier, New York, 1987.
32 M. Puttermans, L. Dryon and D.L. Massart, *Anal. Chim. Acta*, 165 (1984) 245.
33 M. Puttermans, L. Dryon and D.L. Massart, *Anal. Chim. Acta*, 189 (1985) 15.
34 M. Puttermans, L. Dryon and D.L. Massart, *J. Pharm. Biomed. Anal.*, 3 (1985) 503.

35 V. Graef, T. Banken, E. Furuya and O. Nishikaze, *Fresenius Z. Anal. Chem.*, 324 (1986) 289.
36 A.K. Konturi and G. Sundholm, *Acta Chem. Scand. A*, 40 (1986) 121.
37 R. Swarup and S.K. Patil, *J. Inorg. Nucl. Chem.*, 38 (1976) 1203.
38 A. Helenius and K. Simons, *Biochim. Biophys. Acta*, 415 (1975) 29.
39 M.A. Chaiko, R. Nagarajan and E. Ruckenstein, *J. Coll. Interface Sci.*, 99 (1984) 168.
40 R. Nagarajan and E. Ruckenstein, *Sep. Sci. Technol.*, 20 (1985) 285.
41 P. Plucinski, *J. Heat Mass Transfer*, 28 (1985) 451.
42 A. Mahapatra and M. M. Sharma, *Solvent Extr. Ion. Exch.*, 5 (1987) 781.
43 A.S. Abdul and T.L. Gibson, *Environ. Sci. Technol.*, 25 (1991) 665.
44 C.T. Jafvert, *Environ. Sci. Technol.*, 25 (1991) 1039.
45 M.S. Diallo, L.M. Abriola and W.J. Weber, Jr., *Environ. Sci. Technol.*, 28 (1994) 1829.
46 V. DeGiorgio, in: V. DeGiorgio and M. Corti (Eds.), *Physics of Amphiphiles: Micelles, Vesicles and Microemulsions.* North Holland, Amsterdam, 1985, p. 303.
47 M.J. Rosen, *Surfactants and Interfacial Phenomena.* Wiley, New York, 1987.
48 T. Saitoh and W.L. Hinze, *Anal. Chem.*, 63 (1991) 2521.
49 L.A. Rupert, *J. Coll. Interface Sci.*, 153 (1989) 92.
50 M. Corti, V. Degiorgio, J.B. Hayter and M. Zulauf, *Chem. Phys. Lett.*, 109 (1984) 579.
51 L. Cantu, M. Corti, V. DeGiorgio, H. Hoffmann and W. Ulbricht, *J. Coll. Interface Sci.,* 116 (1987) 384.
52 G.G. Wan, C.J. Drummond, F. Grieser, B.W. Ninham and D.F. Evans, *J. Phys. Chem.,* 90 (1986) 4581.
53 M.F. Borgerding and W.L. Hinze, *Anal. Chem.,* 57 (1985) 2183.
54 M.E.F. Laespada, T.L.P. Pavon and B.M. Cordero, *Analyst,* 118 (1993) 209.
55 B.R. Ganong and C.M. Lu, *Anal. Biochem.,* 179 (1989) 66.
56 M. Brusdeilins and V. Zarybnicky, *J. Chromatogr.,* 287 (1984) 313.
57 S.K. Han, S.M. Lee and H. Schott, *J. Coll. Interface Sci.,* 132 (1989) 444.
58 H. Watanabe and H. Tanaka, *Talanta,* 25 (1978) 585.
59 T. Okada, *Anal. Chem.,* 64 (1992) 2138.
60 P.G. Nilson, B. Lindman and R.G. Laughlin, *J. Phys. Chem.,* 88 (1984) 6357.
61 A.S. Abdul, T.L. Gibson and D.N. Rai, *Groundwater,* 28 (1990) 920.
62 R.C. Chawla, C. Porczucek, J.N. Cannon and J.H. Johnson (Eds.), *Emerging Technologies in Hazardous Waste Management.* ACS Symp. Ser. 468, ACS. Washington DC, 1991. Ch. 16.
63 S. Laha and R.G. Luthy, *Environ. Sci. Technol.,* 25 (1991) 1920.
64 B.N. Aronstein, Y.M. Calvillo and M. Alexander, *Environ. Sci. Technol.,* 25 (1991) 1728.

65 A. Lopez Garcia, E. Blanco Gonzalez, J.L. Garcia Alonso and A. Sanz Medel, *Anal. Chim. Acta,* 264 (1992) 241.
66 W.L. Hinze, H.N. Singh, Z.F. Fu, R. Williams, D.J. Kippenberger, M.D. Morris and F.S. Sadek, in: T. Vo Dinh (Ed.), *Chemical Analysis of Polycyclic Aromatic Compounds.* Wiley, New York, 1989, Ch. 5.
67 E. Pramauro, *Ann. Chim. (Rome)*, 80 (1990) 101.
68 C.G. Pinto, J.L. Pavon and B.M. Cordero, *Anal. Chem.,* 64 (1992) 2334.
69 N.D. Gullickson, J.F. Scamehorn and J.H. Harwell, in: J.F. Scamehorn and J.H. Harwell (Eds.), *Surfactant-Based Separation Processes.* Dekker, New York, 1989, Ch. 6.
70 R.P. Frankewich and W.L. Hinze, *Anal. Chem.,* 66 (1994) 944.
71 C. Tiruppathi, D.H. Alpers and B. Seetharam, *Anal. Biochem.,* 153 (1986) 330.
72 C. Bordier, *J. Biol. Chem.,* 256 (1981) 1604.
73 K.J. Clemetson, D. Bienz, M.L. Zahno and E.F. Luscher, *Biochim. Biophys. Acta,* 778 (1984) 463.
74 W.J. Horvath and C.W. Huie, *Talanta,* 39 (1992) 487.
75 W.J. Horvath and C.W. Huie, *Talanta,* 40 (1993) 245.
76 A. Sanchez Ferrer, R. Bru and F. Garcia Carmona, *Anal. Biochem.,* 184 (1990) 279.
77 T.F. Zhang and L.P. Hager, *Arch. Biochem. Biophys.*, 257 (1987) 485.
78 C. Holm, G. Fredrikson and P. Belfrage, *J. Biol. Chem.,* 261 (1986) 15659.
79 D.A. Edwards, R.G. Luthy and Z. Liu, *Environ. Sci. Technol.,* 25 (1991) 127.
80 A. Bockelen and R. Niessner, *Fresenius J. Anal. Chem.,* 346 (1993) 435.
81 H. Watanabe and H. Tanaka, *Talanta,* 25 (1978) 585.
82 H. Hoshino, T. Saitoh, H. Taketomi and T. Yotsuyanagi, *Anal. Chim. Acta,* 147 (1983) 339.
83 H. Watanabe and K. Haraguchi, *Proc. Symp. Solvent Extr.* (1984) 75.
84 T. Saitoh, H. Hoshino and T. Yotsuyanagi, *Proc. Symp. Solvent Extr.* (1984) 81.
85 S. Kawamorita, H. Watanabe and K. Haraguchi, *Anal. Sci.*, 1 (1985) 41.
86 H. Watanabe, T. Kamidate, S. Kawamorita, K. Haraguchi and M. Miyajima, *Anal. Sci.*, 3 (1987) 433.
87 T. Saitoh, Y. Kimura, T. Kamidate, H. Watanabe and K. Haraguchi, *Anal. Sci.*, 5 (1989) 577.
88 T. Okada, *Anal. Chem.,* 64 (1992) 2138.
89 E. Pramauro, E. Pelizzetti, C. Minero, E. Barni, P. Savarino and G. Viscardi, *Ann. Chim. (Rome)*, 77 (1987) 209.
90 E. Pramauro, C. Minero and E. Pelizzetti, in: W.L. Hinze and D.W. Armstrong (Eds.), *Ordered Media in Chemical Separations.* ACS Symp. Ser. 342, ACS, Washington DC, 1987, p. 152.
91 E. Pelizzetti, E. Pramauro, E. Barni, P. Savarino, M. Corti and V. DeGiorgio, *Ber. Bunsenges. Phys. Chem.*, 86 (1982) 529.

92 F.P. Cavasino, C. Sbriziolo, E. Pelizzetti and E. Pramauro, *J. Phys. Chem.*, 93 (1989) 469.
93 L. Buhai and M. Rigan, *Talanta,* 37 (1990) 885.
94 A.E. Vassiliades, in: E. Jungemann (Ed.), *Cationic Surfactants.* Dekker, New York, 1970. Ch. 12.
95 N.D. Gullickson, J.F. Scamehorn and J.H. Harwell, in: J.F. Scamehorn and J.H. Harwell (Eds.), *Surfactant-Based Separation Processes.* Dekker, New York, 1989, p. 139.
96 P.L. Dubin, T.D. Ross, I. Sharma and B.E. Yegerlehener, in: W.L. Hinze and D.W. Armstrong (Eds.), *Ordered Media in Chemical Separations.* ACS Symp. Ser. 342, ACS, Washington DC, 1987, p. 162.
97 W.A. Charewicz and J. Strzelbicki, *J. Chem. Technol. Biotechnol.*, 29 (1979) 149.
98 D.O. Shah and R.S. Schechter (Eds.), *Improved Oil Recovery by Surfactant and Polymer Flooding.* Academic Press, New York, 1977.
99 D.O. Shah (Ed.), *Surface Phenomena in Enhanced Oil Recovery.* Plenum Press, New York, 1981.
100 E.L. Neustadter, in: Th.F. Tadros (Ed.), *Surfactants.* Academic Press, London, 1984.
101 P. Neogi, in: S.E. Friberg and P. Bothorel (Eds.), *Microemulsions. Structure and Dynamics.* CRC Press, Boca Raton, FL, 1987, p. 197.
102 C. Tondre and R. Zana, *J. Dispersion Sci. Technol.*, 1 (1980) 179.
103 S.E. Friberg and P. Neogi, in: J.F. Scamehorn and J.H. Harwell (Eds.), *Surfactant-Based Separation Processes.* Dekker, New York, 1989, p. 119.
104 N.O. Sonntag, in: D.T. Wasan, M. E. Ginn and D. O. Shah (Eds.), *Surfactants in Chemical/Process Engineering.* Dekker, New York, 1988, p. 169.
105 W.L. Lindenberger, in: M.J. Rosen (Ed.), *Surfactants in Emerging Technologies.* Dekker, New York, 1987, pp. 187–200.
106 N.N. Li, *AIChE J.*, 17 (1971) 459.
107 M. Kirch and J.M. Lehn, *Angew. Chem. Int. Engl. Ed.*, 14 (1975) 555.
108 M.E. Starzak, *The Physical Chemistry of Membranes.* Academic Press, New York, 1984.
109 N.N. Li and J.D. Navratil (Eds.), *Recent Developments in Separation Science.* CRC Press, Boca Raton, FL, 1986, Vol. IX.
110 C. Del Cerro and D. Boey, *Chem. Ind.*, (1988) 681.
111 M. Seno, Y. Shiraishi, S. Takeuchi and J. Otsuki, *J. Phys. Chem.,* 94 (1990) 3776.
112 R.A. Bartsch, W.A. Charewicz, S.I. Kang and W. Walkowiak, in: R.D. Noble and J. Douglas Way (Eds.), *Liquid Membranes: Theory and Applications.* ACS Symp. Ser. 347, ACS, Washington DC, 1987, p. 86.
113 R.M. Izatt, R.L. Bruening and J.J. Christensen, in: R. D. Noble and J. Douglas Way (Eds.), *Liquid Membranes: Theory and Applications.* ACS Symp. Ser. 347, ACS, Washington DC, 1987, p. 98.

114 F. Macasek, P. Rajec, V. Rehacek, T. Vu Ngoc Anh, T. Popovnakova, *J. Radioanal. Nucl. Chem.*, 96 (1985) 529.
115 N. Dickens, M. Rose and G.A. Davies, *Inst. Chem. Eng. Symp. Ser.*, 88 (1984) 69.
116 M. Tromp, M. Burgard, M.J.F. Leroy and M. Prevost, *J. Membrane Sci.*, 38 (1988) 295.
117 T. Delloye, M. Burgard and M.J.F. Leroy, *New. J. Chem.*, 13 (1989) 139.
118 D.T. Wasan, Z.M. Gu and N.N. Li, *Faraday Discuss. Chem. Soc.*, 77 (1984) 67.
119 Z. M. Gu, D.T. Wasan and N.N. Li, in: D.T. Wasan, M.E. Ginn and D.O. Shah (Eds.), *Surfactants in Chemical/Process Engineering.* Dekker, New York, 1988. p. 127.
120 C.F. Reusch and E.L. Cussler, *AIChE J.*, 19 (1973) 736.
121 R.M. Izatt, D.V. Dearden, P.R. Brown, J.S. Bradshaw, J.D. Lamb and J.J. Christensen, *J. Am. Chem. Soc.*, 105 (1983) 1785.
122 J.D. Lamb, J.J. Christensen, J.L. Oscarson, E.L. Nielsen, B.W. Asay and R.M. Izatt, *J. Am. Chem. Soc.*, 102 (1980) 6820.
123 A. Xenakis and C. Tondre, *J. Phys. Chem.*, 87 (1983) 4737.
124 A. Xenakis and C. Tondre, *Faraday Discuss. Chem. Soc.*, 77 (1984) 115.
125 A. Xenakis, C. Selve and C. Tondre, *Talanta*, 34 (1987) 509.
126 F. Nakashio, M. Goto, M. Matsumoto, J. Irie and K. Kondo, *J. Membrane Sci.*, 38 (1988) 249.
127 J. Draxler, W. Furst and R. Marr, *J. Membrane Sci.*, 38 (1988) 281.
128 M.P. Biehl, R.M. Izatt, J.D. Lamb and J.J. Christensen, *Sep. Sci. Technol.*, 17 (1982) 289.
129 R.M. Izatt, R.L. Bruening, W. Geng, M.H. Cho and J.J. Christensen, *Anal. Chem.*, 59 (1987) 2405.
130 M. Teramoto, T. Sakuramoto, T. Koyama, H. Matsuyama and Y. Miyake, *Sep. Sci. Technol.*, 21 (1986) 229.
131 R.M. Izatt, R.L. Bruening, G.A. Clark, J.D. Lamb and J.J. Christensen, *J. Membrane Sci.*, 28 (1986) 77.
132 T. Kumamaru, Y. Okamoto, M. Yamamoto, Y. Obata and K. Onizuka, *Anal. Chim. Acta*, 232 (1990) 389.
133 R.O. Dunn, Jr., J.F. Scamehorn and S.D. Christian, *Sep. Sci. Technol.*, 20 (1985) 257.
134 G.A. Smith, S.D. Christian, E.E. Tucker and J.F. Scamehorn, in: W.L. Hinze and D.W. Armstrong (Eds.), *Ordered Media in Chemical Separations.* ACS Symp. Ser. 342, ACS, Washington DC, 1987, p. 184.
135 L. Lane Gibbs, J.F. Scamehorn and S.D. Christian, *J. Membrane Sci.*, 30 (1987) 67.
136 E. Pramauro, A. Bianco Prevot, P. Savarino, G. Viscardi, M. de la Guardia and E. Peris Cardells, *Analyst*, 118 (1993) 23.
137 S.D. Christian, G.A. Smith, E.E. Tucker and J.F. Scamehorn, *Langmuir*,

1 (1985) 564.
138 K. Kandori and R.S. Schechter, *Sep. Sci. Technol.*, 25 (1990) 83.
139 T. D. Long and J.L. Anderson, *J. Polym. Sci. Polym. Phys. Ed.*, 22 (1984) 1261.
140 J.F. Scamehorn, S.D. Christian and R.T. Ellington, in: J.F. Scamehorn and J.H. Harwell (Eds.), *Surfactant-Based Separation Processes.* Dekker, New York, 1989, Ch. 2.
141 F. Oosawa, *Polyelectrolytes.* Dekker, New York, 1971.
142 S.D. Christian, S.N. Bhat, E.E. Tucker, J.F. Scamehorn and D.A. El-Sayed, *AIChE J.*, 34 (1988) 189.
143 J.F. Scamehorn, in: J.F. Scamehorn (Ed.), *Phenomena in Mixed Surfactant Systems.* ACS Symp. Ser. 311, ACS, Washington DC, 1986, Ch. 1.
144 K.J. Sasaki, S.L. Burnett, S.D. Christian, E.E. Tucker and J.F. Scamehorn, *Langmuir,* 5 (1989) 363.
145 E. Pramauro and E. Pelizzetti, *Trends Anal. Chem.*, 7 (1988) 260.
146 J. Klepac, D.L. Simmons, R.W. Taylor, J.F. Scamehorn and S.D. Christian, *Sep. Sci. Technol.*, 26 (1991) 165.
147 E. Pramauro, A. Bianco, E. Barni, G. Viscardi and W.L. Hinze, *Coll. Surfaces*, 63 (1992) 291.
148 E. Pramauro, A. Bianco Prevot, E. Pelizzetti, R. Marchelli, A. Dossena and A. Biancardi, *Anal. Chim. Acta,* 264 (1992) 303.
149 C. Tondre, S.G. Son, M. Hebrant, P. Scrimin and P. Tecilla, *Langmuir,* 9 (1993) 950.
150 M. de la Guardia, E. Peris Cardells, A. Morales Rubio, A. Bianco Prevot and E. Pramauro, *Anal. Chim. Acta,* 276 (1993) 173.
151 M. Hebrant, A. Bouraine, C. Tondre, A. Brembilla and P. Lochon, *Langmuir,* 10 (1994) 3994.
152 E. Pramauro, A. Bianco Prevot, V. Zelano, W.L. Hinze, G. Viscardi and P. Savarino, *Talanta,* 41 (1994) 1261.
153 M. Ismael and C. Tondre, *Langmuir,* 8 (1992) 1039.
154 M. Ismael and C. Tondre, *J. Coll. Interface Sci.,* 160 (1993) 252.
155 S. G. Son, M. Hebrant, P. Tecilla, P. Scrimin and C. Tondre, *J. Phys. Chem.,* 96 (1992) 11072.
156 C. Tondre and M. Hebrant, *J. Phys. Chem.,* 96 (1992) 11079.
157 U.R. Dharmawardana, S.D. Christian, R.W. Taylor and J.F. Scamehorn, *Langmuir,* 8 (1992) 414.
158 D.W. Fuerstenau, *J. Phys. Chem.,* 60 (1956) 981.
159 J.H. Harwell, J.C. Hoskins, R.S. Schechter and W.H. Wade, *Langmuir,* 1 (1985) 251
160 J.H. Harwell, R.S. Schechter and W.H. Wade, *AIChE J.*, 31 (1985) 415.
161 C.C. Nunn, R.S. Schechter and W.H. Wade, *J. Phys. Chem.,* 86 (1982) 3271.
162 J.W. Barton, T.P. Fitzgerald, C. Lee, E.A. O'Rear and J.H. Harwell, *Sep. Sci. Technol.*, 23 (1988) 637.

163 K. Esumi, Y. Sakamoto, T. Nagahama and K. Meguro, *Bull. Chem. Soc. Jpn.*, 62 (1989) 2502.
164 K. Esumi, M. Shibayama and K. Meguro, *Langmuir,* 6 (1990) 826.
165 J.H. Harwell and E.A. O'Rear, in: J.F. Scamehorn and J.H. Harwell (Eds.), *Surfactant-Based Separation Processes.* Dekker, New York, 1989, p. 155.
166 J. Wu, J.H. Harwell and E.A. O'Rear, *Langmuir,* 3 (1987) 531.
167 J. Wu, C. Lee, J.H. Harwell and E.A. O'Rear, in: J.F. Scamehorn and J.H. Harwell (Eds.), *Surfactant-Based Separation Processes.* Dekker, New York, 1989, p. 173.
168 S.P. Nayyar, D.A. Sabatini and J.H. Harwell, *Environ. Sci. Technol.,* 28 (1994) 1874.
169 P. Somasundaran and B.M. Moudgil (Eds.), *Reagents in Mineral Technology.* Dekker, New York. 1988.
170 J. Leja, *Surface Chemistry of Froth Flotation.* Plenum Press, New York, 1982.
171 D.W. Fuerstenau and R. Herrera-Urbina, in: J.F. Scamehorn and J.H. Harwell (Eds.), *Surfactant-Based Separation Processes.* Dekker, New York, 1989, p. 259.
172 E.J. Chou and Y. Okamoto, *Sep. Sci. Technol.*, 13 (1978) 439.
173 Y. Koide, M. Uchino and K. Yamada, *Bull. Chem. Soc. Jpn.*, 60 (1987) 3477.
174 B.L. Karger, R.B. Grieves and R. Lemlich, *Sep. Sci.*, 3 (1968) 393.
175 F. Sebba, *Ion Flotation.* Elsevier, Amsterdam, 1962.
176 R. Lemlich, *Ind. Eng. Chem.*, 60 (1968) 16.
177 P. Somasundaran, in: E.S. Perry and C.J. Van Oss (Eds.), *Separation and Purification Methods.* Dekker, New York, 1972, Vol. I, p. 117.
178 T. E. Carleson, in: J.F. Scamehorn and J.H. Harwell (Eds.), *Surfactant-Based Separation Processes.* Dekker, New York, 1989, p. 233.
179 R.B. Grieves, W. Charewicz and S.M. Brien, *Anal. Chim. Acta,* 73 (1974) 293.
180 G.A. Nyssen, G.S. Lovell, A.A. Simon, J.G. Smith and B.K. Tolar, *Sep. Sci. Technol.*, 22 (1987) 2127.
181 R.B. Grieves, *AICHE Symp. Ser.*, 71 (1975) 143.
182 P. Somasundaran, *Sep. Sci.*, 10 (1975) 93.
183 K. Jurkewicz, *Sep. Sci. Technol.*, 20 (1985) 1979.
184 F. Sebba, *Sep. Purif. Methods*, 14 (1985) 127.
185 F. Sebba, in: J.F. Scamehorn and J.H. Harwell (Eds.), *Surfactant-based Separation Processes.* Dekker, New York, 1989, p. 91.
186 J.F. Scamehorn and J.H. Harwell, in: D.T. Wasan, M.E. Ginn and D.O. Shah (Eds.), *Surfactants in Chemical/Process Engineering.* Dekker, New York, 1988, p. 77.
187 D.L. Blakeburn and J.F. Scamehorn, in: J.F. Scamehorn and J.H. Harwell (Eds.), *Surfactant-Based Separation Processes.* Dekker, New York, 1989, p. 205.

189 N.J. Turro and B. Kraeutler, *J. Am. Chem. Soc.*, 106 (1978) 7432.
190 N.J. Turro and B. Kraeutler, *Acc. Chem. Res.*, 13 (1980) 369.
191 M.B. Zimmt, C. Doubleday and N.J. Turro, *J. Am. Chem. Soc.*, 106 (1984) 3363.
192 I.R. Gould, N.J. Turro and M.B. Zimmt, *Adv. Phys. Org. Chem.,* 20 (1984) 1.
193 A.L. Buchachenko, V.F. Tarasov and V.I. Maltsev, *J. Phys. Chem. (Russ.)*, 55 (1981) 936.
194 C. Garcia Pinto, J.L. Pèrez Pavòn and B. Moreno Cordero, *Anal. Chem.*, 67 (1995) 2606.

Chapter 8

Use of biomimetic amphiphilic systems in the separation and analysis of biomolecules

The surfactant-modified separation and quantitation methods examined in previous chapters can be generally applied to a great number of analytes without restrictions regarding their type or structure. In some cases, however, the peculiar properties of the amphiphilic aggregates render these systems unique for studying labile substrates, such as many biological molecules, which are strongly affected by modifications of their complex structural conformations.

The availability of quite simple amphiphilic aggregate structures, able to simulate those much more complex biological systems, may allow us to perform studies on solubilization and transport mechanisms which are of fundamental importance for a better understanding of biochemical processes, such as uptake of cell nutrients, transport and release of drugs in body tissues, energy-transfer and harvesting in plant cells, information transmission and cell recognition [1,2].

A great number of investigations on the interactions between solutes of biological concern and amphiphilic aggregates have been reported in the literature. Their applications to the separation and purification of selected biomolecules from complex natural matrices are of relevant interest in analytical chemistry. In particular, the study of biomolecule–surfactant interactions below the cmc and the examination of solubilization processes in aggregates from naturally occurring surfactants are aspects of biomimetic chemistry to be considered prior to the development of any analytical application.

The preparation and characterization of the most important membrane mimetic structures, together with the description of their properties and applications in some bioanalytical schemes are fundamental steps in these studies and will also be examined in the following sections.

8.1. INTERACTIONS BETWEEN BIOMOLECULES AND MICELLE-FORMING SURFACTANTS

8.1.1. Interactions between proteins and common surfactants

The proteins are a very important class of biological compounds which contain both hydrophobic and hydrophilic amino acid units in their long chains. Many of the physicochemical and biological properties of proteins depend on the ratio between the two types of amino acids. For example, water-soluble proteins usually contain up to 50% amino acids with hydrophilic groups, which in aqueous solutions are mainly oriented towards the water interface. The hydrophobic groups, on the contrary, tend to be oriented in the interior of the folded protein structure.

Due to the restricted flexibility of the protein molecules, some of the hydrophilic groups can still be present in a predominantly hydrophobic environment and some hydrophobic groups can be in contact with water. These latter groups usually act as binding sites for amphiphilic molecules present in the solution.

An understanding of the relationship between protein structure and reactivity is essential in order to elucidate the mechanism of several crucial biochemical processes. For example, the interactions between serum proteins and free fatty acids, lipids, hormones, drugs and other molecules present in blood play an important role in biochemistry. For this reason, the study of interactions between serum proteins and amphiphilic molecules has been extensive and, from the analysis of the corresponding binding isotherms, useful information about the protein structure has been obtained.

One of the most investigated protein–surfactant systems has been the bovine serum albumin (BSA)-SDS. This surfactant, which is a strong disorganizer of the protein structure when present in concentrated solutions, can act as a protein stabilizer at low concentrations.

The surfactant–protein complex can solubilize molecules which are not water-soluble, acting similarly to micellar aggregates. The surfactant counterion is still associated with the surfactant-protein complex and the associated fraction can be estimated from potentiometric measurements using the sodium ion-specific electrode [3].

Many membrane proteins can be dissolved in their biologically active form using surfactant solutions, where the amphiphile may replace the lipids with which the proteins are associated in the living structures. The molecular weights and the Stokes radius of the proteins can be experimentally determined in these solutions by using different techniques, for example through sedimentation equilibrium studies [4].

The binding of surfactants (below their cmc) to globular proteins in aqueous solutions is usually pH-dependent and, conversely, pH changes are observed in unbuffered solutions upon protein binding. For example, SDS binding to lysozyme in aqueous solutions can induce protonation of the carboxylic groups in the original protein conformation, with subsequent ionization in the unfolded protein [5].

Micellar chromatography of proteins is probably one of the most interesting applications of protein solubilization in micellar solutions. As previously indicated in Chapter 6, micellar solutions can be exploited to dissolve protein-containing biological samples for direct injection in HPLC columns. In addition, protein mixtures can be separated on a RP column using surfactant solutions as mobile phases [6]. For example, ovalbumine, BSA, thyroglobulin, chymotrypsinogen, β-lactoglobulin and lysozyme can be more or less efficiently separated onto a RP-C_8 column by gradient micellar chromatography, using a C_nE_6 commercial surfactant mixture (n = 9, 10, 11) buffered at pH 7 with phosphate. Separation of other membrane proteins by micellar HPLC are described in Ref. [7].

8.1.2. Interactions of solutes with biosurfactant aggregates

Naturally occurring micellar systems, such as phospholipids and bile salts, are very important components of biological structures and their properties are of great interest in a number of biotechnological and other industrial processes [8]. Among the micelle-forming biosurfactants, the bile salts are an important class of compounds, while phospholipids (fundamental components of biomembranes) give rise to the formation of other type of structures which will be examined below.

Bile salts are biological surfactants that are synthesized in the liver and are capable of forming small aggregates composed of up to 10 bulky monomers. The basic structure of the corresponding bile acids is shown below.

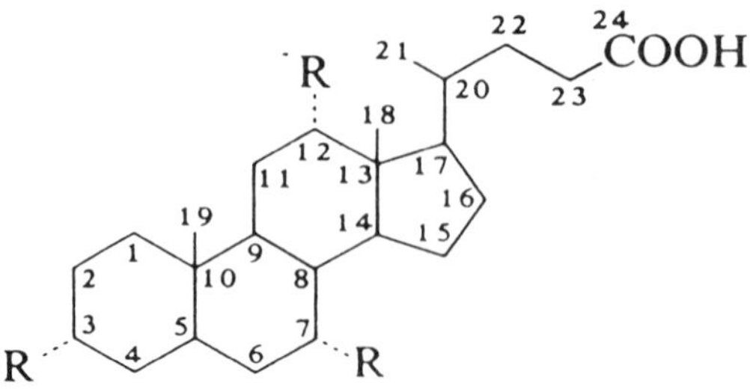

Basic chemical structure of bile acids.

Some examples of the naturally occurring products are the following: cholic acid, which has three OH groups in positions 3,7 and 12, respectively (3α, 7α, 12α-trihydroxy-5β-cholan-24-oic acid); deoxycholic acid has two OH groups in positions 3 and 12 (3α, 12α-dihydroxy-5β-cholan-24-oic acid); litocholic acid has only an OH substituent in position 3 (3α-hydroxy-5β-cholan-24-oic acid); hyodeoxycholic acid is similar to litocholic acid, but has a second OH group in position 6, etc.

The most probable orientation of such molecules at the air–water interface is with the hydrophilic groups towards the aqueous phase. Figure 8.1(A) shows the proposed orientation for cholic acid monomer (OH groups: filled circles, COOH group: open circle), whereas Fig. 8.1(B) shows the structures of the corresponding bile salt micelles, according to Small [9] and Oakenfull and Fisher [10]. The role of hydrogen-bonding in micelle formation from bile salts has been successively questioned [11] and other investigations tended to establish that these aggregates form primarily from hydrophobic bonding, with larger micelles obtained from aggregation of the primary structures. Experimental data (for example, viscosity data obtained from light-scattering measurements) seem to confirm the existence of such structural transitions [12].

The cmc of the bile acids and bile salts depends on the monomer structure. The trihydroxy cholanoic acids exhibit higher cmc values ($3–10\times10^{-3}$ M), whereas the more hydrophobic dihydroxy compounds show cmc values in the range $2–4\times10^{-3}$ M. Their corresponding sodium salts have, as expected, a higher cmc than the acids.

There are some reviews in the literature dealing with the physical and chemical properties of bile salt aggregates [13–16]. Their influence on a number of physiological processes and on drug absorption has also

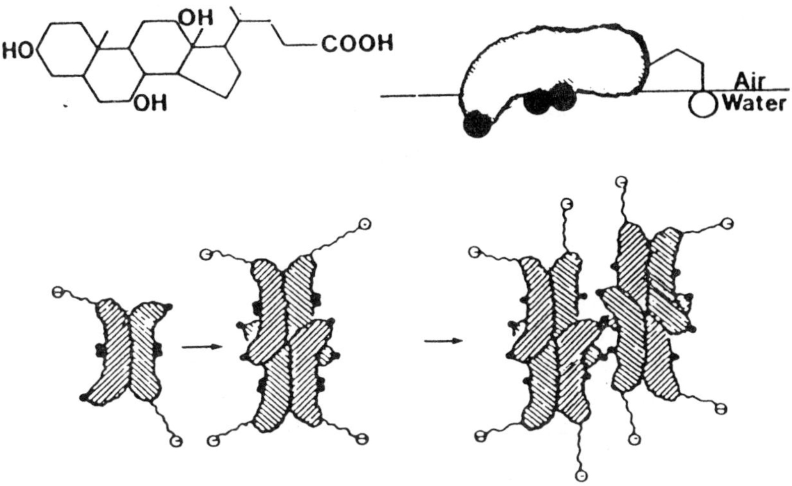

(A)

(B)

Fig. 8.1. (A) Orientation of bile acids in aqueous solutions. (B) Formation of bile salts aggregates.

been examined (see for example Ref. [17] and references cited therein). Among the studies performed, the investigation of different cases of bile-salt-stimulated enzymic activity can be mentioned as an example of the interactions operating between natural surfactants and proteins in biological processes [18–20].

Relevant biological processes involving bile salts, including fat absorption, interactions with protein molecules and interactions with membrane components, have been investigated using simpler model systems. For example, ternary and quaternary systems composed of phospholipids, bile salts and steroids (mainly cholesterol) have been examined in order to gain information about the solubilization of cholesterol [17]. Similarly, the solubilization and diffusion of oleic acid in bile salt–phospholipid mixtures has been investigated at different pH values [21].

Since the main role of bile salts and their micelles is as solubilizers for lecithin, cholesterol, monoglycerides and fatty acids, most of the reported studies concern solubilization of different substrates in these microstructures. The basic aspects of such investigations have been previously discussed in Chapter 2, whereas more detailed information about the phase diagram and the solubilizing power of the mixed aggregates can be obtained from the use of different combined techniques (e.g., NMR, self-diffusion, light scattering, etc.) [22].

8.2. INTERACTIONS OF BIOMOLECULES WITH MEMBRANE-MIMETIC STRUCTURES

Membrane-mimetic chemistry is a rapidly growing field concerned with the development and understanding of transport and separation processes similar to those occurring in biological membranes. Some of the amphiphilic aggregates described in Chapter 1, namely monolayers, organized multilayers (Langmuir–Blodgett films), bilayer lipid membranes and simple or polymerized vesicles, have been used as membrane-mimetic structures in these studies. The fundamental aspects of these investigations, including the preparation and properties of the membrane-mimetic aggregates, the study of their interactions with several biological substrates, and the analysis of their catalytic effects on a great number of chemical reactions, have been examined in detail by Fendler [23–26].

Before we describe the main features and performances of artificial membrane-mimetic aggregates in biomolecules separation, we should take a brief look to the characteristics of natural systems. The most important among them are certainly the biological membranes, which are complex structures capable of providing the compartmentalization and organization of living matter in plant and animal cells, acting as a permeable barrier towards different substrates. They are usually composed of 25–75% lipids, 25–75% proteins and less than 10% carbohydrates.

According to the fluid mosaic model of Singer and Nicolson [27], the structure of a typical biomembrane is composed of a lipid bilayer which incorporates different globular proteins located either at the surface or within the bilayer. In some cases the intercalated protein can fully span the bilayer (transmembrane proteins). A biological membrane structure is represented in Fig. 8.2.

The membrane lipids (phospholipids and/or glycolipids) contain a hydrophobic moiety, which is generally an aliphatic double chain, and phosphate or carboxylate polar head groups. These membrane components are highly mobile, being capable of translocation from one half of the bilayer to the other and giving rise to intermembrane lipid exchange. The important role of the phospholipid head group in membrane interactions has been pointed out by Yeagle [28].

The investigation of selective transport and separation processes in relatively simple amphiphilic structures is among the most important objectives of membrane-mimetic chemistry since many practical

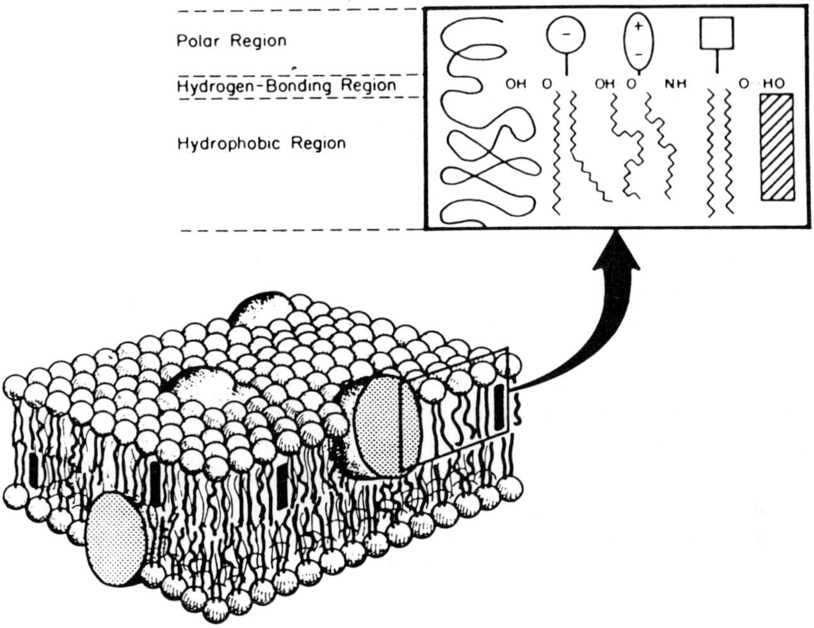

Fig .8.2. Typical structure of a biological membrane. Reproduced with permission from Ref. [23].

applications could arise from this approach. Reactivity control in chemical schemes involving biomolecules, molecular recognition and enrichment of active isomers, studies on photochemical solar energy conversion, drug encapsulation and transport, and development of protected environments for enzymes are some examples of these applications.

The formation and properties of micellar aggregates in water have been extensively examined in Chapter 1, whereas the formation and features of other membrane-mimetic structures, such as monolayers, bilayers and vesicles, have only briefly been described (see Section 1.12.2). The preparation and characterization of suitable amphiphilic microstructures able to interact with common biological molecules and simulate the behaviour of components of living cells will be considered in this section.

8.2.1. Reversed micelles

8..2.1.1. Formation and properties

The formation of ion-pairs and other molecular aggregates in non-polar solvents has long been recognized. The terms reversed micelles and water–oil (w/o) microemulsions have been largely used to designate the more or less spherical amphiphilic aggregates formed in these solvents [29,30], although the term reversed micelle is more often used to indicate microaggregates composed only of surfactant, water and organic solvent (without addition of cosurfactants, such as alcohols). Some of the reversed micelle-forming amphiphiles are listed in Chapter 1 (see Table 1.1). A schematic representation of a typical reversed micelle structure is depicted in Fig. 8.3.

Several applications of reversed micelles as suitable mobile phases for chromatographic separations have been reported and discussed in Chapter 6. The particular properties of these water-containing microstructures, capable of solubilizing and transporting hydrophilic substrates (even macromolecules) within hydrophobic bulk phases, are crucial in other

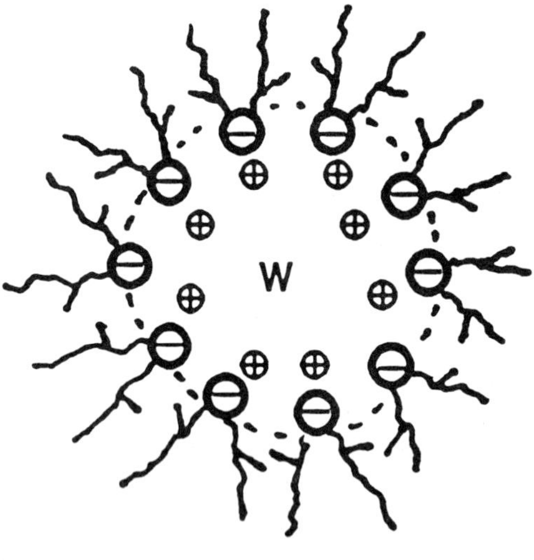

Fig. 8.3 Schematic representation of a sodium bis-2-ethylhexylsulfosuccinate (AOT) reversed micelle. The letter W indicates the inner water pool surrounding the ionized surfactant group (sulfonate). The hydrated counterions (Na^+) present in the water droplet are also indicated. The surfactant branched alkyl chain is oriented towards the bulk non-polar solvent.

important research fields, such as separation and purification of biomolecules and in the modelling of biological membrane permeation.

The solubility and aggregation of amphiphilic molecules in such solvents changes dramatically upon addition of water. Hydrocarbon-soluble surfactants, such as AOT, are able to solubilize a great number of water molecules (up to ca. 60 per amphiphile molecule) in the apolar solvent and, as the amount of entrapped water increases, the size of the aggregate also increases. For example, the aggregation numbers of 0.08 M dodecylammonium propionate (DAP) in cyclohexane vary with the amount of water added to the system, increasing from ca. 5 for 0.093 M water (where there are no free water molecules, all being associated with the surfactant) to ca. 45 for 0.46 M added water (with about 80 free water molecules per aggregate), up to ca. 110 for 0.74 M water (about 600 free water molecules per aggregate) [31].

The radii and sizes of AOT aggregates in different organic solvents have been determined as a function of added water using different experimental techniques, which include ultracentrifugation [32], time-resolved fluorescence [33,34], X-ray scattering [35,36], dynamic light scattering [37–39], neutron scattering [40–43], etc. Figure 8.4 shows the

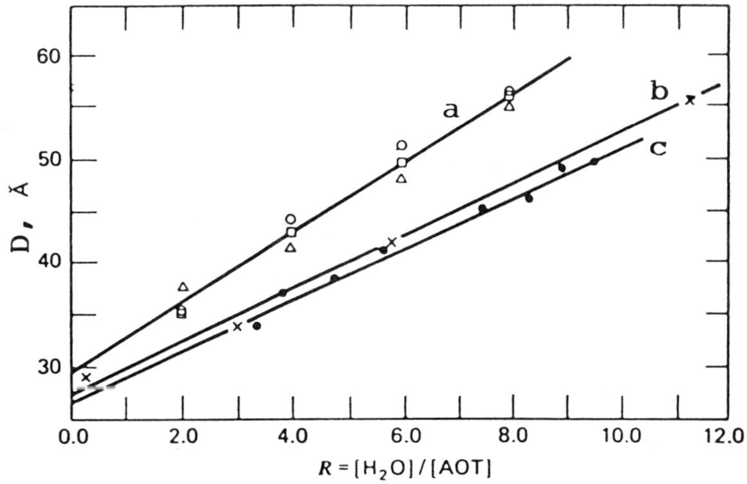

Fig. 8.4 Variation of the droplet diameter of AOT reversed micelles with increasing water content. (a) From viscosity data; (Δ) toluene; (O) chlorobenzene; (□) cyclohexane. (b) From ultracentrifugation data, solvent n-heptane. (c) From dynamic light-scattering in toluene. Reproduced with permission from Ref. [37].

variation of the microdroplet diameter, D, of the AOT reversed micelles in toluene, cyclohexane and chlorobenzene, as a function of the water/AOT ratio, W_0. When molar concentration ratios are used, W_0 and $M_{W/AOT}$ (previously introduced in Chapter 6) are equivalent terms. The symbol R to indicate the molar water/surfactant ratio is also sometimes reported in the literature.

Most of the reported investigations on reversed micelle properties concern the ternary system AOT/water/non-polar solvent. Table 8.1 summarizes some of the studies performed on these systems.

TABLE 8.1

Examples of reported physico-chemical studies on the structure, dynamics and properties of AOT reversed micelles in different organic solvents

Organic phase	Investigated RM parameter	Technique used	Ref.
n-Dodecane	Electrical conductivity	Electrical conductance	44
Isooctane	Electrical conductivity. Development of a charge fluctuation model	Electrical conductance and dynamic light scattering	45
Benzene	Micropolarity and aggregation	UV-VIS absorption of zwitterionic probes	46
Isooctane	AOT hydration	DSC, ESR spin labelling ^2H,NMR	47
n-Heptane	Solubilizate exchange[a] between water pools	Spectrophotometry (normal and continuous-flow)	48
n-Alkanes (C_6 to C_{12})	Intermicellar exchange of solutes	Time-resolved fluorescence probing	49
Isooctane	Ion effects on cation solubilization	Phase-transfer experiments	50
n-Alkanes	Fusion of the RM structures at high pressures	Fluorescence depolarization of solute probes	51
n-Heptane, n-dodecane	Structure of water and RM size. Effect of added solutes	Fluorescence quenching of solute probes	52
Isooctane, $CHCl_3$, n-dodecane	Location of charged solutes within the RM	Probe fluorescence quenching by O_2	53

a: Proton and electron transfer and metal complexation studies.

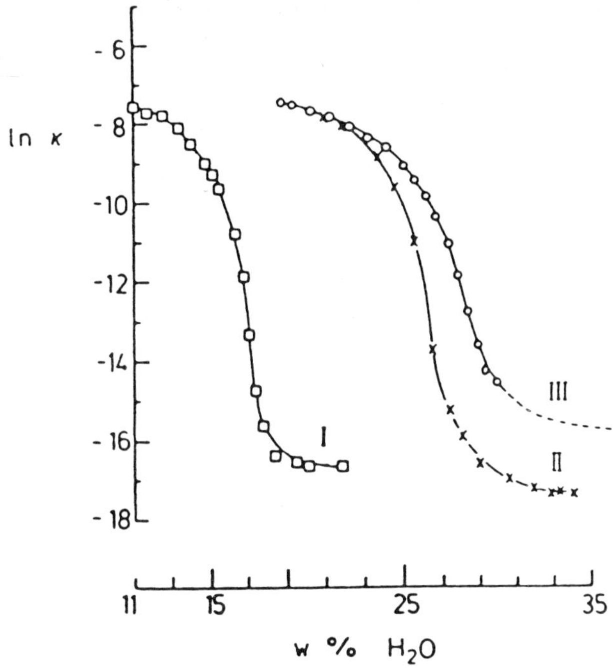

Fig. 8.5 Variation of the electric conductivity of reversed micelles of DDAB as a function of water content. (I) Octane, (II) decane, (III) dodecane. Reproduced with permission from Ref. [55].

The solubilizing power of the aggregates as well as the properties of the entrapped water pools are strongly dependent on the water/surfactant ratios. Different types of water (i.e. water bonded to the surfactant head groups, water bonded to the counterions and unbonded water) have been recognized and, below a critical water/surfactant ratio, only totally bound water seems to exist [54].

Conductivity measurements performed on didodecyldimethylammonium bromide (DDAB) reversed micellar solutions have shown that, in the presence of less than 10% water, the electric conductivity is high. Upon the increase in water content to ca. 40%, the conductivity drops by a number of orders of magnitude (see Fig. 8.5). This unexpected behaviour has been interpreted in terms of the viscoelastic properties of the solution [55], assuming that the water and surfactant components can form continuous channels in the organic phase which allow the transport of the electric current. This network breaks down by the addition of water, because water globules (dispersed into the bulk solvent) form and the continuity of the water domain disappears.

Concerning the thermodynamic aspects of reversed micelle formation, it is useful to remark that recent studies indicate that the uptake of water within the water pools in AOT reversed micelles is an unfavourable process from an enthalpic point of view, whereas the associated entropy change is largely favourable [56,57].

Kinetic analysis of intermicellar exchange of water and electrolytes performed on the same system indicate that these processes are quite fast (in the millisecond range) [58], thus implying a quite fluid microaggregate structure. On the contrary, studies on solubilizates reorganization in reversed micelles of calcium alkylarylsulfonates in heptane revealed a very restricted fluidity. Energy-transfer and FTIR spectroscopy measurements conducted on $TbCl_3$ as probe solute [59] showed a reorganization time in the hour range. These different results confirm that the aggregate properties strongly depend on the surfactant molecule used and cannot be inferred from data obtained when working with other types of reversed micelles.

8.2.1.2. Solubilization in reversed micelles

A common and important feature of the water pools present in these aggregates is that they resemble the polar pockets observed in enzymes, providing a unique microenvironment for interactions with polar solubilizates [60]. A typical location of most solutes in the microstructure is with their polar groups imbedded into the water pools and their hydrophobic moieties placed in the surfactant layer. Studies on the pK_a variation of 4-nitrophenol in AOT/isooctane solutions confirmed such probe location and indicated that the OH group is hydrogen bonded to the sulfonate head groups [61]. However, a less polar environment has been hypothesized for the same solute in AOT/benzene solutions [62], thus indicating that the nature of the organic solvent plays an important role and must be considered

Solubilization in reversed micelles is, as for normal micelles, regulated by the solute–aggregate interactions. Spectrophotometric studies on solubilization of different phenol derivatives in AOT reversed micelles in 2,2,4-trimethylpentane [63] revealed that the strength of binding is here dependent on both the hydrophobicity and the hydrogen-bonding tendency of the solutes.The second factor seems to be more relevant. The values reported for different *para*-substituted phenols (solute concentration ca. 3.5×10^{-4} M) are the following: 4-t-butylphenol

172 M^{-1}, 4-methylphenol 175 M^{-1}, phenol 327 M^{-1}, 4-chlorophenol (5.4×10^{-5} M) 1310 M^{-1}, 4-nitrophenol (ca. 1×10^{-4} M) 3670 M^{-1}. Addition of water to reversed AOT micelles causes a decrease in the binding constants, until the w/o microemulsion transition is observed.

8.2.2. Monolayers

Analytical applications of surfactant monolayers adsorbed onto solid supports, which are used as suitable extraction phases for selective separations of amphiphilic and hydrophobic solutes, have been previously examined when describing the features of admicellar chromatography (see Chapter 7). In order to apply the monolayers as membrane-mimetic media, however, the formation of suitable packings lying on liquid surfaces and/or easily transferred from one solid support to another is necessary. Moreover, the intercalation and careful organization of proper reactive probes within these structures is often required to mimic the biological processes to be investigated.

8.2.2.1. Preparation and characterization of monolayers

Monomolecular layers of amphiphilic molecules (lipids or synthetic surfactants), usually containing long alkyl chains (composed of 12 or more carbon atoms), can be obtained by spreading their organic solutions over water and allowing the evaporation of a solvent. The formation of monolayer films is usually investigated by using the Langmuir film balance, which measures the pressure exerted by a film while the area is reduced using a mobile barrier. Amphiphilic molecules present at the air/water interface are forced to assume a more vertical position, thus reducing the area per molecule when the pressure increases (see Fig. 8.6). The resistance of the film to compression and the area per molecule at a given pressure are the main parameters which characterize the monolayer formation. The thickness of the obtained films can be determined through the measurement of the transmitted and reflected light.

The surface pressure of the monolayer, Π, indicating the area occupied per molecule, is given by:

$$\Pi = \gamma_0 - \gamma \quad (1)$$

where γ_0 is the surface tension of the pure solvent and γ is that of the monolayer-covered liquid surface. The study of spread monolayers is of particular interest since the compression of the structure can alter continuously the interactions among the components, giving rise to neat

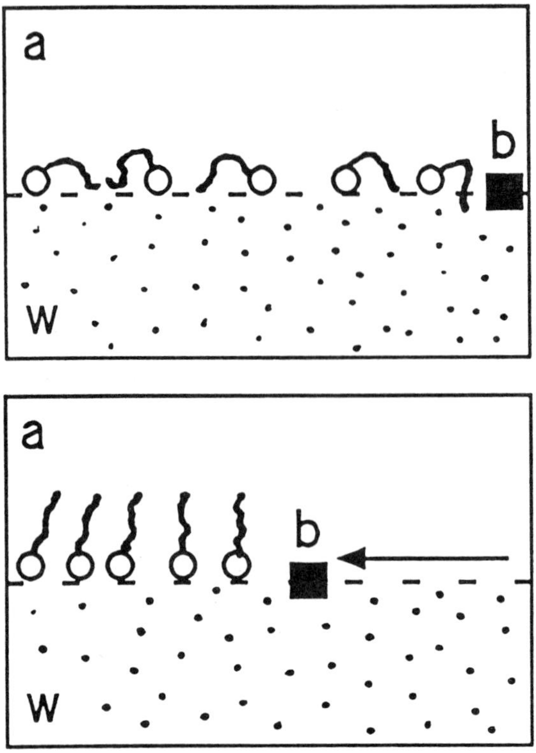

Fig. 8.6. Formation of a monolayer at the air (a) water (w) interface; (b) represents the mobile barrier exerting a variable pressure in the direction indicated by the arrow.

changes in molecular organization and even to phase transitions.

Different subphases can form when monolayers are subjected to increased pressures in the Langmuir trough. The L_1 type, obtained at lower pressures, is composed of molecules having a position more or less perpendicular to the subphase surface. In this state, the intermolecular distances between the monolayer-forming components (their head groups are separated by water molecules) are higher than in bulk liquids.

The compression of L_1 monolayers gives rise to the formation of a more packed structure, termed the liquid condensed phase (L_2), where the molecules are close to each other and tilted with respect to the subphase surface. In the solid state, the amphiphile molecules are very closely packed, perpendicular (or tilted at an angle) to the subphase surface. They are almost incompressible and, under a further increase in pressure, the monolayer breaks into multilayers.

8.2.2.2. Surface potentials and polarity in monolayer structures

Ionic monolayers give rise to the formation of an electrical double layer. For ionizable monolayer-forming amphiphiles, the extent of ionization depends on the surface pH, which in turn is related to the bulk pH by the equation:

$$pH_s = pH_b + e\Psi/2.3\ kT \qquad (2)$$

where Ψ is the interfacial potential and the subscripts s and b indicate the surface and bulk phase, respectively. The amphiphile ionization alters the surface pressure in the monolayer, according to the equation [64]:

$$\Delta\Pi = \Pi - \Pi_0 = -\int_0^{\Psi_0} \varepsilon d\Psi \qquad (3)$$

where Π and Π_0 are the surface pressures of the ionized and unionized monolayer, respectively, and ε is the surface charge density. As expected, counterions largely alter the surface area/surface pressure isotherms and the observed effects are dependent on both the ionized group and the counterion.

As for micelles and vesicles, lipophilic acid–base indicators can be employed to measure the mean field electrostatic potential in surfactant monolayers covering liquid surfaces [65] or deposited onto solid plates [66,67]. Lipoidal eosin and fluorescein derivatives having long alkyl chain substituents (see formulae below) are particularly suitable for probing the electrostatic properties of surfactant monolayers spread at the air/water interface [68].

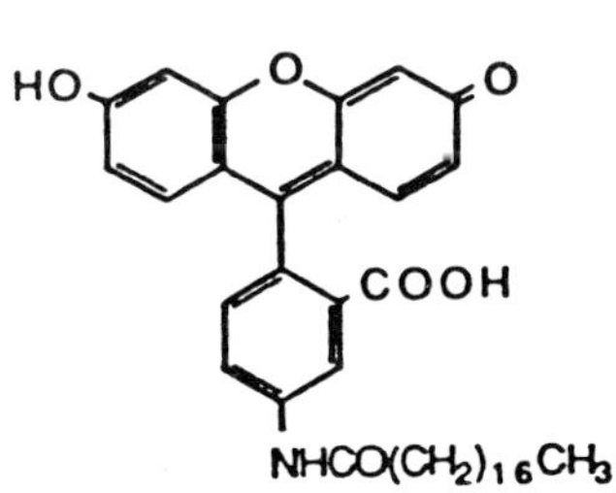

Fluorescein derivative

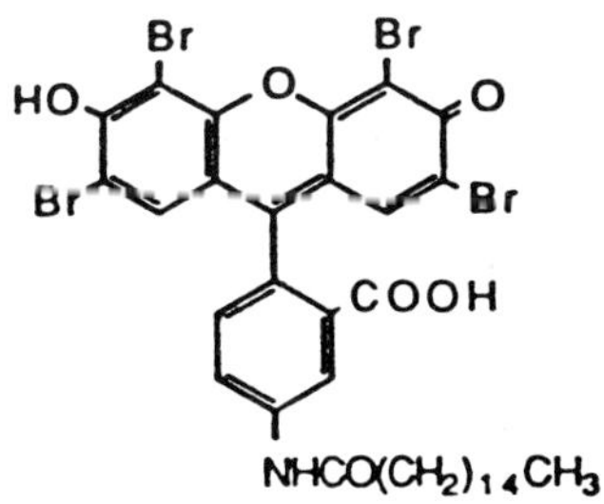

Eosin derivative

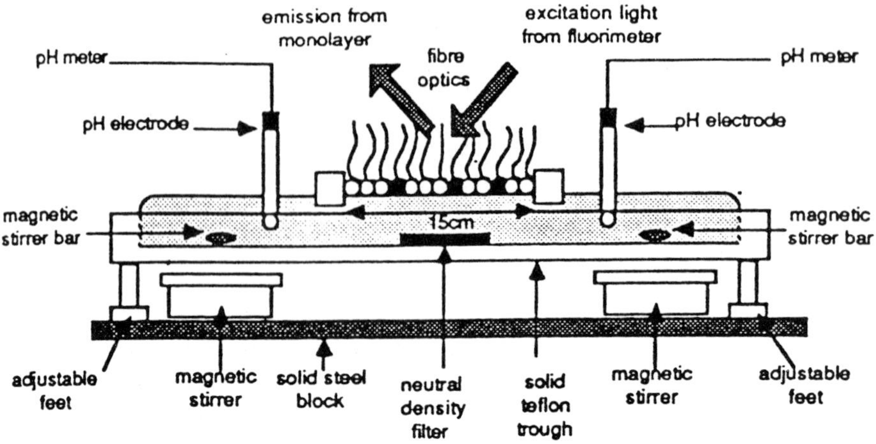

Fig. 8.7. Scheme of the experimental set-up used for the analysis of monolayer properties. The black amphiphiles represent the lipoidal probes. Reproduced with permission from Ref. [68].

The visible absorption spectra, the fluorescence emission spectra and the acid–base properties of these probes, intercalated in the monolayer under observation, have been examined to measure the monolayer/subphase potential. The particular experimental set-up used in these experiments is shown in Fig. 8.7.

The polarity of monolayer structures can also be investigated by exploiting the fluorescence changes of the intercalated probes. For example, the shifts in the fluorescence emission of N-[5-(dimethylamino)naphthalene-1-sulfonyl]dihexadecylamine, a dansyl derivative intercalated in the monolayer, can be correlated with the changes in the conformation of the amphiphile head group and with its compression and hydration in the layer structure [69].

8.2.2.3. Preparation and applications of functionalized monolayers and multilayers

The construction of artificial functional unities simulating more complex structures is one of the most important applications of relatively simple membrane-mimetic aggregates. The intercalation of molecules possessing particular optical properties and/or exhibiting chemical reactivity is a common task in these studies. Suitable techniques for the transport of monolayers onto solid supports and for the formation of

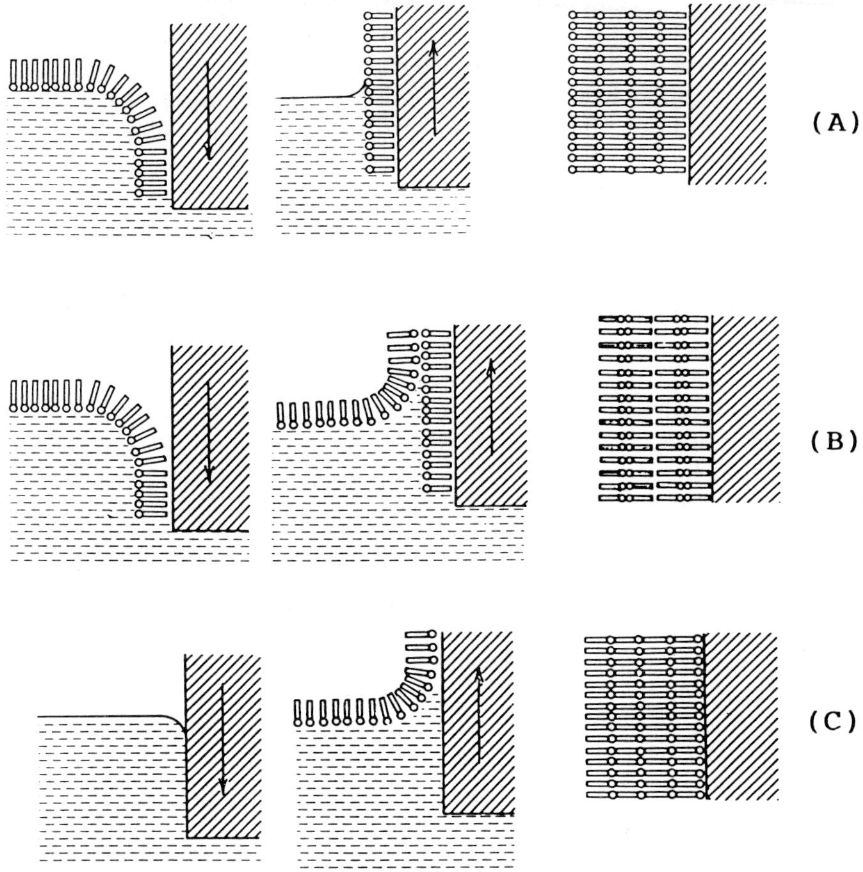

Fig. 8.8. Different structures of fatty acid multilayers. (A) X-deposition; (B) Y-deposition; (C) Z-deposition. Reproduced with permission from Ref. [71].

monolayers and multilayers containing different intercalated components have been developed.

The formation of multilayer structures upon immersion of a metal plate into a monolayer-covered liquid was observed by Blodgett many years ago [70]. Different types of multilayers, characterized by different structures, can be obtained by varying the monolayer-forming molecule and the mode of deposition (see Fig. 8.8). The Y-type multilayers are the most commonly obtained arrangements.

The multilayers can be removed from the solid support by depositing a thin film of PVA onto the external monolayer. After drying the PVA film, it can be removed together with the attached multilayer structure. The good solubility in water of PVA allows the transfer of the removed layers, which lie on the PVA support film, to aqueous solutions. After PVA dissolution, the attached monolayer remains on the solution surface and this approach can be used to prepare monolayers composed of blocks of different molecules lying on the same subphase surface [71].

The formation of monolayers containing (or composed of) indicators and/or reactive molecules has been used to investigate the monolayer packing, to study the solute diffusion across the more or less compressed layers and to simulate certain physico-chemical processes often occurring at the biomembrane level. For example, monolayers composed of molecules containing an azo dye chromophore and a long chain fatty acid linked via ester bonds to one molecule of glycerol have been investigated by Möbius [72]. The examination of different aggregate packings in these monolayers is facilitated by the optical properties of the dye. In fact, different orientations of the chromophore at the monolayer/subphase plane can be inferred from the absorption of linearly polarized light, incident at 45° to this plane. The light having a perpendicular component of the electric vector (in respect to the layer plane) is strongly absorbed, whereas the light with no perpendicular component is not absorbed.

The surface pressure/surface area isotherm of monolayers composed of compound I (see formula below) is shown in Fig. 8.9. An increase of the surface pressure upon reduction of the surface area per molecule to ca. 100 Å^2 is observed, followed by a steep increase at ca. 20 Å^2. The corresponding orientation of the component molecules in these monolayer packings are confirmed by the polarized light absorption measurements.

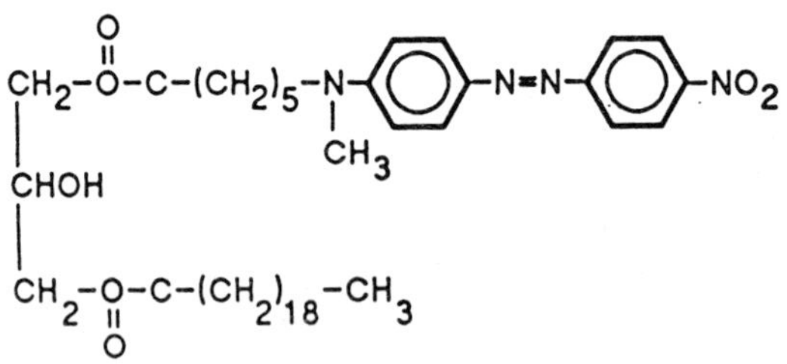

Monolayer-forming dye (compound I).

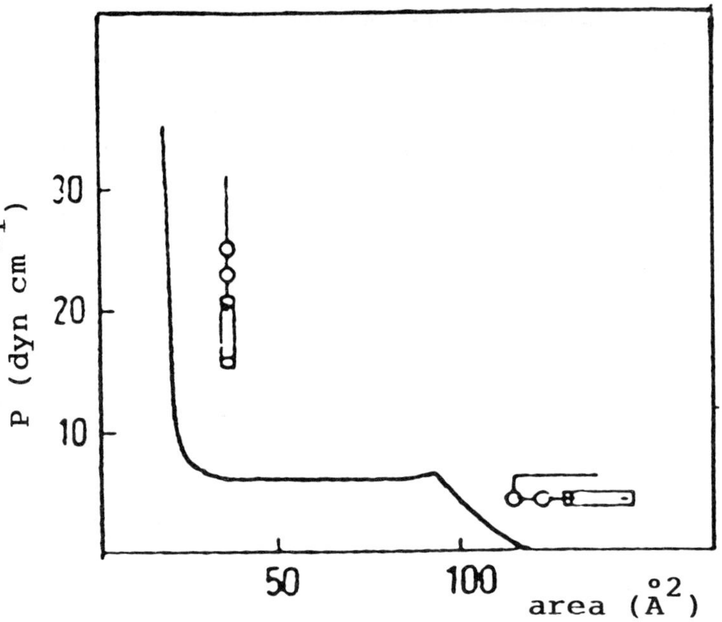

Fig. 8.9. Organization of monolayers of compound I as a function of the area per molecule available. Reproduced with permission from Ref. [72].

Multilayers composed of light-absorbing dyes (sensitizers) and acceptor dyes placed at different distances from each other (using intercalating monolayers of non absorbing compounds), successively irradiated with UV light, show efficient intermolecular energy transfer when the distance is relatively small [71]. An example of these experiments, which can simulate some of the energy transfer processes occurring in plant cells, is the energy transfer from an oxacyanine to a thiacyanine dye in the multilayer assembly depicted in Fig. 8.10.

The dye A is excited by incident UV light and the ratio of emitted fluorescence quanta Q/Q_0 is determined. Q_0 is the number of quanta emitted in multilayer sections where the acceptor B is absent, whereas Q represents the corresponding values in multilayer sections containing B. The ratio Q'/Q, where Q' is the number of quanta of fluorescence of dye B due to the energy transfer from excited A molecules (A^*), is also determined. In the arrangement shown in Fig. 8.10, with a molar ratio of A to arachidic acid of 1:100 and a molar ratio of B to arachidic acid of 1:20 (in their corresponding monolayers), 85% of the fluorescence emitted

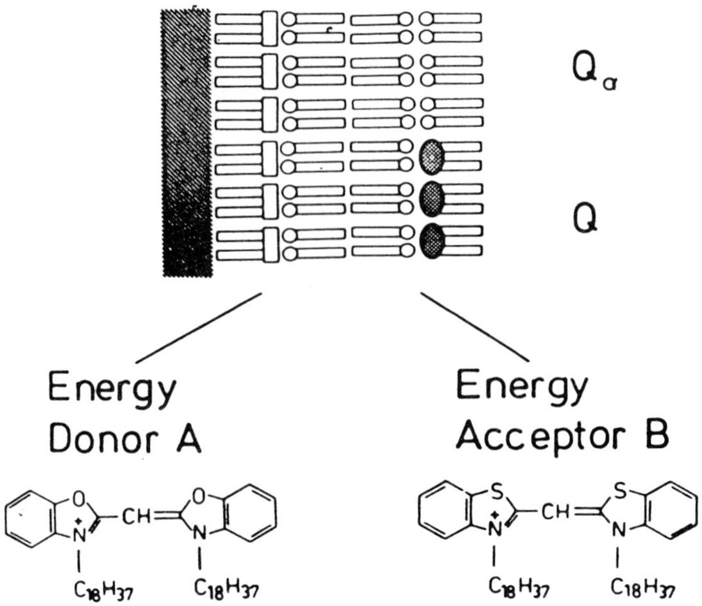

Fig. 8.10. Multilayer assembly containing a donor and an acceptor placed at a fixed distance (54 Å). Two layers of arachidic acid (matrix molecules) separate A from B. Reproduced with permission from Ref. [73].

by A is quenched by B at room temperature ($Q/Q_0 = 0.15$). The intercalation of the same donor–acceptor couple at lower distances (ca. 5 Å) in other bilayer assemblies allows a more effective energy transfer (ca. 50% of the fluorescent emission of A is quenched by B).

The photochemical behaviour of different chromophores incorporated in monolayers has also been investigated by Whitten and coworkers [74]. Cis-trans photoisomerization of stilbazole olefins and thioindigo dyes, ligand photoejection in ruthenium(II) porphyrin complexes, excimer and photodimer formation of long-chain pyrene derivatives, photoinduced electron transfer of hydrophobic ester derivatives of tris(2,2′-bipyridine)ruthenium(II)$^{2+}$ (a sensitizer excited by solar light), are examples of reactions that are very sensitive to the monolayer microenvironment and are used for modelling the solar energy transfer and harvesting processes occurring in biostructures.

The same type of derivatives of tris(2,2′-bipyridine)ruthenium(II)$^{2+}$ has been used to investigate the photoinduced redox processes in monolayers, applied either on aqueous subphases or on solid supports [75]. The hydrolysis of these probes arranged in monolayer films (which is faster than in homogeneous solution) has also been investigated. An efficient analytical control of the chemical process has been achieved by coupling the use of a specially designed multicompartment monolayer trough with accurate HPLC and vapour-phase chromatography determinations of the complexes and their hydrolysis products [76].

The preparation of monolayers formed by $Ru(bpy)_3^{2+}$-based amphiphiles onto semiconductor and metal electrodes is also at the basis of other studies concerning electron transfer processes [77], and further developed in the direction of electrogenerated chemiluminescent emission at monolayer-semiconductor and monolayer–metal electrode interfaces [78].

Other studies have been focused on the analysis of diffusion of small molecules (such as O_2, CO, NO, H_2O) within different multilayers. These investigations clearly indicated that the monolayer structures exhibit a quite relevant permeability to these components [79–81]].

8.2.3. Preparation of bilayer (black) lipid membranes

These microaggregates are formed by brushing an organic solution of the proper amphiphile across a few mm in diameter pinhole separating two aqueous phases. The initially formed thick structure, which shows a grey colour, is transformed (within some minutes) into a thinner one which shows interference colours. Finally, a very thin (40–60 Å thickness) black bilayer membrane is obtained (see Fig. 8.11).

Another method of BLM preparation is based on the preliminary formation of a monolayer onto the liquid subphase. A small teflon loop is then pulled through the monolayer to obtain the membrane. The preparation of these structures require some experience and is, however, difficult independent of the experimental method used.

The particular properties of BLMs have been investigated by several groups, although the difficulties arising from the scarce reproducibility of their preparation limits, at least in part, their application to solute solubilization and transport studies.

Optical techniques (in particular reflectance measurements) and electrical capacitance studies have been used to follow the BLM formation and to estimate the bilayer thickness [82,83]. Studies on selective transport of ions and molecules across BLMs have been performed in order to investigate the solute permeability.

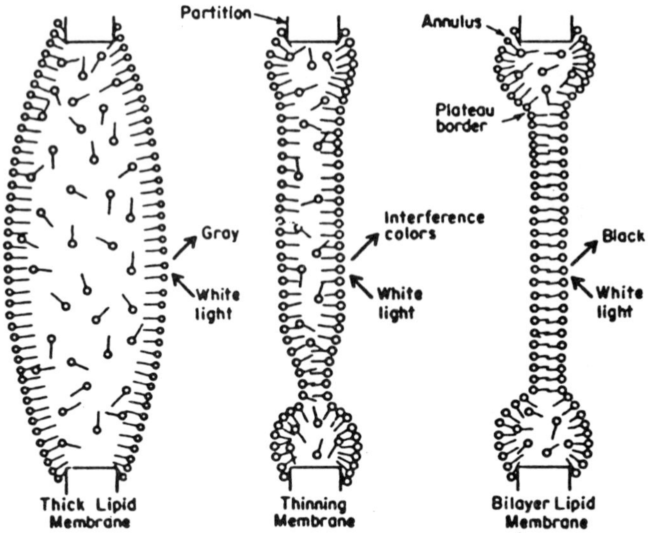

Fig. 8.11. Formation of a BLM. Reproduced with permission from Ref. [24].

8.2.4. Vesicles

8.2.4.1. Preparation and properties

Surfactants having a hydrophilic head group attached to two alkyl chains can spontaneously form bilayer aggregates, which assume the form of a closed spherical structure (vesicle) when sonicated in water. These structures, characterized by an inside region separated from the external one by a more or less hydrophobic barrier (see Fig. 8.12) may be used to simulate the properties of a cell membrane since the introduction of suitable probes (acid–base indicators, fluorescent dyes, drug molecules, redox indicators, etc.) within the vesicle is relatively easy and studies on their transport across the bilayer barrier can be accomplished.

Vesicles composed of naturally occurring (or synthetic) phospholipids are usually termed liposomes, while the term surfactant vesicles is usually reserved for the structures formed from other synthetic amphiphiles. Phospholipids form liposomes having a broad size distribution (100–1800 nm in diameter) which, under ultrasonication, can give rise to small unilamellar vesicles of 30–100 nm diameter and 40–60 Å thickness [84].

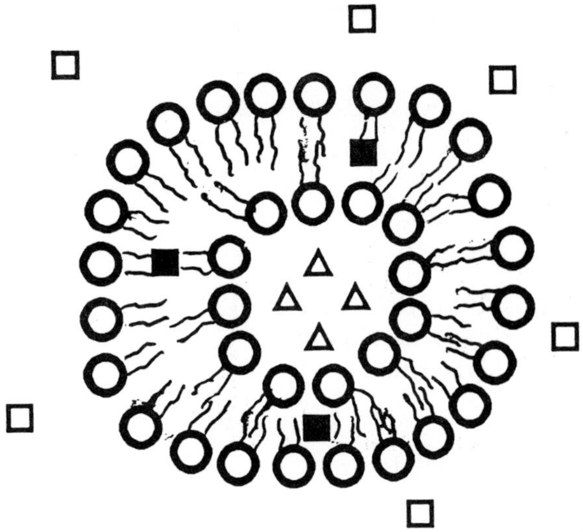

Fig. 8.12. Schematic representation of a single compartment (unilamellar) vesicle. Lipophilic (■) and hydrophilic (□,Δ) probes (or reactants) may be located in different vesicle sites.

Among the vesicle-forming synthetic amphiphiles, the most important are cationic dialkyldimethylammonium salts [85,86] and anionic phosphate derivatives [87], which were proposed several years ago and are still largely used.

The size of the synthetic vesicles can be varied over a wide range (from 25 to 500 nm in diameter) by using different methods of preparation. The injection method is based on a slow injection of a solution of the surfactant in a non-polar solvent (usually chloroform) into the aqueous solution. After solvent evaporation, the formation of large single compartment and multicompartment vesicles is observed. The same result is also obtained by swelling of thin phospholipid films in water.

Sonication of the same surfactants in aqueous solutions yields fairly uniform single bilayer small vesicles (25–60 nm in diameter). Further treatments, such as ultracentrifugation, ultrafiltration or gel permeation, allow more uniform aggregates to be obtained.

The structures of some natural and synthetic vesicle-forming surfactants are shown below.

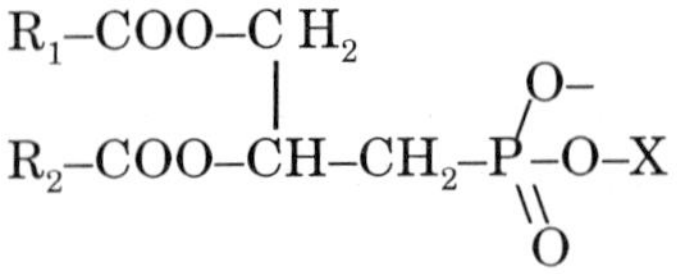

R_1, R_2: normal saturated alkyl chains (C_{11}, C_{13}, C_{15}, C_{17}, C_{19}), oleoyl chain: C_{17}(9-octadecenoic)

X = H phosphotidic acid
= $-CH_2-CH_2-N^+(CH_3)_3$ phosphatidylcholine
= $-CH_2-CH_2-NH_2$ phosphatidylethanolamine
= $-CH_2-CH-NH_2$ phosphatidylserine
|
COOH
= $-CH_2-CH-CH_2OH$ phosphatidylglycerol
|
OH

Liposome-forming phospholipids

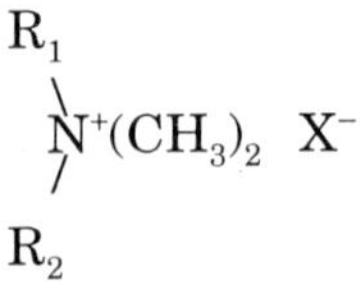

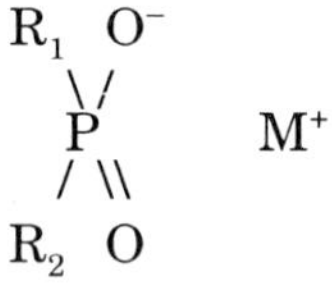

$R_1 = R_2$: saturated alkyl chains (usually $C > C_{10}$)

Vesicle-forming synthetic surfactants

Although most vesicles are obtained with a considerable input of energy (such as sonication) or by particular treatments (including surfactant dialysis or reverse-phase evaporation), in some cases the spontaneous formation of stable vesicles has also been reported [88]. The stability of these vesicles is currently in the range of several weeks or even months.

Pioneering work on characterization and vesicle size determination using electron microscopy and EPR measurements of spin probe-containing didoceyldimethylammonium chloride (DODAC) aggregates has been reported by Fendler and coworkers [89,90]. The incorporation of suitable probe molecules, soluble in different sites of the bilayer vesicles and able to monitor the internal structural features of the microaggregate, is the usual approach followed in most reported studies. Examples of such probes and their solubilization modes in vesicles are given in Ref. [91].

An extended examination of liposome and vesicle properties, together with the examination of their interactions with a large number of different biological substrates can be found in Chapter 6 of Ref. [24]. Some selected examples of other reported physico-chemical investigations, performed using different vesicle-solubilized probes and/or reactions, are listed in Table 8.2.

TABLE 8.2

Examples of studies concerning reactivity, ion-exchange, electron and energy transfer in vesicles

Probe system	Vesicle	Technique used	Information obtained	Ref.
$e^-_{(aq)}$ + Pyrene	DPL[a]	Pulse radiolysis, probe fluorescence, fast kinetic analysis.	Probe location in the bilayer close to the outer surface.	92
Glucose	DDAB[b]	Radioactivity of the probe.	Fast leaking of the probe across the bilayer	93
DPHT[c]		Fluorescent polarization	Vesicle microviscosity higher than micelles and lower than liposomes.	94
MV2+[d] + trans-stilbene or 1,4-diphenyl-butadiene	DHP[e]	Quenching of the *ene* fluorescence.	Location of the hydrophobic probe. Charge-transfer complex formation.	94
trans-4-methoxy-4′- nitrostilbene	DHP, DODAC[f]	Phosphorescence decay (lifetime of the probe triplet).	Polarity and viscosity of the bilayer. Interfacial probe location. Intra-molecular CT complex.	94

continued

TABLE 8.2 *continuation*

Probe system	Vesicle	Technique used	Information obtained	Ref.
Polypeptides + fluorescent probe	DHDAC[g]	Calorimetric and probe fluorescence measurements.	Location of hydrophobic polypeptides in the bilayer forming membrane-spanning helices.	95
Inorganic precipitates: CoS, FeO(OH), Ag, Ag_2O, silicates, etc.	Liposomes	Electron microscopy. NMR measurements.	Nucleation studies in very low volumes. Formation and storage of solids in bio-structures.	96
Pyrene probes + I^-	DODAC, DODAB	Fluorescence quenching.	I^-/Br^- and I^-/Cl^- exchange at the outer VS surface.	97
TP^D + DPO^A	Liposomes	Donor fluorescence quenching	Efficient singlet–singlet energy transfer.	98
DPO^D + Chl a^A			Fast transfer due to probe orientation in the VS bilayer.	
Paramagnetic ions	Liposomes	NMR studies.	Use of trapped probes as contrast agents in NMR imaging.	99
Amino acid esters	Catalytic[h] vesicles	Spectrophotometric kinetic studies.	Stereoselective hydrolysis of enantiomeric amino acid esters.	100

a: DPL = dipalmitoyl lecithin; *b*: DDAB = didodecyldimethylammonium bromide; *c*: DPHT = 1,6-diphenyl-1,3,5-hexatriene; *d*: MV^{2+}= methylviologen derivatives; *e*: DHP = dihexadecylphosphate; *f*: DODAC = dioctadecyldimethylammonium chloride; *g*: DODAB = dioctadecyldimethylammonium bromide. *D*: donor, *A*: acceptor; TP: p-terphenyl; DPO: diphenyloctateraene; Chl *a*: chlorophyll *a*; *h:* synthetic functionalized vesicles containing di- or tripeptide histidine segments.

8.2.4.2. Solubilization in vesicles

The noticeable kinetic stability of most of these structures determines a generally slow permeation rate (or diffusion) of solutes across the bilayer membrane. For example, the timescale for the water transport across the bilayer is in the order of seconds, whereas the diffusion of larger molecules is even slower (minutes, hours or days).

Different solubilization sites have been identified in these aggregates. Generally, the hydrophobic molecules lie among the hydrocarbon bilayers whereas polar molecules can move more freely in the entrapped water pools if they are repelled (electrostatically) from the surfactant head groups. For ionized solutes attracted from the surfactant charged groups, a location at the inner or outer vesicle surface is commonly observed. Amphiphilic polar substrates (i.e., long chain derivatives) are usually placed as the surfactant molecules. The solubilization power of the vesicles is usually dependent on their size and membrane thickness.

It is interesting to note that vesicles can also be formed starting from single-tailed surfactants having oppositely charged head groups. For example, by sonicating mixtures composed of hexadecyltrimethylammonium tosylate and sodium dodecylbenzene sulfonate, vesicles having 30–80 nm radius can be obtained [101]. Parameters such as size, shape, bilayer thickness and sign of the surface charge can be adequately controlled in these vesicles by simply changing the relative amounts of the components and their chain lengths. The stability of such aggregates is enough for most reactivity and solute transfer studies.

The gel–filtration technique can be conveniently used to separate and analyze the free substrates from those bound to the vesicles.

8.2.4.3. Ion-exchange at vesicle interfaces

Depending on the surfactant used and on the experimental conditions chosen, the vesicles are either able to exchange the counterions present at their outer and inner surface [102–104] or to exchange only the outer counterions, keeping a single counterion at the inner surface [105]. The degree of ion dissociation from vesicles has also been determined in experiments performed using free-flow electrophoresis [106].

The charged solutes present at the outer vesicle surface can be removed and/or concentrated by simply using an ion-exchange resin.

8.2.4.4. Investigation of chemical processes occurring across vesicle bilayers

Reactions occurring between reactants separated by the bilayer vesicle can be easily investigated. Sonication of the proper vesicle-forming surfactant in the aqueous solution containing one of the reagents (I) allows us to obtain unilamellar vesicles, which can be separated from the bulk solution (for example, by ultrafiltration) and re-equilibrated with another aqueous solutions containing the external reagent (E). Hydrophobic reaction mediators (M), located in the vesicle membrane, may also be present in the system.

Redox reactions have been the most investigated systems in vesicles because these processes are very important from the biological point of view. In fact, they are at the basis of the respiratory function and are crucial for the capture and storage of solar energy in photosynthetic organisms.

A model cell system composed of phosphatidylcholine vesicles containing $Fe(CN)_6^{4-}$ as an electron acceptor in the inner compartment (I), ascorbate as an electron donor in the external bulk solution (E) and membrane-soluble hydrophobic ferrocene derivatives (M), was built up by Hinkle [107]. Ferrocene acts as an electron carrier between donor and acceptor reagents in this ground state electron transfer reaction and, in the presence of catalytic amounts of carbonylcyanide p-trifluoromethoxyphenylhydrazone (FCCP), the reaction rate largely increases. The role of this compound is to increase the H^+ permeability across the bilayer membrane (see Fig. 8.13).

A large number of studies concerning ground- and excited-state electron-transfer processes occurring in amphiphilic vesicles have appeared in the literature in the last few years. They include the investigation of charge separation and photoassisted electron transfer processes occurring in different types of natural and synthetic vesicles and the introduction of semiconductor particles in the bilayer vesicles for the development of sensitized photoreactions. For a detailed description of these peculiar vesicular redox processes, the readers are addressed to the excellent review of Robinson and Cole-Hamilton [108].

8.2.4.5. Development of polymerized surfactant vesicles

Enhanced stabilities and more reproducible vesicle parameters (including size and shape, membrane fluidity and permeability, etc.) can be obtained using synthetic polymerized vesicles, developed some years ago by Fendler and Tundo [109]. Basically, these aggregates are obtained

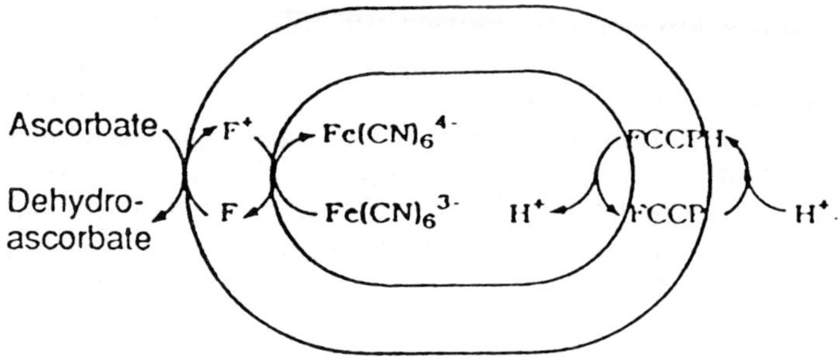

Fig. 8.13 Representation of the ferrocene-mediated transbilayer electron transfer between ascorbate and hexacyanoferrate(II), stimulated by FCCP, in phosphatidylcholine vesicles. Reproduced with permission from Ref. [108].

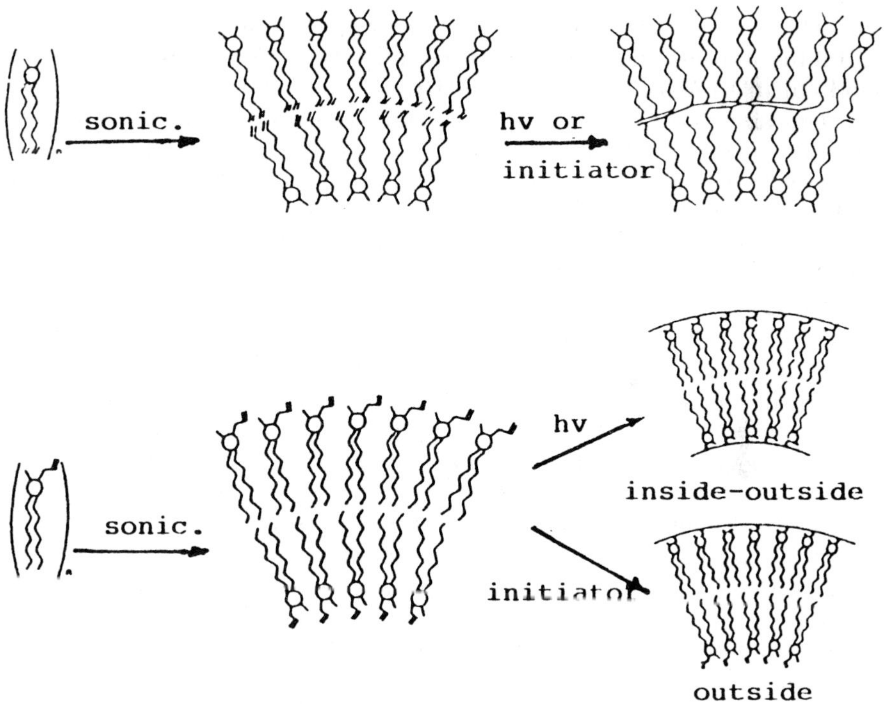

Fig. 8.14. Formation of a polymerized vesicle. Polymerization initiator: azoisobutyronitrile (AIBN), persulfate. Reproduced with permission from Ref. [109].

from vesicle-forming surfactants containing polymerizable double bonds which can be cross-linked adequately. Postvesicle functionalization is usually also possible.

A schematic representation of the synthetic process used can be seen in Fig. 8.14. The enhanced stability of these aggregates suggests their application in photochemical solar energy studies, as a better reaction media (in particular for the formation of small and uniform microparticles) and as drug carriers.

8.3. SEPARATION AND PURIFICATION OF BIOMOLECULES USING MEMBRANE-MIMETIC STRUCTURES

The rapid expansion of genetic engineering and biotechnology in recent years has attracted the interest of many researchers to the development of separation, purification and concentration methods for proteins and other biomolecules, usually formed in fermentation and cell culture media. Due to the fact that many of these molecules are present as polyions under normal working conditions, their liquid–liquid extraction with common organic solvents is usually not possible. Moreover, these solvents can drastically alter the sensitive tertiary structures of the substrates, with a relevant loss of their biological activity.

The use of reversed micelles and w/o microemulsions can help to overcome these inconveniences since these aggregates contain more or less extended water pools which are able to solubilize hydrophilic molecules of different sizes. Moreover, these structures exhibit membrane-mimetic properties [24] and thus can also be used to simulate the uptake of biomolecules by the living cells

8.3.1. Extraction of amino acids in reversed micelles

Amino acids are very important biological molecules as they constitute the basic units of the more complex protein structures. Moreover, these compounds are themselves valuable as essential feed components.

Due to the presence of at least two ionizable groups in their molecules, the amino acids can be present as anions, cations or zwitterions in aqueous solutions, depending on pH. Under physiological conditions both the amino and the carboxylic group are ionized and this zwitterionic character makes the amino acid extraction into organic solvents through conventional ion-pairing or complexation techniques difficult.

Starting from the pioneering work of Fendler et al. on partitioning of amino acids in reversed cationic micelles [110], the main features of amino acids solubilization in these microaggregates have been reported by other research groups. Some of these studies are listed in Table 8.3.

TABLE 8.3

Studies on amino acids solubilization and extraction in reversed micelles

Amino acid	Reversed micelle	Type of investigation/ Technique used	Ref.
Tryptophan	TOMAC	Solubility studies and electrostatic, effects/ Phase-transfer equilibrium method	111
Tryptophan	Phospholecitin	Properties of the water pool/ IR, NMR	112
Tryptophan	AOT	Amino acid location/ Fluorescence quenching	113
Arginine, phenylalanine	AOT/isooctane	Effect of pH and net solute charge	114
Glycine, leucine, phenyl-alanine, arginine, 6-amino-hexanoic acid, tryptophan	AOT/n-heptane	Effect of pH and W_0 on extraction performances. Determination of partition coefficients/ Phase-transfer method	115

a: TOMAC: trioctylmethylammonium chloride.

The extraction curves of arginine and phenylalanine in AOT/isooctane solutions are shown in Fig. 8.15. The solubilization occurs when the amino acid is positively charged and the net positive charge determines the extraction efficiency, thus suggesting a fundamental role of the solute–surfactant electrostatic interactions.

Solubilization in the microaggregates also depends on the amino acid hydrophobicity, being higher for more hydrophobic solutes. In order to explain this observation, simple ion-pairing between the protonated substrate and the anionic surfactant polar head has been assumed. For low pH values this assumption is reasonable, but at intermediate pH's it seems to be of little importance since the deprotonated carboxylic

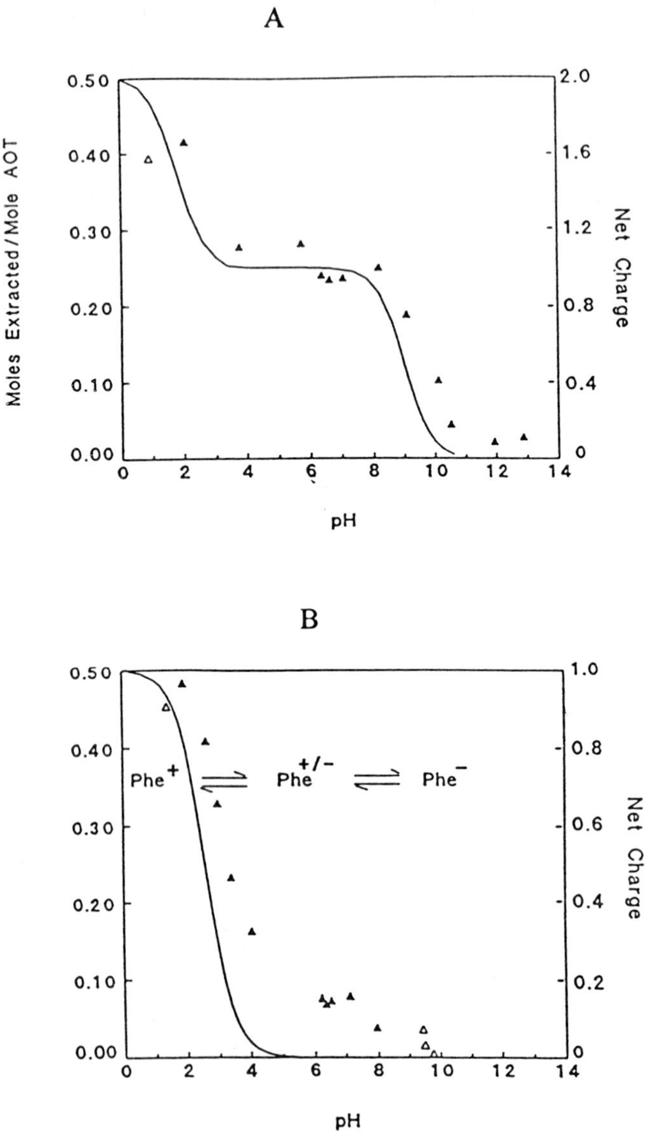

Fig. 8.15. Extraction of arginine (A) and Phenylalanine (B) into reversed AOT micelles in isooctane. Solid lines represent the calculated titration curves. Reproduced with permission from Ref. [114].

groups are highly hydrated, thus facilitating the solute incorporation within the water pool. Other explanations can be given assuming that the non-polar moiety of the amino acid is located within the amphiphile shell, allowing a strong interaction between the protonated amino group and the surfactant charged head group.

An important contribution to the knowledge of amino acids solubilization in AOT reversed micelles was recently given by Leodidis and Hatton [116–119], who examined in detail the effect of different experimental parameters. The crucial role of hydrophobic effect and hydrogen bonding as driving forces for interfacial solubilization of amino acids has been clearly pointed out in these investigations [117].

8.3.2. Extraction of proteins in reversed micelles

Reversed micellar solutions can also offer a suitable extraction medium for proteins since these biomolecules can be transferred from the aqueous matrix to the non-polar extracting environment through their solubilization into the water pools of the microaggregates (see Fig. 8.16).

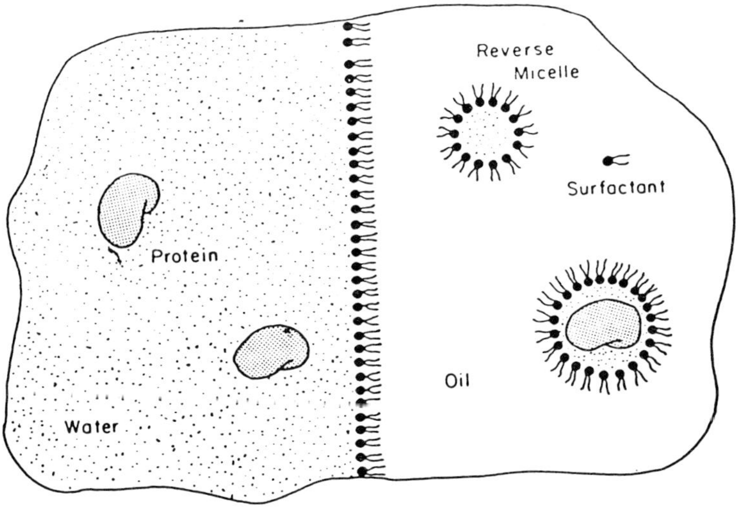

Fig. 8.16. Simplified representation of protein extraction using reversed micelles. Reproduced with permission from Ref. [114].

The potential utility of this approach was first recognized by Luisi and coworkers [120,121], who also demonstrated that several enzymes (such as α-chymotrypsin, ribonuclease, alcohol dehydrogenase, alkaline phosphatase, peroxidase, lysozyme, etc.), can be dissolved in an AOT/isooctane/water solution while still keeping their catalytic activities [122–125]. The practical aspects of such protein extractions have been successively examined in detail by other investigators [126].

From a practical point of view, the preparation of protein-containing reversed micellar solutions for laboratory-scale studies follows two main procedures: (i) the injection technique, based on the injection of low amounts of aqueous buffered protein solutions into the AOT/organic solvent solution; (ii) the phase-transfer technique, in which the two immiscible phases (aqueous protein-containing and organic surfactant-containing) are allowed to stand until the protein partitioning reaches its equilibrium value.

8.3.2.1. Effect of experimental parameters on protein extraction efficiency

Among the factors affecting protein solubilization, the pH effect is obviously relevant since the net charge of the protein is pH-dependent and, moreover, changes in pH can alter the protein conformation. At pH values below the isoelectric point (pI), the protein becomes positively charged, while above the pI it becomes negatively charged. If protein solubilization into reversed micelles is mainly governed by electrostatic interactions, the use of anionic amphiphiles (such as AOT) to form the extracting microaggregate will be possible only at pH values lower than pI. A drop in protein solubility is, however, observed in acid media with the simultaneous formation of precipitates at the interface between the two liquids. This negative effect has been attributed to the protein denaturation.

The above-predicted behaviour has been observed for ribonuclease A, cytochrome c and lysozyme, which are low molecular weight proteins. The marked change in solubilization behaviour around the pI values of each protein is clearly shown in Fig. 8.17.

For larger proteins, such as the enzymes α-chymotrypsin or trypsin, the pH must be reduced well below the pI in order to obtain appreciable solubilization. Interactions other than electrostatic seem to play a very important role here.

The use of cationic amphiphiles that are able to form reversed micelles has also been examined. The extraction of pepsin, trypsin and

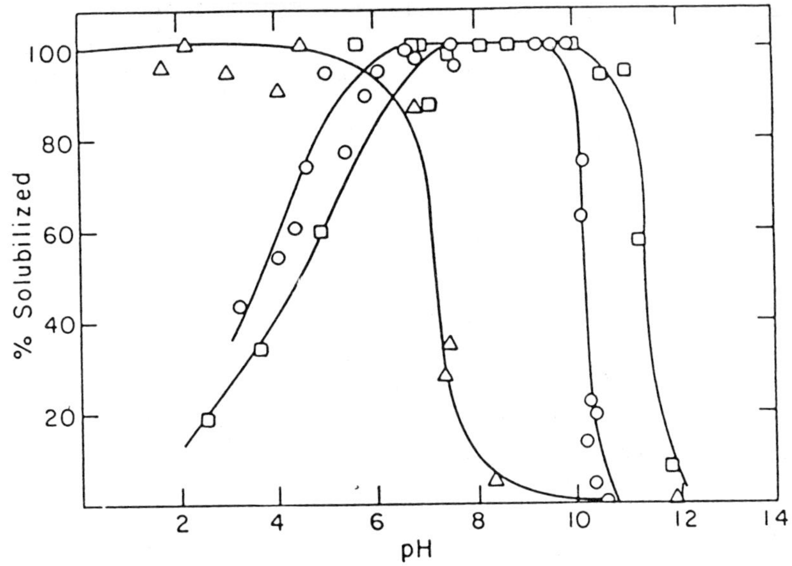

Fig. 8.17. Effect of pH on solubilization of (□) lysozyme, (O) cytochrome c, and (Δ) ribonuclease A in AOT (5×10^{-2} M)-isooctane solutions, in the presence of 0.1 M KCl. Reproduced with permission from Ref. [126].

α-chymotrypsin in reversed micelles of trioctylmethylammonium chloride (TOMAC) in cyclohexane was performed several years ago [120], while more recent studies on the solubilization mechanism of α-amylase in TOMAC/isooctane solutions [127,128] showed the appearance of a narrow solubilization peak for this protein around pH 10, as expected for a negatively charged substrate. Due to the shielding of the charged groups, the addition of salt shifts the extraction maximum towards higher pH values.

Effective separations of proteins are possible by exploiting the ionic strength effects. In fact, the ionic strength of the aqueous solution strongly alter the partitioning of the proteins between this phase and the reversed micellar phase for different reasons: (i) the shielding of the electrostatic interactions between the charged protein sites and the surfactant head groups; (ii) the formation of smaller aggregates through the reduction of the surfactant charged groups repulsion; (iii) the salting-out of the protein from the non-polar phase due to its displacement from the water pools by the salt ions.

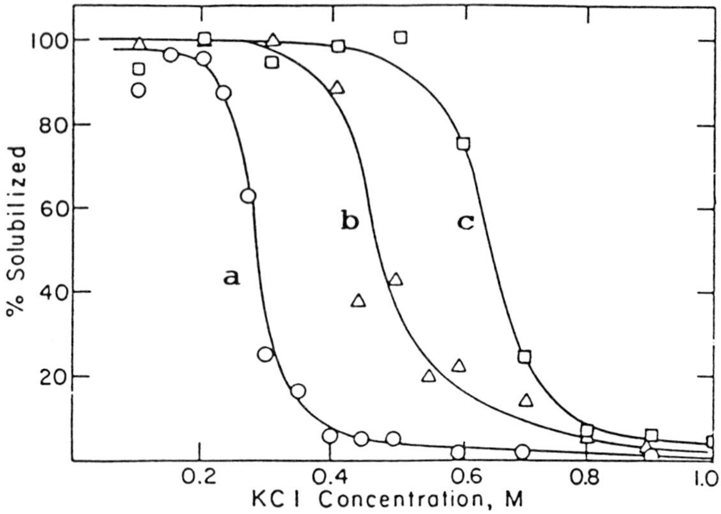

Fig. 8.18. Dependence of protein solubilization in AOT reversed micelles in isooctane on the aqueous solution ionic strength. (a) Cytochrome c; (b) ribonuclease A; (c) lysozyme. Reproduced with permission from Ref. [126].

Figure 8.18 shows the effect of KCl on the solubilization of three different proteins in AOT reversed micelles in isooctane. The solubilization range, which has been assumed to be dependent on the protein species (in particular on the protein surface characteristics), does not seem to be strictly related to the protein charge in aqueous solutions [126].

A practical example of these selective protein separations is given in Ref. [126]. Starting from an aqueous ternary mixture containing ribonuclease a, cytochrome c and lysozyme (proteins having a similar molecular weight) containing 1 mg/ml of total proteins, the components have been efficiently separated using a three-step procedure. After the adjustment of the feed solution pH to ca. 9 (KCL 0.1 M) and treatment with a 50 mM solution of AOT in isooctane, the separation of the unextracted ribonuclease a is obtained in the aqueous phase. The reversed micellar phase containing the other two proteins is then equilibrated with an aqueous solution of higher ionic strength (KCL 0.5 M) to recover cytochrome c and traces of lysozyme. Lysozyme is, finally, back-extracted into a basic aqueous solution (pH 11.5) containing appreciable amounts of added salt (KCl 2.0 M).

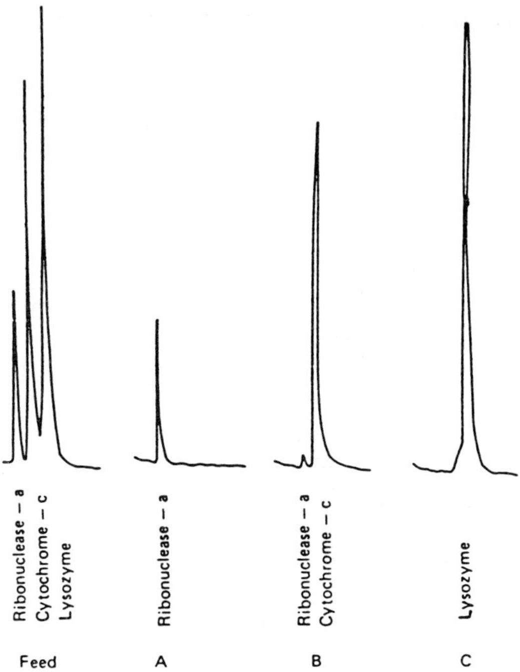

Fig. 8.19. HPLC pattern of the protein-containing aqueous phases after extraction with reversed micellar solutions. Reproduced from Ref. [126], with permission.

The good performances of the above-mentioned separation are confirmed by the HPLC analysis of each aqueous phase (see Fig. 8.19).

The influence of the salt type upon protein extraction has also been examined. For example, the use of salts of multicharged cations (e.g., $CaCl_2$) at constant pH (around 5) strongly increases the protein uptake into the AOT reversed micellar phase and this effect has been attributed to the selective binding of Ca^{2+} to the proteins and to the surfactant [129].

The protein extraction yield can also be increased by increasing the surfactant concentration. The effect of this parameter on solubilization of the proteins ribonuclease A and concanavalin in AOT reversed micelles, performed at a pH below their pI values, has been rationalized by introducing the following equation [130]:

$$\ln P = A + B(pH) + \{D + E(pH)\}\ln C \quad (4)$$

where P is the solute partition coefficient, defined as the ratio between the concentration of the AOT-entrapped and free protein, respectively. C is the AOT concentration, A,B,D,E are parameters depending on the interphase distribution potential, on the slope of the protein titration curve, on the standard state chosen, etc.

8.3.2.2. Selective protein extractions using affinity partitioning

The selectivity of protein extractions in reversed micelles can be further enhanced by exploiting the affinity existing between proteins and specific ligand molecules, already successfully employed in affinity chromatography. The ligand is usually bound to a proper alkyl chain in order to locate the resulting amphiphilic molecule between those of the surfactant, with the affinity group oriented toward the polar core of the reversed micelle.

The selective binding of the ligand to the analyte (protein to be extracted) facilitates its selective extraction into the microaggregate. A simplified scheme of such extractions is depicted in Fig. 8.20.

The dependence of the protein partition coefficient on the ligand concentration can be deduced from a simple thermodynamic treatment of the protein–ligand complex formation equilibrium [130]:

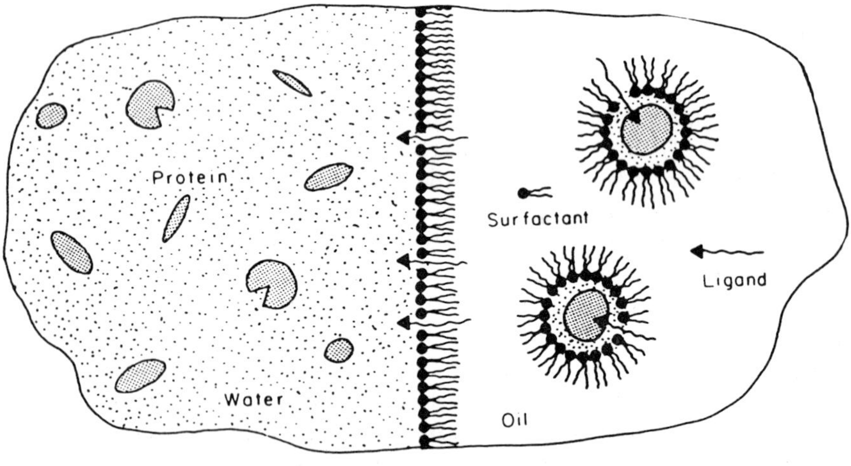

Fig. 8.20. Schematic representation of the selective extraction of proteins into reversed micelles exploiting the affinity partitioning approach. Reproduced with permission from Ref. [114].

$$P = \frac{([P] + [PL])_m}{([P] + [PL])_w} \quad (5)$$

From the above equation, by introducing the dissociation constants of the protein–ligand complex in the two phases and indicating the partition coefficients of the simple protein and of the ligand by P_o and P_L, respectively, we obtain:

$$P = P_o \frac{(1 + [L]/K_{dm})}{(1 + [L]/K_{dw}P_L)} \quad (6)$$

When n-identical binding sites per protein molecule are present, the fraction which multiplies P_o in Eqn. (6) must be raised to the nth power. For example, for concanavalin A (a protein possessing two binding sites per molecule) n = 2 [131]. Assuming a large partition coefficient for the ligand, P_L, it can be demonstrated that the protein partition coefficient depends on both the surfactant and ligand concentrations through the expression:

$$P = P_o'([AOT] + \beta[L])^{\alpha} (1 + [L]/K_{dm})^2 \quad (7)$$

where β is the molar AOT-equivalent solubilizing capacity of the ligand, α is an empirical fitting parameter depending on pH and $P_o = P_o'f(pH)$. A good fit of experimental data for concanavalin A in AOT/isooctane solutions at pH 7.2 has been obtained using Eqn. (7). The effective extraction of this protein (more than 90%) has been achieved in the pH range 6–7.5 using a biological surfactant (octyl-β-D-glucopyranoside) as convenient ligand. The ligand, which was dissolved in 50 mM AOT solution in isooctane, is an essential component since without it the extraction yield drops to less than 15% under the same experimental conditions [132].

8.3.2.3. Stripping of the proteins from the reversed micellar phase

The recovery of the extracted proteins from the organic phase is the final step of the concentration/separation procedure. The retention of the biological activity of the recovered protein is in most cases essential.

Although the back extraction of some proteins may require peculiar and complex manipulations, in several cases the simple adjustment of pH and ionic strength of the aqueous stripping solution may be sufficient.

Cytochrome c, α-chymotrypsin and α-amylase are examples of easily recovered proteins, whereas for other analytes (such as lysozyme) it is necessary to operate under extreme pH and ionic strength conditions.

For proteins such as ribonuclease A and concanavalin, which show strong interactions with the surfactant AOT, the recovery is very difficult unless the reversed micelles are disrupted by adding a more polar water-immiscible solvent [130].

8.3.2.4. Theoretical models for protein uptake by reversed micelles

The prediction of the protein behaviour in the reversed micelle-based extraction systems is of great concern in any practical application of this separation method in biotechnology. Attempts to quantitatively describe the protein uptake using different experimental approaches have been reported in the literature [133–135], although some drastic assumptions (such as the constancy of the water/surfactant ratio) have been introduced in these treatments, thus reducing its general validity.

More recently, a solubilization model developed on a thermodynamic basis, able to fit the data coming from experiments performed using the injection technique and to estimate some important parameters (such as the radii of the filled and empty microaggregates), has been proposed by Caselli et al. [136–138].

The model, which assumes that the reversed micelle can be treated as a microcapacitor (see Fig. 8.21), introduces some parameters necessary to describe the experimental system and basically applies the minimization of the free energy of the system. Other approximations introduced in the treatment are the following: the monodispersity of the reversed micelles (one radius, r_0, before protein uptake and two radii, r_e and r_f, after solubilization, reflecting the presence of filled and empty aggregates); the proteins are considered as rigid spheres and are characterized by their radius r_p, by the average surface charge and by the net average number charge of the ionized groups; no more than one protein molecule is assumed within the extracting aggregate; the interactions occurring between micelles are neglected; a unique dielectric constant for water is considered, whereas a different dielectric constant is assumed at the reversed micelle surface layer.

Without entering into the details of the analytical elaboration of the model, thoroughly described in Ref. [136], it must be emphasized that the calculations of the free energy changes have four main contributions: (i) an electrostatic term related to the interactions between the charged protein and the inner wall of the surfactant polar heads (usually

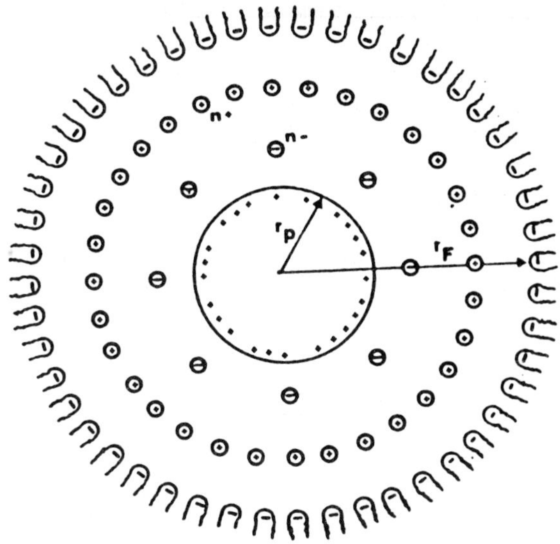

Fig. 8.21. Representation of a protein-containing reversed micelle described in terms of two concentric capacitors. Reproduced with permission from Ref. [138].

estimated by solving the Poisson–Boltzmann equation); (ii) an entropic term arising from these electrostatic interactions; (iii) an entropic term due to the rearrangement of the counterions within the reversed micelle, after the protein uptake; (iv) another entropic contribution coming from changes in concentration and size of microaggregates in the non-polar phase.

The description of the application of the Poisson–Boltzmann equation inside the spherical cavities and the evaluation of the ionic distribution within the reversed micelles in equilibrium with electrolyte solutions can be found in Refs. [139] and [140], respectively.

After the choice of the molecular parameters necessary for the calculations, the radii of empty and filled reversed micelles for different proteins, at different water/surfactant ratios (W_0), have been estimated. For the reported system [136], molecular volumes equal to 29.8, 922 and 274 Å^3 have been assumed for water, AOT and isooctane, respectively. The area of the surfactant polar head (55 Å^2), the r_p values (15.4, 16.0, 17.5, and 22.3 Å for ribonuclease, cytochrome c, myelin basic protein and chymotrypsin, respectively), the mean value of the dielectric constant at the surfactant layer, $\varepsilon = 3$ [141], the pK_a values of the amino

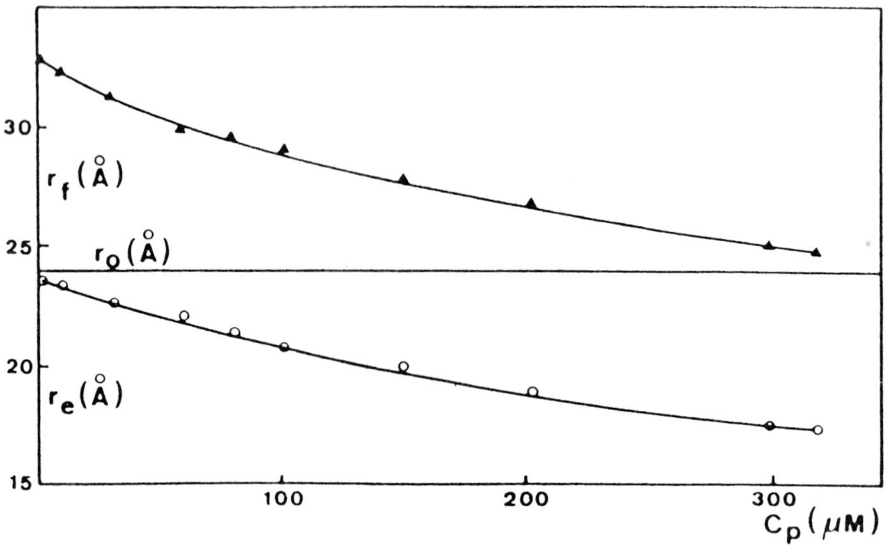

Fig. 8.22. Predicted variation of the radii of empty and filled reversed micelles as a function of MBP concentration. Reproduced from Ref. [136], with permission.

acid residues, and the proper surface charge data at working pH [142], are the other parameters introduced in the model calculations.

A good fit of experimental data with the model predictions has been observed. For example, for a constant protein concentration (60 %M), the predicted r_e and r_f mean values for myelin basic protein (MBP) are 10.7 Å and 25.9 Å, respectively (compared with the corresponding experimental values 12.9 Å and 21.9 Å). For chymotrypsin (CHT) the agreement is even better: 19.1 Å and 29.0 Å are the calculated values of r_e and r_f, respectively, which compare well with those obtained from independent measurements (19.0 Å and 29.2 Å, respectively).

The effect of protein concentration on the reversed micelle dimensions is also relevant. Calculations performed at a fixed water/AOT molar ratio ($W_0 = 13$) for MBP predict a decrease in r_e and r_f (see Fig. 8.22). This behaviour has been interpreted in terms of increased number of filled aggregates, with less water available for the protein molecules.

The excellent agreement between the predicted micellar parameters and the experimental data arising from neutron scattering measurements (SANS) further confirms the validity of the proposed model [137].

8.4. ENZYME SOLUBILIZATION AND REACTIVITY IN REVERSED MICELLES

The interest in applications of enzymes dissolved in reversed micelles [133,143,144] is due to the possible applications of these hosted biomolecules as catalysts for reactions involving water insoluble compounds. The RM aggregate can be viewed as a microreactor whose dimensions can be properly varied by simply controlling the water/surfactant ratio.

Three examples of reactant compartmentalization in enzyme-containing reversed micelles have been considered [145,146]: (i) two reagents (A_1 and B_1) preferentially soluble in water and their reaction product, P_1, soluble in hydrocarbons; (ii) one of the two reagents, A_2, also soluble in non-polar solvents; (iii) the formation of an overwhelming hydrocarbon insoluble product, P_3, entrapped in the water pools (see Fig. 8.23).

Some studies on the synthesis of peptides using the enzyme α-chymotrypsin solubilized in AOT RM are examples of case (ii) [146]:

$$\underset{A}{\text{Z–Ala–Phe–OMe}} + \underset{B}{\text{H–Leu–NH}_2} \rightarrow \underset{C}{\text{Z–Ala–Phe–Leu–NH}_2} + \text{MeOH} \quad (8)$$

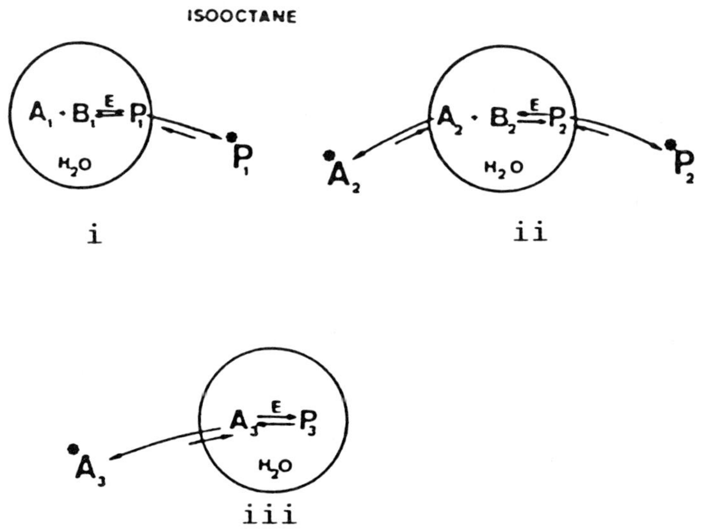

Fig. 8.23. Examples of reagents and products compartmentalization in reactions driven in enzyme-containing reversed micelles. Reproduced with permission from Ref. [146].

Z is a protecting benzyloxycarbonyl group. The reaction yield of the water-insoluble peptide obtained is between 40 and 60% for W_0 = 5–30. One of the starting reagents is water soluble, the other partitions between the RM and the hydrocarbon. The enzyme is water soluble and the product C (water insoluble) is transferred to the external bulk phase. It should be noted that the above reaction cannot be driven in aqueous solutions due to the poor solubility of reagent A.

Another example of a reaction catalyzed by α-chymotrypsin, which corresponds to case (iii), is the following:

$$\underset{A}{\text{Z–Phe–OHex}} \rightarrow \underset{B}{\text{Z–Phe–OH}} + \text{hexanol} \tag{9}$$

The n-hexylester of the N-protected phenylalanine (reagent A) is soluble in the oil phase and insoluble in water. After the hydrolysis catalyzed by the enzyme, the reaction products remain in the water pool. Yields of hydrolysis of ca. 40% have been determined in experiments performed in 0.1 M AOT/isooctane, with W_0 = 12.5 and an enzyme concentration in the μM range [146].

Further contributions to the analysis of the possible stages in the formation of peptide bonds between hydrophobic and polar amino acids (reacting at the interface of AOT/isooctane reversed micelles) can be found in Ref. [147].

A number of kinetic studies on enzymatic activity in non-polar media have also been performed. A typical experimental approach in these investigations has been the analysis of product formation using spectrophotometric techniques. As an example, the study of the hydrolysis of p-nitrophenylesters in the presence of suitable enzymes (mostly α-chymotrypsin or lipases), can be mentioned.

Details concerning the kinetic equations applied, either in the case of competitive inhibition by the surfactant and when this component has no effect on the enzymatic reaction, are given in ref.148. The hydrolysis of N-glutaryl-L-phenylalanine-p-nitroanilide by α-chymotrypsin in AOT reversed micelles showed, for example, a higher value of the Michaelis constant (about 100 times higher than in water) [148].

The denaturation of some enzymes (e.g., lysozyme) in AOT/isooctane reversed micelles has been observed [149]. Since the stability of the solubilized proteins in these systems is crucial for any practical application, studies on guest molecule behaviour have been performed using DSC in a wide temperature range (from –16 to 80°C). It has been found, for example, that the temperature-induced unfolding of the protein molecules within the microaggregates is a two-state process. Moreover, the use of dc-spectroscopy can give useful information about the protein conformational changes [150].

The presence of reversed micelles (or w/o microemulsion droplets) of well defined minimum size seems to be essential in order to obtain high catalytic activities. Fluorescence anisotropy decay measurements of tryptophan in trypsin dissolved in detergentless microemulsions were used to evaluate the proper size of the enzyme-containing microemulsion droplets [151]. A radius in the range 27–37 Å can accommodate the guest macromolecule which, according to the proposed model, is surrounded by a water layer of 7–10 Å thickness. The retention of the enzyme activity within the microaggregate seems to be due to the protective function of the water film against the denaturing contact with the organic phase.

Recently, the structure and reactivity of two types of the widely investigated enzyme α-chymotrypsin, solubilized in AOT/isooctane reversed micelles, have been studied by coupling synchrotron radiation X-ray small-angle scattering measurements with enzymatic activity determinations [152]. The scattering experiments demonstrated the location of the enzyme in the water pools, and the activity measurements performed on two different substrates (p-nitrophenyl acetate and p-nitrophenyl caprylate) again revealed a maximum of the enzymatic activity related to the presence of a critical amount of water ($W_0 = 12$) and indicated that the catalyzed ester hydrolysis is enhanced at the internal microaggregate interface.

The effect of added salts upon protein solubilization has already been examined in previous sections. A specific investigation of ionic strength effects on enzyme solubilization using the shell-and-core model, which allows the prediction of the fraction of the enzyme-containing reversed micelles as a function of the added salt, can be found in Ref. [153].

Finally, studies on solubilization and activity of living structures, such as yeast cells, in water–oil microemulsion systems, should be mentioned as pioneering work in the development of microbiology in non-polar solvents [154]. The application of the polarographic method to estimate the oxygen consumption of the encapsulated cells after the addition of nutrients (glucose) seems to be an adequate analytical choice for monitoring the corresponding biological processes.

REFERENCES

1 M.K. Jain and R.C. Wagner, *Introduction to Biological Membranes*. Wiley-Interscience, New York, 1980.
2 N. Robertson, *The Lively Membranes*. Cambridge University Press, Cambridge, 1983.
3 S. Backlund and K. Rundt, *Acta Chem. Scand. A*, 35 (1981) 785.
4 C. Tanford, Y. Nozaki, J.A. Reynolds and S. Makino, *Biochemistry*, 11 (1974) 2369.
5 M.N. Jones and P. Manley, *J. Chem. Soc. Faraday Trans. I*, 77 (1981) 827.
6 R.A. Barford and B.J. Sliwinski, *Anal. Chem.*, 56 (1984) 1554.
7 R.S. Matson and S.C. Goheen. *LC-GC*, 2 (1984) 536.
8 N. Kosaric, W.L. Cairns and N.C.C. Gray, *Biosurfactants and Biotechnology*. Dekker, New York, 1987.
9 D.M. Small. *Adv. Chem. Ser.*, 84 (1968) 31.
10 D.G. Oakenfull and L.R. Fisher, *J. Phys. Chem.*, 82 (1978) 2443.
11 R. Zana, *J. Phys. Chem.*, 82 (1978) 2440.
12 D.E. Güveli, *J. Chem. Soc. Faraday Trans. I*, 81 (1985) 2103.
13 D.M. Small, in: P.P. Nair and D. Kritchevsky (Eds.), *The Bile Acids*. Plenum Press, New York, 1971, Vol. 1, p. 249.
14 M.C. Carey, in: H. Danielsson and J. Sjövall (Eds.), *Sterols and Bile Acids*. Elsevier, Amsterdam, 1985, p. 345.
15 C.J. O'Connor and R.G. Wallace, *Adv. Coll. Interface Sci.*, 22 (1985) 1.
16 D. Oakenfull, *Aust. J. Chem.*, 39 (1986) 1671.
17 D. Attwood and A.T. Florence, *Surfactant Systems. Their Chemistry, Pharmacy and Biology*. Chapman and Hall, London, 1985. Ch. 4. pp. 124–228.
18 C.J. O'Connor, A.S.H. Mitha and P. Walde, *Aust. J. Chem.*, 39 (1986) 249.
19 C.J. O'Connor and A.S.H. Mitha, *Aust. J. Chem.*, 39 (1986) 259.
20 C.J. O'Connor and P. Walde, *Langmuir*, 2 (1986) 139.
21 M. Lindheimer, J.C. Montet, J. Molenat, A.M. Montet and B. Brun, *J. Chim. Phys.*, 84 (1987) 947.
22 M. Svärd, P. Schurtenberger, K. Fontell, B. Jönsson and B. Lindman, *J. Phys. Chem.*, 92 (1988) 2261.
23 J.H. Fendler, *Acc. Chem. Res.*, 13 (1980) 7.
24 J.H. Fendler, *Membrane Mimetic Chemistry*. Wiley, New York, 1982.
25 J.H. Fendler, in: W.L. Hinze and D.W. Armstrong (Eds.), *Ordered Media in Chemical Separations*. ACS Symp. Ser. 342. Am. Chem. Soc. Washington DC, 1987, p. 83.
26 J.H. Fendler, *Chem. Rev.*, 87 (1987) 877.
27 S.J. Singer and G.L. Nicolson, *Science*, 175 (1972) 720.
28 P.L. Yeagle, *Acc. Chem. Res.*, 11 (1978) 321.
29 P.L. Luisi and B. Straub (Eds.), *Reverse Micelles*. Plenum Press, London, 1984.

30 M.P. Pileni (Ed.). *Structure and Reactivity in Reverse Micelles*. Elsevier. Amsterdam, 1989.
31 G.D. Correll, R.N. Cheser, F. Nome and J.H. Fendler, *J. Am. Chem. Soc.*, 100 (1978) 1254.
32 B.H. Robinson, D.C. Steyler and R.D. Tack, *J. Chem. Soc. Faraday Trans. I*, 75 (1979) 481.
33 N.J. Bridge and P.D.I. Fletcher, *J. Chem. Soc. Faraday Trans. I*, 79 (1983) 2161.
34 A.M. Ganz and B.E. Boeger, *J. Coll. Interface Sci.*, 109 (1986) 504.
35 C. Cabos and J. Marignan, *J. Phys. Lett.*, 46 (1985) L-267.
36 M.P. Pileni, T. Zemb and C. Petit, *Chem. Phys. Lett.*, 118 (1985) 414.
37 R.A. Day, B.H. Robinson, J.H.R. Clarke and J.V. Doherty, *J. Chem. Soc. Faraday Trans. I*, 75 (1979) 132.
38 J.D. Nicholson and J.H.R. Clarke, in: K. Mittal and B. Lindman (Eds.), *Surfactants in Solution*. Plenum Press, New York, 1984, Vol. 3, p. 1663.
39 J.H.R. Clarke, J.D. Nicholson and K.N. Regan, *J. Chem. Soc. Faraday Trans. I*, 81 (1985) 1173.
40 M. Kotlarchyk, S.H. Chen and J.S. Huang, *J. Phys. Chem.*, 86 (1982) 3273.
41 C. Toprakcioglu, J.C. Dore, B.H. Robinson and A. Howe, *J. Chem. Soc. Faraday Trans. I*, 80 (1984) 413.
42 M. Kotlarchyk, J. S. Huang and S. H. Chen, *J. Phys. Chem.*, 89 (1985) 4382.
43 P.D.I. Fletcher, B.H. Robinson and J. Tabony, *J. Chem. Soc. Faraday Trans. I*, 82 (1986) 2311.
44 E. Dutkiewicz and B.H. Robinson, *J. Electroanal. Chem.*, 251 (1988) 11.
45 H.F. Eicke, M. Borkovec and B. Das-Gupta, *J. Phys. Chem.*, 93 (1989) 314.
46 M. Ueda and Z.A. Schelly, *Langmuir*, 5 (1989) 1005.
47 H. Hauser, G. Haering, A. Pande and P.L. Luisi, *J. Phys. Chem.*, 93 (1989) 7869.
48 P.D.I. Fletcher, A.M. Howe and B.H. Robinson, *J. Chem. Soc. Faraday Trans. I*, 83 (1987) 985.
49 J. Lang, A. Jada and A. Malliaris, *J. Phys. Chem.*, 92 (1988) 1946.
50 E.B. Leodidis and T.A. Hatton, *Langmuir*, 5 (1989) 741.
51 K. Tamura and N. Nii, *J. Phys. Chem.*, 93 (1989) 4825.
52 A.M. Howe, J.A. McDonald and B.H. Robinson, *J. Chem. Soc. Faraday Trans. I*, 83 (1987) 1007.
53 M. Sácz, E.A. Abuin and E.A. Lissi, *Langmuir*, 5 (1989) 942.
54 C. Boned, J. Peyrelasse and N. Moha-Onchane, *J. Phys. Chem.*, 90 (1986) 631.
55 H. Hoffmann and G. Ebert, *Angew. Chem. Int. Ed. Engl.*, 27 (1988) 902.
56 D.F. Evans, D.J. Mitchell and B.W. Ninham, *J. Phys. Chem.*, 90 (1986) 2817.
57 A. D'Aprano, A. Lizzio and V. Turco Liveri, *J. Phys. Chem.*, 91 (1987) 4749.
58 H.F. Eicke, J.C.W. Shepherd and A. Steinemann, *J. Coll. Interface Sci.*, 56 (1976) 168.

59 J.B. Nagy, Y. Yuan, T.C. Jao and J. H. Fendler, *J. Phys. Chem.*, 94 (1990) 863.
60 P.L. Luisi and L. Magid, *CRC Crit. Rev. Biochem.*, 20 (1984) 409.
61 F.M. Menger and G. Saito, *J. Am. Chem. Soc.*, 100 (1978) 4376.
62 Y. Jean and H. Ache, *J. Am. Chem. Soc.*, 100 (1978) 6320.
63 L.J. Magid, K. Kon-no and C.A. Martin, *J. Phys. Chem.*, 85 (1981) 1434.
64 J.T. Davies and E.K. Rideal, *Interfacial Phenomena*. Academic Press, New York, 1963.
65 P. Fromherz and R. Kotulla, *Ber. Bunsenges. Phys. Chem.*, 88 (1984) 1106.
66 P. Fromherz and B. Masters, *Biochim. Biophys. Acta*, 356 (1974) 270.
67 B. Lovelock, F. Grieser and T.W. Healy, *Langmuir*, 2 (1986) 443.
68 J. Kibblewhite, C.J. Drummond, F. Grieser and P.J. Thistlethwaite, *J. Phys. Chem.*, 93 (1989) 7464.
69 F. Grieser, P. Thistlethwaite, R. Urquhart and L.K. Patterson, *J. Phys. Chem.*, 91 (1987) 5286.
70 K.B. Blodgett, *J. Am. Chem. Soc.*, 57 (1935) 1007.
71 H. Kuhn, D. Möbius and H. Bücher, in: A. Weissberger and B.W. Rossiter (Eds.), *Physical Methods for Chemistry*. Wiley, New York, 1972, Vol. 1, pp. 577–701.
72 D. Möbius, *Ber. Bunsenges. Phys. Chem.*, 82 (1978) 848.
73 D. Möbius and H. Kuhn, *Israel J. Chem.*, 18 (1979) 375.
74 D.G. Whitten, F.R. Hopf, F.H. Quina, G. Sprintschnik and H.W. Sprintschnik, *Pure Appl. Chem.*, 49 (1977) 379.
75 G.L. Gaines, Jr., P.E. Behnken and S.J. Valenty, *J. Am. Chem. Soc.*, 100 (1978) 6549
76 S.J. Valenty, *J. Am. Chem. Soc.*, 101 (1979) 1.
77 R. Memming and F. Schroppel, *Chem. Phys. Lett.*, 62 (1979) 207.
78 X. Zhang and A.J. Bard, *J. Phys. Chem.*, 92 (1988) 5566.
79 B.E. Horsey and D.G. Whitten, *J. Am. Chem. Soc.*, 100 (1978) 1293.
80 J.A. Mercer-Smith and D.G. Whitten, *J. Am. Chem. Soc.*, 101 (1979) 6620.
81 S. Windreich and A. Silberberg, *J. Coll. Interface Sci.*, 77 (1980) 427.
82 H.T. Tien, *Bilayer Lipid Membranes (BLM) Theory and Practice*. Dekker, New York, 1974.
83 P. Schoch, D.F. Sargent and R. Schwyzer, *J. Membr. Biol.*, 46 (1979) 71.
84 C.D. Tran, P.L. Klahn, A. Romero and J.H. Fendler, *J. Am. Chem. Soc.*, 100 (1978) 1622.
85 T. Kunitake, Y. Okahata, K. Tamaki, F. Kumamuru and M. Takayanagi, *Chem. Lett.* (1977) 387.
86 T. Kunitake, *J. Macromol. Sci. Chem.*, A13 (1979) 587.
87 R.A. Mortara, F.H. Quina and H. Chaimovich, *Biochem. Biophys. Res. Commun.*, 81 (1978) 1080.
88 S. Lacouturiere and S. Piekarski, *J. Chim. Phys.*, 84 (1987) 941.
89 Y.Y. Lim and J.H. Fendler, *J. Am. Chem. Soc.*, 101 (1979) 4023.

90 K. Kano, A. Romero, B. Djermouni, H.J. Ache and J.H. Fendler, *J. Am. Chem. Soc.*, 101 (1979) 4030.
91 J. Shobha, V. Srinivas and D. Balasubramanian, *J. Phys. Chem.*, 93 (1989) 17.
92 W. Schnecke, M. Grätzel and A. Henglein. *Ber. Bunsenges. Phys. Chem.*, 81 (1978) 821.
93 R. McNeil and J.K. Thomas, *J. Coll. Interface Sci.*, 73 (1980) 522.
94 D.M. Shin, K.S. Schanze, J.P. Otruba, P.E. Brown and D.G. Whitten, *Israel J. Chem.*, 28 (1987/88) 37.
95 A. Takizawa, M. Higuchi, T. Kinoshita and Y. Tsujita, *Coll. Polym. Sci.*, 265 (1987) 31.
96 S. Mann, J. Skarnulis and R.J.P. Williams, *Israel J. Chem.*, 21 (1981) 3.
97 I.M. Cuccovia, H. Chaimovich, E. Lissi and E. Abuin, *Langmuir*, 6 (1990) 1601.
98 A. Mehreteab and G.S. Strauss, *Photochem. Photobiol.*, 28 (1978) 369.
99 T. Parasassi, G. Bombieri, F. Conti and U. Croatto, *Inorg. Chim. Acta*, 106 (1985) 135.
100 K. Ohkubo and S. Miyake, *J. Chem. Soc. Perkin Trans II* (1987) 995.
101 E.W. Kaler, A.K. Murthy, B.E. Rodriguez and J.A.N. Zasadzinski, *Science*, 245 (1989) 1371.
102 J.H. Fendler and W.L. Hinze, *J. Am. Chem. Soc.*, 103 (1981) 5439.
103 I.M. Cuccovia, F.H. Quina and H. Chaimovich, *Tetrahedron Lett.*, 38 (1982) 917.
104 R.A. Moss, Y. Ihara and G.O. Bizzigotti, *J. Am. Chem. Soc.*, 104 (1982) 7476.
105 E.A. Lissi, E.B. Abuin, A. Zanocco, C.A. Backer and D.G. Whitten, *J. Phys. Chem.*, 93 (1989) 4886.
106 I.M. Cuccovia, E. Feitosa, H. Chaimovich, L. Sepulveda and W. Reed, *J. Phys. Chem.*, 94 (1990) 3722.
107 P. Hinkle, *Biochem. Biophys. Res. Commun.*, 41 (1970) 1375.
108 J.N. Robinson and D.J. Cole-Hamilton, *Chem. Soc. Rev.*, 20 (1991) 49.
109 J.H. Fendler and P. Tundo, *Acc. Chem. Res.*, 17 (1984) 17.
110 J.H. Fendler, F. Nome and J. Nagyvary, *J. Mol. Evol.*, 6 (1975) 216.
111 P.L. Luisi, F.J. Bonner, A. Pellegrini, P. Wiget and R. Wolf, *Helv. Chim. Acta*, 62 (1979) 740.
112 C.A. Boicelli, F. Conti, M. Giomini, A.M. Giuliani, *Chem. Phys. Lett.*, 89 (1982) 490.
113 M.A.J. Rodgers and P.C. Lee, *J. Phys. Chem.*, 88 (1984) 3480.
114 T.A. Hatton, in: W.L. Hinze and D.W. Armstrong (Eds.), *Ordered Media in Chemical Separations*. ACS Symp. Ser. 342. Am. Chem. Soc. Washington DC, 1987, p. 170.
115 M. Adachi, M. Harada, A. Shioi and Y. Sato, *J. Phys. Chem.*, 95 (1991) 7925.

116 E.B. Leodidis and T.A. Hatton, *J. Phys. Chem.*, 94 (1990) 6400.
117 E.B. Leodidis and T.A. Hatton, *J. Phys. Chem.*, 94 (1990) 6411.
118 E.B. Leodidis, A.S. Bommarius and T.A. Hatton, *J. Phys. Chem.*, 95 (1991) 5943.
119 E.B. Leodidis and T.A. Hatton, *J. Phys. Chem.*, 95 (1991) 5957.
120 P.L. Luisi, F.J. Bonner, A. Pellegrini, P. Wiget and R. Wolf, *Helv. Chim. Acta*, 62 (1979) 740.
121 P. Meier, V.E. Imre, M. Fleschar and P.L. Luisi, in: K.L. Mittal and B. Lindman (Eds.), *Surfactants in Solution*. Plenum Press, New York, 1984, Vol. 2, p. 999.
122 R. Wolf and P.L. Luisi, *Biochem. Biophys. Res. Commun.*, 89 (1979) 209.
123 P. Meier and P.L. Luisi, *J. Solid Phase Biochem.*, 5 (1980) 269.
124 C. Grandi, R.E. Smith and P.L. Luisi, *J. Biol. Chem.*, 256 (1981) 837.
125 S. Barbaric and P.L. Luisi, *J. Am. Chem. Soc.*, 103 (1981) 4239.
126 K.E. Goklen and T.A. Hatton, *Sep. Sci. Technol.*, 22 (1987) 831.
127 M. Dekker, K. Van't Riet, S.R. Weijers, J.W.A. Baltussen, C. Laane and H. Bijsterbosch, *Chem. Eng. J.*, 33 (1986) B27.
128 M. Dekker, J.W.A. Baltussen, K. Van't Riet, B.H. Bijsterbosch, C. Laane and R. Hilhorst. In: C. Laane, J. Tramper and M.D. Lilly (Eds.), *Biocatalysis in Organic Media*. Elsevier, Amsterdam, 1987, p. 285
129 M.E. Leser, G. Wei, P.L. Luisi and M. Maestro, *Biochem. Biophys. Res. Commun.*, 135 (1986) 629.
130 T.A. Hatton, in: J.F. Scamehorn and J.H. Harwell (Eds.), *Surfactant-Based Separation Processes.* Dekker, New York, 1989. p. 55.
131 J.M. Woll, T.A. Hatton and M.L. Yarmush, *Biotechnol. Prog.*, 5 (1989) 57.
132 J.M. Woll, A.S. Dillon, R.S. Rahman and T.A. Hatton. In: R. Burgess (Ed.), *Protein Purification: Micro to Macro*. Alan R. Liss, New York, 1987, p. 117.
133 F.J. Bonner, R. Wolf and P.L. Luisi, *J. Solid Biochem.*, 5 (1980) 525.
134 D. Chatenay, W. Urbach, A. M. Cazabat, M. Vacher, M. Waks, *Biophys. J.*, 48 (1985) 893.
135 G.G. Zampieri, H. Jaeckle and P.L. Luisi, *J. Phys. Chem.*, 90 (1986) 1849.
136 M. Caselli, P.L. Luisi, M. Maestro and R. Roselli, *J. Phys. Chem.*, 92 (1988) 3899.
137 M. Caselli, M. Maestro and G. Morea, *Biotechnol. Prog.*, 4 (102) 1988.
138 P. Bruno, M. Caselli, P.L. Luisi, M. Maestro and A. Traini, *J. Phys. Chem.*, 94 (1990) 5908.
139 M. Caselli and M. Maestro, *J. Electroanal. Chem.*, 283 (1990) 67.
140 P. Bruno, M. Caselli, M. Maestro and A. Traini, *J. Surface Sci. Technol.*, 5 (1989) 38.
141 M.K. Gibson and B. Honig, *Biopolymers*, 25 (1986) 2097.
142 D.G. Barlow and M. Thronton, *Biopolymers*, 25 (1986) 1717.
143 P.L. Luisi and L.J. Magid, *CRC Crit. Rev. Biochem.*, 20 (1986) 409.
144 K. Martinek, A.V. Levashov, N.L. Klyachko, S.N. Namyotkin and I.V.

Berezin, *Eur. J. Biochem.*, 155 (1986) 453.
145 P. Lüthi and P.L. Luisi , *J. Am. Chem. Soc.*, 106 (1984) 7285
146 M. Leser, G. Wei, P. Lüthi, G. Haering, A. Hochkoeppler, E. Blöchliger and P.L. Luisi, *J. Chim. Phys.*, 84 (1987) 1113.
147 D. Ranganathan, G.P. Singh and S. Ranganathan, *J. Am. Chem. Soc.*, 111 (1989) 1144.
148 R. Schoemaecker, B.H. Robinson and P.D.I. Fletcher, *J. Chem. Soc. Faraday Trans. I*, 84 (1988) 4203.
149 B. Steinmann, H. Jäckle and P.L. Luisi, *Biopolymers*, 25 (1986) 1133.
150 L. Magid, P. Walde, G. Zampieri, E. Battistel, Q. Peng, E. Trotta, M. Maestro and P.L. Luisi, *Coll. Surfaces*, 30 (1988) 193.
151 Y.L. Khlmelnitsky, A. van Hoek, C. Veeger and A.J. Visser, *J. Phys. Chem.*, 93 (1989) 872.
152 M. Hirai, T. Takizawa, S. Yabuki, R. Kawai-Hirai, M. Oya, K. Nakamura, K. Kobashi and Y. Amemiya, *J. Chem. Soc. Faraday Trans.*, 91 (1995) 1081.
153 D. Bratko, A. Luzar and S.H. Chen, *Bioelectrochem. Bioenergetics*, 20 (1988) 291.
154 A. Hochköppler, N. Pfammatter and P.L. Luisi, *Chimia*, 43 (1989) 348.

Subject index